T0302067

Interplanetary Astrodynamics

Focusing on the orbital mechanics tools and techniques necessary to design, predict, and guide a trajectory of a spacecraft traveling between two or more bodies in a Solar System, this book covers the dynamical theory necessary for describing the motion of bodies in space, examines the N-body problem, and shows applications using this theory for designing interplanetary missions. While most orbital mechanics books focus primarily on Earth-orbiting spacecraft, with a brief discussion of interplanetary missions, this book reverses the focus and emphasizes the interplanetary aspects of space missions. Written for instructors, graduate students, and advanced undergraduate students in Aerospace and Mechanical Engineering, this book provides advanced details of interplanetary trajectory design, navigation, and targeting.

Interplanetary Astrodynamics

David B. Spencer and Davide Conte

CRC Press is an imprint of the
Taylor & Francis Group, an **informa** business

MATLAB® is a trademark of The MathWorks, Inc. and is used with permission. The MathWorks does not warrant the accuracy of the text or exercises in this book. This book's use or discussion of MATLAB® software or related products does not constitute endorsement or sponsorship by The MathWorks of a particular pedagogical approach or particular use of the MATLAB® software.

First edition published 2023
by CRC Press
6000 Broken Sound Parkway NW, Suite 300, Boca Raton, FL 33487-2742

and by CRC Press
4 Park Square, Milton Park, Abingdon, Oxon, OX14 4RN

CRC Press is an imprint of Taylor & Francis Group, LLC

ISBN: 978-0-367-75970-4 (hbk)
ISBN: 978-0-367-75999-5 (pbk)
ISBN: 978-1-003-16507-1 (ebk)

DOI: 10.1201/9781003165071

Typeset in Times
by codeMantra

Access the Support Material: www.routledge.com/9780367759704

I'd like to dedicate this book to my parents, E.B. and Ruth Spencer, who first inspired me to follow my dream of being an aerospace engineer. I could also not have persevered in this book were it not for the continual support and encouragement from my wife Carol. Your love and patience with me during the years that went into writing this book made this all possible. Lastly, I'd like to thank my children, Brad, Michelle and Matt, who kept me motivated through this project.

DBS

I would like to dedicate this book to my sister Lucia, her daughter Cecilia, my wife-to-be Julia, and my parents Giovanni and Rossana, who have always loved and supported me in my career and life decisions on either sides of the Atlantic Ocean. I would not be where I am today without your continuous help and encouragement.

DC

Contents

Authors

Dr. David Spencer is a Professor Emeritus in the Department of Aerospace Engineering at The Pennsylvania State University, in University Park, Pennsylvania. His research areas include: spacecraft dynamics and controls, trajectory optimization, theoretical and applied astrodynamics, space systems engineering, space traffic management, orbital debris dynamics, and space technology development.

He began his career as a Member of the Technical Staff in the Astrodynamics Department at The Aerospace Corporation. He then worked for the Air Force Research Laboratory in various positions in the Space Vehicles Directorate. He joined the faculty at The Pennsylvania State University in 1999 and retired as an emeritus professor in 2021. He is now back at The Aerospace Corporation as a Senior Staff member. He is a Full Member of the International Academy of Astronautics, a Fellow of the American Astronautical Society, an Associate Fellow of the American Institute of Aeronautics and Astronautics, and a former Associate Editor for the *Journal of Spacecraft and Rockets.*

Dr. Spencer received a B.S. in Mechanical Engineering from the University of Kentucky, an M.S. in Aeronautics and Astronautics from Purdue University, an M.B.A. from Penn State, and a Ph.D. in Aerospace Engineering Sciences from the University of Colorado at Boulder.

Dr. Davide Conte is an Assistant Professor at Embry-Riddle Aeronautical University, in Prescott, Arizona. His research areas include: trajectory optimization, theoretical and applied astrodynamics, space systems engineering, space mission design, and proximity operations.

Dr. Conte was born in Genova, Italy, where he lived and attended school until the age of 17. He then received a scholarship to attend the Altoona Area High School in Altoona, Pennsylvania, as an American Field Service (AFS) scholar. He later returned to his hometown in Italy to complete his high school diploma at Liceo Scientific S. Maria ad Nives, in Genova Pegli, Italy. Dr. Conte then moved to the United States to attend The Pennsylvania State University, where he received a B.S. in Aerospace Engineering, a B.S. in Mathematics, an M.S. in Aerospace Engineering, and a Ph.D. in Aerospace Engineering. He joined the faculty in the Department of Aerospace Engineering at

Embry-Riddle Aeronautical University in 2019, where he has been leading a variety of competition-driven space mission design projects and teaches and does research in astrodynamics and space mission design.

He is a Member of the American Astronautical Society and a Member of the American Institute of Aeronautics and Astronautics.

Acknowledgements

We would like to thank all our students, especially those who took Dr. Spencer's Interplanetary Astrodynamics course at Penn State, who helped us flesh out the material in this book. Additionally, we'd like to thank the many colleagues, who, over the years, provided valuable insight and suggestions for the book, including Hal Beck, Jerome Bell, Denise Brown, Shyam Bhaskaran, Dennis Byrnes, Al Cangahuala, Todd Ely, David Folta, Martin Hechler, Kathleen Howell, Jugo Igarashi, Jeannie Johannesen, Martin Lo, Jim McAdams, William Owen, Ryan Park, Jeff Parker, Hank Pernicka, Chris Scott, Jon Sims, Ted Sweetser, Dave Vallado, Stacy Weinstein, Jim Wertz, Hideaki Yamato, and Ken Young.

1 Introduction

1.1 PURPOSE OF BOOK

We begin with a brief history of the pioneers who developed the fundamental dynamics of the motion of natural orbiting bodies, along the mathematical techniques that were applied to describe and characterize this motion. After the start of the space age, these fundamentals were expanded upon to apply to artificial objects, and additional techniques were developed and applied to the new problems in orbital mechanics such as maneuverable objects and objects that communicate with controllers. Applications of how these tools and techniques have been applied to interplanetary missions are also presented through a review of past and current interplanetary missions.

1.2 STRUCTURE OF BOOK

This book assumes that the reader has a background in calculus, physics, and dynamics, but not necessarily has a broad background in orbital mechanics.

Chapter 2 develops the dynamics necessary for describing the motion of bodies in space. Chapter 3 examines the N-body problem and shows applications using theories unique to this problem for designing interplanetary missions. Coordinate systems and time systems are tools necessary to correctly predict and execute missions, and are the topics for Chapter 4, which also outlines the dynamics of planetary motion and how this is applied to interplanetary spacecraft trajectories. Techniques for designing interplanetary orbital transfers are presented in Chapter 5. Navigation and targeting a spacecraft's trajectory, and knowing where a spacecraft is and comparing it to where we want it to be are the subjects of Chapter 6. Five appendices are also presented, showing constants and other data that are used to describe the motion of a spacecraft in orbit about a body or a body about the Sun.

1.3 HISTORY OF INTERPLANETARY ASTRODYNAMICS

In these next sections, we present a brief history of the study of astrodynamics, along with astronomical observations and mathematics, that over the course of many centuries allowed us to understand the motion of celestial objects in the Solar System, and beyond.

1.3.1 History of the Study of Orbital Mechanics

The fundamental understanding of the motion of bodies orbiting the Sun was grounded in the analysis of observational measurements. However, in order to more accurately explain these results, new mathematical methods were needed. This section briefly describes some of the pioneers in these efforts and briefly explains how

DOI: 10.1201/9781003165071-1

their works are used. Throughout this book, the mathematical details are added as we move to the solution of specific problems.

1.3.2 Ancient Astronomers and Mathematicians

Early civilizations, including the Chinese, Egyptians, Babylonians, Mayans, Romans, and Greeks (among others), looked to the sky to better understand nature [Szebehely, 1989]. The stellar constellations were named and studied to give various cultures a better sense of how to integrate their lives into the universe as a whole. Early activities, primarily based on gathering and growing food, required a good knowledge of the seasons, which are related to what was seen in the sky. Additionally, many early civilizations believed that the constellations were where their gods lived, and the need to understand the gods required them to better understand the sky.

Early theories on mathematically describing the motion were first introduced by the Greek philosopher and polymath Aristotle (384–322 B.C.). He believed that the Earth was the center of the universe, and everything moved relative to Earth. While we know this theory to be wrong, it led to another theory about gravity – that a body on the surface of the Earth would be at rest once any applied force is removed. This brought about the concept of "being at rest" [Seitz, 1992]. Later, Ptolemy (151–127 A.D.) first explained the motion of the heavens in Aristotle's Earth-centered universe by a complex mathematical description. He used an epicycle model with the eccentric method that *almost* fit the observed motion of heavenly bodies. Ptolemy's theories were widely held for the next millennium and became the basis of astronomical studies during the Renaissance until they were proven wrong by Copernicus.

1.3.3 Middle Ages Astronomers and Mathematicians

Nicolaus Copernicus (1473–1543) first proposed the theory that the Sun is the center of the Solar System (heliocentric system) and that the orbits of the planets formed circles about the Sun. This theory also accounts for the apparent retrograde motion of the planets beyond the Earth, as well as variations in the brightness of heavenly bodies [Chaisson, 2007]. This radical theory went against thousands of years of Earth-centered (geocentric) theory, and although Copernicus was not censured by the Church in his lifetime for his heliocentric theory, he was dismissed by Catholic and Protestant theologians, and his book was banned by the Catholic Church in 1854. Finally, only in 1993, Pope John Paul II stated that Copernicus' work was "one of the greatest scientific achievements of all time" [Brussels, 1993]. Copernicus' book, "On The Revolution of Celestial Spheres" [Copernicus, reprinted 2002], was published the year of his death and had a great influence on the future work of Galileo and Kepler, among others.

Furthering and expanding upon the heliocentric theory of the Solar System originated by Copernicus was Italian polymath Galileo Galilei (1564–1642) [Galileo, 1638]. Among Galileo's numerous contributions included the first uses of the telescope to study the motion of planets and their moons in the sky. Using his telescope, he discovered the four principal moons of Jupiter: Io, Callisto, Europa, and Ganymede, also known as the Galilean moons. Some of his early observations showed that these moons were moving back and forth relative to the planet in what is now referred to as simple harmonic motion. This simple harmonic motion is explained as uniform circular motion viewed edge-on. Galileo's observations of the phases of Venus showed that it orbited the Sun, and further strengthened the heliocentric theory. Like Copernicus, Galileo was in conflict with the Church, and spent the later years of his life under house arrest.

Tycho Brahe (1546–1601) [Darling, 2006, pp. 68–73], a Danish observational astronomer, was very meticulous about taking astronomical observations. Several of his accomplishments included the creation of several new measurement instruments. He recorded naked-eye observations of supernova and planetary parallax and collected measurements over a period of several years of the motion of the planet Mars and other planets relative to the constellations, making these the most accurate measurements to date (these measurements predated the invention of the telescope). Based on his observations, he rejected the Copernican model and created a new theory that the Earth was at the center of the universe and the Sun and Moon revolved around it. Additionally, the other planets revolved around the Sun. Brahe also found small errors in the apparent motion of Mars that contradicted the theory that all planets are in circular orbits around the Sun. It wasn't until Kepler had access to this data (after Brahe's death) that the theory of planetary circular orbits was replaced by the concept that planets move in elliptical orbits about the Sun.

Johannes Kepler (1571–1630) [Darling, 2006, pp. 73–75] was a German astronomer and mathematician who is arguably the most influential mathematician in the field of celestial mechanics. Using the observation data taken by Tycho Brahe, Kepler developed and validated his three laws of planetary motion, which the foundations of orbital mechanics are based upon. He believed Copernicus' heliocentric theory and wrote books about it. Kepler's three laws included (1) the planets move in elliptical orbits with the Sun at one of the foci, (2) an orbiting object sweeps out an equal area in an equal time, and (3) the square of the period of an orbit is proportional to the cube of its average distance from the Sun (its semimajor axis). While Kepler uncovered the rules in which the planets moved, he did not know how they moved. It took Isaac Newton (1642–1727) to explain how objects move under the force of gravity.

Sir Isaac Newton [Newton, 1687] was an English mathematician and physicist able to put a mathematical spin on the work of Copernicus, Galileo, Kepler, and others. Applying Kepler's third law (the square of the period is equal to the cube of the semimajor axis), Newton formulated the inverse-square law. This law states that the gravitational force between two objects is inversely proportional to the square of the distance between them. Although Newton is better remembered for his three laws of motion, his law of gravitation is one of the fundamental equations of orbital mechanics.

One of the many accomplishments of Swiss-French polymath Johann Lambert (1728–1777) was his development of a theorem for initial orbit determination. Although he did not produce an analytical proof (Lagrange came up with the analytical proof after Lambert's death), he did formulate a geometric proof [Szebehely, 1989, pp. 92–93]. Lambert's theorem provided a relationship between the time of flight between two points on an ellipse and the semimajor axis of the ellipse, the radii at the two points, and the chord length connecting the two points. An analogous derivation for hyperbolic orbits uses hyperbolic trigonometry functions. Lambert's theorem was previously shown for parabolic orbits by Newton and Euler.

1.3.4 18th-, 19th-, and 20th-Century Astronomers and Mathematicians

George William Hill (1838–1914) was an American mathematician who was known for his work in describing the relative motion between two objects in space [Hill, 1878]. This work was focused in the area of celestial mechanics and the relative motion of planets and their moons. It wasn't until the dawn of the space age that these concepts were applied to the relative motion between two artificial objects. William Clohessy and Raymond Wiltshire were two American engineers of The Martin Company who rederived the linearized equations of relative motion and applied them to developing a terminal guidance system to allow two spacecraft to rendezvous in orbit [Clohessy and Wiltshire, 1960]. While these equations were previously derived by Hill and others, they have become known as the Clohessy-Wiltshire or Hill-Clohessy-Wiltshire equations.

Henri Poincaré (1854–1912) was a French mathematician who did fundamental work on the restricted three-body problem. In 1887, Oscar II, King of Sweden and Norway, initiated a mathematical competition and established a prize for anyone who could find a power series solution of the N-body problem. In the event that a solution could not be derived, the prize would be awarded for an important contribution to classical mechanics. Although the problem was not solved, the panel decided to award the prize to Poincaré for his contribution to the understanding of the equations of dynamics (called Hamiltonian systems today), and for the many new ideas, he brought into mathematics and mechanics. This work demonstrated that certain numerical solutions of the three-body problem result in chaotic motion with no obvious sign of a repetitious path. The evolution of these orbits is so sensitive to minor changes in an object's position and velocity with respect to other gravitating bodies that it is essentially unpredictable. Poincaré's manuscript was important, because it laid the foundation of deterministic chaos theory [Diacu, 1996].

Carl Gustav Jacobi (1804–1851) [Bell, 1937] was a Prussian mathematician who, among other discoveries, expanded on the work of Lagrange and found an integral of motion (Jacobi integral) in the restricted three-body problem. The integration constant used in this solution is known as the Jacobi constant and is a fundamental characteristic of an N-body dynamical system.

Joseph-Louis Lagrange (born Giuseppe Luigi Lagrangia) (1736–1813) [Szebehely, 1989] was an Italian mathematician and astronomer, later naturalized French, who made fundamental contributions to the field of celestial mechanics and discovered equilibrium points in the restricted three-body problem. His prediction of the existence of asteroids in the triangular libration points in the Sun-Jupiter system was later proven by observational astronomers, and his mathematical developments laid the foundation of dynamics for the following centuries, into today.

François Félix Tisserand (1845–1896) [Szebehely, 1989] was a French astronomer who bridged the gap between observational and analytical astronomy. Tisserand discovered that when a comet approaches a massive body (Jupiter is commonly used as the massive body), the orbit of the comet is changed. Tisserand developed a mathematical method (known as Tisserand's Criterion) to correlate the orbit before and after the passage. This allowed for observational astronomers to recognize that a comet that approached and passed near Jupiter was indeed the same comet that was observed after passage [Roy, 2005].

A man who contributed to a multitude of fields was the American scientist Josiah Gibbs (1839–1903), who applied vector calculus to the problem of orbit determination of planetary and comet orbits. This method, known as Gibbs' Method, is still used today, primarily for initial orbit determination [Bate et al., pp. 109–116].

Methods of initial orbit determination using angular measurements were introduced by Pierre-Simon Laplace (1749–1827), a French mathematician and astronomer, whose main contribution to the field of celestial mechanics was to lay the foundation to a mathematical basis for the motion of orbiting bodies [Szebehely, 1989, pp. 18–19].

Sir William Rowan Hamilton (1805–1865), an Irish mathematician, astronomer, and physicist, was another giant of mechanics and mathematics, and is credited with developing the subject of Hamiltonian mechanics. In celestial mechanics, this work is used extensively in general perturbation problems [Szebehely, 1989, p. 19].

Carl Friedrich Gauss (1777–1855) was a German mathematician and physicist. Considered to be one of the greatest mathematicians, Gauss was, too, a master of many fields. In the field of orbital mechanics, Gauss developed a method to compute the orbit of an object, given two position vectors at different times and the time of flight between the observations. This was first applied to the orbit of the minor planet Ceres and is widely used in today's modern astrodynamics [Bate et al., 2020].

Leonhard Euler (1707–1783) was a Swiss mathematician who is known for many contributions to the fields of mathematics, physics, astronomy, geography, and engineering. His primary contribution to celestial mechanics was on lunar theory and how this could be used for celestial navigation. The theory was also used to create ephemeris information – predictions of orbital position and velocity – for the Moon [Szebehely, 1989, pp. 17–18].

Before the advent of computers, various perturbation methods had been developed for studying the motion of orbiting objects. In addition to the computational benefits, many perturbation methods provide valuable physical insight into the system's behavior. Special perturbation methods developed by the British astronomer Philip Herbert Cowell (1870–1949), and German astronomers Johann Franz Encke (1791–1865) and Peter Andreas Hansen (1795–1874) formulate the problems differently, in techniques that integrate the differential equations of motion in rectangular coordinates (Cowell), perturbing two-body orbits (Encke), and polar coordinates (Hansen) [Szebehely, 1989].

Expanding on the work done by Poincaré, Finnish mathematician Karl Sundman (1873–1949) used analytical methods to approximate the solution to the restricted three-body problem. Although Sundman's method is not practical for numerical solutions nor does it provide much information about the actual motion, it does show that a previously unsolvable problem actually is solvable [Bell, 1937].

1.3.5 Pre-space Age and Space Age Engineers, Scientists, and Mathematicians

As the space age neared, various 20th-century scientists and engineers developed theories and methods to further understand spaceflight. Walter Hohmann (1880–1945) was a German engineer who became interested in the problems associated with interplanetary spaceflight. He understood that an important factor in making spaceflight practical was to find ways to minimize the propellant usage. Ultimately, he discovered an orbit transfer that bears his name (Hohmann transfer), which is the minimum propellant, two-burn, circle-to-circle, co-planar orbital transfer [Hohmann, 1960].

Another field that is widely published is the field of estimation theory, and in particular, the ability to provide a best fit of a mathematical model through observation data that is "noisy." This is the basis of the Kalman filter, first published by Hungarian-American engineer Rudolf Kalman (1930–2016) [Kalman, 1960], for a mathematical technique to provide a linear least square fit to noisy data. This method was instrumental in the navigation system for the Apollo program and is widely used in many fields today.

While there are many more people in history who have contributed to the evolution of celestial mechanics, and continue this knowledge quest today, the people outlined in this section have developed techniques and methods that are detailed further in this book.

1.4 INTERPLANETARY SPACE MISSIONS

Information on these missions was gathered from websites that are maintained by various national space agencies, and the URL addresses are shown as footnotes throughout this chapter. The material changes over time, and the reader is encouraged to review the websites for the most up-to-date information.

1.4.1 Earth Moon Missions

On May 25, 1961, in his famous "Message to the Congress on Urgent National Needs," United States (U.S.) President John F. Kennedy spoke to the U.S. Congress [Kennedy, 1961], asking them to support his believe that "…this nation should commit itself to achieving the goal, before this decade is out, of landing a man on the Moon and returning him safely to the Earth". Kennedy reaffirmed this goal in a September 12, 1962 speech at Rice University in Houston, TX [Kennedy, 1962], in which he said

> We choose to go to the Moon in this decade and do the other things, not because they are easy, but because they are hard, because that goal will serve to organize and measure the best of our energies and skills, because that challenge is one that we are willing to accept, one we are unwilling to postpone, and one which we intend to win, and the others, too.

This goal was charged to the U.S. National Aeronautics and Space Administration (NASA). This program was one battle of the Cold War between the U.S. and the Union of Soviet Socialist Republics (U.S.S.R.) – also simply known as Soviet Union. However, regardless of the reason behind landing on the Moon, this program was probably one of the most technologically challenging programs ever undertaken. When Kennedy gave his speech to Congress, the U.S. had yet to orbit an astronaut. NASA faced many challenges, many of which involved developing new technologies, but others, related to the topic of this book, involved development of new flight dynamics tools and techniques.

Before NASA could send humans to the Moon, several precursor missions were necessary. At this time, the Soviet Union was working toward landing on the Moon as well. The first successful missions for both the Soviet and U.S. programs involved flybys of the Moon as well as impacts on its surface. In early 1959, the Soviet Luna 1[1] spacecraft, also known as Mechta (Russian: Мечта; lit. Dream), flew by the Moon at an altitude of between 5,000 and 6,000 km. The spacecraft is now in a heliocentric orbit with a period of 450 days. Shortly after Luna 1, the U.S. successfully launched the Pioneer 4 spacecraft. This spacecraft passed within 60,000 km of the Moon and is also now in a heliocentric orbit with a period of 398 days. Later in 1959, the Luna 2[2] spacecraft (Russian: Луна-2) discovered that the Moon had no appreciable magnetic field and no radiation belts, and its mission ended by impacting the spacecraft on the Moon. In 1964, the first successful U.S. spacecraft to the Moon, Ranger 7,[3] took several pictures before it impacted the Moon. The first successful "hard landing" on the Moon was done by the Luna 9[4] spacecraft (Russian: Луна-9) in 1966. The spacecraft impacted at approximately 6 m/s and survived impact. It transmitted pictures for 3 days and demonstrated that the surface of the Moon could support the weight of a foreign body. Later in 1966, the Luna 10[5] spacecraft (Russian: Луна-10) was inserted into orbit around the Moon and operated for 56 days. Additionally, in 1966, the U.S. Lunar Orbiter 1[6] spacecraft orbited the Moon for about 2 months. Several more Lunar Orbiter spacecraft orbited the Moon. The first "soft landing" on the Moon occurred with the U.S. spacecraft, Surveyor 1,[7] also in 1966. The Surveyor 1 spacecraft continued to operate for about a month and transmitted over 11,000 pictures back to Earth. Additional Surveyor spacecraft were flown, with several additional firsts accomplished, such as the first lunar regolith scoop (Surveyor 3,[8] in 1967), and the first "lift off" from the Moon (Surveyor 6,[9] in 1967). The Surveyor 3 spacecraft was later visited by the Apollo 12 crew, and several parts of the spacecraft were returned to Earth. Other interesting firsts included the Soviet Zond 5[10] spacecraft (Russian: Зонд 5, lit. *Probe 5*) that in 1968 performed a circumlunar mission by flying several biological samples (turtles, flies, worms, plants, seeds, and bacteria) on a trip around the Moon and a successful reentry and recovery back on Earth. These missions laid the groundwork for the first attempt of a human mission to the Moon in 1968.

The U.S. Apollo program was the culmination of a fairly successful series of crewed flights. When President Kennedy announced that the U.S. would send a man to the Moon and return them safely, the U.S. had not even sent a human into Earth orbit. The Soviets, however, had orbited several humans and maintained an early lead on the U.S. in heading to the Moon. After six crewed flights (each with one astronaut) of the U.S. Mercury program [Swenson et al., 1989], the technology demonstration

[1] https://solarsystem.nasa.gov/missions/luna-01/in-depth/
[2] https://solarsystem.nasa.gov/missions/luna-02/in-depth/
[3] https://solarsystem.nasa.gov/missions/ranger-7/in-depth/
[4] https://solarsystem.nasa.gov/missions/luna-09/in-depth/
[5] https://solarsystem.nasa.gov/missions/luna-10/in-depth/
[6] https://solarsystem.nasa.gov/missions/lunar-orbiter-1/in-depth/
[7] https://solarsystem.nasa.gov/missions/surveyor-1/in-depth/
[8] https://solarsystem.nasa.gov/missions/surveyor-3/in-depth/
[9] https://solarsystem.nasa.gov/missions/surveyor-6/in-depth/
[10] https://solarsystem.nasa.gov/missions/zond-5/in-depth/

Gemini series of missions were flown [Hacker and Grimwood, 1977]. These 12 flights (2 were unmanned, 10 had crews of two) demonstrated several technologies that were needed for the ultimate Moon landing, including extravehicular activity, long-duration spaceflight, rendezvous and maneuver of the combined system with propulsion from one vehicle, docking, rerendezvous (undock then rendezvous again), and precision reentry.

Leading up to human lunar landings, the biggest flight dynamics decision that needed to be made was exactly how to safely get the crew to the Moon and back. Ultimately, the direct ascent (taking the whole system to the Moon and back), Earth orbit rendezvous (docking two vehicles in Earth orbit and having that system land on the Moon and returning as a whole unit), and Lunar Surface Rendezvous (multiple vehicles landing on the Moon to support the crew) were rejected, and the Lunar Orbit Rendezvous concept was chosen [Brooks et al., 1979]. This concept allowed for one large vehicle (what ultimately became the Saturn V booster) to launch the entire mission. The crewed system would go into orbit around the Moon, and a smaller vehicle would take humans to the surface and return them to rendezvous with the vehicle that would take the entire crew back to Earth.

While the Gemini program was winding down, uncrewed tests of various Apollo systems were undertaken. Several uncrewed flight tests were conducted leading up to the first crewed test. Tragically, in January 1967, a flash fire occurred during a test of the Apollo 1 command module, killing three astronauts. After a thorough investigation, several improvements to the Apollo system were made, and the first crewed launch (Apollo 7) occurred in October 1968. This was followed up with the first circumlunar mission (Apollo 8) in December 1968, a flight test in Earth orbit of the Lunar Module (Apollo 9) in March 1969, and a lunar orbit test of the Apollo system (Apollo 10) in May 1969. On July 21, 1969 (less than 6 months before President Kennedy's deadline), Apollo 11 successfully landed two astronauts on the surface of the Moon. After a stay of a little more than 21 hours on the surface (of that, 2½ hours were spent outside the lunar module), the crew successfully launched from the Moon and rendezvoused with the command module. The three astronauts then successfully returned to Earth. Six more missions (with five successful landings) followed, and the program ended in 1972. These missions landed at various locations on the near side of the Moon – the landing sites are shown in Figure 1.1.

Various new flight dynamics techniques were developed for the Apollo mission. Many of these were done due to the technological limitations of the Apollo system (e.g., simple computer algorithms for complex dynamical systems), although other techniques, such as free-return trajectories and rendezvous about a distant body, remain timeless. Several of these methods are outlined later in this book.

At the same time, the Soviet Union was working on its own lunar landing program [Johnson, 1979]. During the Apollo program and for a few years after, the Soviet Union also launched several uncrewed missions to the Moon as part of their Luna program, including orbiters, rovers and sample-return missions. Some were successful and some failed. This program ended in 1976.

After the Apollo and Soviet lunar programs ended, there was a hiatus in spacecraft exploration of the Moon. In 1990, the Japanese Space Agency launched a technology demonstration satellite named Muses-A, which was later renamed Hiten

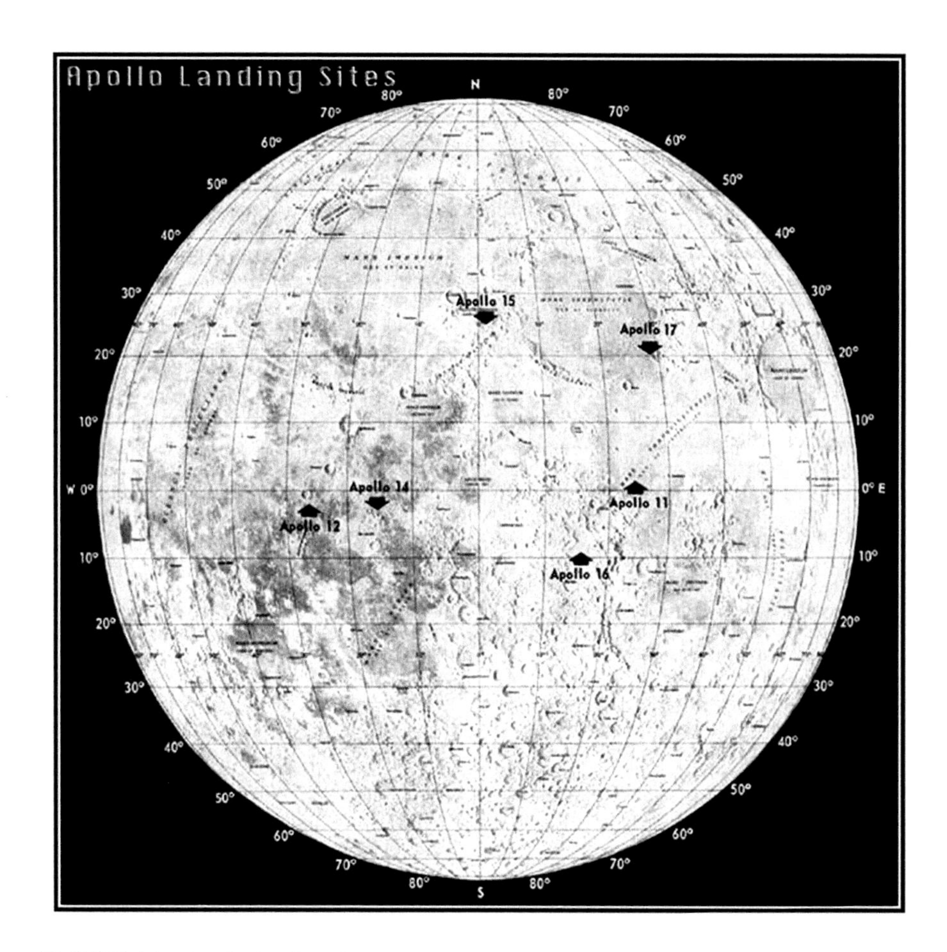

FIGURE 1.1 Apollo landing sites. (Courtesy of NASA.)

(Japanese: ひてん). Among the various experiments of this satellite included flight dynamics techniques of multiple gravity assists, aerobraking, lunar orbiter (a subsatellite was released from the spacecraft), and lunar impact. The spacecraft ultimately depleted its propellant supply and crashed into the Moon in 1993.

Also in 1990 (and again in 1992), NASA's Galileo spacecraft system flew by the Moon on its way to Jupiter. While these were minor goals for the mission, these flybys imaged the Moon's surface at various wavelengths. Although the trajectory is known as a VEEGA (Venus-Earth-Earth Gravity Assist), the gravity of the Moon also contributed to Galileo's gravity assist as it passed the Earth on two separate gravity assists.

In 1994, the U.S. Department of Defense and NASA jointly flew a technology demonstration satellite named Clementine.[11] This mission carried several imaging and mapping instruments, and tested these instruments as the spacecraft passed the Moon. After about 75 days, Clementine left lunar orbit on its way to fly by the asteroid

[11] https://solarsystem.nasa.gov/missions/clementine/in-depth/

Geographos, but a computer failure caused the spacecraft to lose spin control which resulted in no ability to collect data when it flew by Geographos.

In search for lunar ice, NASA launched the Lunar Prospector[12] mission in 1998. The spacecraft flew in a polar orbit for 19 months and was used for magnetic and gravity mapping of the Moon, and the mission ended when the spacecraft was purposely impacted into a crater in the lunar south pole to search for water vapor that may get kicked up (none was found).

The European Space Agency (ESA) launched its first mission to the Moon in September 2003. Named SMART-1[13] (Small Missions for Advanced Research in Technology 1), this technology demonstration satellite was launched as a secondary payload on an ESA Ariane 5 rocket. One of the technologies demonstrated was a low-thrust propulsion system, and the spacecraft eventually entered lunar orbit around 14 months after launch. The mission mapped the surface of the Moon, while the spacecraft's ion engine performed orbit adjustment maneuvers. The spacecraft was intentionally crashed into the Moon in 2006. The Kaguya[14] (Japanese: かぐや) mission (formerly called SELENE, Selenological and Engineering Explorer) was launched by the Japanese Aerospace Exploration Agency (JAXA) in September 2007 with the goal to study the selenology (geology of the Moon). In October 2007, the Chinese National Space Administration (CNSA) launched its Chang'e 1 spacecraft (simplified Chinese: 嫦娥一号). This was the first of a series of missions to study the Moon. This first mission ended in March 2009 when the spacecraft was crashed into the Moon. Chandrayaan-1[15] (Hindi: चन्द्रयान-1) was India's first lunar spacecraft. It was launched in October 2008 and released an impactor (that included an Indian flag) that crashed into the Moon. This made India the fourth nation to place a flag on the Moon. The mission ended in September 2009 after accomplishing 95% of its mission objectives.

NASA's Lunar Reconnaissance Orbiter (LRO)[16] was launched in 2009 and provided scientists with detailed global data about the Moon, such as day/night temperature profiles, geodetic data, and high-resolution imaging and ultraviolet albedo characteristics. This mission's priority was to focus on the polar region due to the solar illumination properties and the possibility of finding water in permanently shaded areas. Launched with the LRO, the Lunar Crater Observation and Sensing Satellite (LCROSS)[17] spacecraft was a short duration experiment. This spacecraft was designed to observe the impact into the Moon of the Centaur upper stage that delivered both payloads to the Moon. The LCROSS spacecraft observed the dust plume from the Centaur impact and transmitted that data back to the Earth before the LCROSS impacted the moon about 5 minutes after the Centaur impact.

NASA's Gravity Recovery and Interior Laboratory (GRAIL)[18] was a lunar orbiter whose observations enabled the creation of a high-quality gravitational field mapping of the Moon. This also allows us to determine the interior structure of the Moon and study it. Launched in September 2011, GRAIL was comprised of two twin spacecraft:

---

[12] https://solarsystem.nasa.gov/missions/lunar-prospector/in-depth/
[13] https://www.esa.int/Enabling_Support/Operations/SMART-1
[14] https://solarsystem.nasa.gov/missions/kaguya/in-depth/
[15] https://solarsystem.nasa.gov/missions/chandrayaan-1/in-depth/
[16] http://lunar.gsfc.nasa.gov/mission.html
[17] http://lcross.arc.nasa.gov/
[18] https://solarsystem.nasa.gov/missions/grail/in-depth/

GRAIL-A (Ebb) and GRAIL-B (Flow). Both spacecraft orbited the Moon in similar orbits in order to map the gravity field of Moon until they were purposely crashed into the lunar surface in December 2012.

As part of the Chinese Chang'e[19] series, a second mission was flown in 2010. This mission was a test vehicle for future lunar missions. Following its lunar mission, the spacecraft left lunar orbit and transferred into a Sun-Earth libration point (L_2). It later left this orbit and performed a flyby of asteroid 4179 Toutatis. The Chang'e 3[20] (simplified Chinese: 嫦娥三号) mission, launched in 2013, successfully landed in the Moon's northern hemisphere, making China the third country to have landed on the lunar surface. China's Chang'e 4[21] (simplified Chinese: 嫦娥四号) was the first mission to land on the far side of the Moon. Launched in May 2018, the lander and rover touched down in the Von Kármán crater in January 2019. Chang'e 4 was tasked to collect lunar samples of the far-side basaltic rocks. Chang'e 5 (simplified Chinese: 嫦娥五号) was China's first lunar sample-return mission – and the first mission to return samples from the Moon since the Soviet Luna 24 in 1976. Launched in November 2020, Chang'e 5 landed on the Moon in December 2020. The spacecraft collected 1.7 kg of lunar samples and returned them to the Earth approximately 2 weeks later.

The Korea Pathfinder Lunar Orbiter (KPLO),[22] also known as Danuri (Korean: 다누리), was South Korea's first orbiting mission to the Moon – a technology demonstration mission – operated by the Korea Aerospace Research Institute (KARI). Launched in August 2022, Danuri's objectives were to scout the lunar surface from orbit in search for water ice, silicon, and aluminum, and elements that are rare on Earth, such as uranium and helium-3.

In January 2004, U.S. President George W. Bush announced a new direction for NASA to develop a new spacecraft with the goal of returning humans to the Moon as part of his Vision for Space Exploration. This plan was perceived as a response to the Space Shuttle Columbia disaster and a way to regain public enthusiasm for space exploration. NASA began undertaking the Constellation Program, with its Ares launch vehicles, its Orion crew module, and its Altair lunar lander. The goals of the Constellation Program were to return humans to the Moon in a sustained and affordable way and to extend human presence across the Solar System, starting with crewed lunar missions by the year 2020 and pave the way for human exploration of Mars and beyond. This led to the development of innovative technologies, knowledge, and infrastructures required for Constellation. A test launch of the Ares I rocket (the Ares I-X mission) was successfully completed in October 2009.

In 2010, U.S. President Barack Obama canceled the Constellation Program after the Review of United States Human Space Flight Plans Committee deemed the program infeasible. While some of the assets being developed by NASA were kept, such as the Orion capsule, others were converted to be used on other proposed missions, such as the Space Launch System (SLS), which began development in 2011. During the Obama administration, NASA changed its focus from missions to the lunar surface to missions in the vicinity of the Moon. In fact, in 2013, the Asteroid Redirect

[19] http://nssdc.gsfc.nasa.gov/nmc/masterCatalog.do?sc=2010–050A
[20] http://nssdc.gsfc.nasa.gov/nmc/spacecraftDisplay.do?id=2013–070A
[21] https://solarsystem.nasa.gov/missions/change-4/in-depth/
[22] https://www.kari.re.kr/eng/sub03_07_01.do

Mission (ARM) was announced. ARM would be comprised of two main phases: the Asteroid Retrieval Robotic Mission (ARRM) spacecraft would rendezvous with a large near-Earth asteroid, retrieve a 4-m boulder from it, and return it to a lunar distant retrograde orbit. The second phase would see astronauts launch from Earth and rendezvous with the retrieved asteroid to collect samples and perform in-situ resource utilization experiments. In 2018, ARM was canceled, although many of its key technologies being developed continued, including SLS, Orion, and the ion thruster propulsion system that would have been flown on ARRM.

In 2019, under U.S. President Donald J. Trump, plans for the Lunar Gateway were laid out. The Lunar Gateway, or simply "Gateway," is a planned space station in a lunar near-rectilinear halo orbit intended to serve both human and robotic missions to the Moon and its vicinity. One of the key aspects of Gateway is that it will be constructed and serviced by collaborations of the public and private sectors of participating nations, including NASA, ESA, JAXA, the Canadian Space Agency (CSA), SpaceX, and Blue Origin. Gateway is planned to be assembled and used for missions orbiting and landing on the Moon, and to eventually enable and facilitate longer human spaceflight missions to Mars.

NASA's Artemis I[23] mission was an uncrewed test flight for NASA's Artemis program. This mission served as an uncrewed technology demonstration for various systems, including NASA's Space Launch System (SLS), a super heavy-lift launch vehicle, and the Orion spacecraft, a crew capsule. Launched in November 2022, the mission consisted in sending the uncrewed Orion capsule to a lunar distant retrograde orbit before returning to Earth.

Each of the programs discussed here has contributed to the next, and along with each of these missions, new flight dynamics techniques have been developed. One goal of this book is to outline these timeless techniques. New discoveries continue to be introduced, and it is expected that future editions of this book will cover many of these new methods.

1.4.2 Planetary Missions

During the early space program, the Moon was not the only destination for spacecraft. Mars and Venus, being the two closest planets to Earth, were the logical destinations for the first interplanetary spacecraft. Early in the space age, the U.S. and U.S.S.R. were the only two countries that attempted interplanetary missions. Later, these countries were joined by countries in ESA,[24] Japan, India, China, and the United Arab Emirates (UAE).

Three different types of interplanetary missions with different objectives were undertaken: planetary orbiters, planetary landers, and planetary flybys. Each has different technical challenges and has been applied to a variety of destinations in the Solar System.

---

[23] https://www.nasa.gov/specials/artemis-i/

[24] As of 2023, ESA is comprised of 22 member states: Austria, Belgium, Czech Republic, Denmark, Estonia, Finland, France, Germany, Greece, Hungary, Ireland, Italy, Luxembourg, The Netherlands, Norway, Poland, Portugal, Romania, Spain, Sweden, Switzerland, and the United Kingdom..

1.4.2.1 Planetary Orbiters

Planetary orbiters were the first interplanetary spacecraft. Sending a spacecraft into orbit around another planet is a complicated endeavor, and, early in the space age, there was a lot of learning needed to accomplish this. Many early missions failed to reach their destination, either missing a planet completely or failing to insert themselves into orbit. However, the space agencies learned from their failures and eventually succeeded in inserting spacecraft into orbit around other planets in our Solar System.

1.4.2.1.1 Mercury

MESSENGER[25] (MErcury Surface, Space ENvironment, GEochemistry and Ranging) was the first spacecraft to orbit Mercury. The probe was launched on August 3, 2004 from Cape Canaveral on a Delta II rocket. After launch, MESSENGER underwent an EVVMMM (Earth-Venus-Venus-Mercury-Mercury-Mercury) gravity assist sequence before performing its final orbit insertion maneuver at Mercury on March 18, 2011 entering a 9,300 × 200 km altitude orbit with a period of 12 hours. Figure 1.2 shows the trajectory of MESSENGER from Earth launch to Mercury arrival.

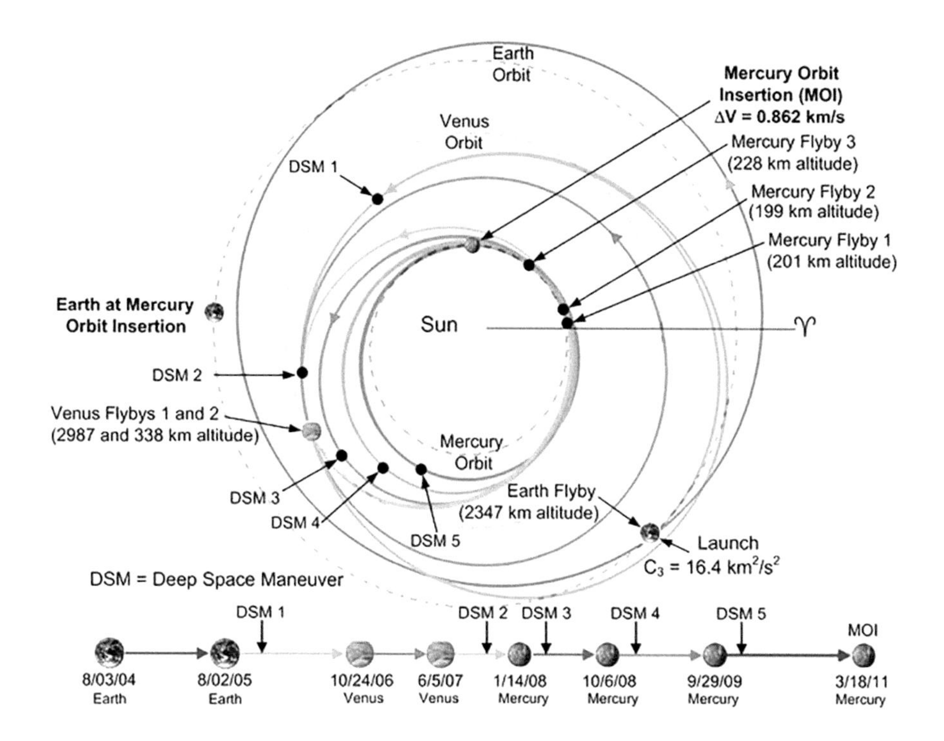

FIGURE 1.2 MESSENGER's trajectory to Mercury. (Courtesy of NASA.)

[25] https://solarsystem.nasa.gov/missions/messenger/

During its 4-year mission at Mercury, MESSENGER determined the composition of Mercury's surface, revealed its geological history, discovered details about its internal magnetic field, and verified that its polar deposits are predominantly water ice. The mission ended when MESSENGER was purposely deorbited and crashed into Mercury's surface on April 30, 2015.

BepiColombo[26] is a joint ESA-JAXA mission comprised of two spacecraft that launched together in October 2018 and are expected to arrive at Mercury in 2025. ESA's Mercury Planetary Orbiter (MPO) aims to study the planet's surface and interior while JAXA's Mercury Magnetospheric Orbiter (MIO) has the objective to study the planet's magnetic field.

1.4.2.1.2 Venus

The Soviet Venera 9 spacecraft (Russian: Вене́ра-9; lit. *Venus-9*) was the first probe to orbit Venus. Launched in June 1975, it consisted of an orbiter and a lander. The Venera 9 orbiter was in a 1,510 × 112,200 km altitude orbit around Venus at a 34-degree inclination functioning as a communications relay for the lander. We will discuss the "firsts" of the Venera program in more details in Section 1.4.2.2.

The Magellan spacecraft[27] was launched in May 1989 by NASA with the goal of mapping the surface and atmosphere of Venus. Named after the famous 15th–16th-century Portuguese explorer Ferdinand Magellan (Portuguese: *Fernão de Magalhães*), the spacecraft represents the first interplanetary mission to be launched by the Space Shuttle. Magellan was also the first mission to use aerobraking maneuvers as means of orbit circularization upon arriving at its destination (more on this topic is discussed in Section 5.14). During its nearly five and a half years of operations, Magellan was the first Venus probe to successfully image the entire surface of Venus while in a 297 × 8,463 km altitude Venusian orbit at a 85.5-degree inclination. On October 13, 1994, the spacecraft was commanded to plunge into the atmosphere of Venus to gather aerodynamic data. Contact with the probe was lost approximately 10 hours later as the spacecraft burned up in the Venusian atmosphere, completing one of the most successful deep space missions. Magellan found that at least 85% of the Venusian surface is covered with volcanic flows. Despite the high surface temperatures (475°C) and high atmospheric pressures (92 atmospheres), the spacecraft's observations suggested that the complete lack of water makes erosion an extremely slow process on the planet. As a result, surface features can persist for hundreds of millions of years. Furthermore, Magellan found that continental drift is not evident on Venus. Magellan's imagery still represents the best high-resolution radar maps of Venus' surface.

1.4.2.1.3 Mars

Here, we briefly present the most significant planetary orbiters that have successfully entered orbit around Mars. The missions to Mars as planned through the 2020s are listed on Figure 1.3.

---

[26] https://www.esa.int/Science_Exploration/Space_Science/BepiColombo

[27] https://solarsystem.nasa.gov/missions/magellan

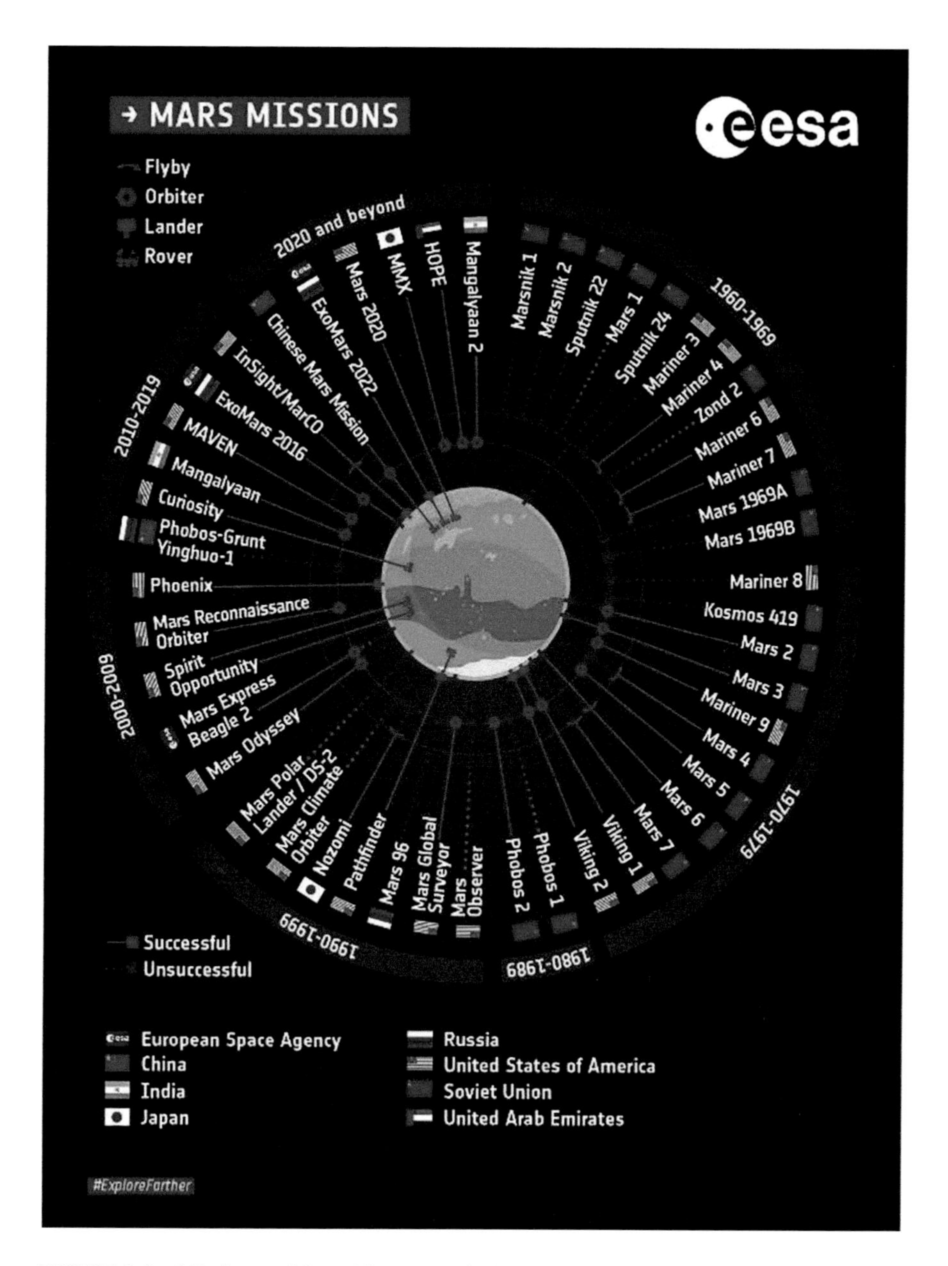

FIGURE 1.3 Missions to Mars. (Courtesy of ESA.)

The Mariner 9[28] was part of the NASA Mariner program and the first spacecraft to orbit another planet. It was launched on May 30, 1971, arriving at Mars on November 14, 1971. During its mission, it mapped 85% of the Martian surface, and transmitted over 7,000 images, including the first detailed views of the Solar System's

[28] https://www.jpl.nasa.gov/missions/mariner-9-mariner-i

largest volcano, Olympus Mons, which has a height of approximately 21.9 km (13.6 mi). Mariner 9 was deactivated on October 27, 1972, remaining a derelict satellite in Mars orbit until the spacecraft will eventually enter Mars and crash into its surface. However, the position of Mariner 9 is currently unknown: it is either still orbiting Mars or it has already crashed into its surface.

Mars 2[29] (Russian: Марс-2) was part of the Mars program, a series of uncrewed landers and orbiters launched by the Soviet Union on May 19, 1971. It included an orbiter and a lander module. Mars 2 reached Mars on November 27, 1971, when it started collecting imagery of the Martian surface and clouds, and topographical data. Part of its primary objectives also included determining the temperature of the planet and the composition and physical properties of its surface and atmosphere. It was also set to monitor the solar wind and the interplanetary and Martian magnetic fields, and transmit signals from the lander to Earth. Mars 2 completed its mission by August 22, 1972, after completing 362 orbits and sending back a total of 60 pictures. The data gathered by Mars 2 supplied the creation of surface relief maps and provided information on the Martian gravity and magnetic fields.

Phobos 2[30] (Russian: Фобос-2) was a Soviet mission launched on July 12, 1988 to explore Mars' largest moon, Phobos. Secondary objectives included studying the Sun, Mars, the interplanetary medium, and gamma-ray burst sources. The spacecraft took 37 pictures of Phobos, which collectively featured 80% of the moon. Additionally, it failed to find signs of water through its infrared spectrometer. The final phase of the mission included the release of two landers on the surface of Phobos (one mobile and a stationary platform); however, the mission was abruptly terminated due to a failure of the on-board computer on March 27, 1989, preventing spacecraft signals to be reacquired. To date, no mission has successfully landed on Phobos.

The Mars Global Surveyor (MGS)[31] was a robotic space probe launched by NASA JPL in November 1996. During its mission around Mars from September 1997 to November 2006, MGS provided information suggesting the presence of past and currently active water on the surface of the planet. Also, part of its original objectives, MGS sought to characterize the surface and geology of Mars, determine its global topography, shape, and magnetic and gravitational fields, monitor the planet's weather and the thermal structure of the atmosphere, and study the seasonal cycles in Mars' surface and atmosphere. MGS's initially unexpected longevity was accomplished through an ingenious strategy developed by the MGS team, the "angular momentum management" strategy, which minimized the need for thrusters to stabilize and balance the spacecraft. In summary, the approach aimed at tipping the spacecraft by 16° backward. By doing so, it reduced the daily use of propellant by 800%. The extension of MGS's mission allowed NASA to receive continuous weather monitoring data with the subsequent arrival of the Mars Reconnaissance Orbiter that took over this task after MGS was discontinued. It also allowed it to record imagery of potential landing sites for future missions (including the 2007 Phoenix spacecraft and the 2011 Curiosity rover), observe and analyze key scientifically relevant sites,

[29] https://nssdc.gsfc.nasa.gov/nmc/spacecraft/display.action?id=1971-045A
[30] https://heasarc.gsfc.nasa.gov/docs/heasarc/missions/phobos2.html
[31] https://mars.nasa.gov/mgs/overview/

and continue monitoring the surface's response to the pressures promoted by wind and ice. Although no longer in use, the spacecraft is still currently orbiting Mars and is expected to crash onto its surface at some point after 2047.

The Mars Atmosphere and Volatile Evolution (MAVEN)[32] mission was selected as part of NASA's now-canceled Mars Scout Program. MAVEN was launched on November 18, 2013 on an Atlas V rocket from Cape Canaveral Air Force Station, in Florida. The spacecraft performed its Mars orbit insertion maneuver on September 21, 2014, after which it entered a four and a half-hour period elliptical orbit around Mars. MAVEN is the first spacecraft whose objectives were to explore the atmosphere and ionosphere of the red planet and understand how they interact with the Sun and solar wind. Data collected from this mission might reveal how the loss of volatiles from the Martian atmosphere has affected the Martian climate over time, and thus contribute to a better understanding of terrestrial climatology. As of 2022, the spacecraft is still operational.

The ExoMars Trace Gas Orbiter (TGO or ExoMars Orbiter)[33] was a collaborative mission between ESA and the Russian space agency, Roscosmos, that sent an atmospheric research orbiter and the Schiaparelli demonstration lander to Mars in 2016 as part of the ExoMars program. Launched on March 14, 2016, TGO arrived at Mars on October 19, 2016 entering a 2-hour period circular orbit at an altitude of 400 km. One of TGO's main mission objectives was to gain a better understanding of methane and other atmospheric gasses that are present in small concentrations (less than 1% of the atmosphere) but nevertheless could be evidence for possible biological or geological activity. The spacecraft was also tasked with delivering the Schiaparelli lander, which suffered a malfunction during entry into the Martian atmosphere and crashed into the surface. Another mission objective of TGO was to serve as a data relay to support communications for the ExoMars 2022 rover and other surface assets.

Although the exploration of Mars has seen primarily the United States, Russia, and recently Europe as protagonists, other countries have started to contribute to the exploration and understanding of the red planet. Here, we list the most recent and successful of these missions.

Launched on November 5, 2013, the Mars Orbiter Mission (MOM; Hindi: मंगल कक्षित्र मशिन),[34] also called Mangalyaan (Hindi: मंगलयान; lit. *Mars-Craft*), was India's first interplanetary mission. This accomplishment made the country's space agency, Indian Space Research Organization (ISRO), the fourth space agency and the first Asian nation to reach Mars orbit. The mission sought to test various technologies that would allow ISRO to develop future interplanetary missions. Some of its primary objectives included to explore Mars' surface features and study its atmosphere, including its molecular constituents and dynamics. As of 2022, the spacecraft is still operational.

The Emirates Mars Mission[35] is UAE Space Agency's first space exploration mission to Mars. Launched on July 19, 2020, the Hope orbiter (Arabic: مسبار الأمل, *Misbar Al-Amal*) reached Mars on February 9, 2021. The objectives of this mission were

---

[32] https://solarsystem.nasa.gov/missions/maven/in-depth/

[33] https://exploration.esa.int/web/mars/-/46475-trace-gas-orbiter

[34] https://solarsystem.nasa.gov/missions/mars-orbiter-mission/in-depth/

[35] https://www.emiratesmarsmission.ae/

to understand climate dynamics and the global weather map by characterizing the lower atmosphere of Mars; explain how the weather affected the loss of hydrogen and oxygen in the Martian atmosphere; and understand the structure and variability of hydrogen and oxygen in the upper atmosphere. As of 2022, the spacecraft is still operational.

Tianwen-1[36] (TW-1; simplified Chinese: 天问; lit. *Heavenly Questions*) was China's first mission to Mars, organized and launched by the China National Space Administration (CNSA). TW-1 consisted of five parts: an orbiter, deployable camera, lander, drop camera, and the Zhurong rover. The spacecraft, launched on July 23, 2020, arrived at Mars on February 10, 2021. With a total mass of nearly 5,000 kg, TW-1 was one of the most massive probes launched to Mars to date. It was the first in a series of planned missions undertaken by CNSA as part of its Planetary Exploration of China program. TW-1's mission objectives are to study the geological structure of Mars; study the surface and underground layers of Martian regolith (including water ice); study the composition and type of geological formations on the Martian surface, including minerals present in ancient lakes and rivers; study the atmosphere of Mars, both in its near-space environment and on its surface; and study the internal structure of Mars, its magnetic field, the history of its geological evolution, the internal distribution of its mass, and its gravitational field. As of 2022, the spacecraft is still operational.

1.4.2.1.4 Jupiter

The Galileo spacecraft,[37] named after the 16th–17th-century Italian astronomer, physicist and engineer Galileo Galilei, was launched on October 18, 1989 from Kennedy Space Center, FL, aboard Space Shuttle Atlantis. Galileo's trajectory is explained in Section 1.4.2.3. Galileo orbited Jupiter for nearly 8 years while making several close approaches of its major moons, primarily the Galilean moons – Io, Europa, Ganymede, and Callisto. Galileo's camera and nine other instruments sent back reports that allowed scientists to determine, among other things, that Jupiter's icy moon Europa probably has a subsurface ocean with more water than the total amount found on Earth. They discovered that the volcanoes of the moon Io repeatedly and rapidly resurface the little world. They found that the giant moon Ganymede possesses its own magnetic field. Galileo achieved a number of firsts in space exploration, including being the first spacecraft to orbit an outer planet, to deploy an entry probe into an outer planet's atmosphere, and to complete the first flyby and imaging of an asteroid (Gaspra and Ida). Additionally, Galileo was the first (and so far only) spacecraft to perform a direct observation of a comet colliding with a planet's atmosphere (Shoemaker-Levy 9). Galileo was also the first spacecraft to operate in a giant planet magnetosphere long enough to identify its global structure and to investigate its dynamics.

1.4.2.1.5 Saturn

The Cassini-Huygens spacecraft[38] was the first probe to orbit Saturn. It was launched on October 15, 1997 on a Titan IV rocket, and it arrived at Saturn on July

[36] http://www.cnsa.gov.cn/english/

[37] https://solarsystem.nasa.gov/missions/galileo

[38] https://solarsystem.nasa.gov/missions/cassini/overview/

1, 2004 after a series of planetary flybys, as explained in Section 1.4.2.3. The mission consisted of two distinct spacecraft: the Italian Space Agency (Italian: *Agenzia Spaziale Italiana*; ASI) and NASA's Cassini orbiter, named after the 17th–18th-century Italian mathematician, astronomer, and engineer Giovanni Domenico Cassini, discoverer of Saturn's ring divisions and four of its satellites; and ESA's Huygens probe, named for the 17th-century Dutch astronomer, mathematician, and physicist Christiaan Huygens, discoverer of Titan. The Cassini spacecraft acquired scientific data of Saturn and its moons, including Titan (primarily through the Huygens lander, as explained in Section 1.4.2.2) and Enceladus. Cassini performed a series of 162 planned flybys of the moons of Saturn, during which it acquired and transmitted nearly 635 Gb of data, together with almost half a million images. Thanks to its 13 years spent in orbit around Saturn and its moons, Cassini was able to analyze the seasonal changes taking place on Saturn's surface and its moons. Additionally, seven new moons of Saturn were discovered thanks to Cassini's observations. One of these (named Pallene) was photographed during the Voyager 2's flyby of Saturn in 1981, but remained unrecognized until decades later. Several flybys of Enceladus, some as close as 48 km from its surface, led to the discovery of water vapor with traces of simple hydrocarbons venting from the geologically active south polar region of the icy moon. Cassini has provided strong evidence that Enceladus has an ocean with an energy source, nutrients, and organic molecules, making Enceladus one of the best places for the study of potentially habitable environments for extraterrestrial life.

1.4.2.1.6 Vesta and Ceres

Dawn[39] was the first mission to visit Ceres and Vesta, which are the two largest objects in the main asteroid belt between Mars and Jupiter. Launched on September 27, 2007 on a Delta II rocket from Cape Canaveral, Dawn arrived at Vesta on July 16, 2011 after performing a gravity assist of Mars, as shown in Figure 1.4. When Dawn arrived at Vesta, it became the first spacecraft to orbit an object in the main asteroid belt. On September 5, 2012, Dawn departed Vesta and arrived at Ceres on March 6, 2015. Dawn was the first spacecraft to orbit two extraterrestrial targets and the first spacecraft to visit a dwarf planet. Dawn used solar-electric propulsion for all of its in-space maneuvers, totaling to a cumulative velocity change of nearly 11.5 km/s, 2.7 times any prior spacecraft, and nearly equal to the velocity provided by Dawn's Delta II launch vehicle from the surface of Earth. Dawn's active powered flight totaled 5.9 years (54% of the time in space). Dawn's observations confirmed that Vesta is the parent of the HED (howardites, eucrites, and diogenites) meteorites, which Dawn connected to Vesta's large south polar basin. Dawn discovered that Ceres, the inner Solar System's only dwarf planet, was an ocean world where water and ammonia reacted with silicate rocks. As the ocean froze, salts and other minerals concentrated into deposits that are now exposed in many locations across the surface. Dawn also found organic compounds in several locations on Ceres' surface. After the end of its mission in 2018, Dawn was placed in a derelict stable orbit around Ceres, where it still remains to this day.

[39] https://solarsystem.nasa.gov/missions/dawn/overview/

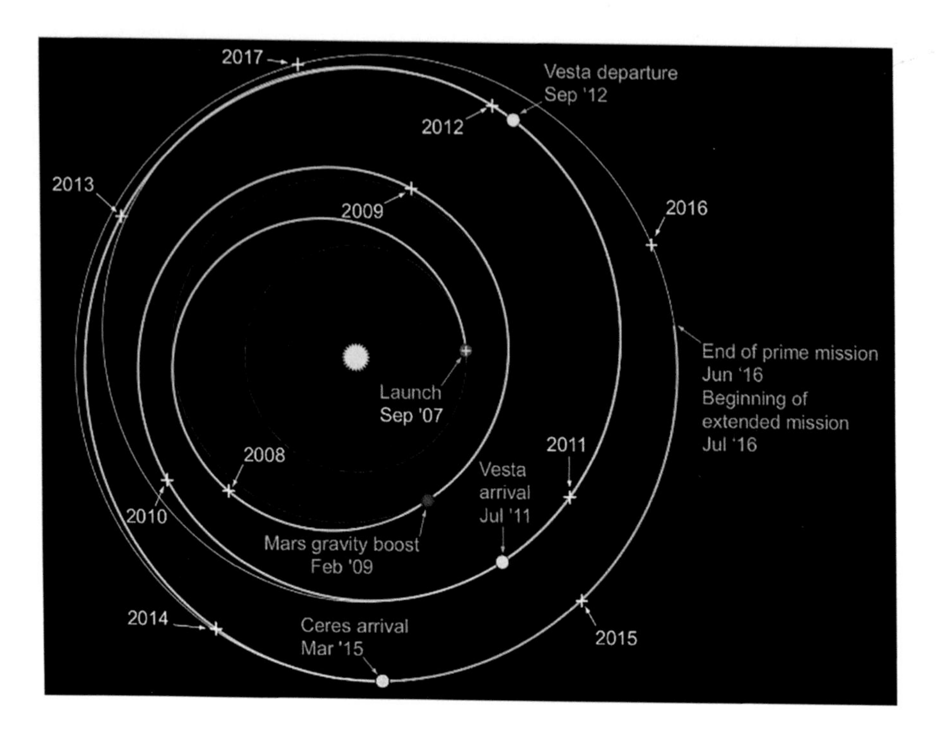

FIGURE 1.4 Dawn's trajectory. (Courtesy of NASA/JPL-Caltech.)

1.4.2.2 Planetary Landers

In this section, we briefly present the most significant planetary landers. Some of the most complex interplanetary missions are those that aim at landing on the surface of a far-away celestial body. This is because due to the time delay in communications, spacecraft need to be designed to complete their entry, descent, and landing (EDL) procedures autonomously through extraterrestrial atmospheres, among other challenges.

1.4.2.2.1 Venus

The Venera program[40] was a series of Soviet Venus probes designed by the Soviet Union to explore the atmosphere and surface of Venus. The program ran from February 1961 to July 1984. Ten probes successfully landed on the surface of the planet, while 13 probes successfully entered the Venusian atmosphere. Because of Venus' extreme surface conditions, the probes could only survive for a short period on the surface, with times ranging from 23 minutes to 2 hours. The Venera program established a number of precedents in space exploration, including being the first human-made objects to enter the atmosphere of another planet (Venera 3 on March 1, 1966), the first to make a soft landing on another planet (Venera 7 on December 15, 1970), the first to return images from another planet's surface (Venera 9 on June 8, 1975), the first to record sounds on another planet (Venera 13 on October 30, 1981), and the first to perform high-resolution radar mapping scans of Venus (Venera 15 on June 2, 1983).

[40] https://nssdc.gsfc.nasa.gov/planetary/venera.html

1.4.2.2.2 Mars

A multitude of missions were directed to exploring the surface of Mars, as shown in Figure 1.5 by the numerous planetary landers that, over the decades, have arrived onto the surface of Mars to study the geology of the planet and try to discover if life ever existed on the red planet.

NASA's Viking 1 mission made the first truly successful landing on Mars. The Soviet Mars 3 lander claimed a technical first with a survivable landing in 1971, but contact was lost seconds after it touched down. Both NASA Viking 1 and 2 missions used a combination of orbiter and lander to explore Mars in unprecedented detail. The Viking 1 lander touched down on Mars on July 20, 1976, while the Viking 2 lander arrived on September 3, 1976, about 6,500 km from the Viking 1. In total, the two Viking orbiters returned more than 50,000 images of Mars and mapped about 97% of its surface at a resolution of 300 m. The landers returned 4,500 photos of the two landing sites. The last contacts received from Viking 1 and 2 were in November 1982 and April 1980, respectively.

NASA's Spirit rover[41] and its twin Opportunity[42] comprised NASA's Mars Exploration Rover (MER) mission. Arriving on Mars on January 4, 2004 and January 25, 2004, respectively, Spirit and Opportunity studied the history of climate and water near their landing sites on Mars where it is believed that conditions may once have been favorable to life. Spirit uncovered strong evidence that Mars once had significantly more water than it does now. Spirit operated for 6 years, 2 months, and 19 days, more than 25 times its original intended lifetime, and traveled 7.73 km across the Martian plains. Opportunity operated almost 15 years, exceeding its life expectancy by 60 times, and had traveled more than 45 km by the time it reached its final resting location in Perseverance Valley. Opportunity stopped communicating with Earth after a severe Mars-wide dust storm blanketed its location in June 2018. Opportunity found evidence that Mars may once have been able to sustain microbial life.

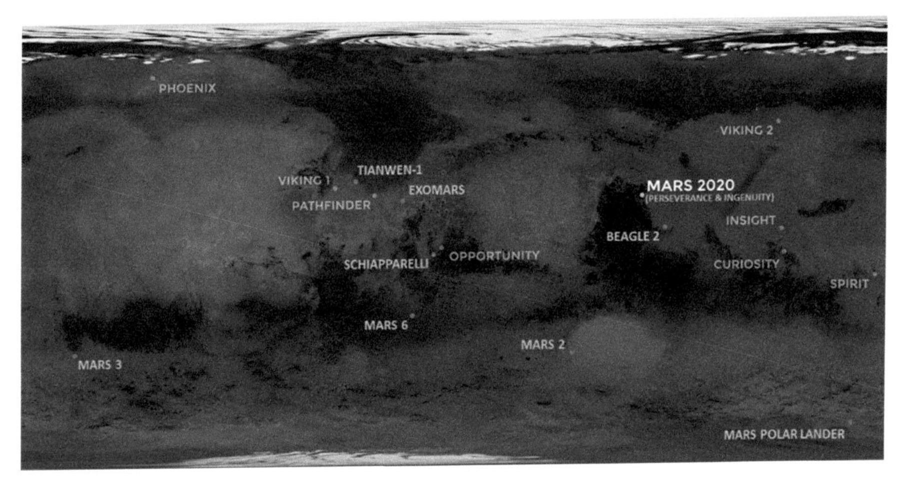

FIGURE 1.5 Map of all Mars landers and rovers, failed and successful. (Courtesy of NASA.)

[41] https://solarsystem.nasa.gov/missions/spirit/in-depth/

[42] https://solarsystem.nasa.gov/missions/opportunity/in-depth/

NASA's Phoenix lander[43] was launched on August 4, 2007 on a Delta II rocket from Cape Canaveral. It arrived and touched down on Mars on May 25, 2008, and remained active until November 2, 2008. Phoenix landed farther north on Mars than any previous spacecraft to date. During its surface mission, the lander verified the presence of water ice in the Martian subsurface, which NASA's Mars Odyssey orbiter first detected remotely in 2002.

NASA's Mars Science Laboratory (MSL) Curiosity[44] was a 900kg car-sized rover aimed at studying the ancient habitability and the potential for life on Mars. Curiosity was launched on November 26, 2011 on an Atlas V from Cape Canaveral. It landed on Mars on August 6, 2012, and started operations on August 29, 2012. The rover's goals included an investigation of the Martian climate and geology, studying if selected sites inside the Gale crater have ever offered environmental conditions favorable for microbial life (including investigation of the role of water), and planetary habitability studies in preparation for human exploration. In December 2012, Curiosity's 2-year mission was extended indefinitely. As of 2022, Curiosity is still operational.

NASA's Mars 2020 Perseverance Rover[45] is searching for signs of ancient microbial life in order to understand the past habitability of Mars. Perseverance launched on July 30, 2020 on an Atlas V from Cape Canaveral, and landed on Mars on February 18, 2021. The rover has a drill to collect samples of Martian rock and regolith, with the ability to store them for future sample-return missions. Perseverance is testing technologies to help pave the way for human exploration of Mars. Among them, the Mars Helicopter, Ingenuity, which landed on Mars stowed inside Perseverance, was the first rotary aircraft on a planet beyond Earth to perform controlled flight in the Martian atmosphere. Perseverance is seeking signs of ancient life by studying Martian terrain that is now inhospitable, but once held flowing rivers and lakes.

1.4.2.2.3 Asteroids and Comets

NASA's Near-Earth Asteroid Rendezvous (NEAR)[46] was the first spacecraft to orbit an asteroid and land on one. NEAR, which was later renamed NEAR Shoemaker, in honor of renowned geologist Eugene M. Shoemaker (1928–1997), was launched on February 17, 1996 on a Delta II from Cape Canaveral. NEAR arrived at its targeted asteroid, 433 Eros, on February 14, 2000. After performing 230 orbits of the asteroid, the spacecraft, which was not intended to be used as a lander, managed to survive touchdown on the asteroid. While in close proximity to the asteroid's surface, the spacecraft returned valuable data for about 2 weeks before mission control lost contact with it, on February 28, 2001, due to the extreme cold.

ESA's Rosetta[47] was the first spacecraft to orbit a comet, comet 67P/Churyumov-Gerasimenko. Rosetta was launched on March 2, 2004 on an Ariane 5 from Europe's Spaceport, ESA's primary launch site, located in French Guiana. Rosetta performed an orbit insertion maneuver at its targeted comet on August 6, 2014. Rosetta carried a lander with it, named Philae. On November 12, 2014, Philae detached from Rosetta

---

[43] https://solarsystem.nasa.gov/missions/phoenix/in-depth/
[44] https://solarsystem.nasa.gov/missions/curiosity-msl/in-depth/
[45] https://solarsystem.nasa.gov/missions/mars-2020-rover/in-depth/
[46] https://solarsystem.nasa.gov/missions/near-shoemaker/in-depth/
[47] https://www.esa.int/Enabling_Support/Operations/Rosetta

and landed on the comet, becoming the first successful landing on the surface of a comet. Although Philae landed in a shaded region of the surface of the comet, it was able to complete its primary mission objectives of sending images and data back to Earth. While in orbit around the comet, Rosetta monitored comet 67P/Churyumov-Gerasimenko's evolution during its closest approach to the Sun and beyond. The mission ended with a controlled impact of Rosetta on the comet on September 30, 2016. Both Rosetta and Philae remain on the surface of the comet.

1.4.2.2.4 Jupiter and Saturn Probes

The Galileo entry probe[48] was carried by the Galileo spacecraft, and it was deployed on July 13, 1995 and sent on a trajectory toward the atmosphere of Jupiter on July 13, 1995. On December 7, 1995, the 337-kg probe entered the Jovian atmosphere tasked with taking readings for nearly 1 hour before it was crushed by Jupiter's overwhelming atmospheric pressure. The Galileo entry probe became the first human-made object to enter the Jovian atmosphere. The entry probe found less lightning, less water vapor, and half the helium than had been expected in the upper atmosphere of Jupiter.

ESA's Huygens[49] was a probe that traveled to Saturn with the Cassini spacecraft. It was released from Cassini on December 25, 2004 targeting Saturn's largest moon, Titan. Huygens landed on Titan on January 14, 2005, becoming the first landing in the outer Solar System. Huygens was also the first spacecraft to sample an extraterrestrial ocean, revealing Titan to be one of the most Earth-like worlds we've encountered. Huygens was designed to study the smog-like atmosphere of Titan as it parachuted to the surface. It also carried cameras to photograph the moon's surface.

1.4.2.3 Planetary Flybys and Orbiters

What made various missions possible where propulsion limitations would make a direct trajectory impossible is the use of single and multiple flyby trajectories of one or several gravitational sources. Gravitational assists were first used for the Grand Tour mission of the Voyager program in the 1970s and 1980s and continue to be used very successfully today. The concept of the Grand Tour began in the 1960s, when JPL engineer Gary Flandro noted that an alignment of the outer planets Jupiter, Saturn, Uranus, and Neptune would occur in the late 1970s and would enable a single spacecraft to visit all of the outer planets by using gravity assists. As computed by Flandro, this planetary alignment repeats approximately every 175 years. JPL promoted the Grand Tour mission since it would allow a complete survey of the outer planets in less time and for less money than sending individual probes to each planet. Two spacecraft were designed and launched as part of the project. Voyager 1's course was planned to perform a flyby of Titan, while Voyager 2's trajectory was optimized for the Grand Tour. Voyager 2 was launched on August 20, 1977 on a Titan III rocket, while Voyager 1 was launched later, on September 5, 1977, also on a Titan III rocket. Voyager 2 would reach Saturn 9 months after Voyager 1, giving ample time to decide if it should proceed with the Grand Tour. Additionally, by launching Voyager 2 first, Voyager 1's launch could be retargeted to perform the Grand Tour if

[48] https://solarsystem.nasa.gov/missions/galileo-probe/in-depth/
[49] https://solarsystem.nasa.gov/missions/huygens/in-depth/

Voyager 2 were lost due to a launch failure. Detailed trajectories of Voyager 1 and 2 are shown in Figure 1.6, including the dates of closest approach during each flyby. Note that Voyager 1 flew-by Saturn at an altitude of approximately 124,000 km and shortly after flew-by Titan as close as 6,400 km. Because of this last planned flyby, Voyager 1 was unable to visit Uranus and Neptune, which were instead visited by Voyager 2. Had Voyager 1 failed or been unable to observe Titan, Voyager 2's trajectory would have been altered to incorporate the Titan flyby. However, doing so would have precluded any visit to Uranus or Neptune.

Voyager 1 and 2 have been flying longer than any other spacecraft in history. These twin spacecraft are helping scientists understand the nature of energy and radiation in space, which is fundamental for protecting future crewed and robotic missions. Voyager 1 was the first spacecraft to cross the heliosphere, the boundary where the influences outside our Solar System are stronger than those from our Sun, and it is thus the first human-made device to reach interstellar space. Voyager 1 discovered a thin ring around Jupiter and two new Jovian moons, Thebe and Metis. When flying-by Saturn, Voyager 1 discovered a new ring of Saturn (the G-ring) and five new Saturnian moons. Voyager 2 is the first (and so far only) spacecraft to study all four of the Solar System's outer planets at close range in one mission. Additionally, Voyager 2 discovered ten new moons of Uranus and two new of its rings. At Neptune, Voyager 2 discovered five moons, four rings, and a "Great Dark Spot" on the surface of the planet.

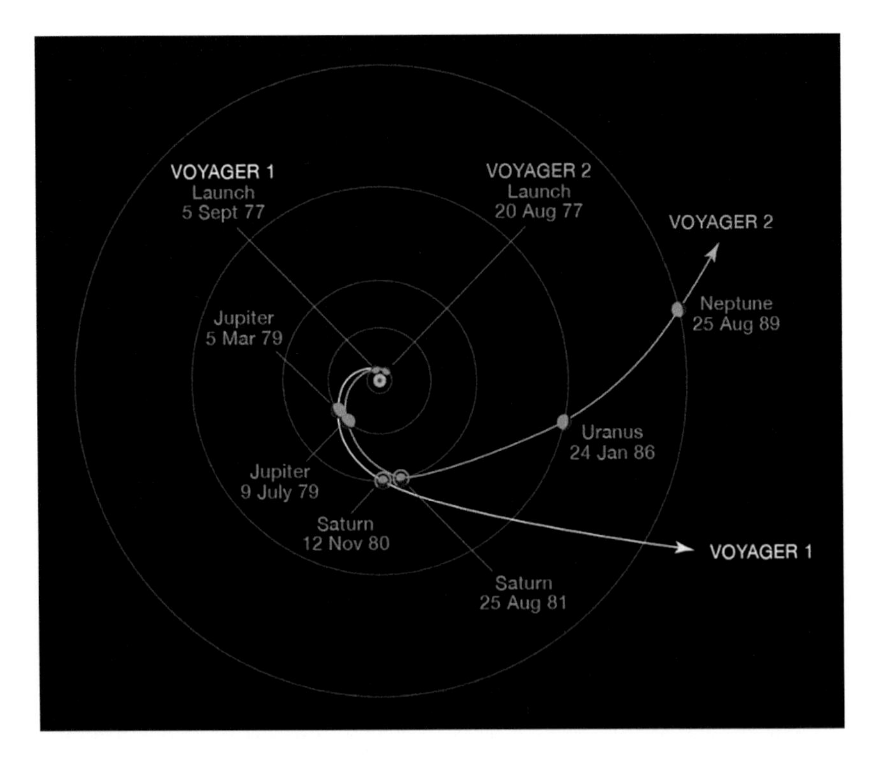

FIGURE 1.6 Voyager 1's and 2's orbital trajectories through the Solar System. (Courtesy of NASA/JPL-Caltech.)

Voyager 1 carries a copy of the well-known Golden Record, which is a message from humanity to the cosmos that includes greetings in 55 languages, pictures of people and places on Earth, and music. The contents of the Voyager's Golden Record were selected by a committee chaired by American scientist Carl Sagan (1934–1996), who noted that "[Voyager] will be encountered and the [Golden] Record played only if there are advanced space-faring civilizations in interstellar space, but the launching of this 'bottle' into the cosmic 'ocean' says something very hopeful about life on this planet".[50]

After Voyager, the next use of multiple gravitational flybys was done for NASA's Galileo mission to Jupiter. Originally, the Galileo spacecraft was to be launched from the Space Shuttle with the liquid-fueled Centaur upper stage. However, after the Space Shuttle Challenger accident in 1986, the decision was made to eliminate the possibility that the shuttle would have to land with a fully fueled spacecraft if, for some reason, the deployment was aborted. Instead, a safer solid propellant rocket, the Inertial Upper Stage (IUS), was to be used. However, the IUS did not have the same propulsive capability as the Centaur, so Galileo could not fly the same trajectory as it would with the Centaur upper stage. Although other options were considered (downsizing the spacecraft or using multiple launches), the final decision was to use gravitational assists from multiple bodies. A series of trajectories was designed to take advantage of a gravity assist from multiple bodies, and over a period of several years, new trajectories were designed to accommodate a continually moving schedule [Diehl et al., 1988; D'Amario et al., 1989].

On October 18, 1989, Galileo was launched on the Space Shuttle Atlantis, and in February 1990, the spacecraft made its first gravitational flyby past Venus at a range of around 16,000 km. Later that year (December 1990), Galileo received a gravitational assist as it made its first Earth flyby at an altitude of 960 km. In October 1991, an opportunity of geometry arose that brought Galileo close to the asteroid 951 Gaspra. While there was no change in the spacecraft's orbit, this gave project scientists an opportunity to test some of the spacecraft's systems in a realistic environment, similar to what the spacecraft would see when it reached Jupiter. A second Earth gravitational assist was completed in December 1992 at an altitude of 305 km. The spacecraft was then on its trajectory to Jupiter.

Another opportunity to fly by another asteroid, 243 Ida, was accomplished in August 1993. The comet Shoemaker-Levy 9 (D/1993 F2), discovered in 1993, was predicted to impact into Jupiter in July 1994. As Galileo was nearing Jupiter, it had a great view of the impact and provided some excellent scientific data. About 5 months before arrival at Jupiter, the spacecraft released a probe that plunged into Jupiter. In order to prevent the Galileo spacecraft from following the probe into Jupiter, a trajectory correction maneuver (TCM-25) was performed 2 weeks after the probe's release. The spacecraft arrived at Jupiter in December 1995 by performing a 49-minute burn, a Jupiter Orbit Insertion (JOI), and went into orbit around Jupiter. Galileo flew close to several moons, including Io (after JOI), Ganymede (G1, G2, G7, G8), Callisto (C3, C9, C10), and Europa (E4, E5, E6, E11). The number following the letter designates the orbit number around Jupiter. The trajectory is shown in Figure 1.7, while Figure 1.8 shows a more detailed orbital plot of the tour of the Jovian moons. On September 21, 2003, the Galileo spacecraft was purposely deorbited into Jupiter.

---

[50] https://voyager.jpl.nasa.gov/golden-record/

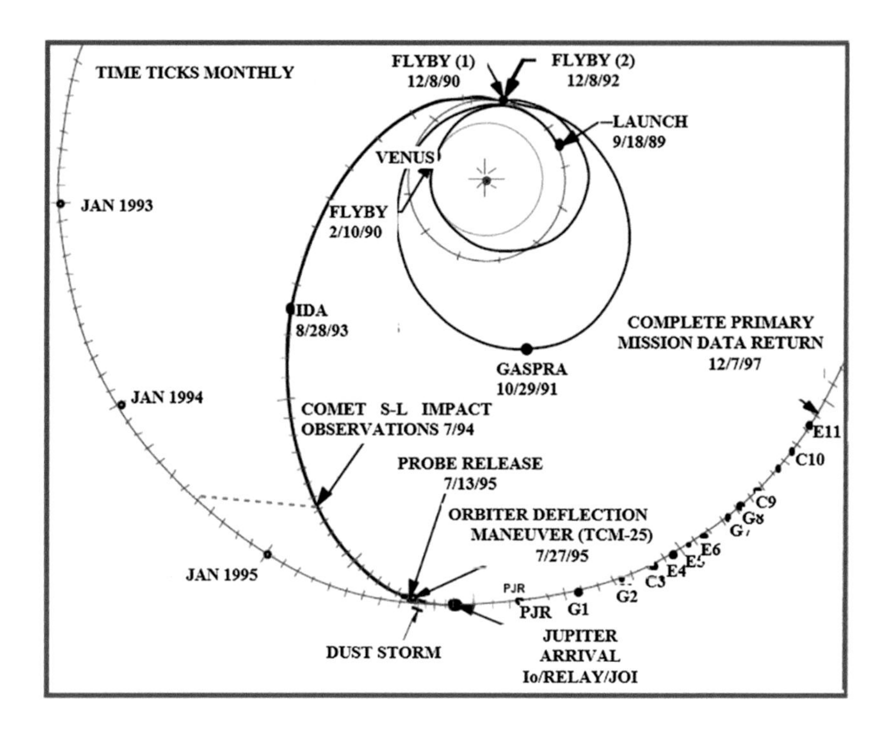

FIGURE 1.7 Orbital path of Galileo spacecraft. (Courtesy of NASA/JPL-Caltech.)

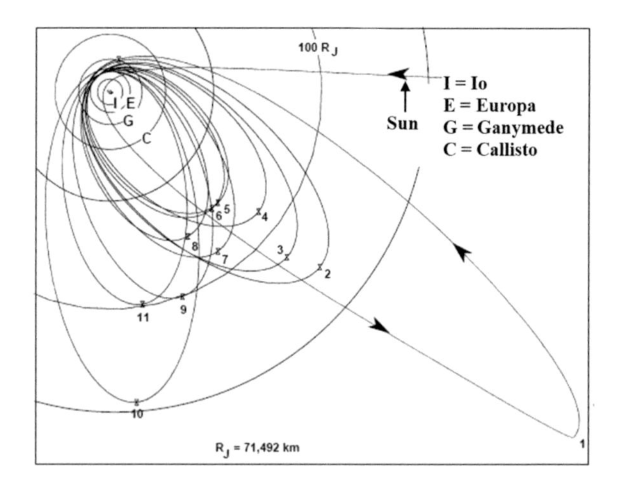

FIGURE 1.8 Jovian Moon tour. (Courtesy of NASA/JPL-Caltech.)

As mentioned previously, Cassini was (and remains) the most influential mission to explore Saturn and its moons. In order to arrive at its planned destination, the spacecraft was launched on October 15, 1997, toward the inner Solar System, to receive a gravity assist from Venus on April 26, 1998 at an altitude of 284 km, equating to approximately 7 km/s of velocity change. On December 3, 1998, Cassini performed a deep space maneuver with its main engines in order to target Venus for a second flyby, which occurred on June 24, 1999 at 623 km from Venus' surface. Then, on August 18, 1999, Cassini flew-by the Earth at an altitude of 1,171 km, resulting in an equivalent velocity change of 5.5 km/s. On its way to Jupiter, Cassini performed a close approach of asteroid 2685 Masursky on January 23, 2000, approximately 1.6 million km from it. On December 30, 2000, Cassini performed a gravity assist of Jupiter, at 9.7 million km. Cassini finally arrived at Saturn on July 1, 2004, when it began its primary scientific operations. Figures 1.9 and 1.10 show the trajectories of Cassini from Earth to Saturn, and during its primary mission phase once it arrived at Saturn, respectively.

Cassini's mission ended with the so-called Grand Finale Orbits (Figure 1.11), during which the spacecraft performed repeated close approaches of Saturn between the planet's upper atmosphere and its closest rings. On September 15, 2017, the Cassini mission ended when the probe was deliberately sent into Saturn's atmosphere ("Final Orbit" on Figure 1.11), concluding its 13 years of exploration and discoveries at Saturn.

Along with outer planet exploration, missions aimed at orbiting the Sun's closest planet, Mercury, are also challenging due to the amount of energy (and thus propellant usage) needed to accomplish. NASA's MESSENGER mission required a long sequence of flyby maneuvers (Figure 1.2) in order to successfully enter orbit around Mercury. After launch from Earth on August 3, 2004, it re-encountered Earth approximately a year later, on August 2, 2005, in order to perform a gravity assist that would decrease the spacecraft's energy with respect to the Sun. MESSENGER performed its Deep Space Maneuver (DSM 1) in order to target two back-to-back flybys of Venus on October 24, 2006, at an altitude of 3,324 km and on June 5, 2007, at an altitude of 300 km, respectively. A series of three flybys of Mercury followed, all at approximately 200 km of altitude, taking place on January 14, 2008, October 6, 2008, and September 29, 2009. On March, 18, 2011, MESSENGER was finally able

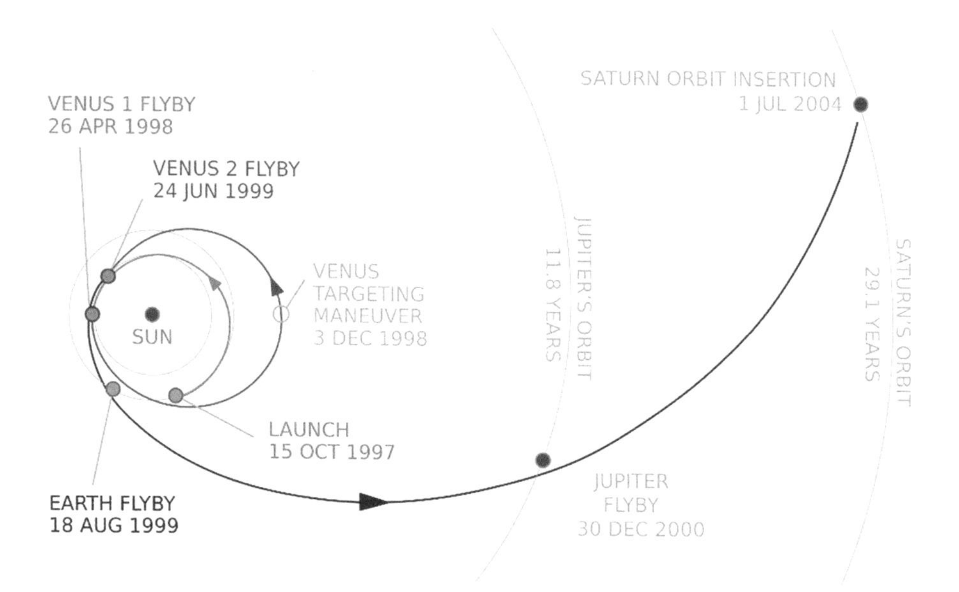

FIGURE 1.9 Cassini's orbital path from Earth to Saturn. (Courtesy of NASA/JPL-Caltech.)

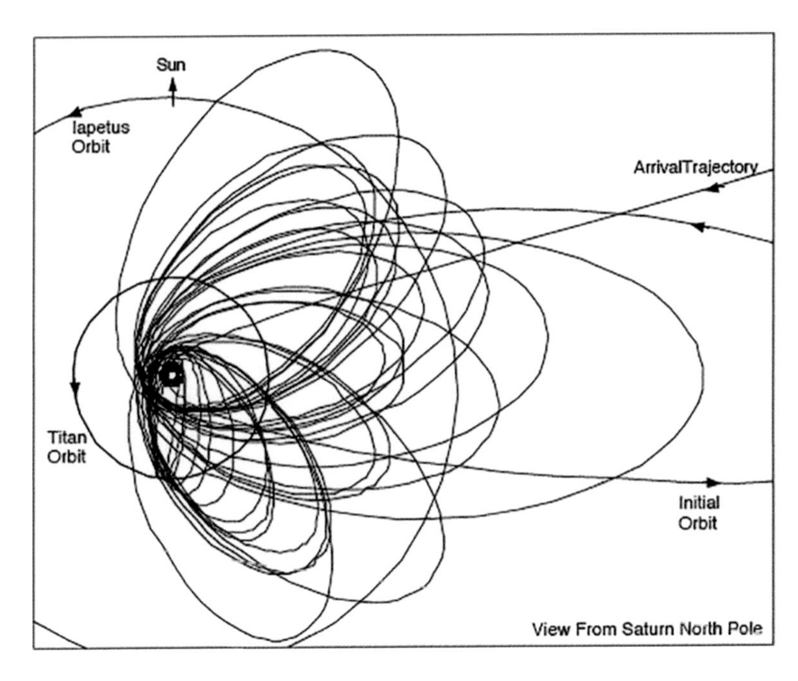

FIGURE 1.10 Cassini's orbital path from arrival at Saturn, including flybys of its Moons. (Courtesy of NASA/JPL-Caltech.)

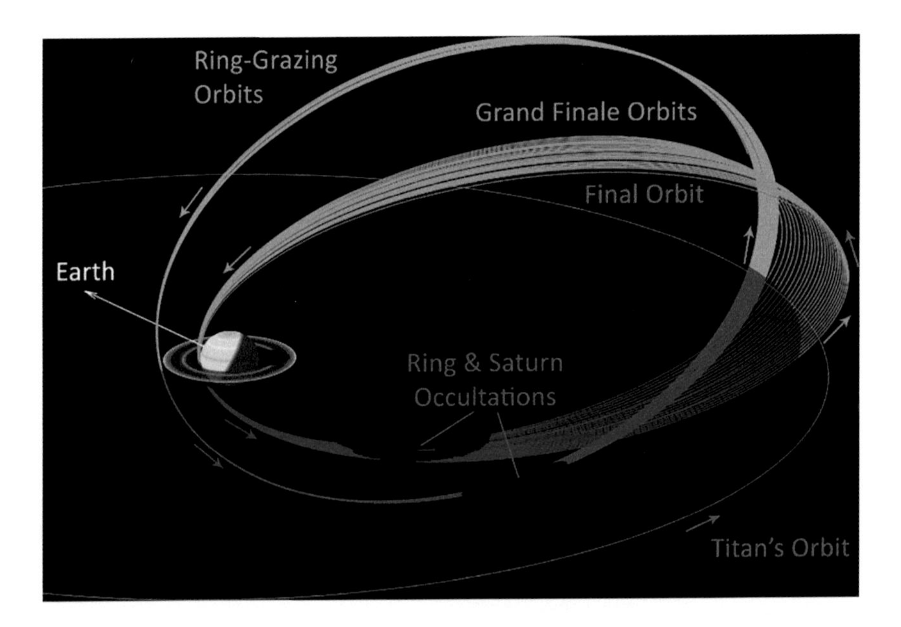

FIGURE 1.11 Cassini's grand finale orbits. (Courtesy of NASA/JPL-Caltech.)

to perform an orbit insertion maneuver at Mercury. Had MESSENGER not performed the last three flybys of Mercury, it would not have had sufficient propellant to enter Mercury's orbit in the required science orbit.

With an average distance from the Sun of 5.9 billion km (~39.5 the distance of Earth from the Sun) and an orbital inclination of 11.88° with respect to the Sun's equator, Pluto represents one of the most difficult objects to explore in the Solar System. Launched on January 19, 2006 with an Atlas V rocket, NASA's New Horizons (whose trajectory is shown in Figure 1.12) probe became the first spacecraft to explore Pluto and its moons up close. New Horizons was also the first spacecraft to explore a second Kuiper Belt Object up close, namely 2014 MU69. Due to the large distance to travel to reach Pluto, the spacecraft was released from the launch vehicle at a speed of 16.26 km/s, making it the fastest object to escape Earth's orbit. Additionally, the spacecraft arrived at and flew-by Jupiter on February 28, 2007, less than a year and a half after launch. With a close approach to Jupiter of 2.3 million km, New Horizons received enough orbital energy from Jupiter's encounter to arrive at and flyby Pluto on July 14, 2015 at an altitude of 12,500 km. New Horizons' objectives included mapping and characterizing the surface of Pluto and its moons, in particular Charon. At the time of writing, New Horizons is still operational and is being used to conduct further investigations of Kuiper Belt objects, such as 486958 Arrokoth, a 36-km-long trans-Neptunian object composed of two planetesimals 21 and 15 km across that are joined along their major axes.

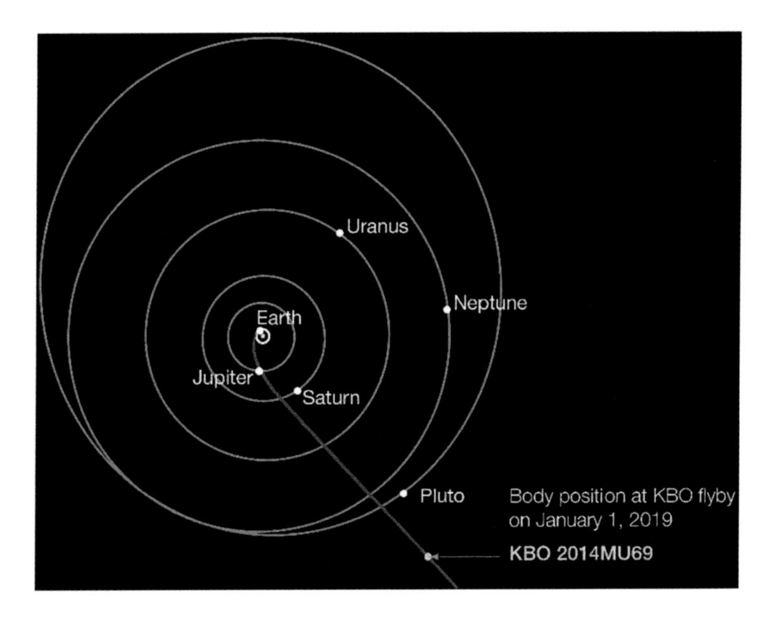

FIGURE 1.12 New Horizon's trajectory. (Courtesy of NASA.)

1.4.3 Future Missions

New space missions are proposed continuously, and throughout the book, we will present the most relevant ones according to the technical topics of discussion. In Chapter 3, we will discuss missions that have used or plan to make use of N-body orbits and how innovative mission design tools, such as dynamical systems theory, are applied for these complex missions. In Chapter 5, we will discuss missions that have used or plan to use a variety of orbital transfers, from the fundamental use of Lambert's problem and its many applications, to more complex mission design strategies, such as gravity assists, atmospheric effects, low-thrust trajectories, and other astrobatic maneuvers.

2 Kinematics, Dynamics, and Astrodynamics

2.1 NEWTON'S LAWS

The many accomplishments of Sir Isaac Newton include the formulation of three laws of motion. These laws state [Hawking, 2002]

1. "Every body perseveres in its state of rest, or of uniform motion in a right line, unless it is compelled to change that state by forces impressed thereon,"
2. "The alteration of motion is ever proportional to the motive force impressed; and is made in the direction of the right line in which that force is impressed,"
3. "To every action there is always opposed an equal reaction: or the mutual actions of two bodies upon each other are always equal, and directed to contrary parts."

These laws govern all motion and are the foundations of all dynamics. The first and third laws are difficult to attach a mathematical equation to, but the second law is commonly written as an equation,

$$\mathbf{F} = \mathrm{m}\mathbf{a} \tag{2.1}$$

where **F** is the net force vector, m is the mass, and **a** is the inertial acceleration.

In astrodynamics, the forces and resulting accelerations we commonly see are thrusting, perturbations, and gravitational. One corollary of Newton's second law is the law of gravity, in which Newton also proved that the gravitational force between two bodies is inversely proportional to the square of the distance between them,

$$\mathrm{F} \propto \frac{\mathrm{m}_1 \mathrm{m}_2}{\mathrm{r}^2} \tag{2.2}$$

The foundations of these laws were based on the work by Kepler. They answered fundamentals of the dynamics of motion, while Kepler's work addressed the kinematics of celestial bodies.

2.2 KEPLER'S LAWS

Johannes Kepler formulated three laws that describe the motion of a body relative to another body in space. While these three laws were not published chronologically, they are:

DOI: 10.1201/9781003165071-2

1. Planetary orbits are ellipses with the Sun at one focus.
2. A body sweeps out an equal area in an equal amount of time.
3. The square of the period is proportional to the cube of the semimajor axis (half of the long axis) of the elliptical orbit.

While each of these laws were first found and applied to natural orbiting objects, they are also applicable to artificial orbiting objects. They form the foundation to interplanetary astrodynamics, and throughout this chapter, the mathematical derivations of these laws will be presented and expanded upon.

2.3 TWO-BODY PROBLEM

The simplest dynamical system used in orbital mechanics is known as the two-body problem. This model describes the orbital motion of a small body with respect to a gravitational source. More complicated models are also used in practice, but generally require numerical methods in order to solve. The distinct advantage of the two-body model is that there is a closed-form, analytical solution. This solution was based on fundamental work done by Sir Isaac Newton, who put a mathematical basis to Kepler's work.

The fundamental physical laws governing the motion of a satellite in a gravity field were formulated by Newton and Kepler. Newton's three laws formed the basis of dynamics, and, along with Kepler, built the foundations of orbital mechanics. In particular, Newton's law of gravitation presented the concept that the gravitational force between two bodies is inversely proportional to the square of the distance between them.

This section shows how these laws were developed to describe the kinematics of planetary motion. In his work, Kepler described the motion of one body with respect to another. A reference frame, fixed in an inertial reference frame that does not move in space, has two moving bodies in this frame. Each body has mass, and this mass causes a gravitational attraction between these two bodies.

As seen in Figure 2.1, these two masses, m_1 and m_2 move in this inertial reference frame with the locations of these two masses as $\mathbf{R}_1$ and $\mathbf{R}_2$, respectively.

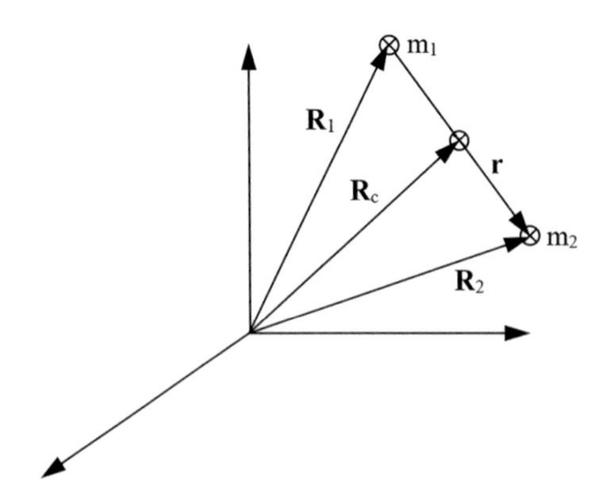

FIGURE 2.1 Two-body problem schematic.

The center of mass of these two bodies is located where the vector $\mathbf{R}_c$ intersects the vector, $\mathbf{r}$, connecting m_1 and m_2. In addition to the two masses moving relative to the inertial frame, their motion relative to each other can also be described.

The location of the center of mass is determined by looking at the mutual gravitational forces, as first developed by Newton. The equations of motion of these two bodies (using Newton's law of gravitation) are

$$m_1 \frac{d^2\mathbf{R}_1}{dt^2} = -\frac{Gm_1m_2}{|\mathbf{R}_1 - \mathbf{R}_2|^3}(\mathbf{R}_1 - \mathbf{R}_2) \tag{2.3}$$

$$m_2 \frac{d^2\mathbf{R}_2}{dt^2} = -\frac{Gm_1m_2}{|\mathbf{R}_2 - \mathbf{R}_1|^3}(\mathbf{R}_2 - \mathbf{R}_1) \tag{2.4}$$

where G is the universal gravitational constant. Adding these two equations together yields

$$m_1 \frac{d^2\mathbf{R}_1}{dt^2} + m_2 \frac{d^2\mathbf{R}_2}{dt^2} = 0 \tag{2.5}$$

We can define the location of the center of mass, $\mathbf{R}_c$, as

$$\mathbf{R}_c = \frac{m_1\mathbf{R}_1 + m_2\mathbf{R}_2}{m_1 + m_2} \tag{2.6}$$

Taking the derivative of Equation (2.6) twice, and inserting Equations (2.3) and (2.4) shows that $\frac{d^2\mathbf{R}_c}{dt^2} = 0$. Integrating this equation twice shows that the location of the center of mass is

$$\mathbf{R}_c = \mathbf{R}_{c0} + \frac{d\mathbf{R}_{c0}}{dt}(t - t_0) \tag{2.7}$$

where the vectors $\mathbf{R}_{c0}$ and $\frac{d\mathbf{R}_{c0}}{dt}$ are the initial position and velocities, with three components each. This yields six constants of motion that fully describes the conservation of linear momentum. This also means that the center of mass can at most move in a straight line with constant speed.

Multiplying Equation (2.3) by m_2 and Equation (2.4) by m_1, and subtracting them produces a new equation that describes the motion of m_2 with respect to m_1, and becomes

$$\begin{aligned} &m_1m_2 \frac{d^2\mathbf{R}_2}{dt^2} - m_2m_1 \frac{d^2\mathbf{R}_1}{dt^2} \\ &= m_1m_2 \frac{d^2\mathbf{r}}{dt^2} = -\frac{Gm_1m_2(m_1 + m_2)}{r^3}\mathbf{r} \end{aligned} \tag{2.8}$$

We can define a gravitational constant, μ, as $\mu = G(m_1 + m_2)$. Dividing by m_1m_2 yields,

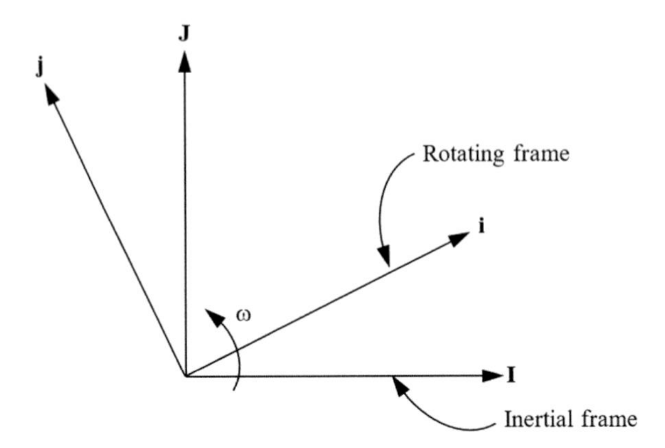

FIGURE 2.2 Rotating and inertial reference frames.

$$\frac{d^2\mathbf{r}}{dt^2} = -\frac{\mu\mathbf{r}}{r^3} \tag{2.9}$$

which gives the non-linear, second-order differential equation that describes the motion of a body relative to another in an inverse-square gravity field.

Before developing the equations of motion, we should review the relationships between the velocity and acceleration in rotating and inertial systems. Define an inertial frame, **IJK** and a rotating frame **ijk** as seen in Figure 2.2. The vectors **K** and **k** align with each other and are out of the page. The rotating frame rotates around the **K** axis at a rate ω.

An arbitrary point can be placed in the rotating frame, as seen in Figure 2.3, and the position vector from the origin of these systems to that point is **r**.

The time rate of change of a vector, $\boldsymbol{\alpha}$, is

$$\frac{d\alpha}{dt}^{(I)} = \frac{d\alpha}{dt}^{(R)} + \omega^{(R/I)} \times \alpha \tag{2.10}$$

where $\boldsymbol{\alpha}$ is a generic vector, the superscripts (I), (R), and (R/I) represent the inertial, rotating, and rotating with respect to the inertial frames, respectively. The vector $\boldsymbol{\omega}^{(R/I)}$ describes how the rotating frame moves around the inertial frame – the magnitude gives the rate in which the frames move with respect to each other and the vector direction describes the direction of the rotation axis, and is

$$\omega^{(R/I)} = \omega\mathbf{K} = \omega\mathbf{k} \tag{2.11}$$

Replacing the vector $\boldsymbol{\alpha}$ with the position vector **r**, the velocity in the inertial frame is

$$\frac{d\mathbf{r}}{dt}^{(I)} = \frac{d\mathbf{r}}{dt}^{(R)} + \omega^{(R/I)} \times \mathbf{r} \tag{2.12}$$

Applying Equation (2.10) to Equation (2.12) yields the acceleration in the inertial frame,

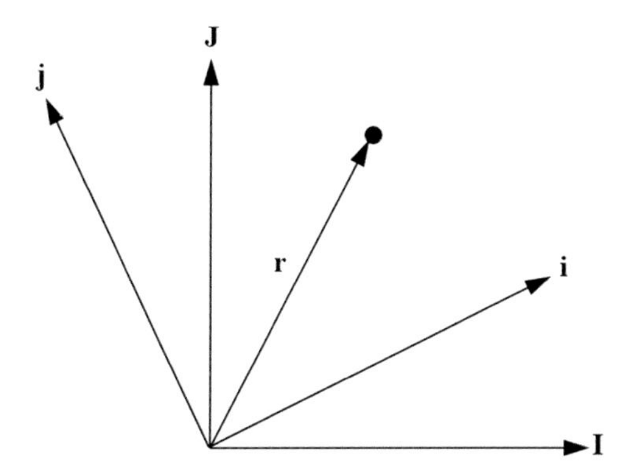

FIGURE 2.3 Position vector in the rotating and inertial frames.

$$\frac{d^2\mathbf{r}}{dt^2}^{(I)} = \frac{d^2\mathbf{r}}{dt^2}^{(R)} + 2\boldsymbol{\omega}^{(R/I)} \times \frac{d\mathbf{r}}{dt}^{(R)} + \frac{d\boldsymbol{\omega}}{dt}^{(R/I)} \times \mathbf{r} + \boldsymbol{\omega}^{(R/I)} \times \left(\boldsymbol{\omega}^{(R/I)} \times \mathbf{r}\right) \tag{2.13}$$

The left-hand side of Equation (2.13) is the acceleration in the inertial frame (the same acceleration in Newton's second law, $\mathbf{F} = m\mathbf{a} = m\frac{d^2\mathbf{r}^{(I)}}{dt^2}$).

Since the dynamical system describing the position of m_2 with respect to m_1 is not subject to external forces, both energy and angular momentum are conserved. Taking the dot product of the velocity vector, $\frac{d\mathbf{r}}{dt}$, with Equation (2.9) becomes

$$\frac{d\mathbf{r}}{dt} \cdot \frac{d^2\mathbf{r}}{dt^2} = -\frac{\mu}{r^3}\frac{d\mathbf{r}}{dt} \cdot \mathbf{r} \tag{2.14}$$

A closer examination of Equation (2.14) shows that the left-hand side of this equation is the time derivative of the kinetic energy per unit mass,

$$\frac{d\mathbf{r}}{dt} \cdot \frac{d^2\mathbf{r}}{dt^2} = \frac{1}{2}\frac{d}{dt}\left(\frac{d\mathbf{r}}{dt} \cdot \frac{d\mathbf{r}}{dt}\right) = \frac{1}{2}\frac{d}{dt}\left(v^2\right) \tag{2.15}$$

The velocity vector can be written as a linear combination of components in the radial and tangential directions as

$$\frac{d\mathbf{r}}{dt} = \mathbf{v} = \frac{dr}{dt}\mathbf{e}_r + r\frac{d\theta}{dt}\mathbf{e}_\theta \tag{2.16}$$

where $\mathbf{e}_r$ and $\mathbf{e}_\theta$ are unit vectors in the radial and tangential directions, respectively. The third component, $\mathbf{e}_h$, is perpendicular to the orbital plane. This geometry is shown in Figure 2.4.

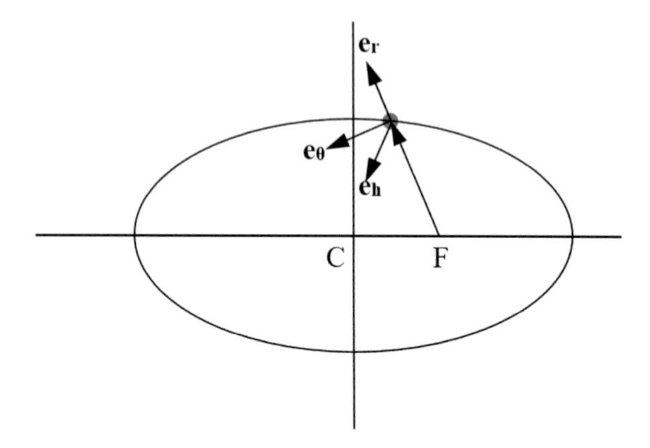

FIGURE 2.4 Rotating coordinate frame.

Taking the dot product of Equation (2.16) with the position vector, $\mathbf{r} = r\mathbf{e}_r$, shows that

$$-\frac{\mu}{r^3}\mathbf{r}\cdot\frac{d\mathbf{r}}{dt} = -\frac{\mu}{r^2}\frac{dr}{dt} = -\frac{d}{dt}\left(\frac{\mu}{r}\right) \tag{2.17}$$

which is the time derivative of the specific gravitational potential energy. Inserting Equations (2.15) and (2.17) into Equation (2.14) and integrating yields,

$$\frac{v^2}{2} - \frac{\mu}{r} = E \tag{2.18}$$

where E is the specific mechanical energy. This equation shows that the sum of the specific kinetic energy and the specific gravitational potential energy is a constant.

Equation (2.9) conserves angular momentum as well. Taking the cross-product of the position vector with Equation (2.9) gives

$$\mathbf{r}\times\frac{d^2\mathbf{r}}{dt^2} = -\frac{\mu}{r^3}\mathbf{r}\times\mathbf{r} = 0 \tag{2.19}$$

as the cross-product of a vector with itself is zero. The left side of Equation (2.19) is also the time derivative of $\mathbf{r}\times\frac{d\mathbf{r}}{dt}$,

$$\mathbf{r}\times\frac{d^2\mathbf{r}}{dt^2} = \frac{d}{dt}\left(\mathbf{r}\times\frac{d\mathbf{r}}{dt}\right) = \frac{d}{dt}\mathbf{h} = 0 \tag{2.20}$$

where $\mathbf{h}$ is the specific angular momentum $\left(\mathbf{r}\times\frac{d\mathbf{r}}{dt}\right)$. Integrating Equation (2.20) shows that

$$\mathbf{h} = \text{constant vector} \tag{2.21}$$

Thus, ten constants of motion are found (six from conservation of linear momentum, three from conservation of angular momentum, and one from conservation of energy). In order to integrate the equations of motion of each mass with respect to the inertial reference frame, a total of 12 constants of motion are needed. Since there are only ten constants of motion, the solution of Equations (2.3) and (2.4) cannot be found. Thus, only the motion of one body with respect to the other (Equation 2.9) can be solved.

In order to solve Equation (2.9), take its cross-product with the specific angular momentum,

$$\frac{d^2\mathbf{r}}{dt^2}\times\mathbf{h} = -\frac{\mu}{r^3}\mathbf{r}\times\mathbf{h} = -\frac{\mu}{r^3}\mathbf{r}\times\left(\mathbf{r}\times\frac{d\mathbf{r}}{dt}\right) \tag{2.22}$$

Applying the vector triple product identity, $\mathbf{a}\times(\mathbf{b}\times\mathbf{c}) = (\mathbf{a}\cdot\mathbf{c})\mathbf{b} - (\mathbf{a}\cdot\mathbf{b})\mathbf{c}$, the right-hand side of Equation (2.22) becomes

$$-\frac{\mu}{r^3}\mathbf{r}\times\mathbf{h} = -\frac{\mu}{r^3}\left(\underbrace{\mathbf{r}\cdot\frac{d\mathbf{r}}{dt}}_{r\frac{dr}{dt}}\right)\mathbf{r} + \frac{\mu}{r^3}\left(\underbrace{\mathbf{r}\cdot\mathbf{r}}_{r^2}\right)\frac{d\mathbf{r}}{dt} = \frac{d}{dt}\left(\frac{\mu\mathbf{r}}{r}\right) \tag{2.23}$$

Since $\mathbf{h}$ is a constant vector, the left-hand side of Equation (2.22) is

$$\frac{d^2\mathbf{r}}{dt^2}\times\mathbf{h} = \frac{d}{dt}\left(\frac{d\mathbf{r}}{dt}\times\mathbf{h}\right) \tag{2.24}$$

Inserting Equations (2.23) and (2.24) into Equation (2.22) results in

$$\frac{d}{dt}\left(\frac{d\mathbf{r}}{dt}\times\mathbf{h} - \frac{\mu\mathbf{r}}{r}\right) = 0 \tag{2.25}$$

Integrating once shows that

$$\mathbf{r}\times\mathbf{h} - \frac{\mu\mathbf{r}}{r} = \mathbf{C}\ (\text{constant vector}) \tag{2.26}$$

Arbitrarily, this constant of integration $\mathbf{C}$ can be replaced by $\mu\mathbf{e}$,

$$\frac{d\mathbf{r}}{dt}\times\mathbf{h} - \frac{\mu\mathbf{r}}{r} = \mu\mathbf{e} \tag{2.27}$$

The vectors $\mathbf{r}$ and $\mathbf{e}$ are always in the same plane (the orbital plane). By taking the dot product of the radius vector with Equation (2.27), and applying vector identities $\mathbf{a}\cdot\mathbf{b}\times\mathbf{c} = \mathbf{a}\times\mathbf{b}\cdot\mathbf{c}$ and $\mathbf{a}\cdot\mathbf{b} = ab\cos(\text{angle between } \mathbf{a} \text{ and } \mathbf{b})$,

$$\mathbf{r}\cdot\frac{d\mathbf{r}}{dt}\times\mathbf{h} = \left(\underbrace{\mathbf{r}\times\frac{d\mathbf{r}}{dt}}_{\mathbf{h}}\right)\cdot\mathbf{h} = h^2 = \frac{\mu}{r}\left(\underbrace{\mathbf{r}\cdot\mathbf{r}}_{r^2}\right) + \mu\left(\underbrace{\mathbf{r}\cdot\mathbf{e}}_{re\cos\theta}\right) \tag{2.28}$$

where θ is the angle between the radius vector and the **e** vector, and e is known as the eccentricity, which is the magnitude of the eccentricity vector. Rearranging this equation gives

$$r = \frac{h^2/\mu}{1 + e\cos\theta} \tag{2.29}$$

This equation, also known as the orbit equation, is the polar form of a conic section with the origin at one focus. While this is not an explicit representation of position as a function of time, the angle θ (also known as the true anomaly) is a function of time. The relationship between the true anomaly and time will be shown later.

Several physical characteristics can be determined using this geometry. First, note the term in the numerator should be a physical length (the term in the denominator is dimensionless). We can define a distance parameter when the true anomaly is equal to 90° as the semilatus rectum (also known as the parameter) and is generally designated as p. For a true anomaly of 90°, $r = p$, and inserting this definition into Equation (2.29), we find that

$$p = \frac{h^2}{\mu} \tag{2.30}$$

Variations in this equation give us another useful relationship between the angular momentum scalar value and a physical dimension of the orbit.

Second, note that at true anomalies of $\theta = 0°$ and 180°, the radius is equal to its smallest value, r_p – also known as periapsis – and its largest value, r_a – also known as apoapsis – respectively. Inserting these true anomaly values into Equation (2.29) gives

$$r_p = \frac{h^2/\mu}{1 - e} \tag{2.31}$$

and

$$r_a = \frac{h^2/\mu}{1 + e} \tag{2.32}$$

Also,

$$r_a + r_p = 2a \tag{2.33}$$

Substituting Equations (2.31) and (2.32) into Equation (2.33) shows that

$$h^2/\mu = a\left(1 - e^2\right) \tag{2.34}$$

which gives a further relationship between angular momentum and the size and shape of the orbit,

$$p = \frac{h^2}{\mu} = a\left(1 - e^2\right) \tag{2.35}$$

Inserting this into the orbit equation, Equation (2.29) gives another valuable interpretation of the parameters in the orbit equation,

$$r = \frac{a\left(1 - e^2\right)}{1 + e\cos\theta} \tag{2.36}$$

Thus, the apoapse and periapse radii are

$$r_a = a(1 + e) \qquad r_p = a(1 - e) \tag{2.37}$$

The second part of defining the physical characteristics of the orbit is the eccentricity. While semimajor axis gives the size of the orbit, eccentricity defines the shape. The eccentricity can be found by examining the difference between the apoapse and periapse radii,

$$r_a - r_p = a\left[1 + e - (1 - e)\right] = 2ae \tag{2.38}$$

Rearranging this equation and solving for the eccentricity, it is found that the eccentricity is

$$e = \frac{r_a - r_p}{2a} = \frac{r_a - r_p}{r_a + r_p} \tag{2.39}$$

A special case of an elliptical orbit is a circular orbit. In a circular orbit, the two foci collapse on the center of the conic section, making the periapse and apoapse equal. This means that a circular orbit has a zero value for eccentricity. While a truly circular orbit is not physically possible, there are many mathematical simplifications in orbital mechanics that can be made by assuming a circular orbit.

2.4 CONIC SECTIONS

The distinguishing features of various conic sections are characterized by several sets of parameters. The three descriptors of a conic section include: (1) is the conic section open or closed, (2) what is the size of the conic section, and (3) what is the shape of the conic section.

The first characteristic describes whether an object moving on the conic section returns to its original location periodically. For example, an object moving on a circle or ellipse returns to its origin and continues to repeat itself. An object moving on an open conic section does not repeat its location – as the object moves closer to the focus and then moves away, it starts at an infinite distance, approaches the focus, and moves away to an infinite distance. Parabolas and hyperbolas are two examples of open orbits.

The second and third characteristics are coupled. In planar geometry, a closed conic section can be described by its major and minor axes. The origin of these axes is the geometrical center of the conic section. In a coordinate system where the origin is located at the focus, a more common set of descriptors is the major axis (or half of the major axis, also known as the semimajor axis) and the eccentricity. In an open

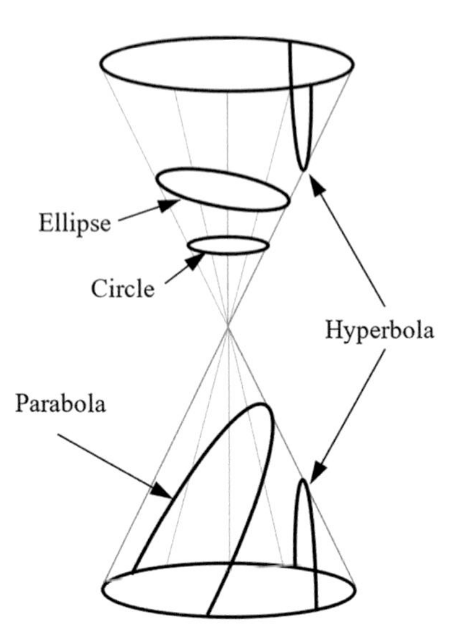

FIGURE 2.5 Conic section cuts.

conic section, a common size and shape descriptor is the distance from the closest point to the focus and the "turning angle" of the path.

A common geometric descriptor of conic sections is the inverted circular cone model, as seen in Figure 2.5. A plane slices the cones at various angles, and the cross section shows various conic sections. Should the slicing plane pass through the cone parallel to either base the cross section that results is a circle. Should the slicing plane pass through the cone at a slight angle relative to the base, as long as the slice doesn't intersect with the base of either cone, the resulting cross section is an ellipse. If the slicing plane intersects the base of one of the cones (an extension of the ellipse case), the resulting slice is a parabola. If the slicing plane passes through the cones perpendicular to the base, the resulting slices are the two mirror images of a hyperbola.

2.4.1 Elliptical Orbits

Ellipses in the two-body problem are closed orbits and repeat themselves. The size of the ellipse is defined by its major and minor axes. Here, the major axis is the length of the long axis, and the minor axis is the length of the short axis. A further definition is the semimajor (a) and semiminor (b) axes. These values are the distances from the center point of the ellipse along the major and minor axes, respectively. Two foci are characteristic of an ellipse, and both lie along the major axis. The gravitational source is at one focus (the occupied focus), while the other focus is empty (known as the vacant focus). Two locations of particular note are the apoapse (furthest point to the occupied focus) and the periapse (nearest point to the occupied focus), represented as r_a and r_p, respectively. This elliptical orbit is shown in Figure 2.6. An additional term that is shown in this figure is the angle between the location of the periapse and the radius vector, and is known as the true anomaly, θ.

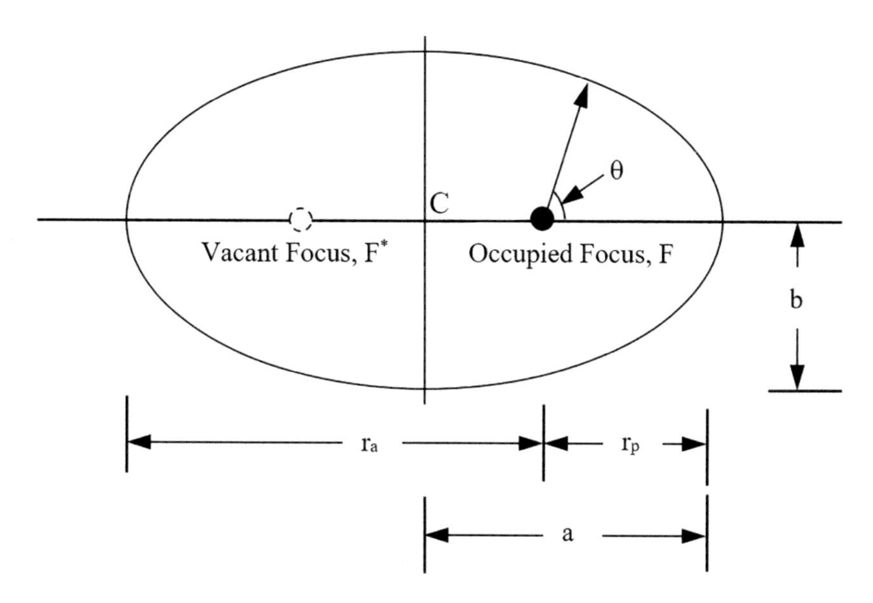

FIGURE 2.6 Example elliptical orbit.

2.4.2 Circular Orbits

The simplest of all conic sections is the circle. A circle is simply a special case of an ellipse, albeit one where the radius is constant and eccentricity is equal to zero. No real orbit is truly circular – there are always slight deviations (and it is impossible to have an infinite number of zeros after the decimal point in the eccentricity term). All of the concepts of an ellipse hold true for circles. All of the equations describing the radius, velocity, angular momentum, and energy (although, in many cases, are trivial).

Many simplifications in astrodynamics use the assumption of a circular orbit. This assumption makes otherwise complex, non-linear problems into simple, linear problems many times, with analytical solutions. While this assumption induces some error in the solution, it can provide interesting insight into the behavior of various problems.

2.4.3 Parabolic Orbits

A parabolic orbit is a physically impossible but mathematically important orbit. Like an exact circular orbit ($e = 0.000\ldots$), a parabolic orbit has an eccentricity of 1 ($e = 1.000\ldots$). Many of the orbit equations are similar to those of elliptical orbits, with the exception of semimajor axis. In a parabolic orbit, the semimajor axis is undefined (it is common to see $a \to \infty$). Instead, the common distance parameter that is used is the semilatus rectum (or parameter), p. Kepler's problem (to be introduced later) also has a slightly different formulation for parabolic orbits.

An example parabolic orbit, also showing a radius vector and true anomaly, is shown in Figure 2.7.

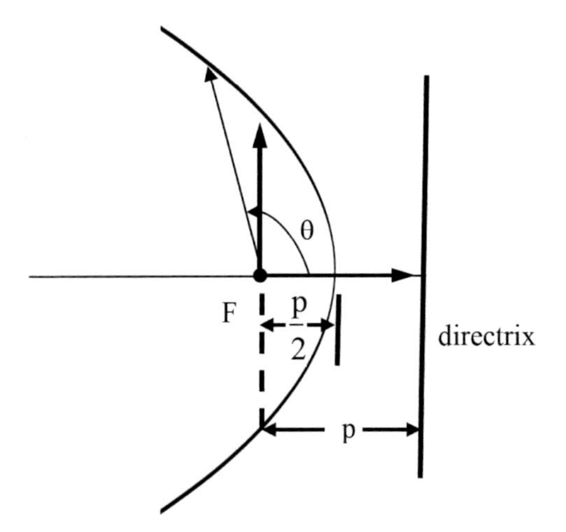

FIGURE 2.7 Example parabolic orbit.

Here, the distance between the origin of the coordinate frame and the periapse point is $p/2$. A directrix is located at a distance of p from the origin of this coordinate frame. Since the eccentricity of the orbit is 1, the orbit equation, Equation (2.29) becomes

$$r = \frac{h^2/\mu}{1+\cos\theta} \tag{2.40}$$

What makes this orbit significant is that it is a transition orbit between a closed (elliptical) orbit and an open (hyperbolic) orbit. A spacecraft leaving orbit around a planet transitions from a closed orbit to an escape orbit. In order to make this jump, a spacecraft instantaneously moves into a parabolic orbit on its way to a hyperbolic orbit.

Another use for a parabolic orbit is for computing the minimum escape velocity. This is the velocity necessary to move from a closed orbit to an open orbit. All spacecraft never simply maneuver to reach exactly the escape velocity; instead, it is common to use this concept for analytical insight into the characteristics of an orbit.

A special case of a parabolic orbit is a rectilinear orbit. Rectilinear motion is defined as motion in a straight line, so a rectilinear orbit is an open orbit where motion is in a straight line. This is not a physical orbit, but a mathematical one that is used for various approximations, including maneuvers, cometary motion, and ballistic trajectories [Vallado, 2013, p. 12]. With this type of orbit, the position and velocity vectors are in the same direction, thus the angular momentum is

$$\mathbf{h} = \mathbf{r} \times \mathbf{v} = r\mathbf{e}_r \times v\mathbf{e}_r = 0 \tag{2.41}$$

Thus, we can see the importance of these types of orbits, if simply to maintain continuity between closed and open orbits.

2.4.4 Hyperbolic Orbits

The hyperbolic orbit is an open orbit and is very commonly used in interplanetary missions (e.g., escape from orbit, gravitational flybys). In this orbit, the eccentricity is greater than 1 ($e > 1$). An example hyperbolic orbit (and its mirror image) is shown in Figure 2.8. Like the elliptical orbit, there is an occupied focus and a vacant focus. A hyperbola is formed around each of these foci (as you can see from Figure 2.8, this conic section has a very symmetrical representation).

Let's focus on the hyperbolic segment around the occupied focus. In a space mission, this occupied focus is the location of the gravitational source. The location of the closest approach is the periapse point (similar to the ellipse and parabola), but the apoapse point is undefined ($r_a \rightarrow \infty$). The semimajor axis, a, is presented as negative or positive values in various texts. The reader is reminded that the radius is always positive and the eccentricity in a hyperbolic orbit is greater than 1, so how Equation (2.36) is presented in texts indirectly defines the sign of the semimajor axis.

2.5 ORBITAL ELEMENTS

There are many ways to describe the characteristics of an orbit. Since the dynamical equations of motion are represented as three coupled, second-order differential equations, they can be recast as six first-order equations of motion. The solution of

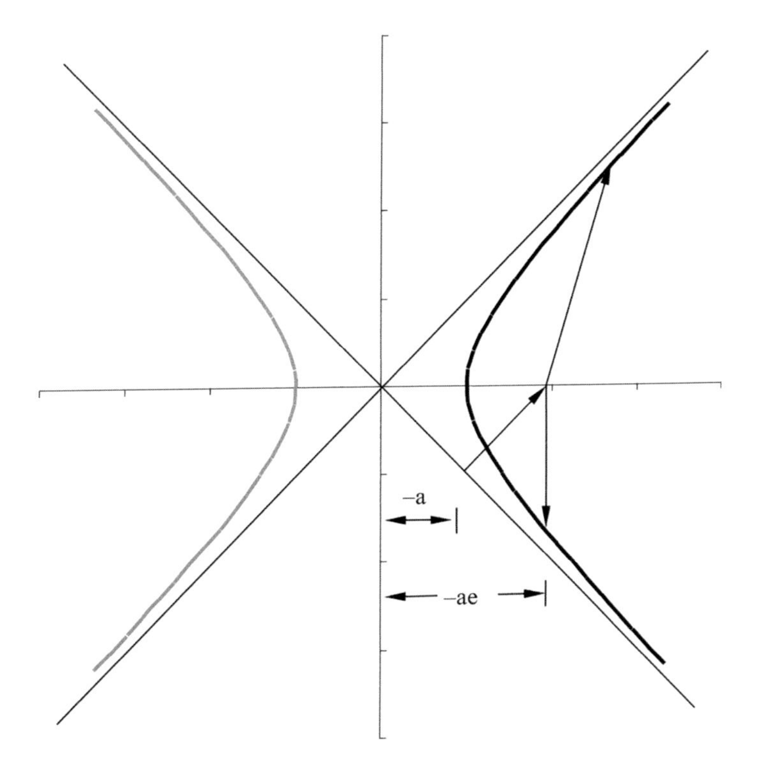

FIGURE 2.8 Hyperbola (and its mirror image).

these six equations of motion fully describes the position and velocity of an orbiting object. Various sets of variables can be used to represent this motion and are commonly known as orbital elements. Here, we present several commonly used orbital elements to describe the orbital motion.

2.5.1 Describing Orbits in Two Dimensions

An unperturbed orbit moves on a fundamental plane and defining the correct plane allows for describing the motion in the orbit as a two-dimensional problem. This geometry is shown in Figure 2.9. A third dimension, perpendicular to the orbit plane, is also defined. In planar geometry, it is common to place the origin of this coordinate frame at the center of the conic section (C). However, when describing the motion in a two-dimensional orbit, we commonly use the occupied focus (F) as the origin of the coordinate frame. With one vector pointing from the origin through the periapse point (**p**), the angular momentum vector pointing out of the plane (**w**) and the vector (**q**) completes the right-handed system ($\mathbf{q} = \mathbf{w} \times \mathbf{p}$). This coordinate frame is known as the perifocal reference frame.

The distance between the center of the ellipse and the occupied focus is found as

$$d_{C \to F} = a - r_p = a - a(1 - e) = ae \tag{2.42}$$

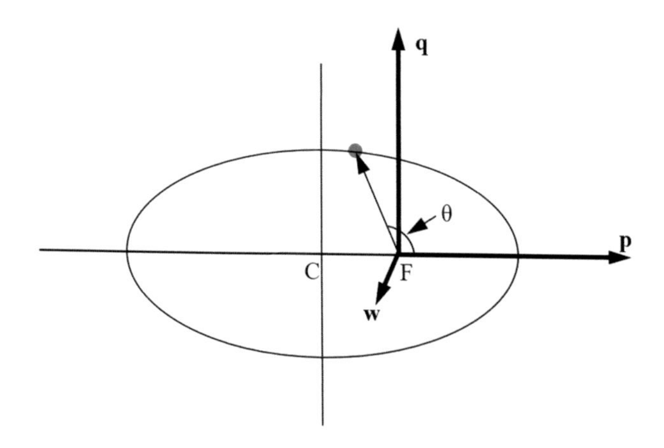

FIGURE 2.9 Perifocal reference frame.

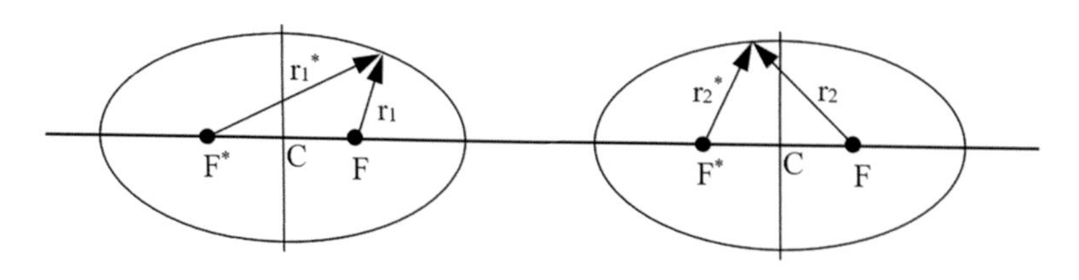

FIGURE 2.10 Graphical representation of similar triangles.

One characteristic of an ellipse is that the sum of the distances between each focus to a common point on the ellipse is equal to twice the semimajor axis of the ellipse for any point on the ellipse. Two different points on an ellipse are shown in Figure 2.10. Here,

$$r_i + r_i^* = 2a, \quad i = 1,2 \tag{2.43}$$

At the point where the radius vector intersects the minor axis of the ellipse, a right triangle is formed with the base, $d_{C \to F}$ (as seen in Equation 2.42), and the hypotenuse value is the semimajor axis of the elliptical orbit (due to the symmetry of the figure and applying Equation 2.43). This is seen in Figure 2.11.

Using the Pythagorean theorem, the semiminor axis, b, can be found as

$$b = a\sqrt{1 - e^2} \tag{2.44}$$

Let's superimpose a rotating reference frame on Figure 2.11, which is shown in Figure 2.12, and examine the angular momentum and energy of the system. From basic dynamics, the angular momentum is equal to the cross-product of the position with the velocity. The velocity vector is the time rate of change in the radius vector. Since this is a rotating coordinate frame, both the radius and the unit vector are differentiable with respect to time. Thus,

$$\mathbf{v} = \frac{d\mathbf{r}}{dt} = \frac{d}{dt}(r\mathbf{e}_r) = \frac{dr}{dt}\mathbf{e}_r + r\frac{d}{dt}(\mathbf{e}_r) = \frac{dr}{dt}\mathbf{e}_r + \boldsymbol{\omega} \times \mathbf{e}_r \tag{2.45}$$

where $\boldsymbol{\omega} = \frac{d\theta}{dt}\mathbf{e}_h$ and $\mathbf{v} = v_r\mathbf{e}_r + v_\theta\mathbf{e}_\theta$.

The radial velocity, v_r, is simply the time rate of change in radius. The tangential velocity, v_θ is the radius, r, times the angular rate, $\frac{d\theta}{dt}$. Using these definitions for radius and velocity, the energy equation becomes

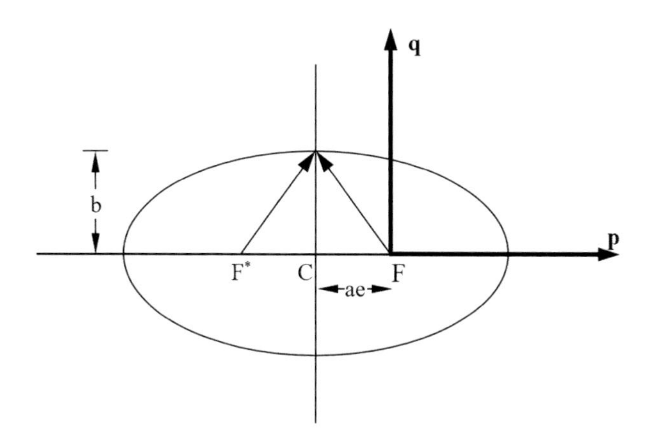

FIGURE 2.11 Mirror image triangle – definition of semiminor axis.

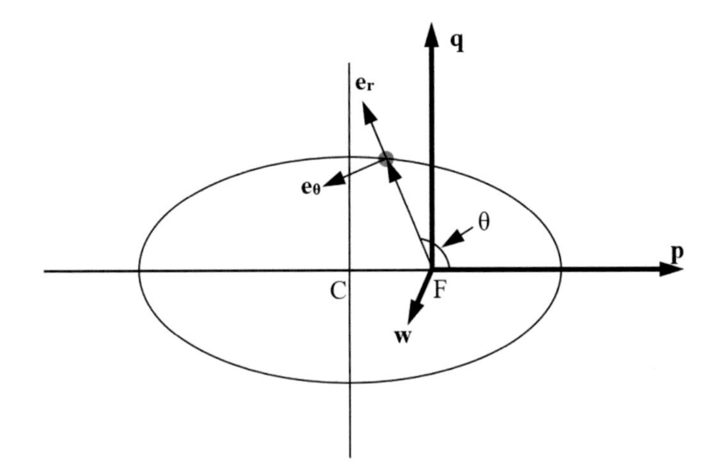

FIGURE 2.12 Superimposed rotating reference frame on perifocal reference frame.

$$E = \frac{\left[(dr/dt)^2 + r^2 (d\theta/dt)^2\right]}{2} - \frac{\mu}{r} = \frac{v^2}{2} - \frac{\mu}{r} \tag{2.46}$$

Recall that the cross-product of the radius and the velocity is the specific angular momentum. Inserting the definitions for the radius and velocity vectors gives the scalar value of the specific angular momentum as

$$h = r^2 \frac{d\theta}{dt} \tag{2.47}$$

Solving Equation (2.47) for $\frac{d\theta}{dt}$ and inserting this into Equation (2.46) shows that the energy is

$$E = \frac{1}{2}\left[(dr/dt)^2 + \frac{h^2}{r^2}\right] - \frac{\mu}{r} \tag{2.48}$$

At both periapse and apoapse, the velocity is made up only of tangential velocity – the radial velocity is equal to zero. This is easily seen by noting that as a body approaches periapse, its radius is decreasing ($dr/dt < 0$). Once it passes periapse, its radius is increasing ($dr/dt > 0$). At periapse, the sign changes, and instantaneously, the radial velocity is equal to zero ($dr/dt = 0$). Therefore, when the radius is equal to r_p, the velocity is purely tangential. Likewise, when a body is approaching apoapse, its radius is increasing ($dr/dt > 0$). After it passes apoapse, its radius starts to decrease ($dr/dt < 0$). Thus, just like at periapse, at apoapse, the radial component of velocity is instantaneously equal to zero ($dr/dt = 0$).

Arbitrarily choosing apoapse (this process works just as well for periapse), inserting the definitions of angular momentum (Equation 2.34) and the definition for apoapse radius, inserting Equation (2.37) into Equation (2.48), noting that $dr/dt = 0$ at apoapse, solving for E yield

$$E = -\frac{\mu}{2a} \tag{2.49}$$

Thus, the energy equation becomes

$$\frac{v^2}{2} - \frac{\mu}{r} = -\frac{\mu}{2a} \tag{2.50}$$

This equation is commonly referred to as the *vis viva* equation (Latin for *living force*, referring to the energy of a dynamical system) and is one of the fundamental equations of orbital mechanics. We will make use of this equation several times throughout this book.

As a spacecraft traces out its orbit, an area of an ellipse is swept out. This geometry is shown in Figure 2.13.

Approximating this area swept out as a triangle, its area is

$$dA = \frac{1}{2} r \times (r d\theta) = \frac{1}{2} r^2 d\theta \tag{2.51}$$

Dividing both sides by a differential in time becomes,

$$\frac{dA}{dt} = \frac{1}{2} r^2 \frac{d\theta}{dt} \tag{2.52}$$

Inserting Equation (2.47) into Equation (2.52) shows

$$\frac{dA}{dt} = \frac{1}{2} h = \text{constant} \tag{2.53}$$

Thus, the area that is swept out over time is proportional to the elapsed time, which proves Kepler's second law.

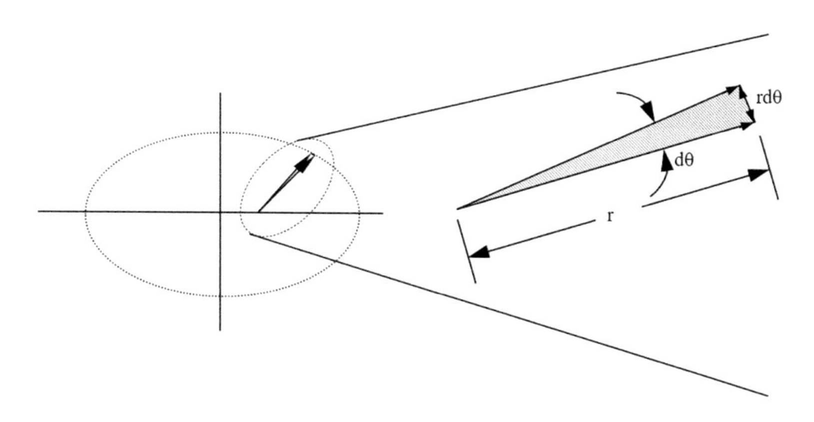

FIGURE 2.13 Area swept out of ellipse segment.

Another important parameter is the period of an orbit. Period is defined as the time it takes to complete one orbit. Since the area swept out over time is a constant, the ratio of the area of an ellipse, A, to the period of the orbit, T, is equal to the time rate of change of the area, Equation (2.53). Thus,

$$\frac{A}{T} = \frac{dA}{dt} \tag{2.54}$$

Using Equation (2.54), and the definition of the area of an ellipse as

$$A = \pi ab = \pi a\left(a\sqrt{1-e^2}\right) \tag{2.55}$$

Solving for the period,

$$T = \frac{A}{dA/dt} \tag{2.56}$$

Inserting Equations (2.44), (2.53), and (2.55) into Equation (2.56) provides the definition of the period,

$$T = \frac{\pi ab}{h/2} = \frac{2\pi a^2\sqrt{1-e^2}}{\sqrt{\mu a\left(1-e^2\right)}} = \frac{2\pi}{\sqrt{\mu/a^3}} = \frac{2\pi}{n} \tag{2.57}$$

where $n = \sqrt{\frac{\mu}{a^3}}$ and is also known as the mean motion, which physically is the average angular velocity in the satellite's orbit.

The above derivations are generic equations for all conic sections. While elliptical orbits are probably the most common conic section, there are other conic sections that are also used. More details on other conic sections can be found in many books, including Vallado [2013] and Curtis [2019].

Ideally, it is desired to find the time history of the orbital radius and velocity. Equation (2.40) gives the radius as a function of the true anomaly, θ. The velocity can be determined via the *vis viva* equation. In order to determine how the radius is a function of time, determination of the true anomaly as a function of time will be used as a bridge to determine this formulation.

2.5.2 Describing Orbits in Three-Dimensions

In order to describe fully the position and velocity of an object in an orbit, a total of six quantities are needed. A common set of elements used are Cartesian coordinates (three elements of position, three elements of velocity). However, the X, Y, and Z coordinates of position and velocity do not provide any physical insight as to the size, shape, and orientation of the object's orbit. To give better physical insight, other coordinate elements have been developed. One common element set is the classical orbital elements. Classical elements provide physical insight on the size and shape of the orbit (two elements), the orientation of the orbit in space (two elements), the orientation of the orbit in the orbital plane (one element), and the location of the object on the orbit (one element). There are other coordinate frames and element sets that are also used, but many are only modifications to these two systems.

2.5.3 Cartesian Orbital Elements

Since the **I-J-K** coordinates of position and velocity do not provide much in the way of a physical insight into the size, shape, and orientation of the object in space, they do provide an easy coordinate frame to numerically integrate the equations of motion. In this inertial **I-J-K** reference frame, the position vector of an object can be written as

$$\mathbf{r} = \xi\mathbf{I} + \eta\mathbf{J} + \zeta\mathbf{K} \tag{2.58}$$

The magnitude of this vector is

$$r = \sqrt{\mathbf{r}\cdot\mathbf{r}} = \sqrt{\xi^2 + \eta^2 + \zeta^2} \tag{2.59}$$

Taking the derivative of Equation (2.58) twice, and inserting this along with Equation (2.59) into Equation (2.9) yields a set of three coupled, second-ordered differential equations,

$$\begin{aligned} \frac{d^2\xi}{dt^2} &= -\frac{\mu\xi}{r^3} \\ \frac{d^2\eta}{dt^2} &= -\frac{\mu\eta}{r^3} \\ \frac{d^2\zeta}{dt^2} &= -\frac{\mu\zeta}{r^3} \end{aligned} \tag{2.60}$$

While it is relatively easy to numerically integrate these equations of motion and obtain valid solutions for all types of orbits, the physical insight into the size, shape, and orientation is still missing. To gain this insight, it is useful to convert these Cartesian elements into classical orbital elements.

2.5.4 Cylindrical and Spherical Orbital Elements

Other sets of variables that find their origins in astronomy are cylindrical and spherical orbital elements. These element sets are a combination of a radial distance, one angle (generally true anomaly), and an out-of-plane distance for cylindrical coordinates, and a radial distance and two angles for spherical coordinates, along with the time rates-of-change of each quantity. We discuss applications of cylindrical and spherical coordinates in more detail in the next chapters. Furthermore, general information about spherical trigonometry (which is trigonometry applied to spherical geometry) can be found in Appendix B.

2.5.5 Classical Orbital Elements

We have previously introduced the concept of semimajor axis and eccentricity. Semimajor axis describes the size of the orbit. Eccentricity describes the shape of the orbit. The magnitudes of the position and velocity vectors can be found as the square root of the dot product of the vector with itself. This yields the scalar radius and velocity. Inserting these quantities into the *vis viva* equation, the semimajor axis can be found as

$$a = \frac{1}{\left(\frac{2}{r} - \frac{v^2}{\mu}\right)} \tag{2.61}$$

where r is defined in Equation (2.59) and $v = \sqrt{\left(\frac{d\xi}{dt}\right)^2 + \left(\frac{d\eta}{dt}\right)^2 + \left(\frac{d\zeta}{dt}\right)^2}$.

Rearranging Equation (2.35) provides an expression for the eccentricity,

$$e = \sqrt{1 - \frac{\mathbf{h} \cdot \mathbf{h}}{\mu a}} = \sqrt{1 - \frac{h^2}{\mu a}} \tag{2.62}$$

In describing the orientation of the orbit in inertial space, it is necessary to introduce some new terms. Commonly used is an inertial reference frame. For example, in the Earth-centered inertial (ECI) frame, the inertial **I–J** plane is the equatorial plane of the Earth. The **I**-axis points in the direction of an inertially fixed constellation Aries, with a symbol of the ram's horn (♈), and is often referred to the "first point of Aries." The **K**-axis aligns with the spin axis of the Earth, and the **J**-axis completes the right-handed system. An illustration of this inertial coordinate frame and the classical orbital elements are shown in Figure 2.14. We will discuss coordinate systems for other planets and interplanetary trajectories in Chapter 4.

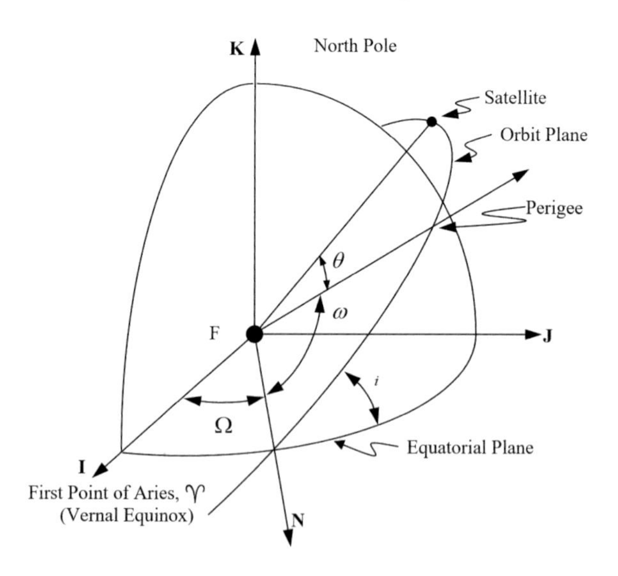

FIGURE 2.14 Definition of Earth-centered inertial frame and classical orbital elements.

An orbit around the Earth breaks the plane formed by the equatorial plane in two places along a nodal vector (**N**), as the object heads north (ascending) and as the object heads south (descending). The angle formed between the first point of Aries and the nodal vector is defined as the right ascension of the ascending node (RAAN) and commonly uses the symbol Ω. The second angle which defines the orientation is the inclination, i. The inclination of the orbit plane is the angle between the equatorial plane and the orbital plane.

Next, we need to describe the orientation of the orbit on the orbit plane. The angle ω, known as the argument of periapsis, is defined as the angle between where the orbit plane crosses the equatorial plane and the vector passing through periapse. The cross-product of the position vector with the radius vector yields the angular momentum vector,

$$\begin{aligned}\mathbf{h} &= \mathbf{r}\times\mathbf{v}\\ &= \left(\eta\frac{d\zeta}{dt}-\zeta\frac{d\eta}{dt}\right)\mathbf{I}+\left(\zeta\frac{d\xi}{dt}-\xi\frac{d\zeta}{dt}\right)\mathbf{J}+\left(\xi\frac{d\eta}{dt}-\eta\frac{d\xi}{dt}\right)\mathbf{K}\end{aligned} \tag{2.63}$$

Several of the equations already presented have not shown a physical representation of the orbital characteristics. For example, looking at Equation (2.47), there isn't a physical "feel" of what numerical size of the angular momentum scalar (h) translates to a specific size orbit. However, an important note is that the direction of the angular momentum vector is normal to the orbit plane.

The inclination is the angle between the orbit plane and the equatorial plane. It is also the angle between the vector perpendicular to the orbit plane and vector perpendicular to the equatorial plane. The vector perpendicular to the orbit plane is the angular momentum vector, while the vector perpendicular to the equatorial plane is the **K**-component of the angular momentum vector,

$$\cos i = \frac{\mathbf{h}}{h}\cdot\mathbf{K} \tag{2.64}$$

Substituting Equation (2.61) into Equation (2.63),

$$\begin{aligned}\cos i &= \frac{\xi\frac{d\eta}{dt}-\eta\frac{d\xi}{dt}}{\sqrt{\left(\eta\frac{d\zeta}{dt}-\zeta\frac{d\eta}{dt}\right)^2+\left(\zeta\frac{d\xi}{dt}-\xi\frac{d\zeta}{dt}\right)^2+\left(\xi\frac{d\eta}{dt}-\eta\frac{d\xi}{dt}\right)^2}}\\ &= \frac{h_\zeta}{h}\end{aligned} \tag{2.65}$$

Since the inclination is defined to be between 0° and 180°, quadrant checks are unnecessary.

The right ascension of the ascending node is the angle between the **I** vector and the unit-node vector, **N**. This node vector is perpendicular to both the angular momentum vector, **h** and the **K** vector, so

$$\mathbf{N}=\frac{\mathbf{K}\times\mathbf{h}}{h} \tag{2.66}$$

The node unit vector is also

$$\mathbf{N}=\cos\Omega\mathbf{I}+\sin\Omega\mathbf{J} \tag{2.67}$$

Equating Equations (2.65) and (2.66), the value of Ω must satisfy both

$$\cos\Omega=\mathbf{I}\cdot\frac{\mathbf{K}\,\Omega\,\mathbf{h}}{h} \tag{2.68}$$

and

$$\sin\Omega=\mathbf{J}\cdot\frac{\mathbf{K}\times\mathbf{h}}{h} \tag{2.69}$$

which provides both a numerical value and quadrant resolution.

The argument of periapse, ω, is the angle between the node vector and the eccentricity vector **e** (found from Equation 2.27),

$$\cos\omega=\frac{\mathbf{N}\cdot\mathbf{e}}{e}=\frac{\mathbf{N}\cdot\left(\dfrac{d\mathbf{r}}{dt}\times\mathbf{h}-\mu\mathbf{r}/r\right)}{e} \tag{2.70}$$

The quadrants can be resolved by examining the eccentricity vector. If the eccentricity vector has a component pointing in the positive **K** direction, the periapse is in the northern hemisphere and vice versa. Thus, if $\mathbf{e}\cdot\mathbf{K}>0$, then

$$\omega=\cos^{-1}\left(\frac{\mathbf{N}\cdot\mathbf{e}}{e}\right) \tag{2.71}$$

otherwise, if $\mathbf{e}\cdot\mathbf{K}<0$, then

$$\omega=-\cos^{-1}\left(\frac{\mathbf{N}\cdot\mathbf{e}}{e}\right) \tag{2.72}$$

Lastly, the fast variable (the variable that moves though its entire limit every orbit) is found in various ways. From Equation (2.36), the true anomaly can be found as

$$\theta=\cos^{-1}\left[\frac{a\left(1-e^2\right)-r}{er}\right] \tag{2.73}$$

The quadrant can be determined by examining the radial velocity. The radial component of velocity is $\frac{dr}{dt}=\frac{\mathbf{r}}{r}\cdot\frac{d\mathbf{r}}{dt}$ and since the radial velocity is positive when moving from periapse to apoapse and negative when moving from apoapse to periapse, then

$$\frac{\mathbf{r}}{r}\cdot\frac{d\mathbf{r}}{dt} > 0, \quad 0^\circ < \theta < 180^\circ$$
$$\frac{\mathbf{r}}{r}\cdot\frac{d\mathbf{r}}{dt} < 0, \quad 180^\circ < \theta < 360^\circ \tag{2.74}$$
$$\frac{\mathbf{r}}{r}\cdot\frac{d\mathbf{r}}{dt} = 0, \quad \theta = 0^\circ \text{ or } \theta = 180^\circ$$

Lastly, we need to describe where the object is in the orbit. Any of the anomalies previously defined as well as a time after the periapsis passage (t_p) can be used to define where the object is relative to the periapsis point. This variable is also known as the "fast" variable, as it moves through its range of values during every orbit while the other elements remain constant.

While the use of classical orbital elements provides significant physical insight to the geometry of an orbit in space, there are two singularities that exist. The first singularity exists for an exactly circular orbit. Recall that the right ascension of the ascending node (Ω) is the angle between the vernal equinox and the ascending crossing of the orbit plane's intersection with the equator. However, when the orbit plane is coincident with the equatorial plane, there is never a crossing. Therefore, Ω is undefined. The second singularity occurs for an exactly circular orbit. In this situation, recall that the argument of periapse (ω) is the angle in the orbit plane between the equatorial crossing and the periapse. However, in a circular orbit, there is no periapse. Also, since the time of periapsis passage is measured with respect to the periapse, there is no time of periapsis passage for the same reason. One way to address this problem is to define the argument of latitude, u,[1] as the angle measured between where the orbit plane crosses the equatorial plane and the location of the satellite ($u = \omega + \theta$). This situation is further exacerbated by a combination of a circular and equatorial orbit (such as seen with a geosynchronous Earth orbit). While this situation is only a mathematical occurrence (getting an exactly circular or equatorial orbit is not an operational possibility), the closer to a zero value causes numerical problems when integrating the equations of motion. Non-singular elements, known as equinoctial elements, are used many times to alleviate these singularities. Additionally, although they provide relatively little physical insight, Cartesian elements have no singularities and can be easily integrated using common numerical integration methods.

The classical elements of a two-body orbit are constants except for the fast variable. If no other forces are present, the description of the orbital motion is complete. However, many forces perturb the motion and make the solution quite complex. The two-body solution is useful only for the simplest models and simulations. This set of orbital elements completely describes the instantaneous location of an object in inertial space.

2.5.6 Other Types of Orbital Elements to Mitigate Singularities

Since six parameters are needed to uniquely define an orbit, the classical orbital elements are usually the standard way to describe an orbit. However, they present singularities, especially in cases involving nearly equatorial and nearly circular orbits, as

[1] Sometime ϖ is used to denote this quantity.

we previously described. Sets of orbital elements aside from those already discussed have been proposed. Here, we're going to briefly present the equinoctial, Delaunay, and Poincaré elements.

The equinoctial elements are often utilized when studying the effects of perturbations, as introduced in Section 2.7. Additionally, these elements make use of the longitude of periapsis, which is the sum of ω and Ω. Although these angles are generally not measured in the same plane and they may individually be undefined, their sum $u = \omega + \Omega$, which is known as the argument of latitude, is defined for non-circular orbits since it is the angle between the **I** vector and the location of periapsis. The equinoctial elements, written as a function of the classical orbital elements, are

$$\begin{aligned}
a_f &= e\cos(\omega+\Omega) \\
&n \\
a_g &= e\sin(\omega+\Omega) \\
L &= M+\omega+\Omega \\
\chi &= \tan\left(\frac{i}{2}\right)\sin(\Omega) \\
\psi &= \tan\left(\frac{i}{2}\right)\cos(\Omega)
\end{aligned} \tag{2.75}$$

The equinoctial elements are valid for circular, elliptic, and hyperbolic orbits. They exhibit no singularity for zero eccentricity and orbital inclinations equal to 0° and 90° although two of the elements, χ and ψ, are singular for equatorial retrograde orbits, which have an orbital inclination of 180°.

Other two useful sets of orbital elements are the Delaunay elements and the Poincaré elements. These are both canonical elements since they are generally used in the Lagrangian and Hamiltonian forms of dynamical systems.

The Delaunay elements are the canonical equivalent of the classical orbital elements since they both present singularities for small values of eccentricity and inclination. The Delaunay elements can be written as a function of the classical orbital elements as follows:

$$\begin{aligned}
&M \\
&\omega \\
&\Omega \\
L_d &= \sqrt{\mu a} \\
&h \\
H_d &= \sqrt{\mu a\left(1-e^2\right)}\cos(i)
\end{aligned} \tag{2.76}$$

Note that in Equation (2.76), the quantity H_d is the z-component of the angular momentum vector.

The Poincaré elements are the canonical equivalent of the equinoctial elements and are therefore non-singular for small values of eccentricities and inclinations. The Poincaré elements can be written as a function of the classical orbital elements as follows:

$$\begin{aligned}
\lambda_M &= M + \omega + \Omega \\
g_p &= \sqrt{2\left[\sqrt{\mu a}\left(1-\sqrt{1-e^2}\right)\right]}\cos(\omega+\Omega) \\
h_p &= \sqrt{-2\sqrt{\mu a\left(1-e^2\right)}\left(\cos(i)-1\right)}\cos(\Omega) \\
L_p &= \sqrt{\mu a} \\
G_p &= \sqrt{2\left[\sqrt{\mu a}\left(1-\sqrt{1-e^2}\right)\right]}\sin(\omega+\Omega) \\
H_p &= \sqrt{-2\sqrt{\mu a\left(1-e^2\right)}\left(\cos(i)-1\right)}\sin(\Omega)
\end{aligned} \tag{2.77}$$

Note the parallelisms between elements g_p and G_p as well as h_p and H_p. These elements are usually used in Hamiltonian dynamics.

The choice of which element set to use is based on various considerations posed by the application. Cartesian elements are generally the easiest to solve numerically, but they don't readily provide the physical insight that other elements provide. The analyst is left to decide the advantages and disadvantages of these and many more element sets needed to solve their problem.

2.6 KEPLER'S PROBLEM

Already introduced was the equation for a whole orbit, but what about a fraction of an orbit? In this section we will develop formulations to describe the relationship between the angular change in location of an object in elliptical, parabolic, and hyperbolic orbits, as a function of time. We will also introduce a formulation using universal variables, which is applicable to all types of conic section orbits.

2.6.1 Kepler's Equation for Elliptical Orbits

For an elliptical orbit we start by introducing an intermediate anomaly, known as the eccentric anomaly, E. A schematic of this anomaly is shown in Figure 2.15. Superimpose an auxiliary circle of radius a around the ellipse, and use the perifocal reference frame, **p** and **q** as was previously defined. A vector from the center of the circle (and ellipse) to a point on the circle has a magnitude of a and is rotated through an angle E, where E = 0 is defined to be in the positive **p**-direction. The position vector of the object on its orbit has components (x, y) in the **p**- and **q**-directions as

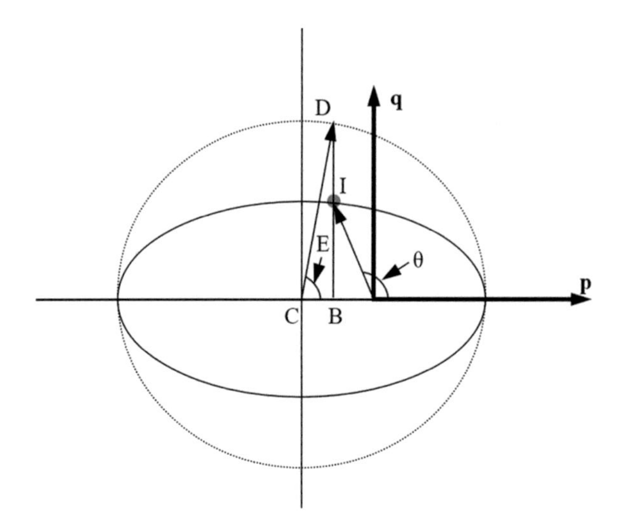

FIGURE 2.15 Geometry relating eccentric anomaly and true anomaly.

$$x = r\cos\theta \tag{2.78}$$

$$y = r\sin\theta \tag{2.79}$$

The **p-** and **q**-frame (perifocal reference frame) has components of the point D on the superimposed circle (x_c and y_c) equal to

$$x_c = a\cos E \tag{2.80}$$

$$y_c = a\sin E \tag{2.81}$$

Two right triangles are formed (B–C–D) and (B–F–I). The distances from C–B and I–D, respectively, are

$$d_{C-B} = ae + r\cos\theta \tag{2.82}$$

$$d_{I-D} = a\sin E - r\sin\theta \tag{2.83}$$

Using the definition of cosine for a right triangle,

$$\cos E = \frac{d_{C-B}}{a} = \frac{ae + r\cos\theta}{a} \tag{2.84}$$

Inserting Equation (2.73) into Equation (2.84) and through some algebraic manipulation, the radius can be found as

$$r = a(1 - e\cos E) \tag{2.85}$$

Through similar algebraic and geometric manipulation, the **p**- and **q**-components of the position as a function of true and eccentric anomalies become

$$x = r\cos\theta = a(\cos E - e) \tag{2.86}$$

$$y = r\sin\theta = a\sqrt{1-e^2}\,\sin E \tag{2.87}$$

In order to find the relationship between eccentric anomaly and true anomaly, insert Equation (2.36) into Equations (2.86) and (2.87) and solving for eccentric anomaly yields

$$\cos E = \frac{e + \cos\theta}{1 + e\cos\theta} \tag{2.88}$$

$$\sin E = \frac{\sqrt{1-e^2}\,\sin\theta}{1 + e\cos\theta} \tag{2.89}$$

Using the tangent half-angle identity,

$$\tan^2\frac{E}{2} = \frac{1-\cos E}{1+\cos E} \tag{2.90}$$

and substituting Equation (2.88) into Equation (2.90), along with applying the tangent half-angle identity for true anomaly, yields,

$$\tan\frac{\theta}{2} = \sqrt{\frac{1+e}{1-e}}\tan\frac{E}{2} \tag{2.91}$$

Here, a quadrant check is unnecessary, since the half angles are both always in the same half-plane.

Recall that the angular momentum is $\mathbf{h} = \mathbf{r} \times \frac{d\mathbf{r}}{dt}$. Since the **p–q** coordinate frame is fixed inertially, $\frac{d\mathbf{r}}{dt} = \frac{dx}{dt}\mathbf{p} + \frac{dy}{dt}\mathbf{q}$. Taking the derivatives of Equations (2.86) and (2.87), and inserting these definitions of **r** and $\frac{d\mathbf{r}}{dt}$ into the angular momentum definition, yields a scalar value for angular momentum of

$$h = ra\sqrt{1-e^2}\,\frac{dE}{dt} \tag{2.92}$$

Inserting Equation (2.40) into Equation (2.92) and solving for the time derivative of the eccentric anomaly,

$$\frac{dE}{dt} = \frac{1}{r}\sqrt{\frac{\mu}{a}} \rightarrow \frac{dE}{dt}(1 - e\cos E) = \sqrt{\frac{\mu}{a^3}} \tag{2.93}$$

Grouping terms with E, integration of Equation (2.93) allows a solution for E as a function of time,

$$\int_0^E (1 - e\cos\zeta)\,d\zeta = \sqrt{\frac{\mu}{a^3}}\int_{t_p}^{t} dt \tag{2.94}$$

where ζ is a dummy variable used for this integration.

The limits on the integration of the left-hand side are the initial eccentric anomaly ($E=0$, periapse) to the final eccentric anomaly, E. The limits on the integration on the right-hand side are the time at periapse passage, t_p, to the time, t, when the eccentric anomaly is equal to E. Solving Equation (2.94),

$$E - e\sin E = \sqrt{\frac{\mu}{a^3}}\left(t - t_p\right) \tag{2.95}$$

Define yet another anomaly, the mean anomaly, M, as

$$M = \sqrt{\frac{\mu}{a^3}}\left(t - t_p\right) \tag{2.96}$$

Thus, Kepler's equation becomes

$$E - e\sin E = M \tag{2.97}$$

Time since periapse passage can be found using Equation (2.96), and mean anomaly can be found using Equation (2.97). This equation is known as a transcendental equation. Specifically, given the mean anomaly, it is an iterative process to find the eccentric anomaly. Conversely, given the eccentric anomaly, the mean anomaly is simply solved. One warning in using this equation is that the eccentric and mean anomalies must be in radians, so if the given E or M are given in degrees, the first step is to convert E or M to radians.

Equation (2.95) gives the eccentric anomaly from the time of periapse passage. However, there are times when it is desired to determine the time of flight between some value not at periapse to another value. In order to accomplish this, the eccentric anomaly at both locations (E_2 and E_1) must be found and subtracted from each other,

$$E_2 - E_1 - e\left(\sin E_2 - \sin E_1\right) = M_2 - M_1 = n\left(t_2 - t_1\right) \tag{2.98}$$

A typical type of problem is to find the radius of an orbit at several times. Equation (2.95) is used to solve the initial eccentric anomaly. Once the initial eccentric anomaly is found, it can be advanced to the next time via Equation (2.98), which is used to find the eccentric anomaly at some different time. Equation (2.85) is then used to find the radius at the different time.

2.6.1.1 Solution to Kepler's Problem for Elliptical Orbits

Although there is no closed-form solution to Kepler's equation, it may be solved numerically. Rearranging Equation (2.97) and equating it to a function $f(E)$ shows that a solution exists at the roots of $f(E)$,

$$f(E) = E - e\sin E - M = 0 \tag{2.99}$$

In order to implement a Newton-Raphson method, the derivative of Equation (2.99) with respect to the desired variable is found,

$$f'(E) = \frac{df}{dE} = 1 - e\cos E \tag{2.100}$$

This Newton-Raphson method iterates on an initial guess on the eccentric anomaly until $|f(E)|$ becomes sufficiently small, and thus

$$E_{k+1} = E_k - \frac{f(E_k)}{f'(E_k)} \tag{2.101}$$

Generally, this is a well-behaved function that converges quickly.

2.6.2 Kepler's Equation for Parabolic Orbits

Just like any other conic section, we can describe the angular momentum, as seen in Equation (2.46). Inserting the definition for the radius from Equation (2.40) into Equation (2.46), recalling that the eccentricity is equal to 1,

$$h = \left(\frac{p}{1+\cos\theta}\right)^2 \frac{d\theta}{dt} \tag{2.102}$$

From a trigonometric identity, we know that

$$1 + \cos\theta = 2\cos^2\frac{\theta}{2} \tag{2.103}$$

Thus, the angular momentum becomes

$$h = \left(\frac{p}{2\cos^2\frac{\theta}{2}}\right)^2 \frac{d\theta}{dt} \tag{2.104}$$

Rearranging this equation (including $\frac{d\theta}{dt}$), and integrating,

$$\int_{t_p}^{t} \frac{4h}{p^2} d\tau = \int_0^{\theta} \sec^4\frac{\Theta}{2} d\Theta \tag{2.105}$$

where τ and Θ are integration dummy variables for t and θ, respectively. Integrating,

$$\frac{2h}{p^2}(t - t_p) = \tan\frac{\theta}{2} + \frac{1}{3}\tan^3\frac{\theta}{2} \tag{2.106}$$

Recalling that $h = \sqrt{\mu p}$, and defining $n_p = \sqrt{\frac{\mu}{p^3}}$ and $B = \tan\frac{\theta}{2}$, Kepler's equation for a parabolic orbit (known in the literature as Barker's equation, named after Thomas Barker, 1722–1809, [Herrick, 1971, p. 191]) becomes

$$M_p = 2n_p\left(t - t_p\right) = B + \frac{B^3}{3} \tag{2.107}$$

where B is known as the parabolic anomaly.

The general solution, when we are not measuring the time swept from periapse, is similar to Equation (2.98),

$$2n_p\left(t_2 - t_1\right) = \left(B_2 - B_1\right) + \frac{1}{3}\left(B_2^3 - B_1^3\right) \tag{2.108}$$

2.6.2.1 Solution to Kepler's Problem for Parabolic Orbits

Unlike the Newton-Raphson method used for the transcendental form used for the solution of Kepler's problem for elliptical orbits, the parabolic orbit equation is not a transcendental equation – it is a cubic equation in parabolic anomaly. The solution to a cubic equation in the form,

$$B^3 + 3B - 3M_p = 0 \tag{2.109}$$

has the solution,

$$\begin{aligned} B &= C_1 + C_2 \\ B &= -\frac{1}{2}(C_1 + C_2) + \frac{\sqrt{-3}}{2}(C_1 - C_2) \\ B &= -\frac{1}{2}(C_1 + C_2) - \frac{\sqrt{-3}}{2}(C_1 - C_2) \end{aligned} \tag{2.110}$$

where

$$C_1 = \sqrt[3]{\frac{3}{2}M_p + \frac{1}{2}\sqrt{9M_p + 4}} \tag{2.111}$$

$$C_2 = -\sqrt[3]{-\frac{3}{2}M_p + \frac{1}{2}\sqrt{9M_p + 4}} \tag{2.112}$$

The solution has one real root (the value of the parabolic anomaly we will use) and two complex conjugate roots.

2.6.3 Kepler's Equation for Hyperbolic Orbits

For closed circular and elliptical orbits, the orbit equations use trigonometric functions. For open hyperbolic orbits, many of these trigonometric functions are replaced with hyperbolic functions. Battin [1987, pp. 165–173] develops the hyperbolic analogy of the elliptical auxiliary circle by replacing it with an equilateral hyperbola having the same semimajor axis as the hyperbolic orbit being studied.

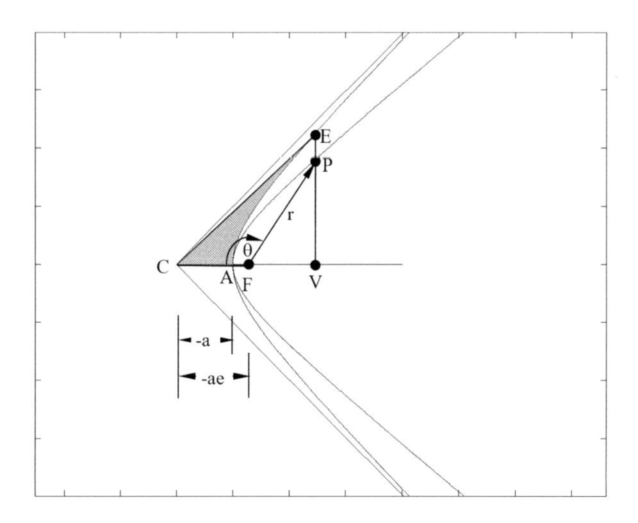

FIGURE 2.16 Geometry of hyperbolic orbits.

Also analogous to the eccentric anomaly, we define the hyperbolic anomaly. In Figure 2.16, the points C and F are the center and occupied focus of the hyperbolic orbit. Point A is the periapse of the orbit. The point P is the point in question on the hyperbolic orbit. We then draw a line perpendicular to the major axis, through the point on the orbit and intersecting an equilateral hyperbola $\left(e = \sqrt{2}\right)$ at point E. The distance between C and A is the negative of the semimajor axis (remember that for a hyperbolic orbit, the semimajor axis is a negative value), and the distance between C and F is –ae.

The hyperbolic anomaly is a measure of area, as opposed to the eccentric anomaly, which is an angle. If we let x be the distance from the center of the hyperbola to the intersection of the line perpendicular (V), and y as the distance from the major axis (V) to the point on the hyperbola (P), the description of the hyperbola relative to the center, C, is

$$\left(\frac{x}{a}\right)^2 - \left(\frac{y}{b}\right)^2 = 1 \tag{2.113}$$

The solution to Equation (2.113) is

$$x = a \cosh H \tag{2.114}$$

$$y = b \sinh H \tag{2.115}$$

Recall that $r = \sqrt{x^2 + y^2}$ and noting that the relationship between the semimajor (a) and semiminor (b) axes, is $\frac{b}{a} = \sqrt{e^2 - 1}$, we find that the radius is

$$r = a\left(1 - e \cosh H\right) \tag{2.116}$$

The ratio of semiminor to semimajor axes is also related to the turning angle (how much the orbit bends as it makes a hyperbolic passage of the planet), ϕ,

$$\tan\phi = \frac{b}{a} \tag{2.117}$$

This relationship will be used later when we discuss gravitational assists and associated problems.

Just like elliptical orbits, hyperbolic orbits have true anomalies as well. However, since these orbits are open, the true anomaly asymptotically approaches a minimum or maximum value. Using the orbit equation, and noting that as r gets large, the true anomaly approaches its minimum or maximum. Inverting Equation (2.116) and taking the limit

$$\lim_{r\to\infty}\frac{1}{r} = 0 = \frac{1+e\cos\theta}{a\left(1-e^2\right)} \tag{2.118}$$

we find the maximum of true anomaly is

$$\theta_{max} = \cos^{-1}\left(-\frac{1}{e}\right) \tag{2.119}$$

and the minimum is

$$\theta_{min} = -\cos^{-1}\left(-\frac{1}{e}\right) \tag{2.120}$$

Thus, the range of true anomalies is

$$-\cos^{-1}\left(-\frac{1}{e}\right) \le \theta \le \cos^{-1}\left(-\frac{1}{e}\right) \tag{2.121}$$

Relationships between true anomaly and hyperbolic anomaly can be found by equating the two forms of the orbit equation, Equations (2.36) and (2.116), as

$$r = a(1-e\cosh H) = \frac{a\left(1-e^2\right)}{1+e\cos\theta} \tag{2.122}$$

Solving Equation (2.122) for true anomaly,

$$\cos\theta = \frac{\cosh H - e}{1-e\cosh H} \tag{2.123}$$

Likewise, the hyperbolic anomaly can be found as

$$\cosh H = \frac{e+\cos\theta}{1+e\cos\theta} \tag{2.124}$$

Using Equation (2.115),

$$y = b\sinh H = r\sin\theta \tag{2.125}$$

Solving for sinhH as a function of true anomaly,

$$\sinh H = \frac{\sqrt{e^2 - 1}\sin\theta}{1 + e\cos\theta} \tag{2.126}$$

and using the relationship in Equations (2.123) and (2.125),

$$\sin\theta = \frac{\sqrt{e^2 - 1}\sinh H}{e\cosh H - 1} \tag{2.127}$$

To move between true anomaly and hyperbolic anomaly requires solving both the sine and cosine equations (which resolves the quadrant ambiguity). Another solution method would be to perform a half-angle solution, similar to Equation (2.91). We find that

$$\tan\left(\frac{\theta}{2}\right) = \sqrt{\frac{e+1}{e-1}}\tanh\left(\frac{H}{2}\right) \tag{2.128}$$

Kepler's equation for hyperbolic orbits follows a similar approach to that of elliptical orbits. In the **p–q** perifocal reference frame (centered at the focus), we again find the angular momentum, $\mathbf{h} = \mathbf{r} \times \frac{d\mathbf{r}}{dt}$. The velocity term is $\frac{d\mathbf{r}}{dt} = \frac{dx_f}{dt}\mathbf{p} + \frac{dy_f}{dt}\mathbf{q}$, where x_f and y_f are the x- and y-coordinates of point P relative to point F, also seen in Figure 2.16.

The perifocal coordinates are

$$\begin{aligned} x_f &= -a(\cosh H - e) \\ y_f &= -a\sinh H\sqrt{e^2 - 1} \end{aligned} \tag{2.129}$$

The derivatives of Equation (2.129) give us the velocity,

$$\frac{d\mathbf{r}}{dt} = -a\sinh H\mathbf{p} + a\cosh H\sqrt{e^2 - 1}\mathbf{q} \tag{2.130}$$

The angular momentum becomes

$$h = a^2\sqrt{e^2 - 1}(1 - e\cosh H)\frac{dH}{dt} \tag{2.131}$$

Rearranging Equation (2.131) and integrating,

$$\int_{t_p}^{t} h\,d\tau = \int_0^H a^2\sqrt{e^2 - 1}(1 - e\cosh\Theta)d\Theta \tag{2.132}$$

Kepler's equation for hyperbolic orbits becomes

$$M = \sqrt{\frac{\mu}{-a^3}}(t - t_p) = (e\sinh H - H) \tag{2.133}$$

To describe the elapsed time between two points that don't include the periapse, Kepler's equation in its general form is

$$\sqrt{\frac{\mu}{-a^3}}(t_2 - t_1) = \left[e(\sinh H_2 - \sinh H_1) - (H_2 - H_1)\right] \quad (2.134)$$

2.6.3.1 Solution to Kepler's Problem for Hyperbolic Orbits

As with Kepler's solution for elliptical orbits, there is no closed-form solution to Kepler's equation and this transcendental equation must be solved numerically. A similar process to solve for the hyperbolic analogy of eccentric anomaly, the hyperbolic anomaly, is used. A function, f, is created such that the goal is to find the value of hyperbolic anomaly, H, that causes this function to be equal to zero,

$$f(H) = e\sinh H - H - M = 0 \quad (2.135)$$

Taking the derivative of Equation (2.135) with respect to the hyperbolic anomaly,

$$f'(H) = \frac{df}{dH} = e\cosh H - 1 \quad (2.136)$$

And inserting this into a similar Newton-Raphson algorithm, similar to what was shown in Equation (2.101),

$$H_{k+1} = H_k - \frac{f(H_k)}{f'(H_k)} \quad (2.137)$$

This equation is iterated upon until $|f(H)|$ becomes sufficiently small. Like the eccentric anomaly, this is generally a well-behaved function and converges after a small number of iterations.

2.6.4 Kepler's Problem Using Universal Variables

In the previous sections, Kepler's problem was solved for the specific case of elliptical, parabolic, and hyperbolic orbits. Each conic section type had similar approaches, but they require different auxiliary angles for the different conic geometries, and depend on different formulations of Kepler's equation.

Rather than taking special approaches to solving Kepler's problem for each type of trajectory, Battin [1987] introduced a universal method that works regardless of trajectory type, needing only an initial $\mathbf{r}_0$ and $\mathbf{v}_0$. Position and velocity are established as functions of a universal anomaly, χ. Kepler's equation appears quite differently in the universal formulation,

$$\sqrt{\mu}(t - t_0) = \sigma_0\chi^2 C_2(\alpha\chi^2) + (1 - r_0\alpha)\chi^3 C_3(\alpha\chi^2) + r_0\chi \quad (2.138)$$

where

$$\sigma_0 \equiv \frac{\mathbf{r}_0 \cdot \mathbf{v}_0}{\sqrt{\mu}} \quad (2.139)$$

$$\alpha \equiv a^{-1} = \frac{2}{r_0} - \frac{v_0}{\mu} \tag{2.140}$$

The functions C_k, known as Stumpff functions [Curtis, 2019], are initially defined here in some detail as they will again be useful in solving Lambert's problem,

$$C_k(z) \equiv \frac{1}{k!} - \frac{z}{(k+2)!} + \frac{z^2}{(k+4)!} - \cdots \tag{2.141}$$

$$C_2(z) = \frac{1}{2!} - \frac{z}{4!} + \frac{z^2}{6!} - \cdots = \begin{cases} \dfrac{1-\cos\sqrt{z}}{z}, & z > 0 \\ \dfrac{1}{2}, & z = 0 \\ \dfrac{\cosh\sqrt{-z}-1}{-z}, & z < 0 \end{cases} \tag{2.142}$$

$$C_3(z) = \frac{1}{3!} - \frac{z}{5!} + \frac{z^2}{7!} - \cdots = \begin{cases} \dfrac{\sqrt{z}-\sin\sqrt{z}}{\sqrt{z^3}}, & z > 0 \\ \dfrac{1}{6}, & z = 0 \\ \dfrac{\sinh\sqrt{-z}-\sqrt{-z}}{\sqrt{-z^3}}, & z < 0 \end{cases} \tag{2.143}$$

2.6.4.1 Solution to Kepler's Problem Using Universal Variables

Similar to the numerical method for solving the special elliptical case, the universal Kepler's equation may also be solved by rearranging and minimizing via Newton-Raphson iteration. It is interesting to find that the first derivative of f with respect to universal anomaly $\left(f' = \frac{df}{d\chi}\right)$ is always equal to the magnitude of the radius,

$$\begin{aligned} f(\chi) &= \sigma_0\chi^2 C_2\left(\alpha\chi^2\right) + (1 - r_0\alpha)\chi^3 C_3\left(\alpha\chi^2\right) \\ &\quad + r_0\chi - \sqrt{\mu}\,(t - t_0) = 0 \end{aligned} \tag{2.144}$$

$$f'(\chi) = \sigma_0\chi\left[1 - \alpha\chi^2 C_2\left(\alpha\chi^2\right)\right] + (1 - r_0\alpha)\chi^2 C_3\left(\alpha\chi^2\right) + r_0$$

$$f'(\chi) = r(\chi)$$

where $r(\chi)$ is the trajectory radius as a function of universal anomaly.

Iteratively, we can solve for the universal anomaly

$$\chi_{k+1} = \chi_k - \frac{f(\chi_k)}{f'(\chi_k)} \tag{2.145}$$

and continue iterating until the universal anomaly doesn't change (within some tolerance) from one iteration to the next. Here, too, the function generally converges after only a few iterations.

The elegance of the universal variable approach is seen when, *a priori*, we do not know what type of conic section we are using. This will be further demonstrated in the discussion of Lambert's problem in Chapter 5.

2.6.5 Lagrange Coefficients

An analytical method to propagate a two-body problem can be applied. Using our definition of the position and velocity vectors in the perifocal reference frame,

$$\begin{aligned} \mathbf{r} &= x\mathbf{p} + y\mathbf{q} \\ \mathbf{v} &= \dot{x}\mathbf{p} + \dot{y}\mathbf{q} \end{aligned} \tag{2.146}$$

where x and y are the **p**- and **q**-components of position, and $\dot{x}$ and $\dot{y}$ are the **p**- and **q**-components of velocity. In matrix form, this can be written as

$$\begin{Bmatrix} \mathbf{r} \\ \mathbf{v} \end{Bmatrix} = \begin{bmatrix} x & y \\ \dot{x} & \dot{y} \end{bmatrix} \begin{Bmatrix} \mathbf{p} \\ \mathbf{q} \end{Bmatrix} \tag{2.147}$$

Similarly, the initial position and velocity vectors are

$$\begin{Bmatrix} \mathbf{r}_0 \\ \mathbf{v}_0 \end{Bmatrix} = \begin{bmatrix} x_0 & y_0 \\ \dot{x}_0 & \dot{y}_0 \end{bmatrix} \begin{Bmatrix} \mathbf{p} \\ \mathbf{q} \end{Bmatrix} \tag{2.148}$$

Solving Equation (2.148) for the perifocal coordinate frame unit vectors,

$$\begin{Bmatrix} \mathbf{p} \\ \mathbf{q} \end{Bmatrix} = \begin{bmatrix} x_0 & y_0 \\ \dot{x}_0 & \dot{y}_0 \end{bmatrix}^{-1} \begin{Bmatrix} \mathbf{r}_0 \\ \mathbf{v}_0 \end{Bmatrix} \tag{2.149}$$

and inserting Equation (2.149) into Equation (2.147),

$$\begin{Bmatrix} \mathbf{r} \\ \mathbf{v} \end{Bmatrix} = \begin{bmatrix} x & y \\ \dot{x} & \dot{y} \end{bmatrix} \begin{bmatrix} x_0 & y_0 \\ \dot{x}_0 & \dot{y}_0 \end{bmatrix}^{-1} \begin{Bmatrix} \mathbf{r}_0 \\ \mathbf{v}_0 \end{Bmatrix} \tag{2.150}$$

Simplifying,

$$\begin{Bmatrix} \mathbf{r} \\ \mathbf{v} \end{Bmatrix} = \begin{bmatrix} f & g \\ \dot{f} & \dot{g} \end{bmatrix} \begin{Bmatrix} \mathbf{r}_0 \\ \mathbf{v}_0 \end{Bmatrix} \tag{2.151}$$

Using the initial position and velocity vectors, we can find the semimajor axis (or semilatus rectum), eccentricity and initial anomaly (true, eccentric, parabolic, hyperbolic or universal). Given the time difference, we can then find the final anomaly using the appropriate Kepler's equation (shown earlier in this section). Ultimately, we can determine future or past position and velocity vectors given the initial position and velocity vectors, along with the terms f, g, $\dot{f}$, and $\dot{g}$.

Depending on the anomaly used for the position and velocity vectors (from their components x, y, $\dot{x}$ and $\dot{y}$), specific definitions of the terms f, g, $\dot{f}$ and $\dot{g}$ can be shown. Let's start with the elliptical orbit. Here, we have a choice between using the true anomaly and the eccentric anomaly. If the position and velocity vectors are given using true anomaly,

$$\begin{aligned} \mathbf{r} &= \left[\frac{p\cos\theta}{1+e\cos\theta}\right]\mathbf{p} + \left[\frac{p\sin\theta}{1+e\cos\theta}\right]\mathbf{q} \\ \mathbf{v} &= \left[-\sqrt{\frac{\mu}{p}}\sin\theta\right]\mathbf{p} + \left[\sqrt{\frac{\mu}{p}}(e+\cos\theta)\right]\mathbf{q} \end{aligned} \tag{2.152}$$

the terms f, g, $\dot{f}$ and $\dot{g}$ are

$$\begin{aligned} f &= 1-\left[\frac{1-\cos(\theta-\theta_0)}{1+e\cos\theta}\right] \\ g &= \frac{rr_0\sin(\theta-\theta_0)}{\sqrt{\mu p}} \\ \dot{f} &= \sqrt{\frac{\mu}{p}}\tan\left(\frac{\theta-\theta_0}{2}\right)\left[\frac{1-\cos(\theta-\theta_0)}{p}-\frac{1}{r}-\frac{1}{r_0}\right] \\ \dot{g} &= 1-\left(\frac{r_0}{p}\right)\left[1-\cos(\theta-\theta_0)\right] \end{aligned} \tag{2.153}$$

Similarly, if eccentric anomaly is used, then

$$\begin{aligned} \mathbf{r} &= a(\cos E - e)\mathbf{p} + a\left(\sqrt{1-e^2}\sin E\right)\mathbf{q} \\ \mathbf{v} &= -\sqrt{\frac{\mu}{a}}\left(\frac{\sin E}{1-e\cos E}\right)\mathbf{p} + \sqrt{\frac{\mu}{a}}\left(\frac{\cos E}{1-e\cos E}\right)\left(\sqrt{1-e^2}\right)\mathbf{q} \end{aligned} \tag{2.154}$$

and the coefficients are

$$
\begin{aligned}
f &= 1 - \frac{a}{r_0}\left[1 - \cos(E - E_0)\right] \\
g &= (t - t_0) - \sqrt{\frac{a^3}{\mu}}\left[E - E_0 - \sin(E - E_0)\right] \\
\dot{f} &= -\frac{\sin(E - E_0)\sqrt{\mu a}}{r_0 r} \\
\dot{g} &= 1 - \left(\frac{a}{r}\right)\left[1 - \cos(E - E_0)\right]
\end{aligned}
\tag{2.155}
$$

For the unlikely case of a purely parabolic orbit (included here for completeness), the position and velocity vectors are

$$
\begin{aligned}
\mathbf{r} &= \left(\frac{p}{2}\right)\left(1 - B^2\right)\mathbf{p} + pB\mathbf{q} \\
\mathbf{v} &= \frac{\sqrt{\mu p}}{r} B\mathbf{p} - \frac{\sqrt{\mu p}}{r}\mathbf{q}
\end{aligned}
\tag{2.156}
$$

where B is the parabolic anomaly (defined in Equation 2.107). The coefficients are

$$
\begin{aligned}
f &= \frac{1 - B^2 + 2BB_0}{1 + B_0^2} \\
g &= \frac{p^2(B - B_0)(1 + BB_0)}{2h} \\
\dot{f} &= \frac{4h(B - B_0)}{p^2\left(1 + B^2\right)\left(1 + B_0^2\right)} \\
\dot{g} &= \frac{1 - B_0^2 + 2BB_0}{1 + B^2}
\end{aligned}
\tag{2.157}
$$

For hyperbolic orbits, the position and velocity vectors are

$$
\begin{aligned}
\mathbf{r} &= a(\cosh H - e)\mathbf{p} + \left(-a\sqrt{e^2 - 1}\sinh H\right)\mathbf{q} \\
\mathbf{v} &= \left(-\frac{\sqrt{-\mu a}}{r}\sinh H\right)\mathbf{p} + \left(\frac{\sqrt{\mu p}}{r}\cosh H\right)\mathbf{q}
\end{aligned}
\tag{2.158}
$$

and the coefficients are

$$
\begin{aligned}
f &= 1 - \frac{a}{r_0}\left[1 - \cosh(H - H_0)\right] \\
g &= (t - t_0) - \sqrt{\frac{(-a)^3}{\mu}}\left[\sinh(H - H_0) - H - H_0\right] \\
\dot{f} &= -\frac{\sinh(H - H_0)\sqrt{-\mu a}}{r_0 r} \\
\dot{g} &= 1 - \left(\frac{a}{r}\right)\left[1 - \cosh(H - H_0)\right]
\end{aligned}
\tag{2.159}
$$

A universal variable form of this process is also available, as we have previously defined in Section 2.6.4. This is certainly useful when we don't want to be forced to predetermine what type of a conic section orbit we have. Using the universal variable definitions defined in Equations (2.138) through (2.143), the position and velocity vectors are

$$
\begin{aligned}
\mathbf{r} &= \left[-ae - a\sin\left(\frac{\chi + C_0}{\sqrt{a}}\right)\right]\mathbf{p} + \left[a\sqrt{1-e^2}\cos\left(\frac{\chi + C_0}{\sqrt{a}}\right)\right]\mathbf{q} \\
\mathbf{v} &= \left[-\frac{\sqrt{\mu a}}{r}\cos\left(\frac{\chi + C_0}{\sqrt{a}}\right)\right]\mathbf{p} + \left[-\frac{\sqrt{\mu p}}{r}\sin\left(\frac{\chi + C_0}{\sqrt{a}}\right)\right]\mathbf{q}
\end{aligned}
\tag{2.160}
$$

with coefficients,

$$
\begin{aligned}
f &= 1 - \frac{\chi_0^2}{r_0}C_2 \\
g &= (t - t_0) - \frac{\chi_0^3}{\sqrt{\mu}}C_3 \\
\dot{f} &= \frac{\sqrt{\mu}}{r_0 r}\chi_0\left(\frac{\chi^2}{a}C_3 - 1\right) \\
\dot{g} &= 1 - \frac{\chi_0^2}{r}C_2
\end{aligned}
\tag{2.161}
$$

where the coefficients C_0, C_2, and C_3 are defined in Equations (2.141) through (2.143).

2.7 PERTURBATIONS

As the dynamical system gets more realistic, various perturbations affect the motion of a satellite. Bumps and dips in the gravity field perturb all objects. Drag strongly affects satellites whose orbits are in the atmosphere (up to several hundreds of kilometers in altitude). Solar radiation pressure (the force exerted on a body by the Sun's light) causes the orbit to drift slightly. A planet's reflected light perturbs the motion slightly. A third body's gravitational forces will also perturb the spacecraft's orbit.

Modifying Equation (2.9), the spacecraft's motion in orbit now includes the sum of the perturbing accelerations, **P**, on the right-hand side

$$\frac{d^2\mathbf{r}}{dt^2} + \frac{\mu\mathbf{r}}{r^3} = \mathbf{P} \tag{2.162}$$

The vector notation is arbitrary, so it can be written in any number of coordinate frames.

Selection of the types of perturbations to be included in an analysis depends on multiple things, including the type of mission being operated and the type of accuracy needed. For most simple problems, neglecting perturbations is a reasonable assumption for the first level of analysis. Further details on perturbations can be found in Chao [2005], among others.

2.7.1 Lagrange's Planetary Equations

Lagrange's planetary equations, which are sometimes referred to as the Lagrangian variation of parameter (VOP), are historically the first set of equations written in the form of $\frac{d\mathbf{COE}}{dt} = f(\mathbf{COE}, t)$ where **COE** is a six-element vector composed of all of the classical orbital elements (COE). The derivation of these equations can be found in Vallado [2013]. For the sake of brevity, we report the resulting equations once the variation of parameter has been implemented

$$\begin{aligned}
\frac{da}{dt} &= \frac{2}{na}\frac{\partial F}{\partial M} \\
\frac{de}{dt} &= \frac{1-e^2}{na^2e}\frac{\partial F}{\partial M} - \frac{\left(1-e^2\right)^{1/2}}{na^2e}\frac{\partial F}{\partial \omega} \\
\frac{di}{dt} &= \frac{\cos i}{na^2\left(1-e^2\right)^{1/2}\sin i}\frac{\partial F}{\partial \omega} - \frac{1}{na^2\left(1-e^2\right)^{1/2}\sin i}\frac{\partial F}{\partial \Omega} \\
\frac{d\Omega}{dt} &= \frac{1}{na^2\left(1-e^2\right)^{1/2}\sin i}\frac{\partial F}{\partial i} \\
\frac{d\omega}{dt} &= -\frac{\cos i}{na^2\left(1-e^2\right)^{1/2}\sin i}\frac{\partial F}{\partial i} + \frac{\left(1-e^2\right)^{1/2}}{na^2e}\frac{\partial F}{\partial e} \\
\frac{dM}{dt} &= -\frac{1-e^2}{na^2e}\frac{\partial F}{\partial e} - \frac{2}{na}\frac{\partial F}{\partial a}
\end{aligned} \tag{2.163}$$

where F is the negative of the total energy, $F = V - T = \left(\frac{\mu}{r} + R\right) - \frac{v^2}{2} = \frac{\mu}{2a} + R$ and R is the disturbing potential. Using this disturbing potential, the Lagrange's planetary equations can also be written as

$$
\begin{aligned}
\frac{da}{dt} &= \frac{2}{na}\frac{\partial R}{\partial M} \\
\frac{de}{dt} &= \frac{1-e^2}{na^2 e}\frac{\partial R}{\partial M} - \frac{\left(1-e^2\right)^{1/2}}{na^2 e}\frac{\partial R}{\partial \omega} \\
\frac{di}{dt} &= \frac{\cos i}{na^2\left(1-e^2\right)^{1/2}\sin i}\frac{\partial R}{\partial \omega} - \frac{1}{na^2\left(1-e^2\right)^{1/2}\sin i}\frac{\partial R}{\partial \Omega} \\
\frac{d\Omega}{dt} &= \frac{1}{na^2\left(1-e^2\right)^{1/2}\sin i}\frac{\partial R}{\partial i} \\
\frac{d\omega}{dt} &= -\frac{\cos i}{na^2\left(1-e^2\right)^{1/2}\sin i}\frac{\partial R}{\partial i} + \frac{\left(1-e^2\right)^{1/2}}{na^2 e}\frac{\partial R}{\partial e} \\
\frac{dM}{dt} &= n - \frac{1-e^2}{na^2 e}\frac{\partial R}{\partial e} - \frac{2}{na}\frac{\partial R}{\partial a}
\end{aligned}
\qquad (2.164)
$$

These equations were originally derived by Lagrange to understand the small disturbances of the planets of a given object orbiting the Sun. However, Equation (2.164) provides the time rate of change of all classical orbital elements for any given disturbing potential R. Because of this, Lagrange's planetary equations can give an analytical formulation for various kinds of perturbations, including Earth's oblateness, drag effects, and third-body effects, among others.

2.7.2 Gravitational

An excellent source for modeling the gravitational potential is presented by Kaula [1966]. He expands the geopotential using spherical harmonic functions, a method that can apply to the potential of any massive body. Any planet's perturbing gravitational potential at distance r, latitude δ, and east longitude λ is

$$
U = \frac{\mu}{r}\sum_{n=2}^{\infty}\left(\frac{a_c}{r}\right)^n \sum_{m=0}^{n} P_{nm}\left(\sin\delta\right)\left[C_{nm}\cos(m\lambda) + S_{nm}\sin(m\lambda)\right] \qquad (2.165)
$$

where a_e is the equatorial radius, P_{nm} is the associated Legendre function of the first kind of degree n and order m, and C_{nm} and S_{nm} are the harmonic coefficients of degree n and order m. The harmonic coefficients contain the geopotential model for any celestial body of interest and are experimentally determined by satellite geodesy. While any depth into this topic is beyond the scope of this book, one term in this gravitational potential (for Earth) is discussed.

The Earth's dominating gravitational perturbation results from its oblateness – a symmetrical bulge caused by its rotation (n = 2, m = 0). The Earth's radius at the equator is about 21.38 km greater than at the poles. The oblateness term takes the designation J_2, because that's how it appears in this harmonic expansion of the Earth's gravitational field. Although the J_2 effect is about 1,000 times smaller than the two-body force, it dominates all other gravitational perturbations from the central body.

The J_2 term affects several orbital elements. Two in particular – the right ascension of the ascending node and the argument of periapse – can produce a significant deviation from the two-body motion over time. These effects can be incorporated into controlling a spacecraft's motion. The oblateness term causes the orbital plane to precess in inertial space; the orbit's right ascension of the ascending node undergoes a constant change. This nodal precession rate depends on the orbit's semimajor axis, eccentricity, and inclination. The average time rate of change of the right ascension of ascending node due to J_2 is

$$\frac{d\Omega}{dt} = -\frac{3nR_{\otimes}^2 J_2 \cos i}{2a^2\left(1-e^2\right)^2} \tag{2.166}$$

where $R_{\otimes}$ is the planet's radius. The secular J_2 nodal regression vanishes for polar orbits ($i = 90°$) and reaches its maximum value for a given semimajor axis and eccentricity when the inclination is zero. This is a significant effect, as the orbital plane may rotate in inertial space by several degrees per day.

Familiar missions that benefit from this perturbation are satellites placed in Sun-synchronous orbits. The semimajor axis, eccentricity, and inclination are chosen so that the satellite's nodal regression matches the planet's rotation about the Sun. This arrangement fixes the line of nodes relative to the Sun, giving constant lighting conditions for observing objects on the ground or consistent illumination for solar panels. Remote-sensing spacecraft commonly use this type of orbit. A schematic of this behavior is shown in Figure 2.17.

Without this J_2 perturbation, the orbit plane would not rotate as the planet goes around the Sun. By judicious choice of semimajor axis, eccentricity, and inclination,

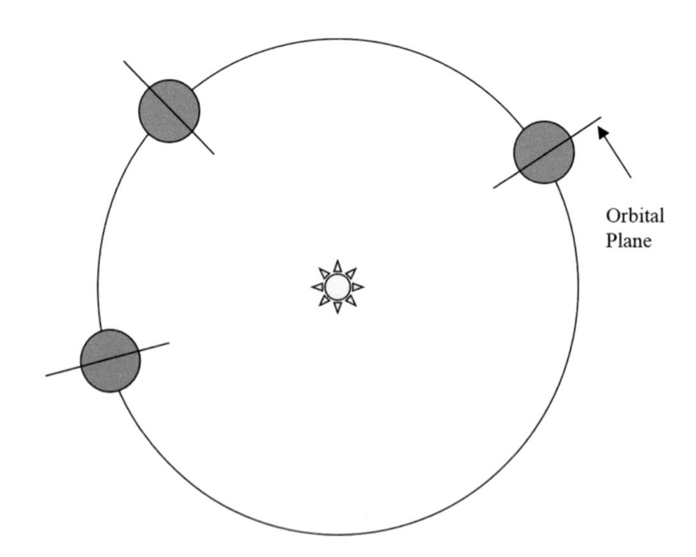

FIGURE 2.17 Schematic of nodal regression for Sun-synchronous orbit.

a satellite's orbit plane can rotate at the rate of 360°/year, exactly matching the angular rotation of the planet around the Sun.

The oblateness term also affects the argument of perigee's time rate of change, which, too, is a function of the orbit's semimajor axis, eccentricity, and inclination. Its average time rate of change is

$$\frac{d\omega}{dt} = \frac{3nR_{\otimes}^2 J_2}{2a^2\left(1-e^2\right)^2}\left(2-\frac{5}{2}\sin^2 i\right) \quad (2.167)$$

where all terms represent the same values as in Equation (2.163). Minimizing the effects given by Equation (2.167) using gravitational perturbations can help control the spacecraft's apsidal rotation. The shape and apsidal rotation rate of such an orbit remain nearly constant over the mission lifetime and are created by selection of the orbit's semimajor axis, eccentricity, inclination, and argument of perigee. In this type of orbit, the perigee point is fixed over a latitude on the planet's surface. Because of this geometry, the spacecraft's orbit repeats at a fixed altitude over any location on the planet. Applications include oceanographic, geodetic, and remote-sensing spacecraft.

As the spacecraft's orbit moves further from the non-spherical planet and closer to other gravitational sources, the gravitational perturbation from a third body becomes a larger relative perturbation. Usually, the perturbing body is treated as a point mass. The equations of motion in a reference frame whose origin is at the center of mass of the two constant massive bodies is

$$\frac{d^2\mathbf{r}_3}{dt^2} = \frac{Gm_1}{r_{31}^3}\left(\mathbf{r}_1-\mathbf{r}_3\right)+\frac{Gm_2}{r_{32}^3}\left(\mathbf{r}_2-\mathbf{r}_3\right) \quad (2.168)$$

where $\mathbf{r}_3$ is the radius vector from the center of mass of the two primary bodies (m_1 and m_2) to the third body, $\mathbf{r}_1$ and $\mathbf{r}_2$ are the radius vectors from the center of mass to m_1 and m_2, respectively, and $\mathbf{r}_{ij}$ is the vector from body i to body j. Additional gravitational sources can be added in a similar way.

The influence of other planets is generally considered negligible for satellites orbiting the Earth. They are, of course, significant for interplanetary trajectories. However, for satellites orbiting the Earth, the Earth's Moon's gravity is minor, but non-negligible, for higher altitude spacecraft (such as geosynchronous spacecraft). Third-body forces influence studies of long-term orbits for satellites in deep space. Third-body perturbations can affect eccentric orbits by changing the eccentricity enough to send the perigee into the drag regime and accelerating the object's decay. Further details of third-body gravitational effects are covered in Chapter 3.

2.7.3 Non-Gravitational

Atmospheric drag is the other main perturbation for spacecraft in a low-altitude orbit. Motion of a body through an atmosphere (even one with very low-density) is subject to drag caused when the body interacts with the atmosphere's molecules. The atmosphere's density is a complex and dynamic function of altitude, latitude, time of day, and the solar radiation hitting the planet. Other factors that affect the drag forces are the satellite's mass, projected cross-sectional area, satellite velocity, and drag coefficient.

Atmospheric drag produces a force opposite to the instantaneous velocity vector, expressed as

$$\mathbf{F} = m\frac{d^2\mathbf{r}}{dt^2} = -\frac{C_D A\rho v\mathbf{v}}{2} \tag{2.169}$$

where C_D is the drag coefficient, A is the cross-sectional area in the direction of motion, ρ is the instantaneous atmospheric density, **v** is the spacecraft's instantaneous velocity vector, and m is the spacecraft's mass. The perturbation applies in the in-track direction (the velocity's direction). Over time, the semimajor axis gets smaller, and the orbital period decreases. The orbit's eccentricity also decreases, which makes the orbit nearly circular. Eventually, a spacecraft's orbit will decay so far that the satellite reenters. This is what happened to the U.S. Skylab space station in 1979.

Atmospheric drag from a planet can also be used for the entry, descent, and landing (EDL) portion of a mission. Here, drag is used purposely to remove energy from a spacecraft's orbit, thus reducing the amount of propellant needed for a soft landing. More details on atmospheric effects and EDL are found in Chapter 5.

The perturbation from solar radiation pressure is similar in many ways to the drag perturbation. The force's magnitude is proportional to the satellite's area-to-mass ratio and depends on the albedo, a parameter analogous to the drag coefficient. In this case, however, the force acts along the line from the Sun to the satellite. As this force doesn't have a constant direction with respect to the orbit (unlike drag), the effects on the elements are more diverse.

2.8 NUMERICAL INTEGRATION

The perturbed two-body equations of motion generally do not have analytical solutions, so it is necessary to numerically integrate the equations of motion in order to find the solution. These are coupled, non-linear, second-order equations. This section discusses analytical and numerical integration methods used.

While analytical methods work for the simple two-body motion, anything more complex generally requires the use of a numerical integration. The ordinary differential equation with perturbations for the two-body problem is

$$\ddot{\mathbf{r}} = -\frac{\mu\mathbf{r}}{r^3} + \mathbf{P} \tag{2.170}$$

where **P** is the perturbation vector. This perturbation vector is what generally requires this equation to be solved numerically.

Regardless of what the perturbation vector represents, it can be written as a three-dimensional vector, $\mathbf{P} = \begin{Bmatrix} P_x & P_y & P_z \end{Bmatrix}^T$, where the subscripts x, y, and z represent Cartesian coordinates, although other sets of three-element basis (such as cylindrical and spherical) can be used. To solve this problem numerically, the three-dimensional second-order ordinary differential equation needs to be converted into a six-dimensional set of first-order ordinary differential equations. In vector form, Equation (2.170) can be written in first-order form as

$$\frac{d}{dt}\begin{Bmatrix} \mathbf{r} \\ \dot{\mathbf{r}} \end{Bmatrix} = \frac{d}{dt}\begin{Bmatrix} \mathbf{r} \\ \mathbf{v} \end{Bmatrix} = \begin{Bmatrix} \mathbf{v} \\ -\dfrac{\mu\mathbf{r}}{r^3} \end{Bmatrix} + \begin{Bmatrix} 0 \\ 0 \\ 0 \\ P_x \\ P_y \\ P_z \end{Bmatrix} \tag{2.171}$$

We then define a state vector, **X**, as

$$\mathbf{X} = \begin{Bmatrix} \mathbf{r} \\ \mathbf{v} \end{Bmatrix} \tag{2.172}$$

where the vectors, **r** and **v,** can be written in array format as

$$\mathbf{r} = \begin{Bmatrix} X(1) \\ X(2) \\ X(3) \end{Bmatrix} \tag{2.173}$$

and

$$\dot{\mathbf{r}} = \mathbf{v} = \begin{Bmatrix} X(4) \\ X(5) \\ X(6) \end{Bmatrix} \tag{2.174}$$

where X(1), …, X(6) are the elements of the state vector.

The perturbation vector can be written as

$$\mathbf{P} = \begin{Bmatrix} P(1) \\ P(2) \\ P(3) \end{Bmatrix} \tag{2.175}$$

where P(1), P(2), and P(3) are the elements of the perturbation vector.

In this array form, Equation (2.170) becomes

$$\frac{dX(1)}{dt} = X(4)$$
$$\frac{dX(2)}{dt} = X(5)$$
$$\frac{dX(3)}{dt} = X(6)$$
$$\frac{dX(4)}{dt} = -\frac{\mu X(1)}{r^3} + P(1) \tag{2.176}$$
$$\frac{dX(5)}{dt} = -\frac{\mu X(2)}{r^3} + P(2)$$
$$\frac{dX(6)}{dt} = -\frac{\mu X(3)}{r^3} + P(3)$$

where $r = \sqrt{[X(1)]^2 + [X(2)]^2 + [X(3)]^2}$. This formulation of the equations of motion is called state-space form. In the following sections, we go over different integration schemes that can be used to numerically integrate these equations of motion in state-space form.

2.8.1 Cowell's Method

The simplest numerical integration method used when dealing with perturbations is Cowell's method, named after 19th–20th-century British astronomer Philip Herbert Cowell. With the help of contemporary colleague Andrew Claude de la Cherois Crommelin, Cowell was able to apply this method to predict the return of Halley's Comet in 1910 [Brouwer and Clemence, 1961]. Cowell's method is rather straightforward:

1. Write all equations of motion in state-space form including all perturbing accelerations
2. Numerically integrate the equations of motion step by step

For the perturbed two-body problem, as we saw, the equations of motion are

$$\frac{d^2\mathbf{r}}{dt^2} = -\frac{\mu\mathbf{r}}{r^3} + \mathbf{P} \tag{2.177}$$

where **P** is the perturbation vector. For example, for a spacecraft orbiting close to Mars, **P** may represent the gravitational acceleration due to its moons, Phobos and Deimos. Representing Equation (2.177) in state-space form gives

$$\dot{\mathbf{r}} = \mathbf{v}$$
$$\dot{\mathbf{v}} = -\frac{\mu\mathbf{r}}{r^3} + \mathbf{P} \tag{2.178}$$

where **r** and **v** are the position and velocity of the spacecraft with respect to the central main body. In Cartesian coordinates, $\mathbf{r} = [x, y, z]$ and $\mathbf{v} = [\dot{x}, \dot{y}, \dot{z}]$ so that Equation (2.178) becomes

$$
\begin{aligned}
\dot{x} &= v_x \\
\dot{y} &= v_y \\
\dot{z} &= v_z \\
\dot{v}_x &= -\frac{\mu x}{r^3} + P_x \\
\dot{v}_y &= -\frac{\mu y}{r^3} + P_y \\
\dot{v}_z &= -\frac{\mu z}{r^3} + P_z
\end{aligned}
\tag{2.179}
$$

where $\mathbf{P} = \left[P_x, P_y, P_z\right]$ is the perturbation vector in Cartesian coordinates. Note that, therefore, $[\mathbf{r}, \mathbf{v}]$ is the six-element state vector. In the example of the spacecraft orbiting Mars, if we only consider Phobos as a perturbation, we get

$$
\begin{aligned}
\dot{\mathbf{r}} &= \mathbf{v} \\
\dot{\mathbf{v}} &= -\frac{\mu_{\text{Mars}}\mathbf{r}}{r^3} - \mu_{\text{Phobos}}\left(\frac{\mathbf{r}_{\text{SC/Phobos}}}{r^3_{\text{SC/Phobos}}} - \frac{\mathbf{r}_{\text{Phobos/Mars}}}{r^3_{\text{Phobos/Mars}}}\right)
\end{aligned}
\tag{2.180}
$$

where **r** is the position of the spacecraft (SC) with respect to Mars, $\mathbf{r}_{\text{SC/Phobos}}$ is the position of the spacecraft with respect to Phobos, $\mathbf{r}_{\text{Phobos/Mars}}$ is the position of Phobos with respect to Mars.

Once the analytical formulation of the equations of motion is provided in state-space form, such as in Equation (2.180), one can apply any numerical integration scheme of their choosing. Cowell's method is therefore a simple and straightforward method that can handle any perturbation, as long as it can be written in the form of an acceleration vector as in Equation (2.178). However, Cowell's method often results slower than Encke's method (which is approximately ten times faster on average) since the time step taken in the integration can depend greatly. For example, near body with large gravitational attraction, smaller time steps must be used. As suggested in Bate et al. [2020], Cowell's method can be improved if polar or spherical coordinates are used in the formulation. This is due to the fact that orbital motion is inherently curvilinear with continuous changes to the curvature of trajectories. In fact, if using, for example, spherical coordinates, the position vector [r, θ, φ] tends to change slower than its Cartesian counterpart. Thus, in spherical coordinates, the perturbed two-body equations of motion become

$$
\begin{aligned}
&\ddot{r} - r\left(\dot{\theta}^2\cos^2\phi + \dot{\phi}^2\right) = -\frac{\mu}{r^2} + P_r \\
&r\ddot{\theta}\cos\phi + 2\dot{r}\dot{\theta}\cos\phi - 2r\dot{\theta}\dot{\phi}\sin\phi = P_\theta \\
&r\ddot{\phi} + 2\dot{r}\dot{\phi} + r\dot{\theta}^2\sin\phi\cos\phi = P_\phi
\end{aligned}
\tag{2.181}
$$

2.8.2 Encke's Method

Nineteenth-century German astronomer Johann Franz Encke is responsible for a method of numerical integration that is generally more efficient than Cowell's method [Brouwer and Clemence, 1961]. In fact, while Cowell's method considers all acceleration terms at once and numerically integrates them together, Encke's method focuses its attention on the difference between the primary acceleration and the secondary (perturbing) acceleration terms. Note that the equations of motion of this dynamical system are still given by Equation (2.177). Furthermore, it is assumed that the primary acceleration alone would produce a conic section as described in Section 2.4. This resulting conic section is called the osculating (or reference) orbit. "Osculation" originates from the Latin word *osculum*, which means "kiss." Since the reference orbit touches (and thus "kisses") the true (perturbed) trajectory at the starting point, it is therefore called an osculating orbit. The osculating orbit would be equal to the true orbit for all integration times only if all perturbations were removed. However, only at the starting point both the osculating and the true orbits touch, and they soon start deviating as the numerical integration process is carried out. When the osculating orbit has deviated too much, rectification must occur in order for the integration to continue. At the point of rectification, a new osculating orbit is generated, numerical integration continues until the osculating and true orbits deviate, and the procedure is then repeated (Figure 2.18).

The goal of Encke's method is to analytically quantify the difference between the osculating and the true orbit. Thus, defining the difference vector, $\delta\mathbf{r}$, as the difference between the two vectors $\mathbf{r}$ (position vector of the true orbit) and $\boldsymbol{\rho}$ (position vector of the osculating orbit), we have

$$\delta\mathbf{r} = \mathbf{r} - \boldsymbol{\rho} \tag{2.182}$$

as seen in Figure 2.19. The equations of motion that govern the true (perturbed) orbit are

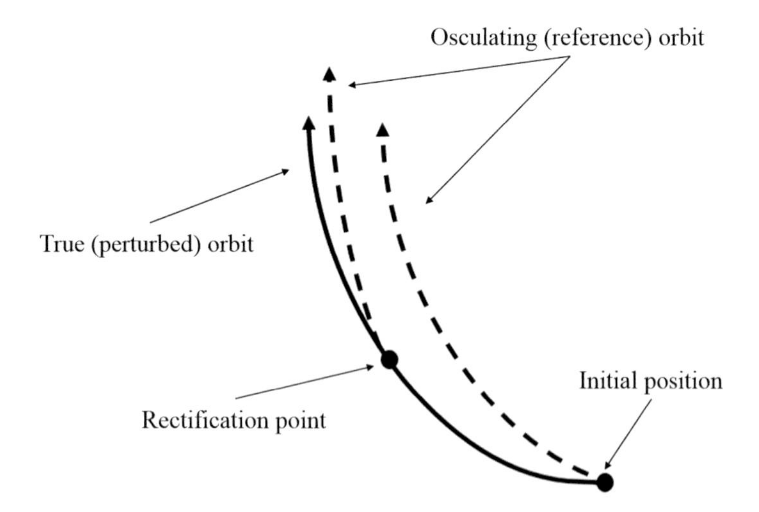

FIGURE 2.18 Representation of the osculating orbit and rectification.

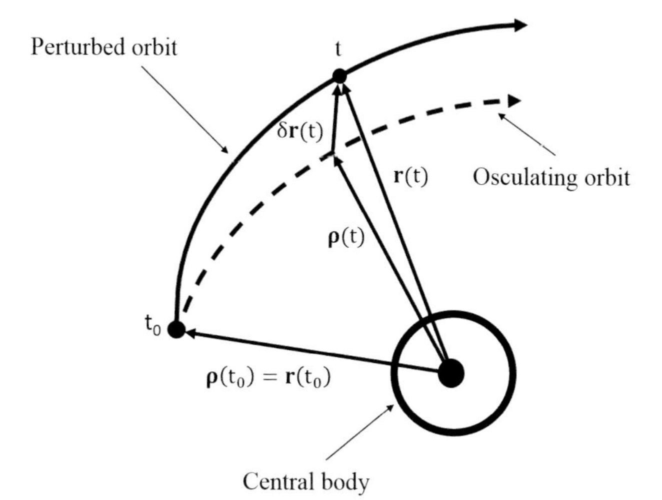

FIGURE 2.19 Osculating and perturbed orbits deviating from each otheR.

$$\frac{d^2\mathbf{r}}{dt^2} = -\frac{\mu\mathbf{r}}{r^3} + \mathbf{P} \tag{2.183}$$

while for the osculating orbit is governed by

$$\frac{d^2\boldsymbol{\rho}}{dt^2} = -\frac{\mu\boldsymbol{\rho}}{\rho^3} \tag{2.184}$$

By definition, we know that at the starting point, $t = t_0$, we have

$$\begin{aligned} \mathbf{r}(t_0) &= \boldsymbol{\rho}(t_0) \\ \mathbf{v}(t_0) &= \dot{\boldsymbol{\rho}}(t_0) \end{aligned} \tag{2.185}$$

Taking two-time derivatives of Equation (2.182) results in

$$\delta\ddot{\mathbf{r}} = \ddot{\mathbf{r}} - \ddot{\boldsymbol{\rho}} \tag{2.186}$$

Combining Equations (2.183), (2.184), and (2.186) and simplifying

$$\delta\ddot{\mathbf{r}} = -\frac{\mu\mathbf{r}}{r^3} + \frac{\mu\boldsymbol{\rho}}{\rho^3} + \mathbf{P} \tag{2.187}$$

Factoring out the two-body solution, $\frac{\mu}{\rho^3}$,

$$\delta\ddot{\mathbf{r}} = \frac{\mu}{\rho^3}\left(\boldsymbol{\rho} - \frac{\rho^3}{r^3}\mathbf{r}\right) + \mathbf{P} \tag{2.188}$$

and substituting Equation (2.182) into Equation (2.188)

$$\delta\ddot{\mathbf{r}} = \frac{\mu}{\rho^3}\left(\mathbf{r} - \delta\mathbf{r} - \frac{\rho^3}{r^3}\mathbf{r}\right) + \mathbf{P} \tag{2.189}$$

and simplifying

$$\delta\ddot{\mathbf{r}} = \frac{\mu}{\rho^3}\left[\left(1 - \frac{\rho^3}{r^3}\right)\mathbf{r} - \delta\mathbf{r}\right] + \mathbf{P} \tag{2.190}$$

Equation (2.190) can be used to numerically compute the deviation between the perturbed and osculating orbit, $\delta\mathbf{r}$, for any future time $t = t_0 + \Delta t$. In fact, $\boldsymbol{\rho}$ can be computed as a function of time (see Section 2.6), and thus $\mathbf{r}$ can be computed using Equation (2.182). One problem when computing $\mathbf{r}$ is the term $\left(1 - \frac{\rho^3}{r^3}\right)$ which represents the difference between two small quantities. This requires a lot of decimal digits for a computer to maintain a reasonable accuracy during the integration. One efficient way to compute the difference between two number which are similar in value to each other is discussed in Battin [1987] and briefly summarized here. Define three vectors $\mathbf{a}$, $\mathbf{b}$, and $\mathbf{c}$, such that $\mathbf{c} = \mathbf{b} - \mathbf{a}$ and $a \ll b$. Clearly, $c \approx b$. Let's further define

$$F = 1 - \frac{c^3}{b^3} \tag{2.191}$$

and

$$q = 1 - \frac{c^2}{b^2} \tag{2.192}$$

Therefore,

$$F = 1 - \left(\frac{c^2}{b^2}\right)^{3/2} = 1 - (1-q)^{3/2} \tag{2.193}$$

and, using some algebraic manipulations,

$$F = \left[1 - (1-q)^{3/2}\right]\frac{1 + (1-q)^{3/2}}{1 + (1-q)^{3/2}} = \frac{1 - (1-q)^3}{1 + \left(\sqrt{1-q}\right)^3} \tag{2.194}$$

from which

$$F(q) = \frac{q^2 - 3q + 3}{1 + (1-q)^{3/2}} \tag{2.195}$$

Note that computing F using Equation (2.195) does not require to compute the difference between nearly equal number like in Equation (2.191). Similarly, we can compute q by noting that

$$q = \frac{b^2 - c^2}{b^2} = \frac{(\mathbf{b} - \mathbf{c}) \cdot (\mathbf{b} + \mathbf{c})}{b^2} \tag{2.196}$$

and since $\mathbf{c} = \mathbf{b} - \mathbf{a}$

$$q = \frac{\mathbf{a} \cdot (2\mathbf{b} - \mathbf{a})}{b^2} \tag{2.197}$$

Thus, Equations (2.195) and (2.197) avoid roundoff errors. Applying these to the orbit problem results in

$$\delta\ddot{\mathbf{r}} = \frac{\mu}{\rho^3}\left[Fq\mathbf{r} - \delta\mathbf{r}\right] + \mathbf{P} \tag{2.198}$$

where F is given by Equation (2.195) and q is given by

$$q = \frac{\delta\mathbf{r} \cdot (2\mathbf{r} - \delta\mathbf{r})}{r^2} \tag{2.199}$$

Using Equation (2.182) and simplifying

$$q = -\frac{\rho_x \delta x + \rho_y \delta y + \rho_z \delta z}{\rho^2} \tag{2.200}$$

which means that now the perturbed trajectory $\mathbf{r}$ can be fully computed numerically.

The procedure for Encke's method is thus as follows:

1. Given the initial conditions $\mathbf{r}(t_0) = \rho(t_0)$ and $\mathbf{v}(t_0) = \dot{\rho}(t_0)$, define the osculating orbit. Note that $\delta\mathbf{r} = \mathbf{0}$ for $t = t_0$. Also, $q(t_0) = 0$.
2. Calculate $\delta\mathbf{r}(t_0 + \Delta t)$ using a user-defined time step Δt.
3. Compute
   a. $\rho(t_0 + \Delta t)$
   b. $q(t_0 + \Delta t)$ using Equation (2.201)
   c. $F(t_0 + \Delta t)$ using Equation (2.195)
4. Integrate another time step Δt to obtain $\delta\mathbf{r}(t_0 + k\Delta t)$
5. If $\delta r/\rho >$ some user-defined constant, rectify and go to step (1). Otherwise, continue.
6. Calculate $\mathbf{r} = \rho + \delta\mathbf{r}$ and $\mathbf{v} = \dot{\rho} + \delta\dot{\mathbf{r}}$.
7. Go to step (3) with Δt replaced by $k\Delta t$, where k is the step number.

In step (5), the user-defined constant is suggested to be on the order of 10^{-2} [Bate et al., 2020]. Note that this algorithm, while generally more accurate than Cowell's method, requires that the accelerations due to perturbation be known in analytic form. Encke's method reduces the number of integration steps since $\delta\mathbf{r}$ changes more slowly than $\mathbf{r}$. However, this is no longer true when $\mathbf{P} > \frac{\mu}{\rho^3}\left[Fq\mathbf{r} - \delta\mathbf{r}\right]$ (i.e., the perturbations are becoming larger than the primary acceleration), or when $\delta r/\rho$ grows large (i.e., when rectification is required). Encke's method is generally faster than Cowell's method, and it was shown to be, on average, ten times faster for interplanetary orbits and approximately three times faster for Earth satellites [Bate et al., 2020].

2.8.3 Euler-Cauchy Methods and Predictor-Corrector Methods

In its simplest form, numerical integration involves solution of the differential equation

$$\frac{dy(x)}{dx} = f(y,x) \tag{2.201}$$

subject to initial conditions, $y(0)=y_0$. Note that Equation (2.202) can be a scalar or vector equation. By definition, the derivative of y with respect to x $\frac{dy}{dx}$ is

$$\frac{dy}{dx} = \lim_{h\to 0}\frac{y(x+h)-y(x)}{h} \tag{2.202}$$

where h is the incremental change in the independent variable x. Rearranging Equation (2.203) and substituting Equation (2.202) one can approximate $y(x+h)$ as

$$y(x+h) \approx y(x) + hy'(x) = y(x) + hf(x,y) \tag{2.203}$$

Equation (2.204) can be used to create a simple first-order integration scheme such that each successive step y_{n+1} can be computed as

$$y_{n+1} = y_n + hf(x_n, y_n) \tag{2.204}$$

where, in order to begin the iteration, the initial condition $y(0)=y_0$ must be used to obtain y_1. This integration scheme is called the Euler-Cauchy method, named after 18th-century Swiss mathematician Leonhard Euler and 18th–19th-century French mathematician Augustin-Louis Cauchy. This "crude" method is almost never used because of its inaccuracy in being able to solve differential equations. The general definition of a Taylor series expansion for the function y is given by

$$y(x+h) = y(x) + h\dot{y}(x) + \frac{1}{2!}h^2\ddot{y}(x) + \frac{1}{3!}h^3\dddot{y}(x) + \cdots \tag{2.205}$$

Generally speaking, a higher order of derivatives used results in higher accuracy in the numerical integration. In fact, an improved Euler-Cauchy method exists, where higher-order derivative of the Taylor series expansion of $y(x+h)$ is used. First, an auxiliary value of y_{n+1} is computed

$$y^*_{n+1} = y_n + hf(x_n, y_n) \tag{2.206}$$

and then a new value is computed

$$y_{n+1} = y_n + \frac{1}{2}h\left[f(x_n, y_n) + f\left(x_{n+1}, y^*_{n+1}\right)\right] \tag{2.207}$$

This method, called improved Euler-Cauchy method, is a predictor-corrector method because in each step the prediction of y_{n+1} is computed (Equation 2.205), and then it is corrected in the next step (Equation 2.206).

2.8.4 Runge-Kutta

A classic method of great practical importance is the Runge-Kutta method of fourth order, or simply Runge-Kutta, named after 19th–20th-century German mathematicians Karl Runge and Wilhelm Kutta. This method makes use of Taylor series expansions, as given in Equation (2.206), to compute the solution of an initial value problem. Given a function f, initial values x_0 and y_0, step size h, and number of steps N, the numerical solution of y(x) can be computed following this algorithm:

For n = 0, 1, … N−1 do:

$$k_1 = hf\left(x_n, y_n\right)$$

$$k_2 = hf\left(x_n + \frac{1}{2}h, y_n + \frac{1}{2}k_1\right)$$

$$k_3 = hf\left(x_n + \frac{1}{2}h, y_n + \frac{1}{2}k_2\right)$$

$$k_4 = hf\left(x_n + h, y_n + k_3\right)$$

$$x_{n+1} = x_n + h$$

$$y_{n+1} = y_n + \frac{1}{6}\left(k_1 + 2k_2 + 2k_3 + k_4\right)$$

Output x_{n+1} and y_{n+1}

End

Stop

This method is recommended for use in computing since it does not require any special procedure to start, demands rather small amounts of computer storage, and makes use of the same straightforward algorithm. Furthermore, the method is numerically stable [Kreyszig, 2021]. However, one major limitation of this method is that it does not have a variable time step, thus resulting in many unnecessary computations, especially if only the final value of y is required. Lastly, it should be noted that if f depends on x alone, this method reduces to Simpson's rule of integration, which is generally discussed in calculus courses.

PROBLEMS

1. Starting with the unperturbed two-body equations of motion, Equation (2.9), derive its state-space form in spherical coordinates.
2. Prove that for the unperturbed two-body problem, orbital energy is constant.

3. Prove that the angular momentum vector and eccentricity vector are orthogonal to each other.
4. Prove that for the unperturbed two-body problem, orbital motion is confined to a plane (this is the orbital plane).
5. Show that the speed of a spacecraft in a circular orbit is

$$v_c = \sqrt{\frac{\mu}{r}}$$

6. A spacecraft in Earth orbit passes through its apogee (apoapsis with respect to Earth) at a radius of $r_a = 9{,}500$ km, and with speed $v_a = 5.95$ km/s. Determine how much time (in hours) elapses before the spacecraft returns to apogee.
7. Starting with any known relationships, show that the orbital eccentricity can be expressed as

$$e = \sqrt{1 + \frac{2pE}{\mu}}$$

   where p is semilatus rectum, E is orbital energy, and μ is the gravitational parameter.
8. Radar measurements give the following data for a spacecraft in Earth orbit at some time t_1:
   $r = 8{,}000$ km
   $\dot{r} = -0.8201$ km/s
   $\dot{\theta} = 5.635 \times 10^{-2}$ deg/s
   Calculate:
   a. Orbital energy, in km^2/s^2
   b. Semimajor axis, in km
   c. Angular momentum, in km^2/s
   d. Semilatus rectum, in km
   e. Eccentricity
   f. Radii of periapsis and apoapsis, in km
   g. Orbital period, in hours
   h. Flight-path angle at time t_1 in rad
9. At time $t = 0$, a spacecraft on an Earth orbit with $a = 50{,}612$ km and $e = 0.63174$ is located at $\theta = 42.319°$. At what time did the satellite previously pass through periapsis?
10. A satellite on an Earth orbit with $a = 12{,}587$ km and $e = 0.42$ passed through periapsis at 10:04:06 (time here is given in the form in 24-hour format). Calculate the time, in 24-hour format, at which the satellite will reach $\theta = 297°$.
11. A satellite in Earth orbit passes through perigee at $T_0 = 978$ seconds on an orbit with $e = 0.7$ and $a = 30{,}500$ km. Using the Newton-Raphson iteration method and an absolute tolerance of $\theta_A = 0.001$, calculate the true anomaly θ at $t = 35{,}000$ seconds. Do not implement a relative tolerance check.
12. Write a computer program that

a. takes as input
   i. A position vector (in km) and a velocity vector (in km/s) in ECI (Earth-centered inertial) coordinates.
   ii. An initial time, t_1
   iii. A later time, t_2
   iv. Tolerances Δ_R (relative tolerance) and Δ_A (absolute tolerance) for the iterative procedure used to solve Kepler's time equation.
b. Outputs the input data, with labels and units, and calculates the following:
c. Classical orbital elements a, e, i, ω, and Ω (*Note that these are five of the six COEs*).
d. The DCM $\underline{C}^{EP}$ that converts from ECI to perifocal coordinates.
e. Position (in km) and velocity (in km/s) vectors in perifocal coordinates at time t_1.
f. True and eccentric anomalies θ_1 and E_1 at time t_1.
g. Time of periapsis passage, T_0.
h. True and eccentric anomalies θ_2 and E_2 at time t_2.
i. Position (in km) and velocity (in km/s) vectors in perifocal coordinates at time t_2.
j. Position (in km) and velocity (in km/s) vectors in ECI coordinates at time t_2.
k. Outputs the values in items (c)–(j) with labels and units.

Run the program with the input values:

$$\mathbf{r}_1 = 5214.1\ \hat{I} + 6322.9\ \hat{J} - 3001.7\ \hat{K}\ (\text{km})$$

$$\mathbf{v}_1 = -4.1549\ \hat{I} + 3.1666\ \hat{J} - 4.5044\ \hat{K}\ (\text{km/s})$$

$$t_1 = 0\ \text{s}$$

$$t_2 = 12{,}500\ \text{s}$$

$$\Delta_R = 10^{-6}$$

$$\Delta_A = 10^{-6}$$

13. Consider a spacecraft in orbit around the Earth with the following initial conditions (ICs; position and velocity vectors) in an Earth-centered inertial (ECI) frame at some time t_0:

$$\mathbf{r}_0 = 666.780\ \mathbf{I} + 7621.32\ \mathbf{J} + 2338.97\ \mathbf{K}\ (\text{km})$$

$$\mathbf{v}_0 = -7.22541\ \mathbf{I} + 0.266320\ \mathbf{J} + 1.52177\ \mathbf{K}\ (\text{km/s})$$

Consider the Cartesian formulation of the J_2-perturbed equations of motion for the two-body problem:

$$\ddot{x} = -\frac{\mu x}{r^3}\left\{1 - \frac{3}{2}J_2\left(\frac{R_e}{r}\right)^2\left[5\frac{z^2}{r^2} - 1\right]\right\}$$

$$\ddot{y} = -\frac{\mu y}{r^3}\left\{1 - \frac{3}{2}J_2\left(\frac{R_e}{r}\right)^2\left[5\frac{z^2}{r^2} - 1\right]\right\}$$

$$\ddot{z} = -\frac{\mu z}{r^3}\left\{1 - \frac{3}{2}J_2\left(\frac{R_e}{r}\right)^2\left[5\frac{z^2}{r^2} - 3\right]\right\}$$

where $\mu = 398{,}600\ \text{km}^3/\text{s}^2$ is the gravitational parameter of the Earth, $R_e = 6{,}378.14$ km is the equatorial radius of the Earth, $J_2 = 1.08263 \times 10^{-3}$ the coefficient of Earth's oblateness, and $r = \sqrt{x^2 + y^2 + z^2}$ is the magnitude of the position of the spacecraft with respect to Earth's center.

Set up a state $\bar{S} = \{x\ \ y\ \ z\ \ \dot{x}\ \ \dot{y}\ \ \dot{z}\} = \{s_1\ \ s_2\ \ s_3\ \ s_4\ \ s_5\ \ s_6\}$ and rewrite the above EOMs in state-space form using these given states. Then, using a computer program such as MATLAB, numerically integrate the ICs of this spacecraft using the above EOMs for 30 days. Provide two figures:

I. Figure 1: a 3D plot (x vs. y vs. z) of the spacecraft orbit (in red) with respect to Earth's center in ECI coordinates. Also, for reference plot the Earth as a blue sphere having radius equal to R_e

II. Figure 2: subplots showing:
   1. Semimajor axis (SMA) in km vs. time
   2. Eccentricity vs. time
   3. Inclination in degrees vs. time
   4. Orbital period in seconds vs. time
   5. Perigee in km vs. time
   6. Apogee in km vs. time

   If using MATLAB, compare the difference between using different integration schemes, such as ode45 vs. ode113. What do you notice? Which one is faster for a given accuracy?

3 N-Body Problem

3.1 FORMULATION OF THE N-BODY PROBLEM

As was shown in Chapter 2, the N-body problem, where $N > 2$, does not have an analytical solution. However, a numerical integration of the equations of motion can be solved to describe the motion of multiple bodies subject to gravitational accelerations from numerous bodies. In the N-body problem, position vectors between the origin and all gravitational sources are shown in Figure 3.1.

Here, the vector $\mathbf{r}_{ij}$ represent the vector between masses m_i to m_j. The acceleration of each body in the inertial frame (which is a generalization of Equations (2.3) and (2.4)) is

$$m_i \frac{d^2\mathbf{R}_i}{dt^2} = G \sum_{\substack{j=1 \\ j \neq i}}^{N} \frac{m_i m_j}{r_{ij}^3} \mathbf{r}_{ij} \tag{3.1}$$

This fundamental equation can be expanded out to include however many gravitational sources are desired. In the following sections, we present the development of the equations of motion for the restricted three-body and restricted four-body problems. One of the masses (usually corresponding to the mass of a spacecraft or a small asteroid) is assumed to have negligible gravitational effects on the other masses as the term restricted means that there are N−1 gravitational sources in the N-body problem. For example, in the restricted three-body problem $N = 3$, so the third body has no gravitational effect on the motion of the other two.

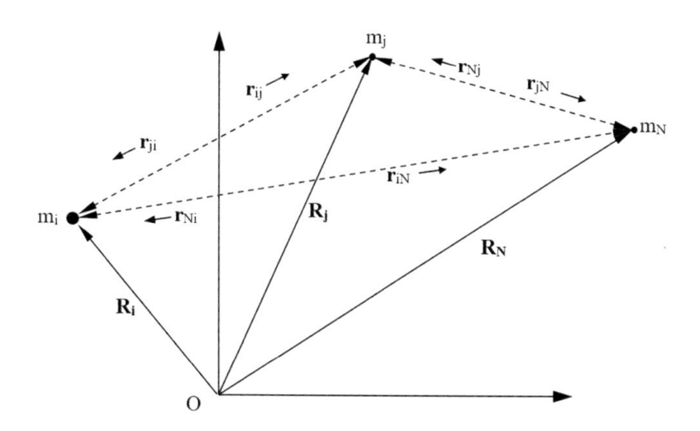

FIGURE 3.1 Position vectors in an inertial reference frame.

DOI: 10.1201/9781003165071-3

During the early stages of mission design, it is often common to approximate complex systems using the three-body and four-body approximations. Sometimes, even the two-body approximation is sufficient to obtain the necessary velocity change (Δv) and time-of-flight values within rather small percentages (<5%) of the actual values calculated when using a full-force model, such as that obtainable through JPL Horizons, as described in Chapter 4. However, having an analytical understanding of these "simpler" dynamical systems can provide significant insight. For example, an Earth-Mars transfer may be approximated using patched conics (two-body problems linked together), as described in Chapter 5, while an Earth-Moon transfer may be approximated using the three-body problem and then refined using a full-ephemeris model during mission planning.

3.2 RESTRICTED THREE-BODY PROBLEM

The restricted three-body problem is one of the simplest unsolvable gravitational problems. The problem consists of three bodies, two with mass and one with negligible mass. The two massive bodies move around their common center of mass (barycenter) and the third body moves relative to the massive bodies.

The concept was first defined by Euler in 1767 [Szebehely, 1967] in his studies of the motion of the Moon around the Earth. In 1772, Lagrange [Szebehely, 1967] also discovered what is now known as the triangular libration points, as we will discuss in a later section in detail. Jacobi [1836] defined one integral of motion, and Poincaré [1899] later proved that this is the only exact constant of motion.

In the inertial reference frame (**IJK**), we can define a rotating reference frame (**ijk**). The **i–j** plane and the **I–J** plane are aligned, with the **i–j** plane moving relative to the **I–J** plane at the rate ω, and sweeps out an angle $\omega(t-t_0)$ with $t-t_0$ being the time since the two frames were aligned. Note that for a circular orbit, ω is constant. The geometry of these two reference frames with the masses m_1 and m_2 with their respective vectors to an infinitesimal mass (m_3) are $\mathbf{r}_1$ and $\mathbf{r}_2$, and are all shown in Figure 3.2.

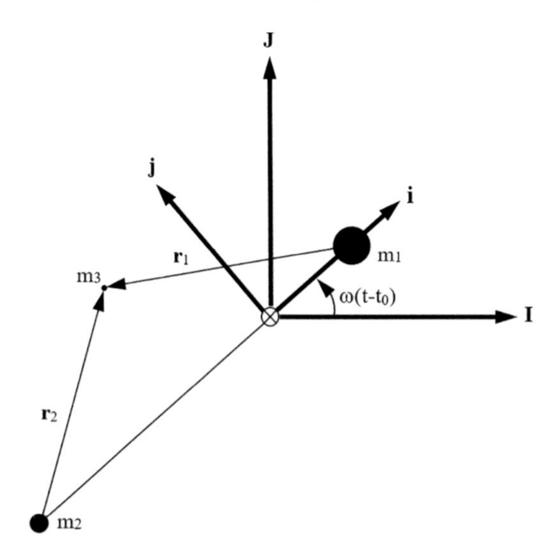

FIGURE 3.2 Geometry of inertial and rotating frames in the restricted three-body problem.

A direction cosine matrix relating these two frames can be developed, as

$$\begin{Bmatrix} \mathbf{I} \\ \mathbf{J} \\ \mathbf{K} \end{Bmatrix} = \begin{bmatrix} \cos\omega(t-t_0) & -\sin\omega(t-t_0) & 0 \\ \sin\omega(t-t_0) & \cos\omega(t-t_0) & 0 \\ 0 & 0 & 1 \end{bmatrix} \begin{Bmatrix} \mathbf{i} \\ \mathbf{j} \\ \mathbf{k} \end{Bmatrix} \tag{3.2}$$

The equations of motion in the inertial frame can be found in either the inertial or rotating frame. Here, we will develop the equations of motion in the rotating frame since the system's equilibrium points are "stationary" (only when ω is constant) and many orbits of interest in this system appear to be periodic only with respect to the rotating frame.

Let's define a state vector, $\mathbf{X}$, of Cartesian coordinates in the rotating frame as $\mathbf{X} = \left(x \;\; y \;\; z \;\; \frac{dx}{dt} \;\; \frac{dy}{dt} \;\; \frac{dz}{dt} \right)^T$. The position vector of an object in the rotating frame is

$$\mathbf{r} = x\mathbf{i} + y\mathbf{j} + z\mathbf{k} \tag{3.3}$$

Using the kinematic formulation shown in Chapter 2, the velocity in the rotating frame is

$$\frac{d\mathbf{r}}{dt}^{(R)} = \frac{dx}{dt}\mathbf{i} + \frac{dy}{dt}\mathbf{j} + \frac{dz}{dt}\mathbf{k} \tag{3.4}$$

and the acceleration is

$$\frac{d^2\mathbf{r}}{dt^2}^{(R)} = \frac{d^2x}{dt^2}\mathbf{i} + \frac{d^2y}{dt^2}\mathbf{j} + \frac{d^2z}{dt^2}\mathbf{k} \tag{3.5}$$

The second term in Equation (2.13) is the Coriolis acceleration, and here simplifies to

$$2\omega^{(R/I)} \times \frac{d\mathbf{r}}{dt}^{(R)} = 2\omega\frac{dx}{dt}\mathbf{j} - 2\omega\frac{dy}{dt}\mathbf{i} \tag{3.6}$$

The third term in Equation (2.13) is the acceleration due to the non-constant movement of the rotating frame with respect to the inertial frame, and here it is

$$\frac{d\boldsymbol{\omega}}{dt}^{(R/I)} \times \mathbf{r} = x\frac{d\omega}{dt}\mathbf{j} - y\frac{d\omega}{dt}\mathbf{i} \tag{3.7}$$

The last term is the centripetal acceleration, which here simplifies to

$$\boldsymbol{\omega}^{(R/I)} \times \left(\boldsymbol{\omega}^{(R/I)} \times \mathbf{r}\right) = -\omega^2\left(x\mathbf{i} + y\mathbf{j}\right) \tag{3.8}$$

Assembling all contributions, the acceleration in the inertial frame is

$$\frac{d^2\mathbf{r}}{dt^2}^{(I)} = \left(\frac{d^2x}{dt^2} - 2\frac{dy}{dt} - y\frac{d\omega}{dt} - \omega^2 x\right)\mathbf{i} + \left(\frac{d^2y}{dt^2} + 2\frac{dx}{dt} + x\frac{d\omega}{dt} - \omega^2 y\right)\mathbf{j} + \frac{d^2z}{dt^2}\mathbf{k} \tag{3.9}$$

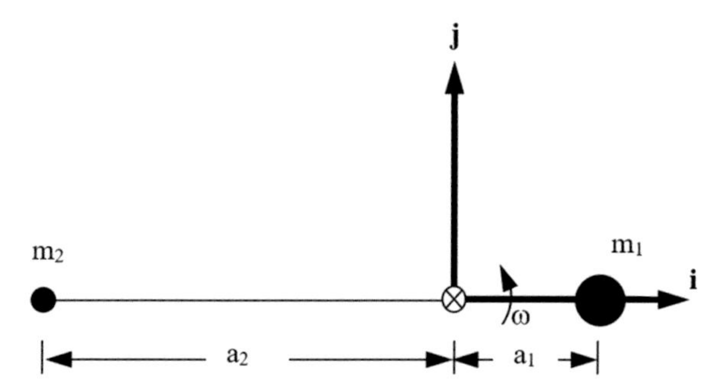

FIGURE 3.3 Two bodies rotating about their barycenter.

In many orbital mechanics problems, we generally work in physical lengths, but the need for multiple variables can make a problem cumbersome. In physical dimensions, the masses of each primary body and the distance between the primary bodies need to be known. However, non-dimensionalization of the equations of motion can reduce the problem inputs to simply a relationship between the two masses. This also means that results obtained in non-dimensional form can be applied to a multitude of problems as long as the constants used to non-dimensionalize the problem are equal. A rotating coordinate frame centered at the center of mass (⊗) of the two primary bodies is shown in Figure 3.3.

The period of the bodies around the barycenter is computed as

$$T = \frac{2\pi}{n} = 2\pi\left[\frac{(a_1 + a_2)^3}{G(m_1 + m_2)}\right]^{1/2} \tag{3.10}$$

Non-dimensionalizing the parameters in the system such that the sum of the masses, the distance between the two bodies and the gravitational constant, G, results in all of these quantities being equal to unity. Consequently, the period becomes

$$T = 2\pi \tag{3.11}$$

If the sum of the two masses is equal to one, we define the mass of the smaller body, m_2, as the proportion of the total mass

$$m_2 = \mu \tag{3.12}$$

Since the normalization of the masses is equal to one, the larger mass is

$$m_1 = 1 - \mu \tag{3.13}$$

Thus, by definition, the maximum value of μ is 0.5 (note that μ here is not the same as the gravitational constant previously defined).

The distance from the center of mass to the primaries can be found by using the definition of the center of mass

$$m_1 a_1 = m_2 a_2 \tag{3.14}$$

Inserting the definitions for m_1 and m_2,

$$\frac{a_1}{a_2} = \frac{\mu}{1-\mu} \tag{3.15}$$

and noting that the distance between m_1 and m_2 is one,

$$a_1 + a_2 = 1 = a_2\left(\frac{\mu}{1-\mu} + 1\right) = a_2\left(\frac{1}{1-\mu}\right) \tag{3.16}$$

Solving Equation (3.16) for a_1 and a_2 shows that $a_1 = \mu$ and $a_2 = 1-\mu$.

Inserting these normalized values into Equation (3.9), the acceleration of the third massless object in the inertial frame becomes

$$\frac{d^2\mathbf{r}}{dt^2}^{(I)} = \left(\frac{d^2x}{dt^2} - 2\frac{dy}{dt} - y\frac{d\omega}{dt} - x\right)\mathbf{i} + \left(\frac{d^2y}{dt^2} + 2\frac{dx}{dt} + x\frac{d\omega}{dt} - y\right)\mathbf{j} + \frac{d^2z}{dt^2}\mathbf{k} \tag{3.17}$$

3.2.1 Circular Restricted Three-Body Problem

The circular restricted three-body problem (CR3BP) is the special case in which the two primaries orbit their barycenter following circular orbits. For this case, the angular rate of the primaries' motion around their barycenter is constant, n, so $\frac{dn}{dt} = \frac{d\omega}{dt} = 0$, and thus the acceleration in the inertial frame becomes

$$\frac{d^2\mathbf{r}}{dt^2}^{(I)} = \left(\frac{d^2x}{dt^2} - 2\frac{dy}{dt} - x\right)\mathbf{i} + \left(\frac{d^2y}{dt^2} + 2\frac{dx}{dt} - y\right)\mathbf{j} + \frac{d^2z}{dt^2}\mathbf{k} \tag{3.18}$$

Position vectors are defined from the primary masses to the infinitesimal body as $\mathbf{r}_1$ and $\mathbf{r}_2$ and are seen in Figure 3.4. In some texts, the locations of m_1 and m_2 are reversed. However, to convert between these two representations, one can simply replace x with −x.

These vectors can be found as

$$\mathbf{r}_1 = \mathbf{r} - a_1\mathbf{i} = (x-\mu)\mathbf{i} + y\mathbf{j} + z\mathbf{k} \tag{3.19}$$

and

$$\mathbf{r}_2 = (x + a_2)\mathbf{i} + y\mathbf{j} + z\mathbf{k} \tag{3.20}$$

The magnitudes of these distances are

$$r_1 = \sqrt{\mathbf{r}_1 \cdot \mathbf{r}_1} = \sqrt{(x-\mu)^2 + y^2 + z^2} \tag{3.21}$$

and

$$r_2 = \sqrt{\mathbf{r}_2 \cdot \mathbf{r}_2} = \sqrt{(x+1-\mu)^2 + y^2 + z^2} \tag{3.22}$$

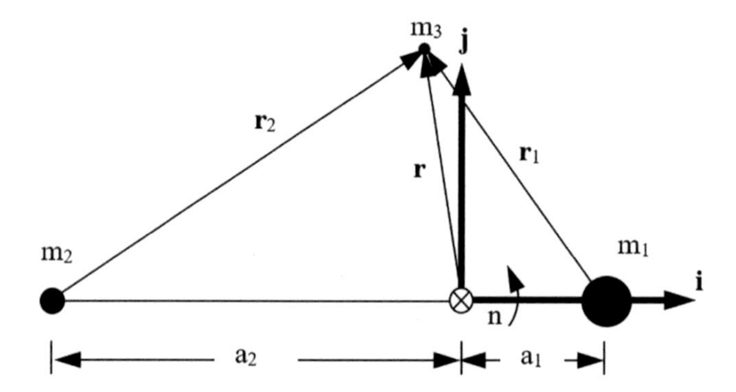

FIGURE 3.4 Basic geometry of the primaries (m_1 and m_2) and an infinitesimal mass (m_3).

The gravitational acceleration on the third, infinitesimal body m_3 is

$$\mathbf{a}_g = -\frac{(1-\mu)\mathbf{r}_1}{r_1^3} - \frac{\mu \mathbf{r}_2}{r_2^3} \tag{3.23}$$

This gravitational acceleration is equal to the inertial acceleration from Equation (3.18).

Inserting Equations (3.19) and (3.20) into Equation (3.23) yields the inertial acceleration as

$$\frac{d^2\mathbf{r}}{dt^2}^{(I)} = -\frac{(1-\mu)}{r_1^3}\left[(x-\mu)\mathbf{i} + y\mathbf{j} + z\mathbf{k}\right] - \frac{\mu}{r_2^3}\left[(x+1-\mu)\mathbf{i} + y\mathbf{j} + z\mathbf{k}\right] \tag{3.24}$$

Equating Equation (3.24) with (3.18) and separating the **ijk** components yields three, second-order, non-linear, coupled differential equations,

$$\begin{aligned}
&\frac{d^2x}{dt^2} - 2\frac{dy}{dt} - x = -\frac{(1-\mu)(x-\mu)}{r_1^3} - \frac{\mu(x+1-\mu)}{r_2^3} \\
&\frac{d^2y}{dt^2} + 2\frac{dx}{dt} - y = -\frac{(1-\mu)y}{r_1^3} - \frac{\mu y}{r_2^3} \\
&\frac{d^2z}{dt^2} = -\frac{(1-\mu)z}{r_1^3} - \frac{\mu z}{r_2^3}
\end{aligned} \tag{3.25}$$

The differential equations of motion in Equation (3.25) are also commonly written as

$$\begin{aligned}
&\frac{d^2x}{dt^2} - 2\frac{dy}{dt} = \frac{\partial U}{\partial x} \\
&\frac{d^2y}{dt^2} + 2\frac{dx}{dt} = \frac{\partial U}{\partial y} \\
&\frac{d^2z}{dt^2} = \frac{\partial U}{\partial z}
\end{aligned} \tag{3.26}$$

where the pseudo-potential, U, is $U = \frac{1}{2}\left(x^2 + y^2\right) + \frac{1-\mu}{r_1} + \frac{\mu}{r_2}$. Note that the non-linearity of these equations of motion comes from the fact that r_1 and r_2 are functions of x, y, and z, as provided in Equations (3.21) and (3.22).

If we want to rewrite Equation (3.26) using Hamiltonian mechanics, we first define the generalized momentum vector $\mathbf{p}$ of m_3 as

$$\mathbf{p} = \left(\frac{\partial L}{\partial \dot{\mathbf{q}}}\right)^T = \begin{bmatrix} \dot{x} - y \\ \dot{y} + x \\ \dot{z} \end{bmatrix} \tag{3.27}$$

where $\mathbf{q}$ is a vector of generalized coordinates and L is the Lagrangian, which is defined as the sum of the kinetic energy, T, and potential energy, U (which here is the pseudo-potential)

$$L = T + U = \frac{1}{2}\left(\dot{X}^2 + \dot{Y}^2 + \dot{Z}^2\right) + U \tag{3.28}$$

The Hamiltonian, H, of a system is given by

$$H = \frac{1}{2}\langle \mathbf{p}, \mathbf{p} \rangle - U \tag{3.29}$$

so the Hamiltonian for the CR3BP becomes

$$\begin{aligned} H &= \frac{1}{2}\langle \mathbf{p}, \mathbf{p} \rangle - \langle \mathbf{p}, \hat{\mathbf{k}} \times \mathbf{r} \rangle - U(\mathbf{r}) \\ &= \frac{1}{2}\left(p_x^2 + p_y^2 + p_z^2\right) + \left(p_x y - p_y x\right) - U(x, y, z) \end{aligned} \tag{3.30}$$

where $\langle \cdot, \cdot \rangle$ denotes the dot product. The equations of motion can thus be written in terms of the Hamiltonian as

$$\dot{\mathbf{q}} = \frac{\partial H}{\partial \mathbf{p}} \qquad \mathbf{p} = -\frac{\partial H}{\partial \mathbf{q}} \tag{3.31}$$

Thus, plugging Equation (3.30) into Equation (3.31) and simplifying gives us the equations of motion of the CR3BP in Hamiltonian form

$$\begin{aligned} \dot{x} &= \frac{\partial H}{\partial p_x} = p_x + y & \dot{p}_x &= -\frac{\partial H}{\partial x} = p_y - x + U_x \\ \dot{y} &= \frac{\partial H}{\partial p_y} = p_y - x & \dot{p}_y &= -\frac{\partial H}{\partial y} = -p_x - y + U_y \\ \dot{z} &= \frac{\partial H}{\partial p_z} = p_z & \dot{p}_z &= -\frac{\partial H}{\partial z} = U_z \end{aligned} \tag{3.32}$$

where the subscripts indicate partial derivatives.

3.2.2 Lagrange Points

Since the equations of motion of the CR3BP are non-linear equations that cannot be analytically solved, the question arises as to whether there are any points where, if an object is placed there, it will not drift away. In order for such equilibrium points to exist, the velocities and accelerations must be equal to zero such that if an object is located at one of these equilibrium points, said object will remain there indefinitely. Setting all velocities and accelerations (first and second derivatives) in Equation (3.25) equal to zero, which is the same as setting $\frac{\partial U}{\partial x} = \frac{\partial U}{\partial y} = \frac{\partial U}{\partial z} = 0$, yields

$$\begin{aligned} x &= \frac{(1-\mu)(x-\mu)}{r_1^3} + \frac{\mu(x+1-\mu)}{r_2^3} \\ y &= \frac{(1-\mu)y}{r_1^3} + \frac{\mu y}{r_2^3} \\ 0 &= -\frac{(1-\mu)z}{r_1^3} - \frac{\mu z}{r_2^3} \end{aligned} \tag{3.33}$$

Equation (3.33) are coupled algebraic equations that, when solved, give the x, y, and z coordinates of the equilibrium points of the CR3BP. However, the third equation reveals that $z=0$ is the only possible solution. This means that all equilibrium points lie in the **i–j** plane, which is the plane of motion of the primaries.

By carefully choosing an equilibrium triangle such that $r_1 = r_2 = a_1 + a_2 = 1$, the so-called equilateral equilibrium points can be found. Equating Equations (3.21) and (3.22) yields the following algebraic equation

$$\sqrt{(x-\mu)^2 + y^2} = \sqrt{(x+1-\mu)^2 + y^2} \tag{3.34}$$

Solving for x, Equation (3.34) becomes

$$x = \mu - \frac{1}{2} \tag{3.35}$$

Any value of y satisfies the y equation, but, since $r_1 = 1$,

$$r_1^2 = (x-\mu)^2 + y^2 = 1 \tag{3.36}$$

Inserting Equation (3.35) into Equation (3.36) and solving for y show that $y = \pm\frac{\sqrt{3}}{2}$. Thus, two equilibrium points are $(x,y) = \left(\mu - \frac{1}{2}, \frac{\sqrt{3}}{2}\right)$ and $(x,y) = \left(\mu - \frac{1}{2}, -\frac{\sqrt{3}}{2}\right)$. These two points are designated as L_4 and L_5, and are known as Lagrange points (named after Lagrange, who first discovered them). These points are called the equilateral points because the location of the primaries and L_4 and the location of the primaries and L_5 form two equilateral triangles.

From the second equation of Equation (3.33), a second group of equilibrium points are where $y=0$. This shows that these equilibrium points are all along the **i**-axis, and are thus designated as the collinear Lagrange points. Since $y=0$, the values of r_1 and r_2 become

$$r_1 = x - \mu \tag{3.37}$$

and

$$r_2 = x + 1 - \mu \tag{3.38}$$

Recognizing that both r_1 and r_2 must be positive, and inserting Equations (3.37) and (3.38) into the first equation of Equation (3.33), we get

$$x = \frac{(1-\mu)(x-\mu)}{|x-\mu|^3} + \frac{\mu(x+1-\mu)}{|x+1-\mu|^3} \tag{3.39}$$

where absolute values are used to enforce the fact that the terms in the denominators represent physical distances, and thus must always be positive. Multiplying this equation out yields a cubic equation in x. The solution to this equation gives three real roots which are the locations of the collinear libration points (L_1, L_2, and L_3). The solution of the libration points, L_1 and L_2 along the non-dimensional x-axis, is shown in Figure 3.5. The locations of all five libration or Lagrange points are shown in Figure 3.6.

In the CR3BP, at each of these five Lagrange points, an object placed **exactly** at one of these locations will remain there forever. However, there are always perturbations in this system, such as the non-homogeneous gravitational field of the primaries, solar radiation pressure, etc. Also, can an object be put exactly at a location in space with exactly a specified velocity? (This falls into the same category as: can you

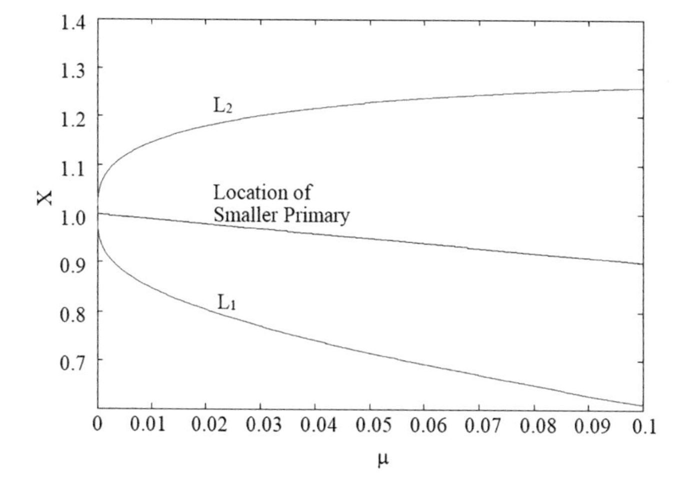

FIGURE 3.5 Distance from barycenter to collinear libration points.

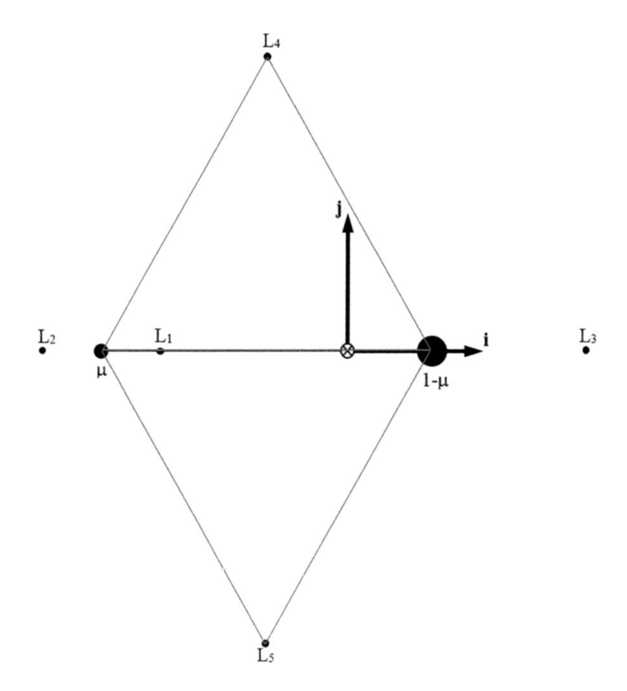

FIGURE 3.6 Five libration points in the restricted three-body problem.

have an exactly circular or parabolic orbit?) Also, planetary orbits are not exactly circular, and there are always other gravitational sources that perturb the system. While putting a spacecraft at exactly one of these locations may not be doable or even useful due to perturbations, periodic orbits **around** the Lagrange points can be found and will be discussed later in this chapter.

3.2.3 Stability of Libration Points

Since the location of the libration points are a mathematical rather than a physical occurrence, the question that needs to be answered is whether these points are stable or not. Stability is defined such that if a particle is slightly perturbed from its equilibrium point, it will not drift away from that point. In order to assess the stability, let's perturb the exact solutions slightly,

$$
\begin{aligned}
x &= x_e + \delta x \\
y &= y_e + \delta y \\
z &= z_e + \delta z = \delta z
\end{aligned}
\tag{3.40}
$$

where x_e, y_e, and z_e are the equilibrium solutions previously found and δx, δy, and δz are small perturbations in the x-, y-, and z-directions, respectively. Note that $z_e = 0$ for all Lagrange points.

First, let's examine the distances from the primaries to the infinitesimal body, which are

$$\frac{1}{r_1^3} = \left[\left(x_e + \delta x - \mu\right)^2 + \left(y_e + \delta y\right)^2 + \delta z^2\right]^{-3/2} \tag{3.41}$$

and

$$\frac{1}{r_2^3} = \left[\left(x_e + \delta x + 1 - \mu\right)^2 + \left(y_e + \delta y\right)^2 + \delta z^2\right]^{-3/2} \tag{3.42}$$

Multiplying everything out and dropping higher-order terms ($\delta x^2, \delta y^2, \delta z^2$) yields

$$r_1^{-3} \approx r_{1e}^{-3} - 3r_{1e}^{-5}\left[\left(x_e - \mu\right)\delta x + y_e \delta y\right] \tag{3.43}$$

and

$$r_2^{-3} \approx r_{2e}^{-3} - 3r_{2e}^{-5}\left[\left(x_e + 1 - \mu\right)\delta x + y_e \delta y\right] \tag{3.44}$$

Taking the first and second derivatives of Equation (3.40), recognizing that the velocity and acceleration of the equilibrium points are zero, and inserting Equations (3.43) and (3.44) into Equation (3.25), the differential equations that govern the motion of the infinitesimal mass near the equilibrium points become

$$\begin{gathered} \delta\ddot{x} - 2\delta\dot{y} - (1 - A)\delta x - B\delta y = 0 \\ \delta\ddot{y} + 2\delta\dot{x} - B\delta x - (1 - C)\delta y = 0 \end{gathered} \tag{3.45}$$

where A, B, and C are constants such that

$$\begin{gathered} A = (1-\mu)\left[\frac{1}{r_{1e}^3} - 3\frac{(x_e - \mu)^2}{r_{1e}^5}\right] + +\mu\left[\frac{1}{r_{2e}^3} - 3\frac{(x_e + 1 - \mu)^2}{r_{2e}^5}\right] \\ B = 3(1-\mu)\frac{(x_e - \mu)y_e}{r_{1e}^5} + +3(1-\mu)\frac{(x_e + 1 - \mu)y_e}{r_{2e}^5} \\ C = (1-\mu)\left[\frac{1}{r_{1e}^3} - 3\frac{y_e^2}{r_{1e}^5}\right] + +\mu\left[\frac{1}{r_{2e}^3} - 3\frac{y_e^2}{r_{2e}^5}\right] \end{gathered} \tag{3.46}$$

We can assume a solution to Equation (3.45) in the form,

$$\begin{gathered} \delta x = c_1 e^{\lambda t} \\ \delta y = c_2 e^{\lambda t} \end{gathered} \tag{3.47}$$

Taking the first and second derivatives of Equation (3.47), inserting this in Equation (3.45), and putting this in matrix form yields,

$$\begin{bmatrix} \lambda^2-(1-A) & -B-2\lambda \\ -B+2\lambda & \lambda^2-(1-C) \end{bmatrix} \begin{Bmatrix} \delta x \\ \delta y \end{Bmatrix} = \begin{Bmatrix} 0 \\ 0 \end{Bmatrix} \tag{3.48}$$

In this eigenvalue problem, the trivial solution, $\delta x = \delta y = 0$, does not provide any insight on the problem. Instead, the more important solution is when the matrix in Equation (3.48) is singular. When singular, this results in a quadratic characteristic equation in λ^2,

$$\lambda^4 + \lambda^2(2+C+A) + \left(1-C-A+AC-B^2\right) = 0 \tag{3.49}$$

The roots of the characteristic equation are

$$\lambda = \pm\sqrt{-2-C-A \pm \sqrt{2C+2A+\frac{1}{4}C^2-\frac{1}{2}AC+\frac{1}{4}A^2+B^2}} \tag{3.50}$$

In order for this system to be stable, each value of λ must be purely imaginary.

For example, let's look at the triangular points. The values of r_{1e} and r_{2e} are both equal to 1 (by definition). The characteristic equation for the triangular points is

$$\lambda^4 + \lambda^2 - \left(\frac{27}{4}\right)\mu(\mu-1) = 0 \tag{3.51}$$

The roots to this equation are

$$\lambda^2 = \frac{1}{2}\left(-1 \pm \sqrt{1-27\mu(1-\mu)}\right) \tag{3.52}$$

For marginal stability, all roots must be purely imaginary. For this to happen,

$$1-27\mu(1-\mu) < 0 \tag{3.53}$$

Simplifying by solving for μ,

$$\mu < \frac{1}{18}\left(9-\sqrt{69}\right) \approx 0.0385209 \tag{3.54}$$

(there are actually two roots, but by definition since $\mu \le 0.5$, we ignore the second root, which is 0.961479). Thus, the L_4 and L_5 points are marginally stable with oscillatory solutions. However, recall the assumptions that were made – the planets are in circular orbits around the Sun (the libration points are no longer defined for elliptical orbits) and we neglect the gravitational influence of all other bodies (adding other gravitational forces destabilizes destroys the marginal stability of this system [Spencer, 1985]). For our Solar System, some typical values for μ are shown in Table 3.1.

Note how these values of μ can vary greatly depending on the primaries that are being considered. Since $\mu < 0.0385209$ for all Sun-planet and planet-moon systems, in theory, in each of these systems, there can be something at L_4 and L_5. An exception

TABLE 3.1
Values of μ for our Solar System

System	μ (Non-dimensional Mass Ratio)
Earth-Moon	1.215060E-02
Sun-Mercury	1.660148E-07
Sun-Venus	2.447835E-06
Sun-Earth	3.003460E-06
Sun-Mars	3.227137E-07
Sun-Jupiter	9.536922E-04
Sun-Saturn	2.857260E-04
Mars-Phobos	1.660513E-08
Mars-Deimos	2.299699E-09
Jupiter-Ganymede	7.803794E-05
Saturn-Titan	2.365667E-04
Pluto-Charon	1.085112E-01

TABLE 3.2
Lagrange point locations in non-dimensional (ND) units and kilometers, and their stability for the Earth-Moon system ($\mu = 0.01215060$)

Lagrange Point	(x, y) Coordinates [ND]		(x, y) Coordinates [km]		Stability
L_1	–0.836915	0	–321,710	0	Unstable
L_2	–1.15568	0	–444,244	0	Unstable
L_3	1.00506	0	386,346	0	Unstable
L_4	0.487849	0.8660254	187,529	332,900	Stable
L_5	0.487849	–0.8660254	187,529	–332,900	Stable

is the Pluto-Charon system, which has $\mu = 0.1085112 > 0.0385209$ and therefore its L_4 and L_5 are unstable.[1] On the other hand, in the Sun-Jupiter system, there are natural asteroids in the vicinity of these points, called Greek and Trojan asteroids. In the Earth-Moon system, these points have no objects permanently around them due to the gravitational perturbation of the Sun.

A similar analysis can be done for the collinear points. Since the roots of the characteristic equation for the collinear points present positive real parts, these points (L_1, L_2, and L_3) are all always unstable. Table 3.2 lists the location of the Earth-Moon Lagrange points with respect to their rotating reference frame, as defined in general in Figure 3.5 (the z-coordinate of all Lagrange points is zero, as demonstrated earlier).

[1] Pluto is no longer considered a planet. However, we included the Pluto-Charon system in this section to show an example of combination of celestial bodies for which L_4 and L_5 are not stable.

Computing the location of the collinear Lagrange points for a given system requires solving algebraic equations that generally do not have a simple analytic solution. However, if the value of μ is very small, Equation (3.39) can be simplified to obtain approximate analytic solutions for the Cartesian coordinates of L_1, L_2, and L_3

$$
\begin{aligned}
L_1 &\approx \left(\left[\frac{\mu}{3}\right]^{1/3} - 1,\ 0,\ 0\right) \\
L_2 &\approx \left(-\left[\frac{\mu}{3}\right]^{1/3} - 1,\ 0,\ 0\right) \\
L_3 &\approx \left(1 + \left[\frac{5\mu}{12}\right],\ 0,\ 0\right)
\end{aligned}
\tag{3.55}
$$

Note that computing the location of the Lagrange points does not take into account the physical sizes of the primaries. Therefore, it is possible to compute Lagrange point locations that are inside the physical size of the smaller of the two primaries, especially if the system has a very small value of μ. Additionally, for certain systems, while the Lagrange points are located above the physical surface of the smaller primary, positioning a spacecraft there would be dangerously close to the surface and would require very frequent stationkeeping maneuvers. One such example is the Mars-Phobos system, where L_1 and L_2 are approximately 3.5 km above Phobos' surface, making it dangerous to position a spacecraft at L_1 or L_2, or even in orbits around these Lagrange points, as discussed in more detail in later sections.

3.2.4 Jacobi Integral

In the two-body problem, we identified ten constants of motion – six from total linear momentum, three from total angular momentum and one from total energy. The restricted three-body problem has these constants of motion, but they tell nothing about the motion of the infinitesimal body with respect to the primaries. The restricted three-body problem has only one exact conservation law which is known as Jacobi integral (or Jacobi constant).

In order to find this parameter, we start with the equations of motion in Equation (3.25). We multiply each equation by the first derivative of the state variable associated with the second derivative of the state variable, and add them together,

$$
\begin{aligned}
&\left[\frac{d^2x}{dt^2} - 2\frac{dy}{dt} - x + \frac{(1-\mu)(x-\mu)}{r_1^3} + \frac{\mu(x+1-\mu)}{r_2^3}\right]\frac{dx}{dt} \\
&+\left[\frac{d^2y}{dt^2} + 2\frac{dx}{dt} - y + \frac{(1-\mu)y}{r_1^3} + \frac{\mu y}{r_2^3}\right]\frac{dy}{dt} \\
&+\left[\frac{d^2z}{dt^2} + \frac{(1-\mu)z}{r_1^3} + \frac{\mu z}{r_2^3}\right]\frac{dz}{dt} = 0
\end{aligned}
\tag{3.56}
$$

Multiplying this out and grouping terms,

$$\frac{dx}{dt}\frac{d^2x}{dt^2}+\frac{dy}{dt}\frac{d^2y}{dt^2}+\frac{dz}{dt}\frac{d^2z}{dt^2}-x\frac{dx}{dt}-y\frac{dy}{dt}=-\frac{(1-\mu)}{r_1^3}\left[(x-\mu)\frac{dx}{dt}+y\frac{dy}{dt}+z\frac{dz}{dt}\right]$$
$$-\frac{\mu}{r_2^3}\left[(x+1-\mu)\frac{dx}{dt}+y\frac{dy}{dt}+z\frac{dz}{dt}\right] \tag{3.57}$$

The first three terms on the left-hand side are the time derivative of the velocity in the rotating frame,

$$\frac{dx}{dt}\frac{d^2x}{dt^2}+\frac{dy}{dt}\frac{d^2y}{dt^2}+\frac{dz}{dt}\frac{d^2z}{dt^2}=\frac{1}{2}\frac{d}{dt}\left(v^2\right) \tag{3.58}$$

and second three terms on the left-hand side is the time derivative of the centrifugal force,

$$-x\frac{dx}{dt}-y\frac{dy}{dt}=-\frac{1}{2}\frac{d}{dt}\left(x^2+y^2\right) \tag{3.59}$$

Each term on the right-hand side is proportional to the time derivative of $1/r$. Combining these terms, Equation (3.57) becomes

$$0=\frac{d}{dt}\left[\frac{1}{2}v^2-\frac{1}{2}\left(x^2+y^2\right)-\frac{(1-\mu)}{r_1}-\frac{\mu}{r_2}\right] \tag{3.60}$$

Integrating this equation,

$$J=\frac{1}{2}\left[\left(\frac{dx}{dt}\right)^2+\left(\frac{dy}{dt}\right)^2+\left(\frac{dz}{dt}\right)^2\right]-\frac{1}{2}\left(x^2+y^2\right)-\frac{1-\mu}{r_1}-\frac{\mu}{r_2} \tag{3.61}$$

where J is a constant of integration and known as the Jacobi constant [Jacobi, 1836]. The value of the constant helps describe the characteristics of the orbit. Note that various texts define the Jacobi constant in different ways – another commonly used definition of the Jacobi constant is $J=\frac{2(1-\mu)}{r_1}+\frac{2\mu}{r_2}+\left(x^2+y^2\right)-\left[\left(\frac{dx}{dt}\right)^2+\left(\frac{dy}{dt}\right)^2+\left(\frac{dz}{dt}\right)^2\right]$, which makes the constant a positive numerical value.

Another way to express the Jacobi integral is in the inertial reference frame. Using the direction cosine matrix in Equation (3.2), and defining a new state vector, **X**, which is the Cartesian coordinates in the inertial reference frame, $X=\left(\xi\ \eta\ \zeta\ \frac{d\xi}{dt}\ \frac{d\eta}{dt}\ \frac{d\zeta}{dt}\right)^T$, the Jacobi constant can be written as

$$J=\frac{1}{2}\left[\left(\frac{d\xi}{dt}\right)^2+\left(\frac{d\eta}{dt}\right)^2+\left(\frac{d\zeta}{dt}\right)^2\right]-\left(\xi\frac{d\eta}{dt}-\eta\frac{d\xi}{dt}\right)-\frac{1-\mu}{r_1}-\frac{\mu}{r_2} \tag{3.62}$$

The numerical value of the Jacobi constant provides bounds on how far the massless body can move away from the primaries. Just like the total energy in a pendulum controls how high the pendulum swings (which is at the point where the velocity of the pendulum becomes zero), so does the amount of energy, as described by the Jacobi constant, which determines when all of the energy is transferred to potential energy and none is left as kinetic energy. Since kinetic energy is directly related to the velocity, the point where the velocity in the rotating frame is equal to zero forms a surface in three-dimensional space where the object cannot cross at its present energy level. This boundary forms what is known as the zero-velocity curve.

The physical coordinates of this boundary are found by solving

$$J = -\frac{1}{2}\left(x^2 + y^2\right) - \frac{1-\mu}{r_1} - \frac{\mu}{r_2} \tag{3.63}$$

for the corresponding x, y, and z coordinates. There are various ways to find these coordinates. Probably the simplest way is to specify two of the three coordinates and solve for the third. The most common curve as seen in the literature is the slice of the three-dimensional surface that exists in the plane of motion of the primaries (the plane where $z=0$). Given a value for x yields multiple values of y. In the restricted three-body problem, this curve is symmetric about the x-axis. Additional methods of solution are presented by Moulton [1970].

The zero-velocity curve forms a boundary where an object does not have enough energy to cross. An example zero-velocity curve is found in Figure 3.7, where the generic distance units (DU) are used. Mathematically, this "inaccessible region" is

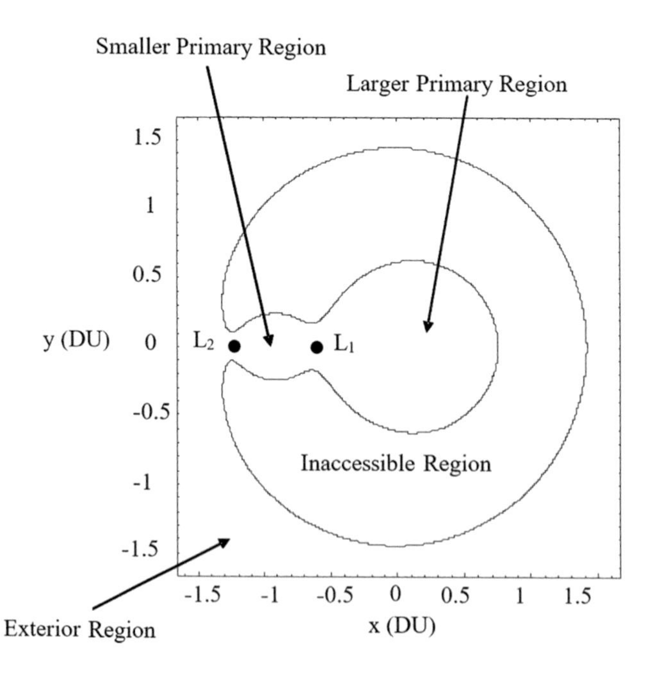

FIGURE 3.7 Regions in zero-velocity curves.

the region where an object inside this region would have an imaginary value for the velocity. In order for an object moving in the area where the larger primary is to move to the exterior region, it must pass through the opening located near the smaller primary region. Likewise, for an object in the exterior region to enter the region near the larger primary, it must pass through this opening near the smaller primary. If it doesn't have enough energy, it remains trapped in the region where it starts.

A series of zero-velocity curves is shown in Figure 3.8. This series is for a mass ratio, $\mu = 0.01215060$ (the value for the Earth-Moon system). In Figure 3.8a, an object with the amount of energy associated with a Jacobi constant of $J = -1.65$ (in non-dimensional units) is shown. An object in the interior region around the larger primary cannot leave the interior region. Likewise, an object in the interior region around the smaller primary cannot leave this region. Physically, this could be thought

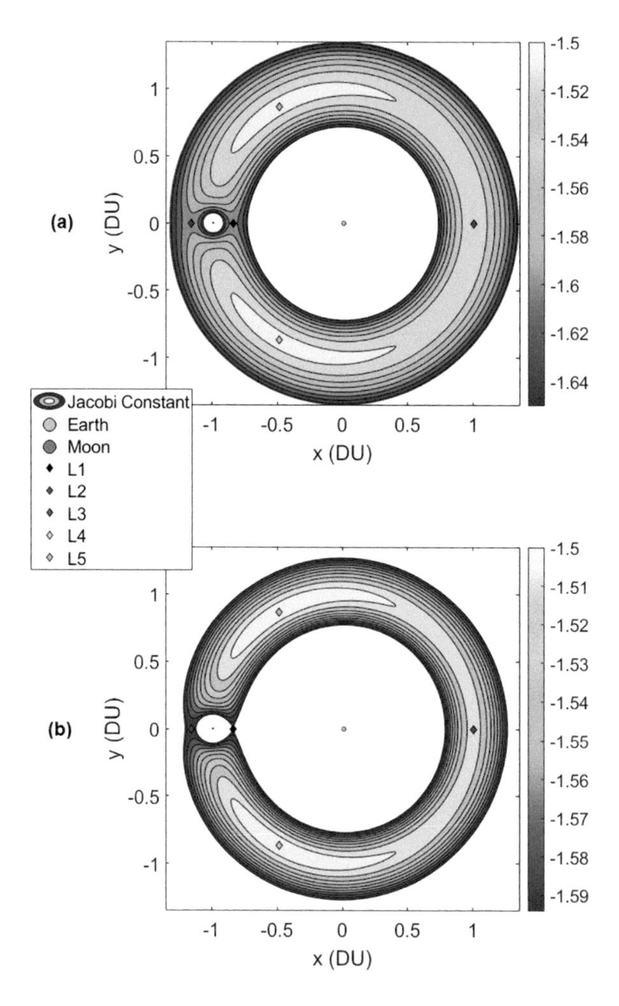

FIGURE 3.8 Zero-velocity curves for the Earth-Moon system for $J = -1.65$ (a) and $J = -1.59411$ (b).

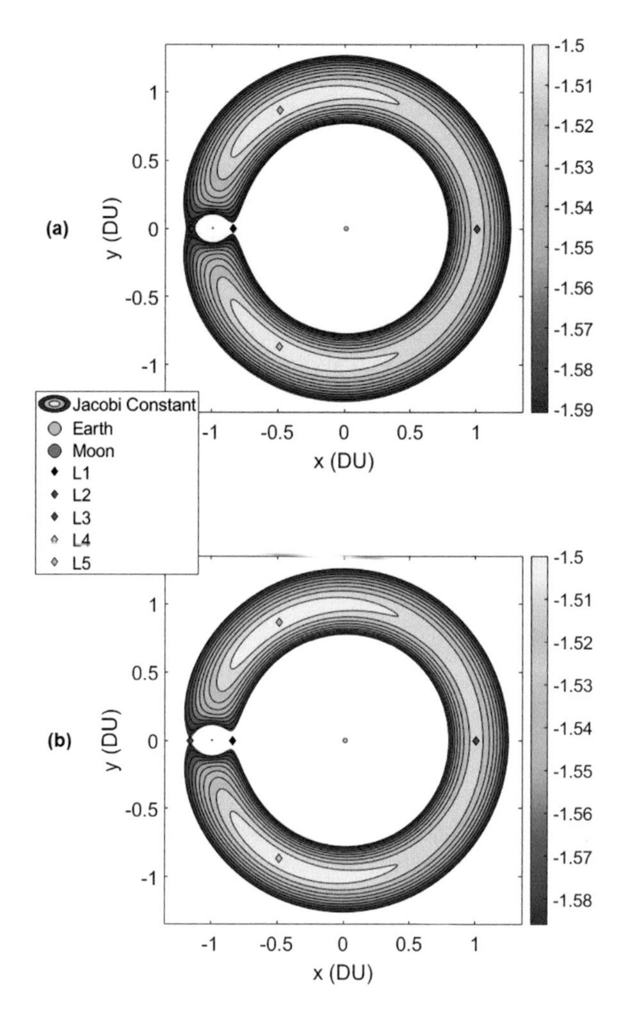

FIGURE 3.9 Zero-velocity curves for the Earth-Moon system for J = −1.59 (a) and J = −1.58603 (b).

of as an object orbiting the Earth with a certain energy level does not have enough energy to reach the Moon simply due to the gravitational effects of the Moon. An object orbiting the Moon does not have enough energy to reach the Earth due to the gravitational effects of the Earth.

As we increase the energy of the system (by increasing the value of the Jacobi constant), the region opens up, and eventually connects these two interior regions. The point where the two regions merge is the collinear libration point, L_1, in this system, as is seen in Figure 3.8b. Here, the value of the Jacobi constant is J = −1.59411. Increasing the Jacobi constant opens up the region between these two interior regions. The Jacobi constant in Figure 3.9a is J = −1.591, and now an object has enough energy to reach the Moon (when moving in the region in the vicinity of the Earth) and passes through the neck at the libration point. The reverse is true when considering an object moving in the vicinity of the Moon heading to the vicinity of Earth due to the Earth's gravitational attraction.

Increasing the value of the Jacobi constant further, as seen in Figure 3.9b, opens up this interior region further and allows an object to escape from orbit around the primaries (it also allows for objects to enter the region around the two primaries). As this region opens up, the opening location corresponds to the collinear libration point, L_2. The Jacobi constant at this point is $J = -1.58603$. Increasing the Jacobi constant opens this region further. The two branches of the upper and lower inaccessible regions in Figure 3.10a touch last at the libration point L_3. At this point, the value of the Jacobi constant is $J = -1.50607$. As the Jacobi constant continues to increase ($J = -1.50$), these two inaccessible regions continue to shrink, as seen in Figure 3.10b, and eventually collapse on the triangular libration points, L_4 and L_5.

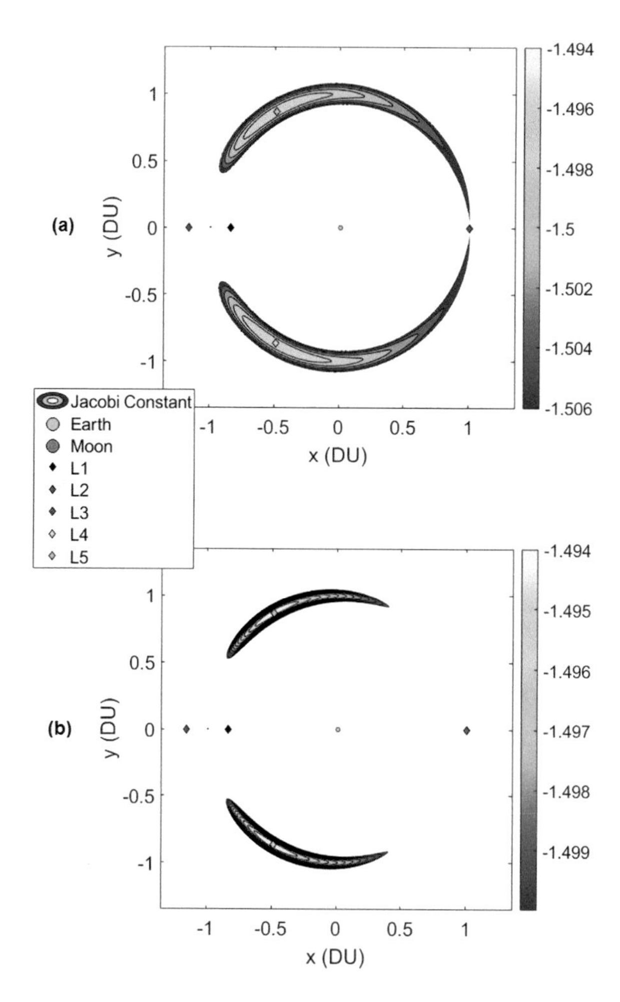

FIGURE 3.10 Zero-velocity curves for the Earth-Moon system for $J = -1.50607$ (a) and $J = -1.50$ (b).

Table 3.3 summarizes the values of the Jacobi constant at the Earth-Moon Lagrange points, including the values for L_4 and L_5. Note that, due to symmetry, these last two values are identical.

If we stack these curves on each other, we get a three-dimensional orthographic projection of the zero-velocity curves as shown in Figure 3.11. We can think of this as sort of a topographic map – each of the libration points are located at various levels in this contour. As the Jacobi constant increases (which is analogous to an increase in energy), we move from the lowest energy libration point (L_1) to the highest energy libration points (L_4 and L_5).

Previously, it was shown that the libration points L_4 and L_5 were marginally stable. However, in order to reach these points, a higher energy level (as characterized by the value of the Jacobi constant) needs to be attained. The libration point requiring the lowest energy to reach is the L_1 point. Ironically, the first space missions to go to any of the libration points went to the Sun-Earth L_1 point.[2] Additional details on libration point space missions are presented later in this chapter.

TABLE 3.3
Values of the Jacobi constant at the Earth-Moon Lagrange points ($\mu = 0.01215060$)

Lagrange point	Jacobi constant (non-dimensional)
L_1	−1.59411
L_2	−1.58603
L_3	−1.50607
L_4	−0.512877
L_5	−0.512877

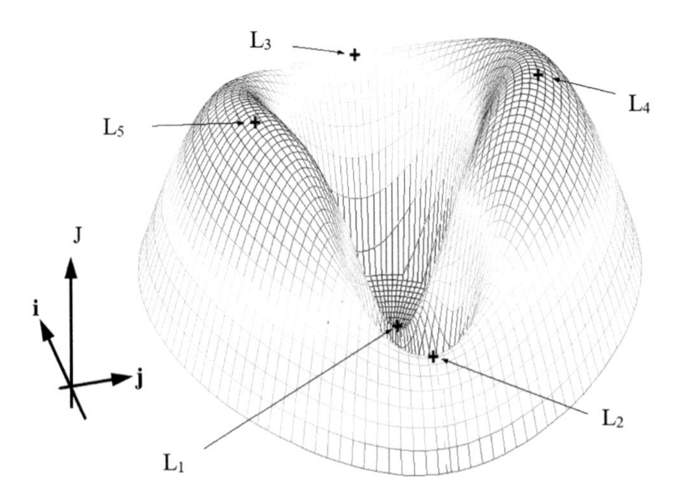

FIGURE 3.11 Contour of zero-velocity curves.

[2] There are scientific reasons to go to the Sun-Earth L_1 point – the fact that it was the lowest energy libration point to get to helped make it more feasible to send spacecraft there.

3.2.5 Tisserand's Criterion

When an object passes near a gravitational source, the orbital elements of the object change. Correlation of the object before it passes the gravitational source and after the pass is important in understanding whether the object entering the close approach is the same one moving away from the gravitational source. The test of whether this object is the same one was developed by François Félix Tisserand who, using the concept of the Jacobi integral, found a relationship between semi-major axis, eccentricity, and inclination of the object before and after the passage of the gravitational source.

Starting with Equation (3.61), we analyze various parts of the equation. First, the velocity in the inertial frame is

$$v^2 = \left(\frac{d\xi}{dt}\right)^2 + \left(\frac{d\eta}{dt}\right)^2 + \left(\frac{d\zeta}{dt}\right)^2 \tag{3.64}$$

Using the vis viva equation,

$$\left(\frac{d\xi}{dt}\right)^2 + \left(\frac{d\eta}{dt}\right)^2 + \left(\frac{d\zeta}{dt}\right)^2 = \frac{2}{r} - \frac{1}{a} \tag{3.65}$$

The angular momentum in inertial frame coordinates is

$$\mathbf{h} = \mathbf{r} \times \mathbf{v} = (\xi\mathbf{I} + \eta\mathbf{J} + \zeta\mathbf{K}) \times \left(\frac{d\xi}{dt}\mathbf{I} + \frac{d\eta}{dt}\mathbf{J} + \frac{d\zeta}{dt}\mathbf{K}\right) \tag{3.66}$$

Expanding this out,

$$\begin{aligned} \mathbf{h} &= h_\xi\mathbf{I} + h_\eta\mathbf{J} + h_\zeta\mathbf{K} \\ &= \left(\eta\frac{d\zeta}{dt} - \zeta\frac{d\eta}{dt}\right)\mathbf{I} + \left(\zeta\frac{d\xi}{dt} - \xi\frac{d\zeta}{dt}\right)\mathbf{J} + \left(\xi\frac{d\eta}{dt} - \eta\frac{d\xi}{dt}\right)\mathbf{K} \end{aligned} \tag{3.67}$$

The second term in Equation (3.67) is the h_ζ term. From Equation (2.65), this second term can also be written as

$$\xi\frac{d\eta}{dt} - \eta\frac{d\xi}{dt} = h\cos i \tag{3.68}$$

Lastly, we can make an assumption that we are examining the passage of an object near a planetary body, and this planetary body is small when compared to the Sun. Thus, the non-dimensional gravitational parameter is approximately equal to zero. Additionally, the distance from the Sun to the object passing by the planet (r_1) is assumed to be very large, so the two terms on the end of Equation (3.63) are approximately equal to zero. Thus, inserting Equations (3.65) and (3.68) into Equation (3.63), and recalling that $h^2 = a\left(1 - e^2\right)$, shows us that the Jacobi constant is approximately the same before the passage as after the passage. Therefore, Tisserand's criterion states that

$$\frac{1}{2a_0}+a_0^{1/2}\left(1-e_0^2\right)^{1/2}\cos i_0=\frac{1}{2a_1}+a_1^{1/2}\left(1-e_1^2\right)^{1/2}\cos i_1 \tag{3.69}$$

While this was originally done for a comet passing a planet, it applies as easily for a spacecraft passing near a planet. Thus, knowing the semimajor axis, eccentricity and inclination before (subscript 0) and after (subscript 1) the planetary passage and testing whether Equation (3.68) holds allows us to determine whether the object seen before the passage is the same one seen after the passage.

3.2.6 The State-Transition Matrix

In order to take advantage of the features of some of the orbital characteristics of libration point orbits, we need to develop the equations of motion for the state-transition matrix. The state-transition matrix is a mapping matrix that maps the linearized state equations from one time to another, and is used for various operations.

Recall that the pseudo-potential, U, is defined as

$$U=\frac{1}{2}\left(x^2+y^2\right)+\frac{1-\mu}{r_1}+\frac{\mu}{r_2} \tag{3.70}$$

A Taylor series expansion of the pseudo-potential, U, can be written as

$$\begin{aligned}\frac{\partial U}{\partial x}&=\frac{\partial}{\partial x}\left(\frac{\partial U}{\partial x}\right)x+\frac{\partial}{\partial y}\left(\frac{\partial U}{\partial x}\right)y+\frac{\partial}{\partial z}\left(\frac{\partial U}{\partial x}\right)z+\frac{\partial}{\partial \dot{x}}\left(\frac{\partial U}{\partial x}\right)\left(\frac{dx}{dt}\right)\\&+\frac{\partial}{\partial \dot{y}}\left(\frac{\partial U}{\partial x}\right)\left(\frac{dy}{dt}\right)+\frac{\partial}{\partial \dot{z}}\left(\frac{\partial U}{\partial x}\right)\left(\frac{dz}{dt}\right)\end{aligned} \tag{3.71}$$

but since U is not a function of the velocities $(\dot{x},\dot{y},\dot{z})$, Equation (3.70) reduces to

$$\begin{aligned}\frac{\partial U}{\partial x}&=\frac{\partial}{\partial x}\left(\frac{\partial U}{\partial x}\right)x+\frac{\partial}{\partial y}\left(\frac{\partial U}{\partial x}\right)y+\frac{\partial}{\partial z}\left(\frac{\partial U}{\partial x}\right)z\\&=\left(\frac{\partial^2 U}{\partial x^2}\right)x+\left(\frac{\partial^2 U}{\partial x\,\partial y}\right)y+\left(\frac{\partial^2 U}{\partial x\,\partial z}\right)z\end{aligned} \tag{3.72}$$

By a similar method,

$$\frac{\partial U}{\partial y}=\left(\frac{\partial^2 U}{\partial x\,\partial y}\right)x+\left(\frac{\partial^2 U}{\partial y^2}\right)y+\left(\frac{\partial^2 U}{\partial y\,\partial z}\right)z \tag{3.73}$$

and

$$\frac{\partial U}{\partial z}=\left(\frac{\partial^2 U}{\partial x\,\partial z}\right)x+\left(\frac{\partial^2 U}{\partial y\,\partial z}\right)y+\left(\frac{\partial^2 U}{\partial z^2}\right)z \tag{3.74}$$

The differential equations can be written as

$$\begin{aligned}
\frac{d^2x}{dt^2} - 2\frac{dy}{dt} &= \left(\frac{\partial^2 U}{\partial x^2}\right)x + \left(\frac{\partial^2 U}{\partial x\,\partial y}\right)y + \left(\frac{\partial^2 U}{\partial x\,\partial z}\right)z \\
\frac{d^2y}{dt^2} + 2\frac{dx}{dt} &= \left(\frac{\partial^2 U}{\partial x\,\partial y}\right)x + \left(\frac{\partial^2 U}{\partial y^2}\right)y + \left(\frac{\partial^2 U}{\partial y\,\partial z}\right)z \\
\frac{d^2z}{dt^2} &= \left(\frac{\partial^2 U}{\partial x\,\partial z}\right)x + \left(\frac{\partial^2 U}{\partial y\,\partial z}\right)y + \left(\frac{\partial^2 U}{\partial z^2}\right)z
\end{aligned} \tag{3.75}$$

where

$$\frac{\partial^2 U}{\partial x^2} = 1 - \frac{(1-\mu)}{r_1^3} - \frac{\mu}{r_2^3} + \frac{3(1-\mu)(x-\mu)^2}{r_1^5} + \frac{3\mu(x+1-\mu)^2}{r_2^5}$$

$$\frac{\partial^2 U}{\partial x\,\partial y} = \frac{3(1-\mu)(x-\mu)y}{r_1^5} + \frac{3\mu(x+1-\mu)y}{r_2^5}$$

$$\frac{\partial^2 U}{\partial x\,\partial z} = \frac{3(1-\mu)(x-\mu)z}{r_1^5} + \frac{3\mu(x+1-\mu)z}{r_2^5}$$

$$\frac{\partial^2 U}{\partial y^2} = 1 - \frac{(1-\mu)}{r_1^3} - \frac{\mu}{r_2^3} + \frac{3(1-\mu)y^2}{r_1^5} + \frac{3\mu y^2}{r_2^5}$$

$$\frac{\partial^2 U}{\partial y\,\partial z} = \frac{3(1-\mu)yz}{r_1^5} + \frac{3\mu yz}{r_2^5}$$

$$\frac{\partial^2 U}{\partial y^2} = -\frac{(1-\mu)}{r_1^3} - \frac{\mu}{r_2^3} + \frac{3(1-\mu)z^2}{r_1^5} + \frac{3\mu z^2}{r_2^5}$$

In matrix form, Equation (3.74) is

$$\frac{d\mathbf{X}}{dt} = \frac{d}{dt}\begin{Bmatrix} x \\ y \\ z \\ \dot{x} \\ \dot{y} \\ \dot{z} \end{Bmatrix} = \begin{bmatrix} 0 & 0 & 0 & 1 & 0 & 0 \\ 0 & 0 & 0 & 0 & 1 & 0 \\ 0 & 0 & 0 & 0 & 0 & 1 \\ \left(\frac{\partial^2 U}{\partial x^2}\right) & \left(\frac{\partial^2 U}{\partial x\,\partial y}\right) & \left(\frac{\partial^2 U}{\partial x\,\partial z}\right) & 0 & 2 & 0 \\ \left(\frac{\partial^2 U}{\partial x\,\partial y}\right) & \left(\frac{\partial^2 U}{\partial y^2}\right) & \left(\frac{\partial^2 U}{\partial y\,\partial z}\right) & -2 & 0 & 0 \\ \left(\frac{\partial^2 U}{\partial x\,\partial z}\right) & \left(\frac{\partial^2 U}{\partial y\,\partial z}\right) & \left(\frac{\partial^2 U}{\partial z^2}\right) & 0 & 0 & 0 \end{bmatrix} \begin{Bmatrix} x \\ y \\ z \\ \dot{x} \\ \dot{y} \\ \dot{z} \end{Bmatrix} \tag{3.76}$$

Writing Equation (3.75) in a more compact form

$$\frac{d\mathbf{X}}{dt} = A(t)\mathbf{X} \tag{3.77}$$

To use a differential correction process (see Section 3.7.1 for details), the differential equations for the state-transition matrix is

$$\frac{d\Phi(t,0)}{dt} = A(t)\Phi(t,0) \tag{3.78}$$

where A(t) was defined in Equation (3.76).

Solution of Equations (3.77) and (3.78) yield a set of 42 coupled, simultaneous differential equations. In order to solve such as set of equations, initial conditions are needed. The initial conditions of the state vector, **X**, are problem specific. The initial conditions for the state-transition matrix equation are, by definition, $\Phi(0,0) = I$, where I is a 6×6 identity matrix.

If a particular solution of the circular restricted three-body problem is known then another solution can be found by considering that the system is invariant (unchanged) if we replace t with −t and y with −y. This is easily verified by making these substitutions into the equations of motion so that $\ddot{x} \to \ddot{x}$, $\dot{y} \to \dot{y}$, $\Omega_x \to \Omega_x$, $\ddot{y} \to -\ddot{y}$, $\dot{x} \to -\dot{x}$, $\Omega_y \to -\Omega_y$, $\ddot{z} \to \ddot{z}$, and $\Omega_z \to \Omega_z$. This property is of particular importance when numerically calculating initial conditions for periodic orbits and when computing trajectories to and from these orbits.

3.3 ELLIPTICAL RESTRICTED THREE-BODY PROBLEM

Simplifications of the motion of the two primary bodies by forcing them to keep a constant distance between them makes for some elegant mathematics and unique mathematical characteristics, such as libration points and the Jacobi constant. However, in reality, these objects are not in circular orbits around their barycenter, but in elliptical orbits. This dynamical system is referred to as the elliptical restricted three-body problem (ER3BP). In the circular restricted three-body problem, the derivation of the equations of motion kept the distance $a_1 + a_2$ constant (as was seen in Figure 3.4). Conversely, in the elliptical case, these values are no longer constant, but oscillate periodically.

The description of the elliptical restricted three-body problem shown here is based on the work of Howell [1983a, 1983b]. Let's describe the distance between the two primary bodies as

$$a_1 + a_2 = 1 + \gamma = \frac{a\left(1 - e^2\right)}{1 + e\cos\omega\left(t - t_p\right)} \tag{3.79}$$

where γ is a periodic function, a is the semimajor axis of the primaries' orbit, e is the eccentricity of the primaries' orbit, ω is the mean motion of the primaries around the inertial frame, and $t-t_p$ is the time since the bodies were at their periapse points around the barycenter.

In non-dimensional terms, the periodic term, γ, is found as

$$\gamma = -\frac{\left[e^2 + e\cos\left(t-t_p\right)\right]}{1+e\cos\left(t-t_p\right)} \tag{3.80}$$

Combining Equations (3.17) and (3.24), letting $n=\omega$, and using the non-dimensional terms, the equations of motion of the elliptical restricted three-body problem become

$$\begin{aligned}
&\frac{d^2x}{dt^2} - 2n\frac{dy}{dt} - y\frac{dn}{dt} - n^2x = -\frac{(1-\mu)\left[x+\mu(1-\gamma)\right]}{r_1^3} - \frac{\mu\left[x-(1-\mu)(1+\gamma)\right]}{r_2^3} \\
&\frac{d^2y}{dt^2} + 2n\frac{dx}{dt} + x\frac{dn}{dt} - n^2y = -\frac{(1-\mu)y}{r_1^3} - \frac{\mu y}{r_2^3} \\
&\frac{d^2z}{dt^2} = -\frac{(1-\mu)z}{r_1^3} - \frac{\mu z}{r_2^3}
\end{aligned} \tag{3.81}$$

and the vectors $\mathbf{r}_1$ and $\mathbf{r}_2$ are

$$\mathbf{r}_1 = \left[x-\mu(1+\gamma)\right]\mathbf{i} + y\mathbf{j} + z\mathbf{k} \tag{3.82}$$

$$\mathbf{r}_2 = \left[x+(1-\mu)(1+\gamma)\right]\mathbf{i} + y\mathbf{j} + z\mathbf{k} \tag{3.83}$$

Like we did for the CR3BP, we can introduce a pseudo-potential, U, such that

$$\begin{aligned}
U = &\frac{1-\mu}{\left\{\left[x+\mu(1+\gamma)\right]^2 + y^2 + z^2\right\}^{1/2}} + \frac{\mu}{\left\{\left[x-(1-\mu)(1+\gamma)\right]^2 + y^2 + z^2\right\}^{1/2}} \\
&+ \frac{1}{2}n^2\left(x^2+y^2\right)
\end{aligned} \tag{3.84}$$

Thus, Equation (3.81) can be simplified to

$$\begin{aligned}
&\frac{d^2x}{dt^2} - 2n\frac{dy}{dt} = \frac{\partial U}{\partial x} y\frac{dn}{dt} \\
&\frac{d^2y}{dt^2} + 2n\frac{dx}{dt} = \frac{\partial U}{\partial y} - x\frac{dn}{dt} \\
&\frac{d^2z}{dt^2} = \frac{\partial U}{\partial z}
\end{aligned} \tag{3.85}$$

3.4 CIRCULAR RESTRICTED FOUR-BODY PROBLEM

While the circular restricted three-body is a mathematical approximation to reality, it does provide insight into behavior of an infinitesimal body orbiting in the vicinity of the primary bodies. However, an extension of the circular restricted three-body problem can be made by adding additional gravitational sources. A good example is the effects of the Sun's gravity on orbits moving in the Earth-Moon restricted three-body problem. This mathematical model is known as the circular restricted four-body problem (CR4BP). This model consists of a space vehicle (m_3), a larger mass m_4, and two smaller masses m_n with $n=1, 2$. It is assumed that both of the smaller bodies are circulating around the barycenter of the m_1–m_2–m_3 system, with the constant radii, D, and angular velocity Ω. The bodies m_1 and m_2 rotate around their common barycenter at an angular velocity ω. These rotations form concentric circles in the same plane and are shown in Figure 3.12.

In this model, the origin of the inertial frame is at the barycenter of the m_1–m_2–m_3 system and the plane of motion is the $\boldsymbol{\xi}$–$\boldsymbol{\eta}$ plane. Similar to the common CR3BP model, the intermediate frame **X–Y** is a rotating frame such that the **X**-axis is in the direction from the center of the inertial frame to the barycenter of the m_1–m_2 system. This frame rotates at a rate of Ω around an axis perpendicular to the plane of motion and sweeps out an angle θ_4. A rotating frame, **i–j** rotates around the barycenter of the m_1–m_2 system at a rate of ω and sweeps out an angle θ with respect to the **X–Y** frame. The vector, $\boldsymbol{\rho}$, has its origin at the barycenter of the system and represents the position of the infinitesimal body with respect to the inertial frame.

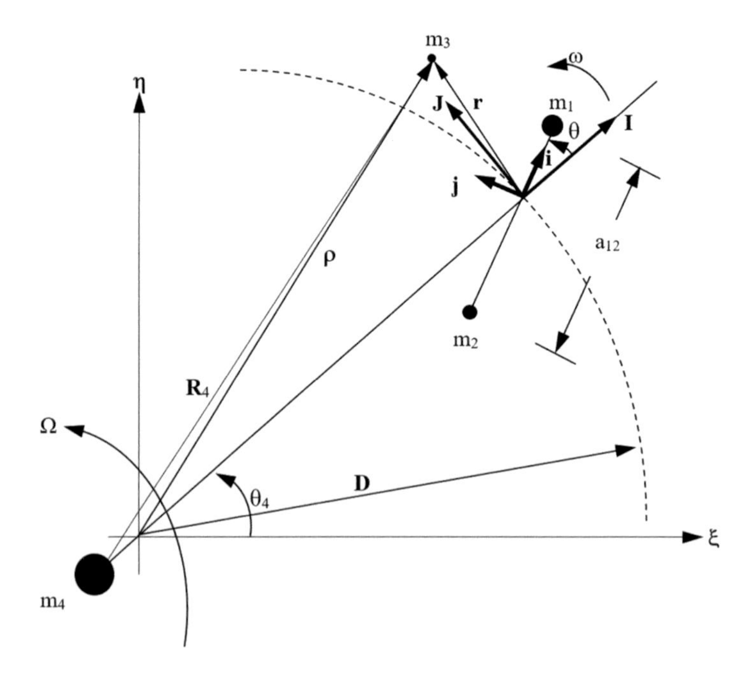

FIGURE 3.12 Geometry of circular restricted four-body problem (CR4BP).

In addition to using the non-dimensionalization that we used for the circular restricted three-body problem, let's define a new normalized parameter that represents the mass of the fourth body in normalized variables,

$$\mu_4 = \frac{m_4}{m_1 + m_2} \tag{3.86}$$

The acceleration of the infinitesimal mass in the inertial frame, from Equation (3.23) is therefore

$$\frac{d^2\boldsymbol{\rho}}{dt^2}^{(I)} = -\frac{(1-\mu)\mathbf{r}_1}{r_1^3} - \frac{\mu\mathbf{r}_2}{r_2^3} - \frac{\mu_4\mathbf{R}_4}{R_4^3} \tag{3.87}$$

where

$$\boldsymbol{\rho} = D\mathbf{I} + \mathbf{r} = X\mathbf{I} + Y\mathbf{J} + Z\mathbf{K} \tag{3.88}$$

where X, Y, and Z are the position components of the infinitesimal mass in the **IJK** frame. Taking two derivatives of Equation (3.88) in the inertial frame, using Equation (2.13),

$$\frac{d^2\boldsymbol{\rho}}{dt^2}^{(I)} = \left(-D\Omega^2\right)\mathbf{I} + \left(\frac{d^2x}{dt^2} - 2\omega\frac{dy}{dt} - \omega^2 x\right)\mathbf{i} + \left(\frac{d^2y}{dt^2} + 2\omega\frac{dx}{dt} - \omega^2 y\right)\mathbf{j} + \frac{d^2z}{dt^2}\mathbf{k} \tag{3.89}$$

The position vectors $\mathbf{r}_1$, $\mathbf{r}_2$, and $\mathbf{R}_4$ are found as

$$\mathbf{r}_1 = (x + a_2)\mathbf{i} + y\mathbf{j} + z\mathbf{k} \tag{3.90}$$

$$\mathbf{r}_2 = (x + a_1)\mathbf{i} + y\mathbf{j} + z\mathbf{k} \tag{3.91}$$

$$\mathbf{R}_4 = \mathbf{r} + \frac{D}{\mu_4}(\mu_4 + 1)\mathbf{I} \tag{3.92}$$

Equations (3.89) and (3.92) are mixes between the coordinate frames **IJK** and **ijk**. In order to convert all vectors into the same frame, it is necessary to find the direction cosine matrix relating the two frames, which can be written as

$$\begin{Bmatrix} \mathbf{I} \\ \mathbf{J} \\ \mathbf{K} \end{Bmatrix} = \begin{bmatrix} \cos\theta & -\sin\theta & 0 \\ \sin\theta & \cos\theta & 0 \\ 0 & 0 & 1 \end{bmatrix} \begin{Bmatrix} \mathbf{i} \\ \mathbf{j} \\ \mathbf{k} \end{Bmatrix} \tag{3.93}$$

Thus, Equation (3.89) becomes

$$\begin{aligned} \frac{d^2\boldsymbol{\rho}}{dt^2}^{(I)} &= \left(\frac{d^2x}{dt^2} - 2\omega\frac{dy}{dt} - \omega^2 x - D\Omega^2\cos\theta\right)\mathbf{i} \\ &+ \left(\frac{d^2y}{dt^2} + 2\omega\frac{dx}{dt} - \omega^2 y + D\Omega^2\sin\theta\right)\mathbf{j} + \frac{d^2z}{dt^2}\mathbf{k} \end{aligned} \tag{3.94}$$

and Equation (3.92) becomes

$$\mathbf{R}_4 = \left(x + \frac{D}{\mu_4}(\mu_4 + 1)\cos\theta \right)\mathbf{i} + \left(y - \frac{D}{\mu_4}(\mu_4 + 1)\sin\theta \right)\mathbf{j} + z\mathbf{k} \tag{3.95}$$

In the circular restricted four-body problem, the distances between the barycenter of the m_1–m_2 frame and the barycenter of the m_1–m_2–m_4 system and the distance between m_1 and m_2 are constant, the time rates of change of the angles θ and θ_4 (ω and Ω) are constant as well. Inserting Equations (3.90), (3.91), and (3.95) into Equation (3.94) yields a set of non-linear second-order differential equations,

$$\frac{d^2x}{dt^2} - 2\omega\frac{dy}{dt} - \omega^2 x - D\Omega^2\cos\theta = -\frac{(1-\mu)(x+a_2)}{r_1^3} - \frac{\mu(x+a_1)}{r_2^3} - \frac{\mu_4\left[x + D(\mu_4+1)\right]}{R_4^3} \tag{3.96}$$

$$\frac{d^2y}{dt^2} + 2\omega\frac{dx}{dt} - \omega^2 y + D\Omega^2\sin\theta = -\frac{(1-\mu)y}{r_1^3} - \frac{\mu y}{r_2^3} - \frac{\mu_4(D\sin\theta - y)}{R_4^3} \tag{3.97}$$

$$\frac{d^2z}{dt^2} = -\frac{(1-\mu)z}{r_1^3} - \frac{\mu z}{r_2^3} - \frac{\mu_4 z}{R_4^3} \tag{3.98}$$

Additional gravitational sources can be added in a similar manner. Concepts such as Lagrange points and the Jacobi constant do not exist outside of the restricted three-body problem, thus there are no equilibrium points or constants of motion for anything more complex than the circular restricted three-body problem.

3.5 ELLIPTICAL RESTRICTED FOUR-BODY PROBLEM

A more complex model with three gravitational sources with all bodies in elliptical orbits about various barycenters can be developed. This model is known as the elliptical restricted four-body problem (ER4BP). The smaller bodies are assumed to be on elliptic orbits around the barycenter of the three bodies. Likewise, the smaller bodies, m_1 and m_2 are on elliptical orbits around their common barycenter. This system is seen in Figure 3.13. Here, the length between the barycenter of the m_1 and m_2 system and the barycenter of the entire system is no longer constant. Additionally, the distance between m_1 and m_2 is also not constant.

The position vectors from the barycenter of the m_1–m_2 system to m_1 and m_2 are found using Equations (3.82) and (3.83). The position vector of the infinitesimal mass to the fourth body, $\mathbf{R}_4$ is

$$\mathbf{R}_4 = \left[\frac{D(1-e_4^2)}{1 + e_4\cos\Omega(t - t_p^*)} \right]\left[\cos\Omega(t - t_p^*)\mathbf{i} + \sin\Omega(t - t_p^*)\mathbf{j} \right] \tag{3.99}$$

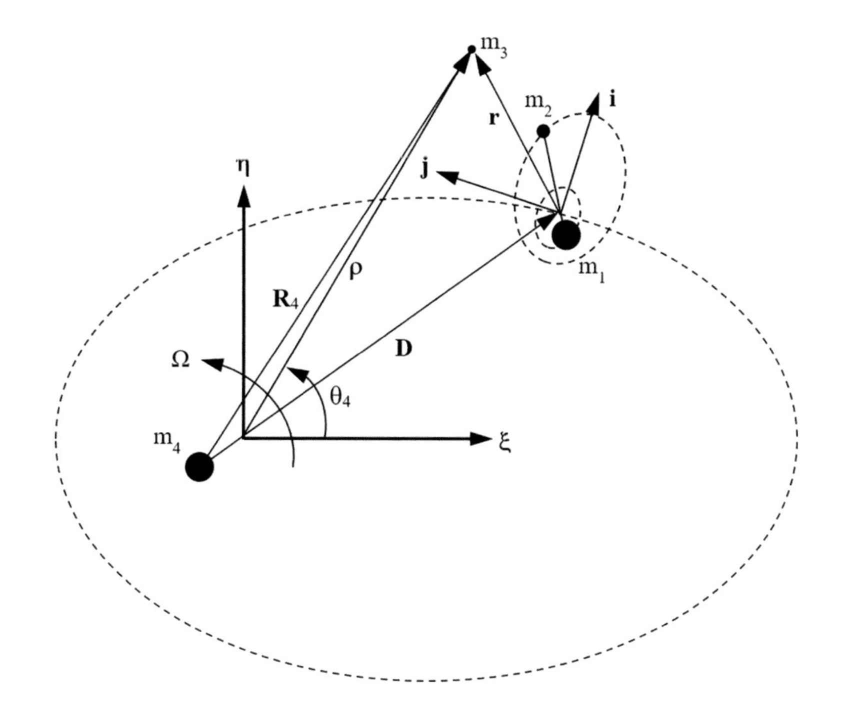

FIGURE 3.13 Geometry of elliptic restricted four-body problem (ER4BP).

where the argument of the trigonometry functions, $\Omega\left(t-t_p^*\right)$, is the true anomaly of the m_1–m_2 system as it moves about the barycenter of the entire system. The term t_p^* is the time of periapse passage, D is the semimajor axis, and e_4 is the eccentricity of the orbit of the barycenter of the m_1–m_2 system while it is orbiting the barycenter of the entire system.

The derivation of the equations of motion follows a similar development to those of the CR4BP. These equations are

$$\frac{d^2x}{dt^2}-2\omega\frac{dy}{dt}-y\frac{d\omega}{dt}-\omega^2x+\left(\frac{d^2D}{dt^2}-D\Omega^2\right)\cos\theta$$
$$+\left(2\Omega\frac{dD}{dt}+D\frac{d\Omega}{dt}\right)\sin\theta=-\frac{(1-\mu)(x+a_2)}{r_1^3}$$
$$-\frac{\mu(x+a_1)}{r_2^3}-\frac{\mu_4\left[x+D(\mu_4+1)\right]}{R_4^3} \tag{3.100}$$

$$\frac{d^2y}{dt^2}+2\omega\frac{dx}{dt}+x\frac{d\omega}{dt}-\omega^2y+\left(D\Omega^2-\frac{d^2D}{dt^2}\right)\sin\theta+\left(2\Omega\frac{dD}{dt}+D\frac{d\Omega}{dt}\right)\cos\theta$$
$$=-\frac{(1-\mu)y}{r_1^3}-\frac{\mu y}{r_2^3}-\frac{\mu_4(D\sin\theta-y)}{R_4^3} \tag{3.101}$$

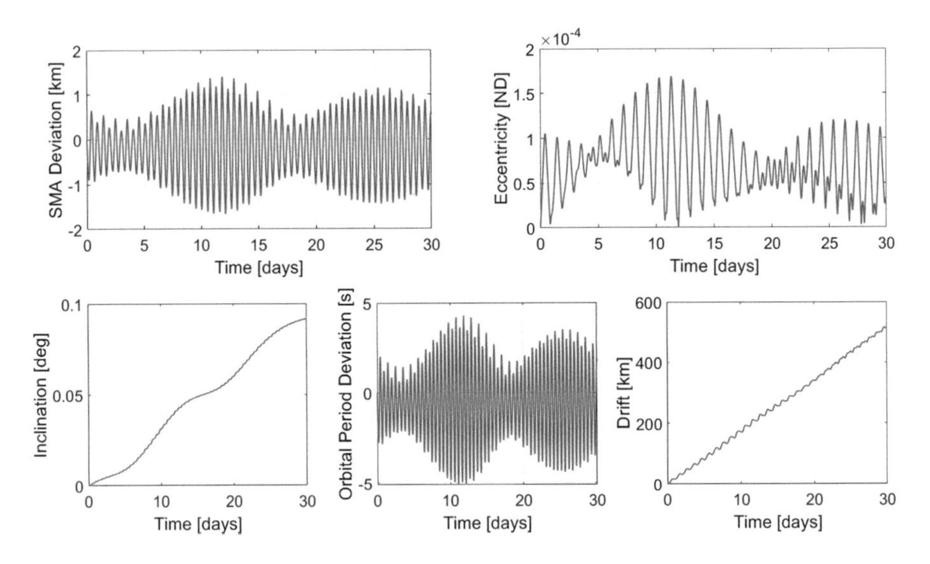

FIGURE 3.14 Deviation of a GEO satellite from its nominal orbit.

$$\frac{d^2z}{dt^2} = -\frac{(1-\mu)z}{r_1^3} - \frac{\mu z}{r_2^3} - \frac{\mu_4 z}{R_4^3} \tag{3.102}$$

Generally, N-body systems with a higher number of bodies are analyzed using approximations, such as assuming that one or more of said bodies are significantly smaller than the others and/or assume that they are perturbations. Alternatively, the "brute force method" to approach N-body systems with $N > 4$ is to numerically integrate Equation (3.1), although this does not give much mathematical insight on the problem.

For example, let's consider a spacecraft in a geostationary orbit, i.e., a spacecraft that has orbital period equal to the rotation of the Earth and orbits the Earth at an altitude of approximately 35,786 km at a 0-degree inclination. If we only consider the point-mass gravitational force of the Earth acting on the spacecraft, the orbital elements would stay constant, as we know from Chapter 2. However, if we introduce the gravitational forces of the Moon and the Sun as point-masses – thus making this into a four-body problem – and integrate the equations of motion, i.e., Equation (3.1), we see that the orbital elements change in time. Because of this, the spacecraft drifts away from its nominal GEO location, as shown in Figure 3.14. Here, we computed the relative positions of the Earth, Moon, and Sun with respect to the spacecraft using the JPL Horizons ephemeris model discussed in Section 4.8, using January 1, 2024 as the starting date and propagating the spacecraft orbit over a period of 30 days using Equation (3.1).

In this example, notice how the semimajor axis (SMA) (and thus orbital period) and eccentricity have a somewhat periodic pattern. However, the spacecraft drifts from its nominal location because the inclination grows over time. This is a typical example of a spacecraft orbit that, when analyzed using the two-body dynamics appears to have a bounded motion. However, when introducing the gravitational disturbances of the Moon and Sun, it is clear that the spacecraft requires stationkeeping in order to avoid drifting from its nominal assigned location in GEO.

In the next section, we cover periodic orbits in the CR3BP that have been extensively used in past missions and that have been proposed for future missions. While most of these orbits still require spacecraft to have active control systems for stationkeeping maneuvers, they exhibit particular characteristics that are useful for missions that require certain lighting conditions, line-of-sight, temperature, distance to certain targets, among other considerations.

3.6 ORBITS IN THE RESTRICTED THREE-BODY PROBLEM

Among the dynamical systems we discussed in previous sections, the most commonly studied problem in astrodynamics is the restricted three-body problem. However, these studies have relatively recently been advanced due to the progress in high-speed computers. With these modern tools, along with continuing analytical work, more thorough, in-depth investigations of the problem are possible. One solution of the restricted three-body problem of particular interest is that of periodic orbits. In the following sections, we will introduce some of the two-dimensional and three-dimensional orbits in the CR3BP that have found applications in past, current, and upcoming space missions. A summary of "useful" orbits in the CR3BP for the Earth-Moon system was compiled by Holzinger et al. [2021]. Maneuvering between CR3BP orbits (or, in general, N-body orbits) can be very challenging due to the chaotic nature of these dynamical systems. Performing these orbital transfers, including station—keeping maneuvers and ballistic maneuvers (as explained in more detail in Section 3.7) is often called astrobatics.[3]

3.6.1 Lyapunov Orbits

A well-researched family of periodic orbits in the vicinity of the Lagrange Points are Lyapunov orbits. Samples of these orbits are shown in Figure 3.15. In red is a sample Lyapunov orbit centered around the Earth-Moon Lagrange Point 1 (EML_1) while in blue is a sample EML_2 orbit. Green arrows point toward the direction of motion. Lyapunov orbits are unstable periodic orbits that have been a fundamental basis for analyzing transfer and capture maneuvers in astrodynamics.

Lyapunov orbits are usually categorized into planar Lyapunov orbits (Figure 3.15) and vertical Lyapunov orbits (Figure 3.16).

The symmetries in Equation (3.25) indicate the existence of planar periodic orbits symmetric with respect to the x-axis, such as Lyapunov orbits, intersect the x-axis perpendicularly. Mathematically, planar Lyapunov orbits and distant retrograde orbits (DROs; we'll discuss them in the next section) have a tangential intersection point at $y = \dot{x} = 0$. Additionally, Lyapunov orbits are prograde near the smaller primary facing the larger primary, but are retrograde on the opposite side [Lam and Whiffen, 2005]. This means that Lyapunov orbits about L_1, L_2, and L_3 cannot have patching points of parallel velocities with the same direction at $y = \dot{x} = 0$ while Lyapunov

[3] Astrobatics: as.tro.bat.ics\as-troe-'ba-tiks\ noun

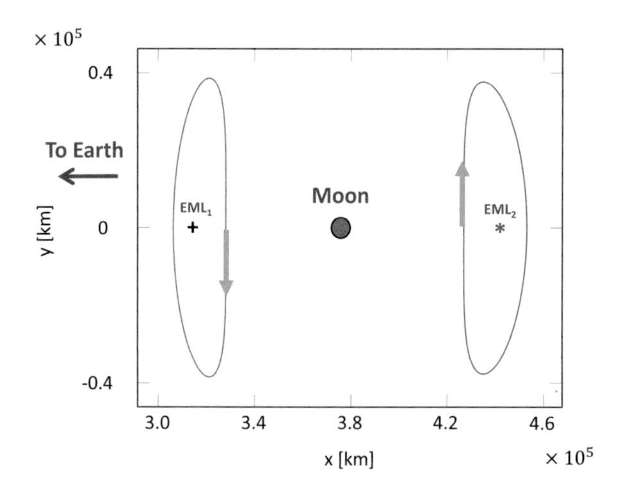

FIGURE 3.15 Sample planar Lyapunov orbits in the Earth-Moon system.

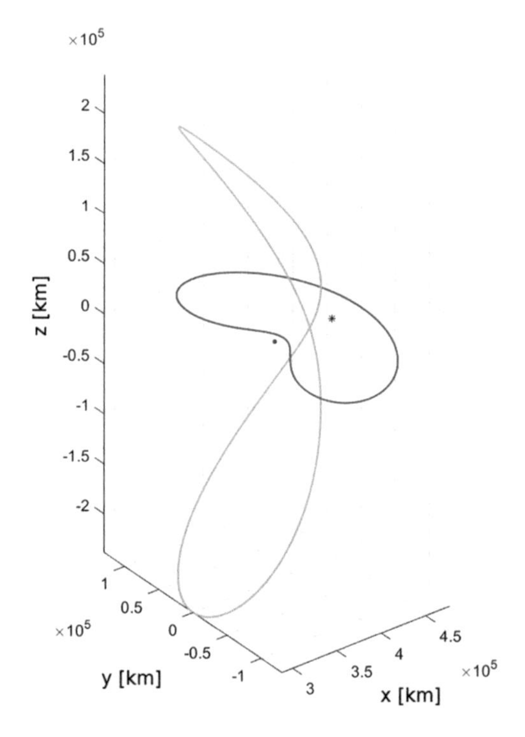

FIGURE 3.16 Vertical (yellow) and planar (blue) Lyapunov orbits generated using the CR3BP model in the Earth-Moon system. The asterisk represents the Earth-Moon L_2 point while the gray sphere represents the Moon (to scale).

orbits about L_4 and L_5 do, as shown in Figure 3.17. In the figure, the Lagrange points are plotted as asterisks while ten sample Lyapunov orbits are plotted about each Lagrange point: blue for L_1 Lyapunov orbits, red for L_2 Lyapunov orbits, green for L_3 Lyapunov orbits, purple for L_4 Lyapunov orbits, orange for L_5 Lyapunov orbits.

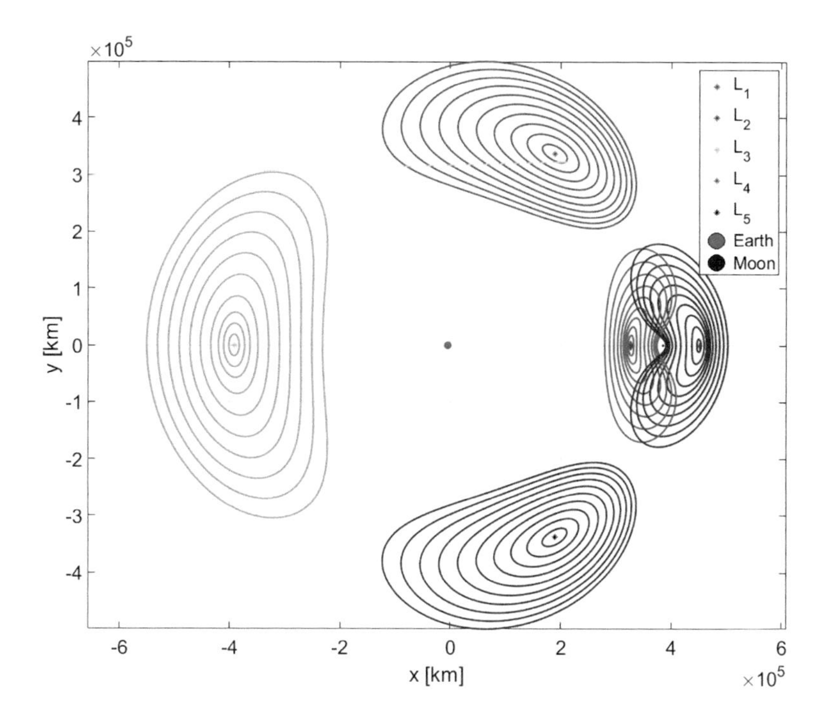

FIGURE 3.17 Planar Lyapunov orbits in the Earth-Moon rotating frame.

Further studies and details on the general background of such orbits and stability of planar and vertical periodic Lyapunov orbits with respect to perturbations perpendicular to the plane can be found in the literature [Hénon, 1973]. A methodology to find accurate initial conditions for such orbits is outlined in Kim and Hall [2002].

3.6.2 Distant Retrograde Orbits

A Distant Retrograde Orbit (DRO) is an orbit that is close to the sphere of influence of the smaller primary and has a retrograde motion when observed from the prospective of the rotating frame. The amplitude of a DRO is defined as the positive x-component of position measured from the second primary's center, known as A_x, where the y- and z-components of position are zero. Such orbits are of particular interest because of their long-term stability [Bezrouk and Parker, 2014; Conte and Spencer, 2018; Hirani and Russell, 2006]. Examples of Earth-Moon DROs are shown in Figure 3.18.

DROs were also considered by NASA for the Asteroid Redirect Mission (ARM) and also for future deep space exploration, starting with the Mars-Phobos and Mars-Deimos systems [Condon and Williams, 2014; Conte and Spencer, 2018]. In fact, a DRO could be used as a parking orbit for a refueling station that could be used as an intermediate staging point for deep space missions, especially for those missions requiring large spacecraft, such as human missions to Mars. The station would be kept operational by bringing resources such as propellant and water that can be harvested from asteroids or from the Moon by performing In-Situ Resource Utilization (ISRU) [Ho et al., 2014; Conte et al., 2017].

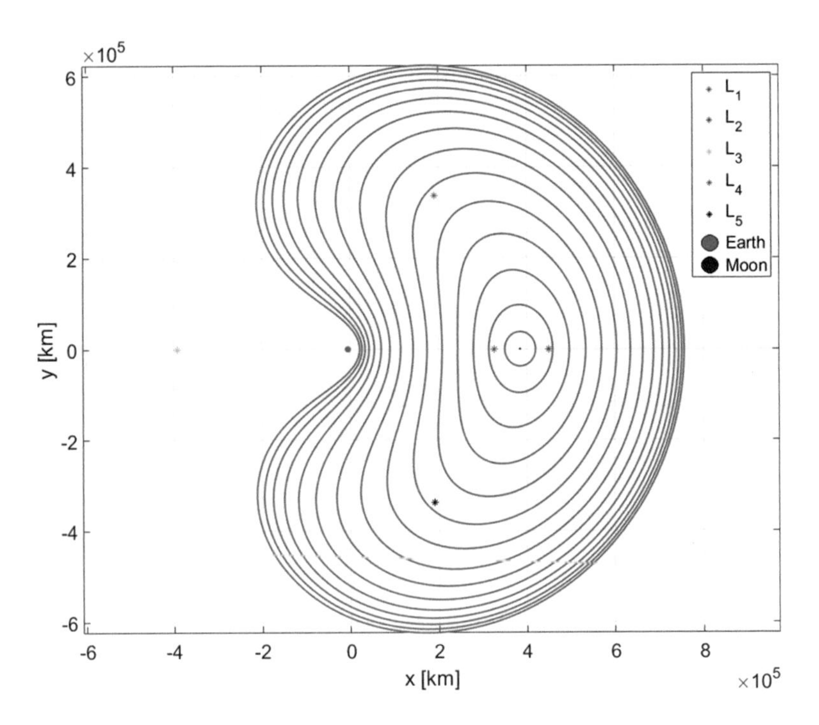

FIGURE 3.18 Sample Earth-Moon DROs.

The lunar DRO that was proposed for NASA's ARM was designed with $A_x = 61{,}500$ km, but the mission ended up being canceled. This orbit, shown in Figure 3.19, was chosen due to its relative vicinity to the Moon and Earth and its long-term stability [Condon and Williams, 2014]. In the figure, a short time-of-flight direct transfer from LEO to the DRO proposed by ARM is also shown. It should be noted that while DROs appear to have a retrograde motion when viewed in the rotating reference frame, they have a prograde motion when seen from an inertial reference frame.

DROs are of interest also for other planetary systems, such as the Mars-Phobos/Deimos system, Jupiter-Europa, among others. Sample Mars-Phobos DROs are shown in Figure 3.20. Additionally, Figure 3.21 shows a sample Mars-Phobos DRO with $A_x = 300$ km in both the Phobos-centered rotating reference frame and Mars-inertial reference frame. In the figure, the red arrows indicate the direction of motion: clockwise (retrograde) on the left and counterclockwise (prograde) on the right. In all both figures, Phobos is depicted as an ellipsoid of size $13.4 \times 11.2 \times 9.2$ km (to scale), while L_1 and L_2 correspond to the Lagrange points 1 and 2 of the Mars-Phobos system.

Most DROs are neutrally stable orbits in the CR3BP. For example, Mars-Phobos DROs with A_x values larger than 20 km are stable for rather long periods of times (over 60 days) when full-force models are utilized for orbit propagation and no control is applied to the spacecraft. On the other hand, smaller DROs require quasi-periodic stationkeeping maneuvers every 1–2 hours in the order of

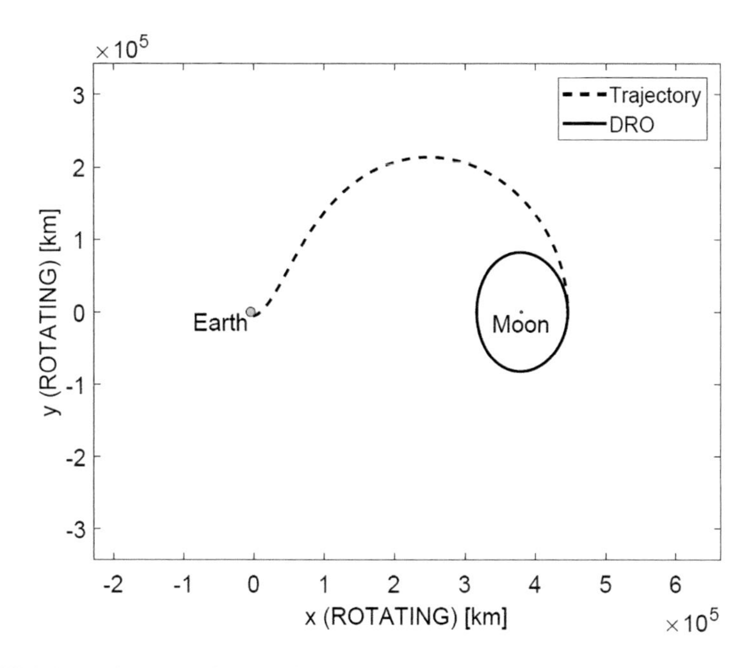

FIGURE 3.19 Direct transfer orbit from LEO to the lunar DRO proposed by ARM [Conte et al., 2017].

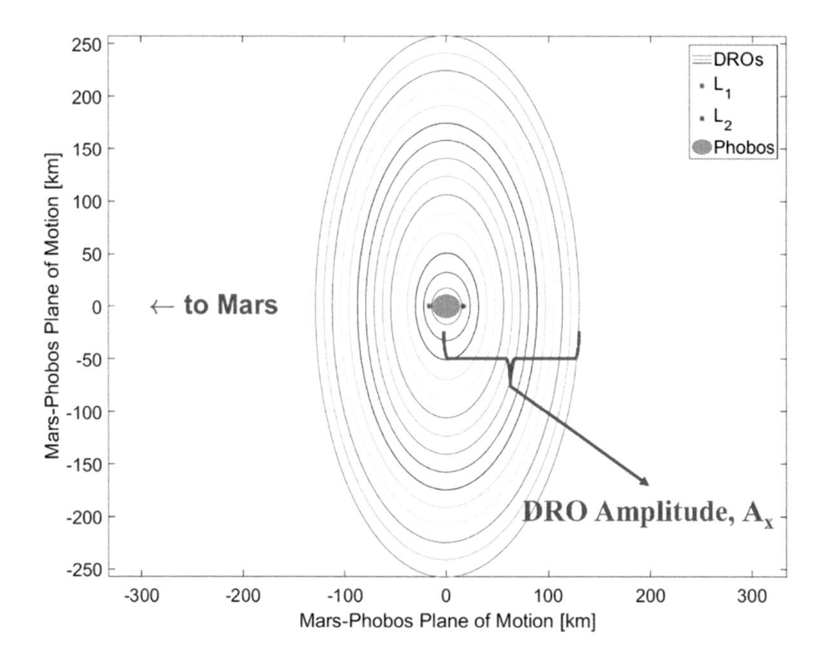

FIGURE 3.20 Sample Mars-Phobos DROs in the Phobos-centered rotating reference frame [Conte and Spencer, 2018].

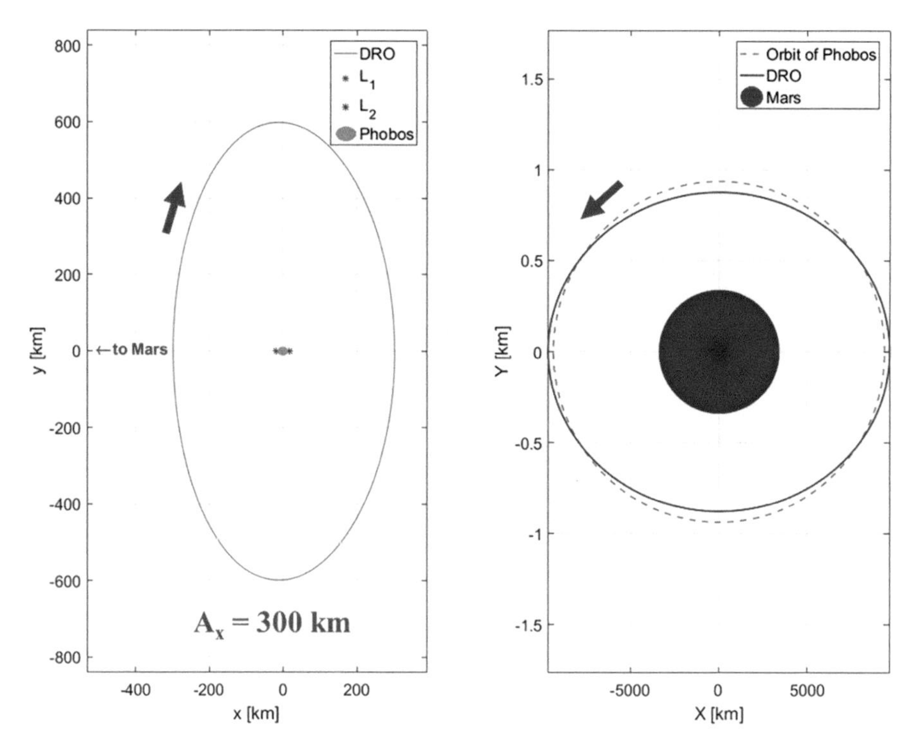

FIGURE 3.21 A sample Mars-Phobos DRO with Ax = 300 km in the xyz Phobos-centered rotating reference frame (a) and XY Z Mars-centered inertial reference frame (b) [Conte and Spencer, 2018].

cm/s to stay on course and avoid crashing onto Phobos' surface [Wallace et al., 2012]. Sample orbit insertion maneuvers required for interplanetary missions to enter such stable Mars-Phobos DROs can be found in the literature [Conte et al., 2017, 2019; Conte and Spencer, 2018]. In other systems where the mass ratio is very small (refer to Table 3.1), similar neutrally stable DROs can be found. These orbits are particularly of interest for exploration missions of, for example, the moons of Jupiter and Saturn, since they allow for long-term stability thus extending the lifetime of such missions. Furthermore, a so-called Δv roadmap is shown in Figure 3.22, where a cislunar DRO is used as the "central hub" between various Earth and lunar orbits of interest, along with other Solar System destinations. The numerical values represent the change in velocity, or Δv, needed to depart and arrive at the desired orbit, assuming a Δv-optimal two-burn maneuver (more on this aspect of orbital transfers is discussed in Chapter 5), unless otherwise specified. "Direct" indicates Δv-optimal two-burn maneuver, but that time of flight is generally very long and thus alternative transfers are usually desirable. This is true especially for far-away destinations such as the gas giants (Jupiter, Saturn, Uranus, and Neptune) and other distant objects. Table 3.4 explains the acronyms used in the figure.

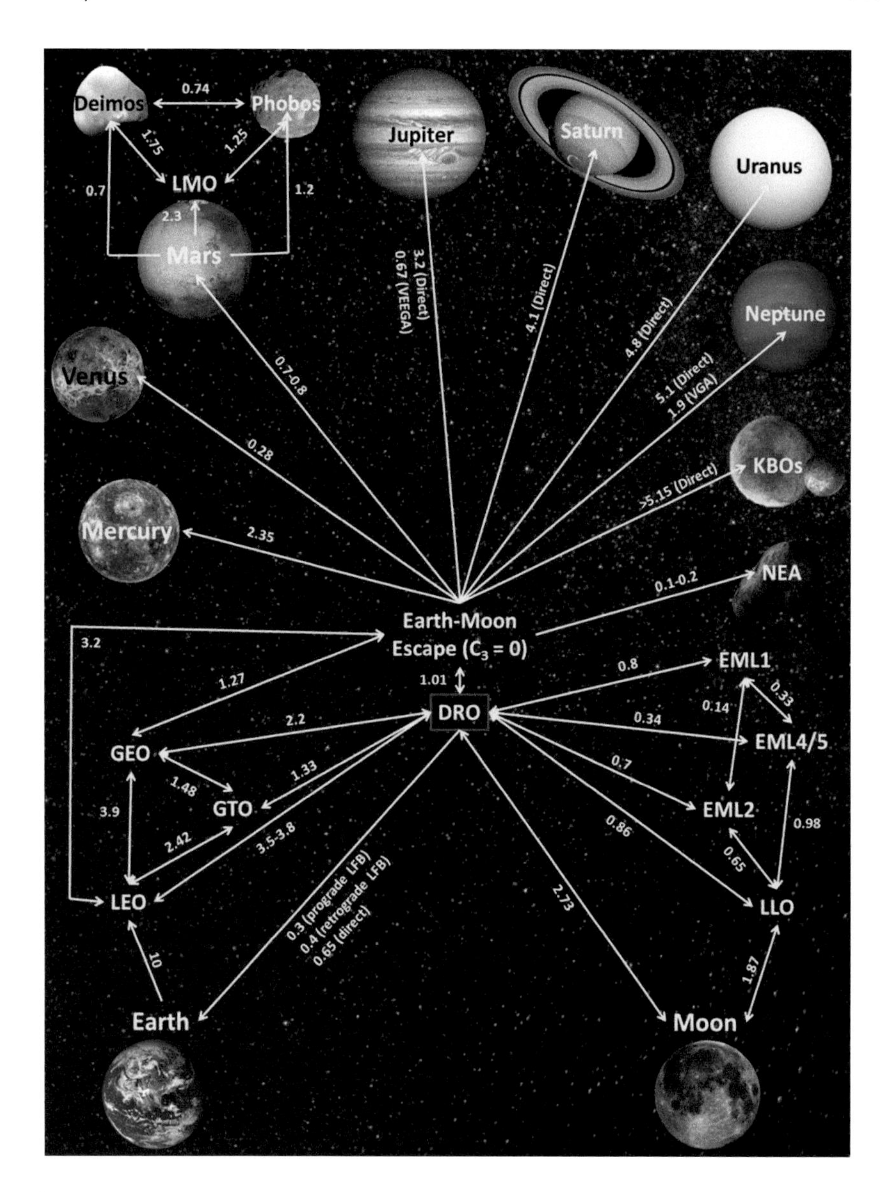

FIGURE 3.22 A Δv roadmap showing the central role of an Earth-Moon DRO with respect to cislunar orbits of interest and the rest of the Solar System [Conte, 2019].

3.6.3 Lissajous Orbits

Analytical approaches have shown that linearization of the equations of motion about any of the collinear libration points include a periodic path in the plane of motion of the primaries and uncoupled out-of-plane motion. The two frequencies of these motions are generally unequal, and for small amplitudes, the paths traced out form Lissajous curves, thus the orbits are known as Lissajous orbits [Farquhar and Kamel, 1973; Richardson and Cary, 1975; Howell and Pernicka, 1988].

TABLE 3.4
List of acronyms used in Figure 3.22

Acronym	Meaning
LEO[a]	Low Earth orbit
GTO[a]	Geosynchronous transfer orbit
GEO[a]	Geosynchronous Earth orbit
EMLX	Earth-Moon Lagrange point number X
LLO[a]	Low lunar orbit
LMO[a]	Low Martian orbit
KBOs	Kuiper Belt objects, i.e., trans-Neptunian objects
NEA	Near-Earth asteroids

[a] These orbits are discussed in detail in Section 5.1.1.

To find the characteristics of these Lissajous trajectories, this problem can be recast by looking at the deviation from a known reference point to the libration point. Let's define three positive deviations, ε_1, ε_2, and ε_3, as seen in Figure 3.23, for each of the collinear points.

The locations of the three collinear points are

$$
\begin{aligned}
&L_1 : x_1 = \mu - 1 + \varepsilon_1, y_1 = z_1 = 0 \\
&L_2 : x_2 = \mu - 1 - \varepsilon_2, y_2 = z_2 = 0 \\
&L_3 : x_3 = \mu + 1 - \varepsilon_3, y_3 = z_3 = 0
\end{aligned}
\tag{3.103}
$$

Inserting the coordinates for the collinear libration points from Equation (3.103) into the equations of motion, Equation (3.25), a set of second-order differential equations can be found as

$$
\begin{aligned}
&\frac{d^2x}{dt^2} - 2\frac{dy}{dt} - (1+2c)x = 0 \\
&\frac{d^2y}{dt^2} + 2\frac{dx}{dt} + (c+1)y = 0 \\
&\frac{d^2z}{dt^2} + cz = 0
\end{aligned}
\tag{3.104}
$$

where

$$
\begin{aligned}
&L_1 : c = \frac{1-\mu}{(1-\varepsilon_1)^3} + \frac{\mu}{\varepsilon_1^3} \\
&L_2 : c = \frac{1-\mu}{(1-\varepsilon_2)^3} + \frac{\mu}{\varepsilon_2^3} \\
&L_3 : c = \frac{1-\mu}{(1+\mu-\varepsilon_3)^3} + \frac{\mu}{(2-\varepsilon_3)^2}
\end{aligned}
$$

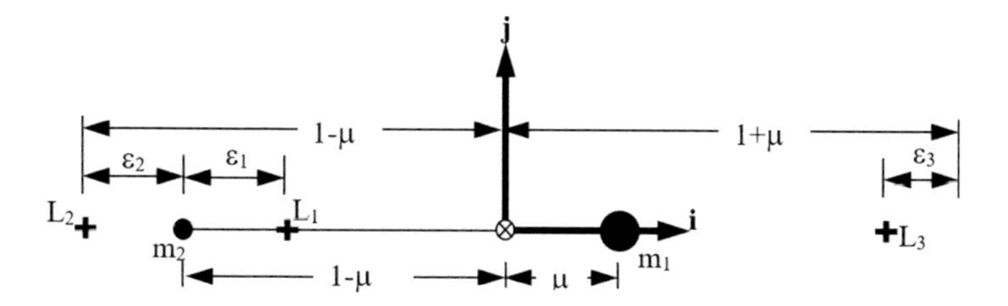

FIGURE 3.23 Locations of deviations from reference points.

The first two equations in Equation (3.104) are coupled, while the third equation (the motion in the z-direction) is uncoupled from the first two. The characteristic equations for the x–y motion and z-motion are then

$$\begin{aligned} &\lambda_{xy}^4 - (c-2)\lambda_{xy}^2 - (2c+1)(c-1) = 0 \\ &\lambda_z^2 + c = 0 \end{aligned} \tag{3.105}$$

The roots of the first equation consist of a pair of equal real roots in magnitude but opposite in sign, and a complex conjugate pair of purely imaginary roots. The characteristic frequencies of x–y motion and z-motion are

$$\begin{aligned} \lambda_{xy} &= \pm u, \pm iv \\ \lambda_z &= \pm iw \end{aligned} \tag{3.106}$$

where $i = \sqrt{-1}$ and u, v, w are positive real values.

The linearized system in Equation (3.104) has the solution,

$$\begin{aligned} x &= \alpha_1 e^{ut} + \alpha_2 e^{-ut} + \beta_1 \cos vt + \beta_2 \sin vt \\ x &= \alpha_3 e^{ut} + \alpha_4 e^{-ut} + \beta_3 \cos vt + \beta_4 \sin vt \\ z &= \gamma_1 \cos wt + \gamma_2 \sin wt \end{aligned} \tag{3.107}$$

where the coefficients in the x- and y-equations are related by a recursion relationship,

$$\begin{aligned} \alpha_3 &= -\frac{\alpha_1}{2u}\left(1 + 2c - u^2\right) \\ \alpha_4 &= \frac{\alpha_2}{2u}\left(1 + 2c - u^2\right) \\ \beta_3 &= \frac{\beta_2}{2v}\left(1 + 2c + v^2\right) \\ \beta_4 &= -\frac{\beta_1}{2v}\left(1 + 2c + v^2\right) \end{aligned} \tag{3.108}$$

With appropriate choices of the initial values ($\alpha_1 = \alpha_2 = 0$), the remaining constants (α_1, α_2, β_1, β_2, γ_1, and γ_2) can be computed using Equation (3.108), and the linearized system produces a bounded solution,

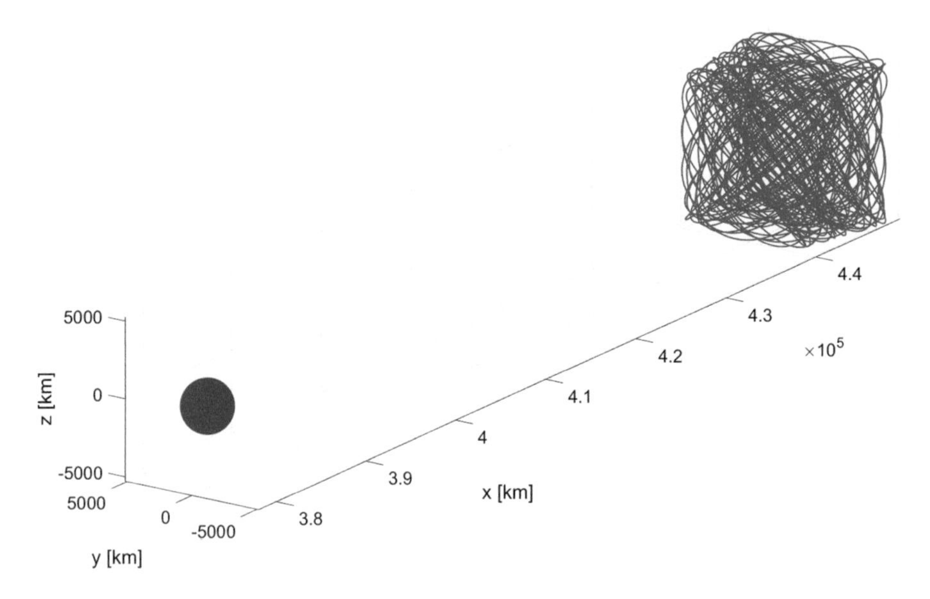

FIGURE 3.24 Example of a Lissajous orbit around the Earth-Moon L_2. The gray sphere represents the Moon (to scale).

$$\begin{aligned} x &= \beta_1 \cos vt + \beta_2 \sin vt = -A_x \cos(vt + \psi) \\ x &= \beta_3 \cos vt + \beta_4 \sin vt = kA_x \sin(vt + \psi) \\ z &= \gamma_1 \cos wt + \gamma_2 \sin wt = A_z \sin(wt + \varphi) \end{aligned} \tag{3.109}$$

where the amplitudes and phase angles are related to the initial conditions, and

$$k = \frac{1 + 2c + v^2}{2v} \tag{3.110}$$

When the two frequencies, v and w, are equal, a halo orbit results,[4] as discussed in the following section. When the two frequencies are not equal (which is the more likely case), a Lissajous orbit results, and the trajectories don't repeat orbit after orbit, but do remain bounded. Figure 3.24 shows an example of a Lissajous orbit in the Earth-Moon system around L_2 which was generated using initial conditions obtained with the equations presented in this section and integrated numerically accounting for the Sun's perturbation and other planetary bodies (obtained through JPL Horizons). Figure 3.25 shows a three-view of the same example orbit, where it can be seen that the motion of the spacecraft remains approximately bounded inside a $5{,}000 \times 5{,}000 \times 5{,}000$ km box centered around L_2. These distances are often referred to as the x-, y-, and z-amplitudes, or A_x, A_y, and A_z, respectively. The motion of a spacecraft in such orbit would not remain bounded indefinitely due to solar and

[4] A halo orbit is computed using the full non-linear CR3BP equations of motion, but the results obtained from setting v = w can be used as initial guesses when searching for initial conditions of halo orbits using differential correction, as discussed in Section 3.7.1.

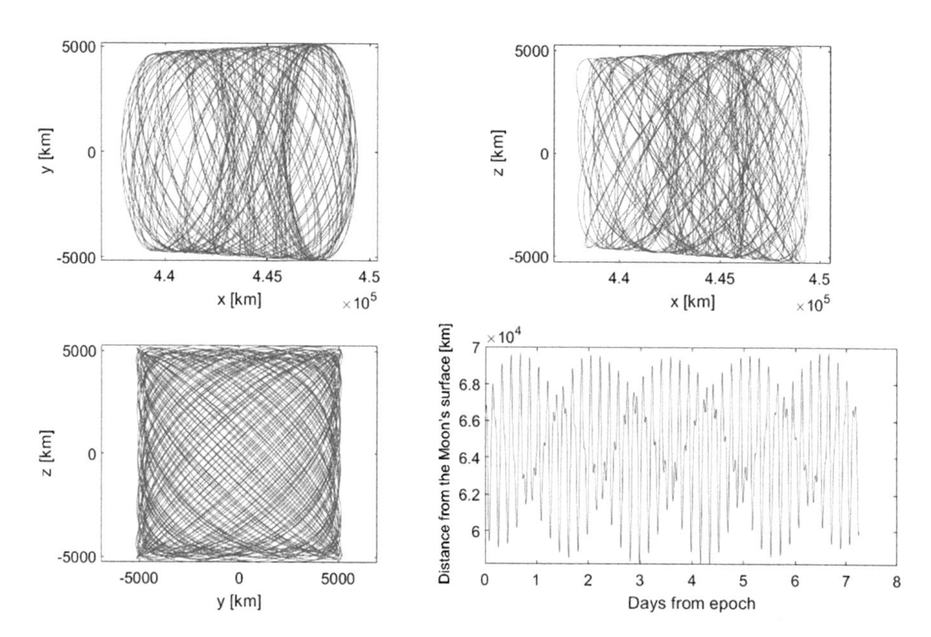

FIGURE 3.25 Three-view of a Lissajous orbit around the Earth-Moon L_2.

planetary perturbations, thus requiring stationkeeping maneuvers over time. For reference, the distance from the lunar surface is also plotted. In fact, due to their bounded motion, Lissajous orbits around the Earth-Moon L_2 could be used for communications spacecraft with the far side of the Moon.

The first mission to successfully make use of a Lissajous orbit was the joint ESA/NASA mission Solar and Heliospheric Observatory (SOHO), which launched in December 1995 and targeted a Lissajous orbit around the Sun-Earth/Moon[5] L_1 [NASA-SOHO, 2019]. Another more recent mission using Lissajous orbits was Gaia, a mission launched in December 2013 by ESA with the purpose of charting a three-dimensional map of the Milky Way, in the process revealing the composition, formation, and evolution of the Galaxy [ESA, 2018]. Figure 3.26 shows a plot of Gaia's nominal science orbit around the Sun-Earth/Moon L_2 as seen from an Earth-centered inertial reference frame. The orbital period of Gaia is about 180 days and the size of the orbit is $A_x \times A_y \times A_z = 263{,}000 \times 707{,}000 \times 370{,}000$ km measured relative to Sun-Earth/Moon L_2. Gaia is expected to be operational until 2025 [ESA, 2018].

3.6.4 Halo Orbits

The term halo orbit was first coined by Farquhar [1973]. In his work, Farquhar used the Lindstedt-Poincaré method to determine quasi-periodic orbits, commonly referred to as Lissajous orbits, in the Earth-Moon system. These orbits are three-dimensional,

[5] We use "Sun-Earth/Moon" to indicate that these Lagrange points correspond to a three-body system in which the first primary is the Sun, and the other primary is the combination of Earth and the Moon considered as a single point mass located at the barycenter of the Earth and Moon system.

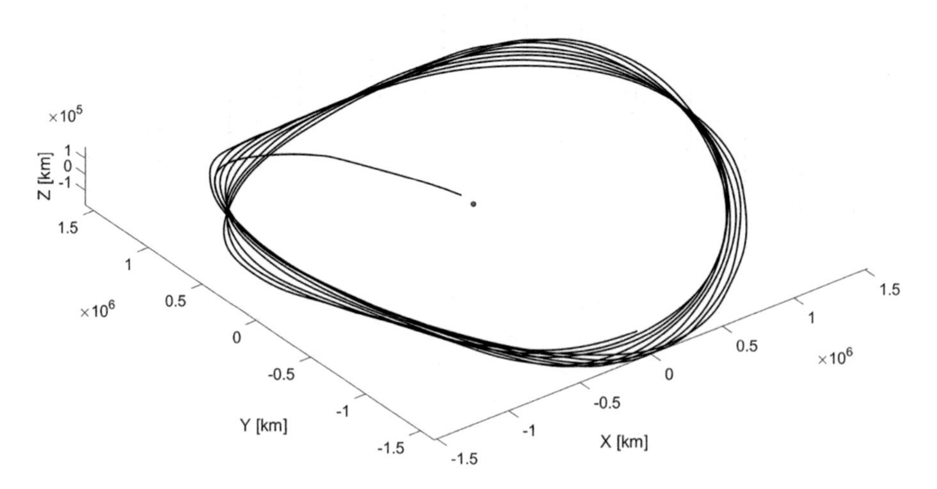

FIGURE 3.26 Gaia orbital plot. The blue dot at the origin represents the Earth. (Orbital data obtained from JPL Horizons.)

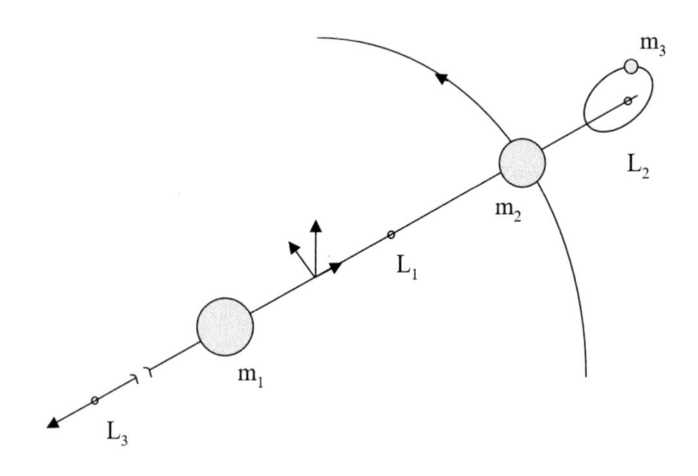

FIGURE 3.27 Schematic of Halo orbit about L_2 libration point.

an example of which is shown in Figure 3.27 in the vicinity of the L_2 libration point. Another view is seen in Figure 3.28. This projection on the y–z plane shows the trajectory as viewed from the larger mass body in the system. This orbit appears to form a "halo" about the smaller body, although it is not in orbit around the smaller body, nor it is restricted to motion in two dimensions (halo orbits look somewhat like hyperbolic paraboloids). It is known that finite, planar periodic orbits exist about the libration points, but halo orbits are of interest because of their three-dimensional characteristics. In fact, unlike Lissajous orbits around L_2, which generally do not provide continuous direct line-of-sight with the far side of the smaller body and the larger primary at the same time, halo orbits that are large enough can provide such non-stop communication link.

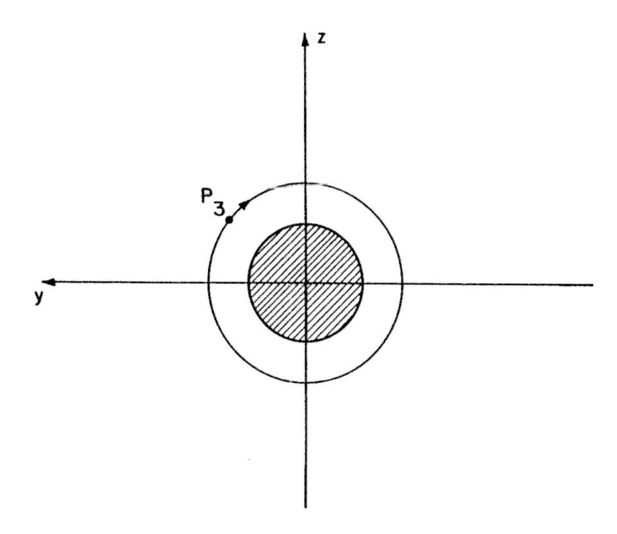

FIGURE 3.28 View of Halo orbit from larger primary – the orbit's name comes from the appearance of a "Halo" about the smaller primary.

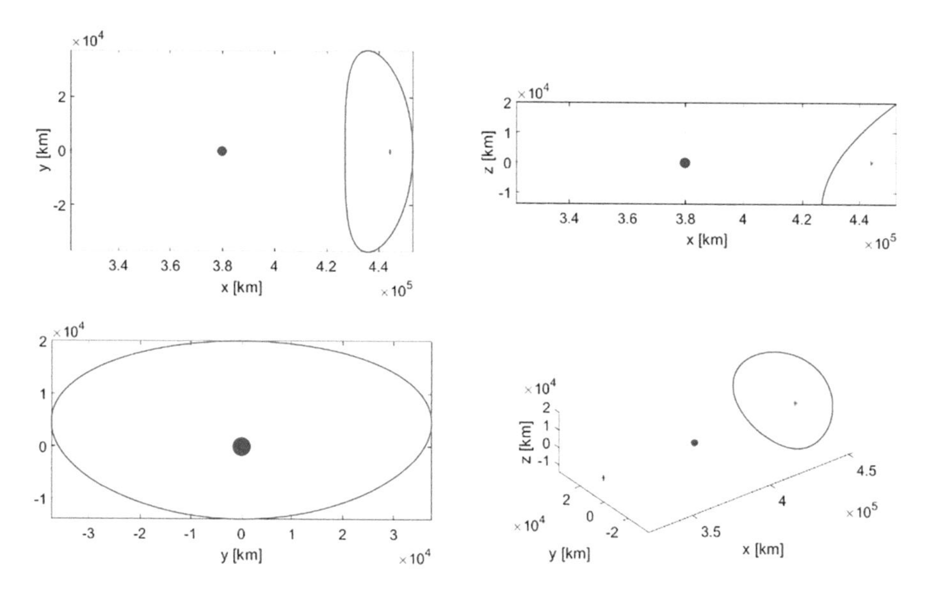

FIGURE 3.29 Example of an L_2 Earth-Moon halo Orbit with $A_z = 20{,}000$ km.

An example of Earth-Moon L_2 halo orbit is shown in Figure 3.29, where a three-view is also shown. The asterisk in red represents the Earth-Moon L_2 location. Usually, halo orbits are characterized by their maximum z-component of position achieved in an orbital period, which is referred to as A_z. For example, in Figure 3.29, the halo shown has an A_z of 20,000 km.

A wide variety of halo orbits exist and are usually categorized based on the Lagrange point (L_1, L_2, or L_3) of origin and their z-orientation (Northern or

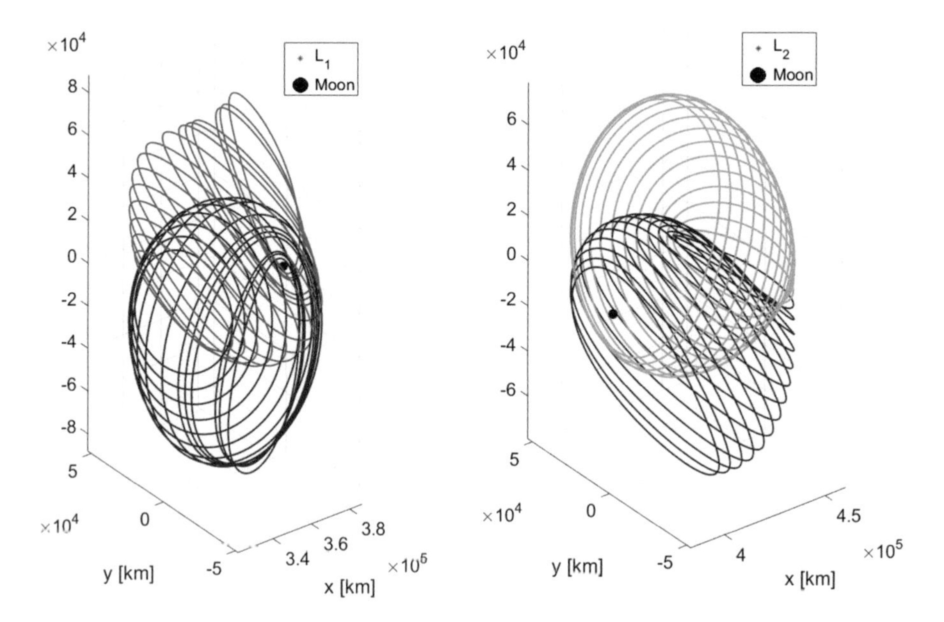

FIGURE 3.30 Examples of L_1 (a) and L_2 (b) Earth-Moon halo orbits.

Southern). For example, halo orbits originating from Earth-Moon L_1 and L_2 are plotted in Figure 3.30. In the figure, L_1 halos are on the left: L_1 Northern halos are in blue and L_1 Southern halos are in red; L_2 halos are on the right: L_2 Northern halos are in green and L_2 Southern halos are in purple.

As they get closer to the smaller mass, halo orbits eventually become elongated and more rectilinear. These last kinds of halo orbits are usually referred to as near-rectilinear halo orbits (NRHO). Examples of Earth-Moon L_2 NRHOs are shown in Figure 3.31, where one can see how close they get to the surface of the Moon.

A mission that will utilize an NRHO is the Lunar Gateway, or simply Gateway, whose diagram is shown in Figure 3.32. Gateway will be a relatively small space station intended to serve as a solar-powered communication hub, science laboratory, short-term habitation module for astronauts. Thanks to its vicinity to the Moon, Gateway can also serve as a staging location for robotic lunar surface assets that are intended to be used for lunar exploration. The Lunar Gateway was envisioned as a multinational collaborative project similar to the International Space Station, except near the Moon. Currently, partner agencies involved in the development of Gateway are NASA, ESA, JAXA, and the Canadian Space Agency (CSA). Other commercial partners are also planned to be involved to facilitate the assembly of the station, the delivery of robotic assets, and astronauts. If successful, Gateway will be both the first space station beyond low Earth orbit and the first space station to orbit close to the Moon. NASA's Cislunar Autonomous Positioning System Technology Operations and Navigation Experiment (CAPSTONE)[6] mission was launched in

[6] https://www.nasa.gov/directorates/spacetech/small_spacecraft/capstone

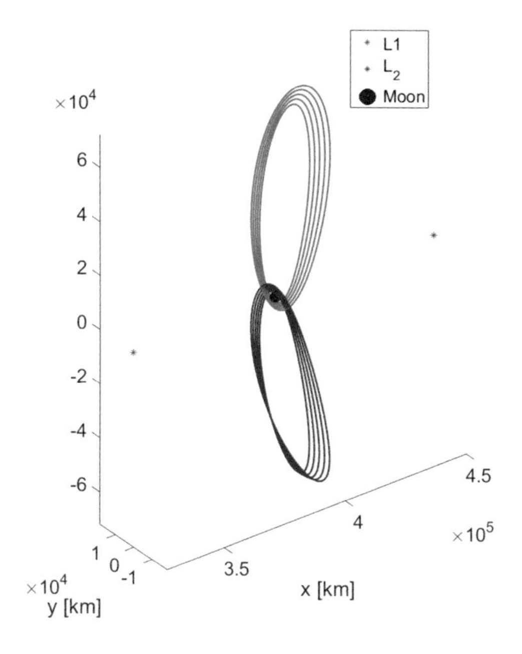

FIGURE 3.31 Earth-Moon L_2 near-rectilinear halo orbits (NRHOs).

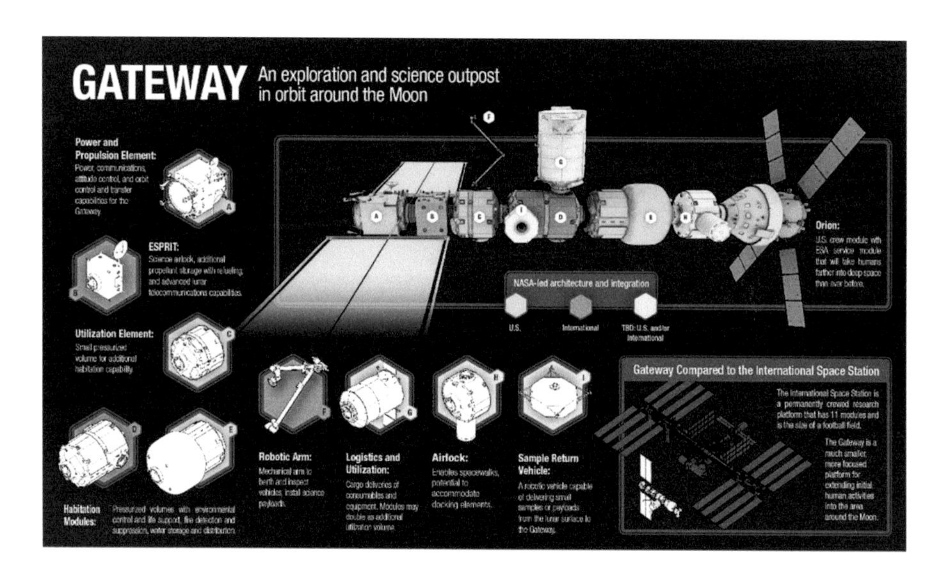

FIGURE 3.32 Diagram of the Lunar Gateway. (Courtesy of NASA.)

June 2022 to test and verify the orbital stability for the Lunar Gateway's planned NRHO, which has periselene and aposelene altitudes of approximately 1,500 and 70,000 km, respectively. CAPSTONE was a 12-unit CubeSat whose mission objectives also included testing navigation systems to be used in tandem with NASA's

Lunar Reconnaissance Orbiter (LRO), which would function as a communications relay with Earth's ground stations.

Figure 3.33 shows a plot of the NRHO chosen for the Lunar Gateway. Notice how the small periselene altitude (1,500 km) gives Gateway an advantageous location for close lunar observations and the ability for assets docked to Gateway to descend to the lunar surface. However, this is also the location in the orbit where speed is the highest, so Δv maneuvers to descend to the lunar surface would likely be initiated around aposelene, at an altitude of approximately 70,000 km. Multiple potential orbits that were analyzed for a trade study aimed at finding the optimal orbit for the Lunar Gateway, which was eventually decided to be an NRHO, as already discussed [Whitley and Martinez, 2016]. A tabular summary of the orbits considered for this trade study is given in Table 3.5.

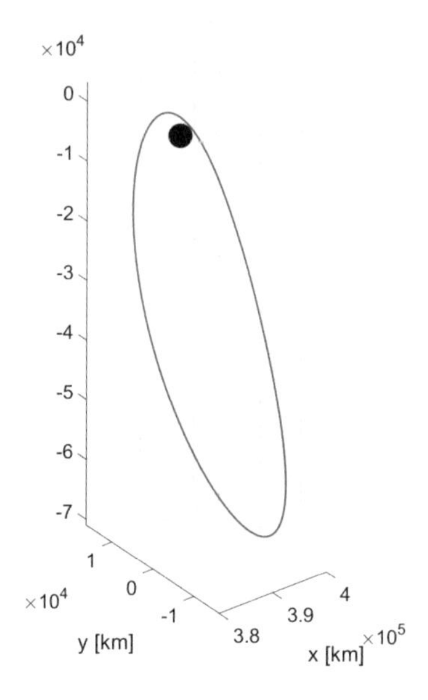

FIGURE 3.33 Lunar Gateway's NRHO.

TABLE 3.5
Characteristics of the periodic cislunar orbits considered for the Lunar Gateway [Whitley and Martinez, 2016]

Orbit type	Orbital characteristic	Value
Distant retrograde orbit (DRO)	A_x	70,000 km
Near-rectilinear halo orbit (NRHO)	Perilune	1,500 km
Prograde circular orbit (PCO)	Perilune	4,000 km
High elliptical lunar orbit (HELO)	Altitude	100 × 10,000 km
Frozen orbit	Inclination	40°
L_2 halo orbit	A_z	60,000 km

3.7 DYNAMICAL SYSTEMS THEORY

Dynamical systems theory (often simply referred to as "DST") was first introduced by the 19th century French mathematician Henri Poincaré [1892–1897], and today is commonly known at the geometric theory of differential equations, or chaos theory. In mathematics, a dynamical system – such as N-body systems – is called chaotic if the system is highly sensitive to the initial conditions. Many apparently simple systems can prove to be chaotic. One of such examples is a double pendulum. In the field of astrodynamics, N-body systems are highly chaotic. In the CR3BP, using slightly altered initial conditions for one of the periodic orbits described in Section 3.6 would result in a drastically different orbit which is unlikely to be periodic. Conversely, in the two-body problem, varying the initial conditions slightly results in a slightly different orbit, whose orbital elements are likely to be numerically close to those of the original orbit. The study of chaotic systems generally requires significant computing power. Before the beginning of the computer age, this was a topic embedded in theoretical mathematics. However, with the development of high-performance computers, this subject was introduced into several practical applications in many fields. The application that is appropriate for this book is the field of complex space mission design.

Many of the tools and techniques developed in the space age have involved analytical methods that allow a mission designer to piece together phases of a mission. The most common examples of this include the method of patched conics and Lambert's problem, both introduced in Chapter 5. These examples take advantage of the analytical nature of these techniques, primarily for the relatively simple two-body problem. However, dynamical systems theory provides an umbrella of techniques that are applicable to many of the complex options available to contemporary space missions in more complex dynamics in highly non-linear, N-body systems.

Dynamical systems theory is capable of inferring the "rules" of a particle that is chaotically moving through space in an N-body system. These "rules" are invariant phase space structures that allow a given particle's orbit to propagate through space and time. Additionally, dynamical systems theory gives a systematic way to discover and classify sets, or families, of orbits that can be used in space missions as transfer orbits and periodic parking orbits.

Most of the methodologies and approaches to orbits in this textbook thus far were developed primarily in the 18th and 19th centuries, or earlier. However, dynamical systems theory represents a more modern approach to orbit design and selection for space missions. Lo [1997] introduced the concept that older methods are limited in helping mission designers, and that newer theories allow newer and potentially more efficient options when designing the orbits for a mission that involves a spacecraft under the influence of several gravitational sources. One of the major disadvantages of dynamical systems theory is that it generally requires significant computing power, as stated previously. However, this limitation is becoming less of a concern in the modern days, especially considering that we have more computational power in our hands through today's cell phone than NASA had in its entirety during the days of the Apollo missions. Even later missions were planned using the patched conics approximations (or similar methods) including the Voyager missions, Galileo, and Cassini, to name a few.

The first missions that made use of dynamical systems theory for their final orbit selection were missions targeting the Sun-Earth Lagrange points which used halo and Lissajous orbits. Notable missions of this kind include ISEE-3, WIND, SOHO, ACE, and MAP which were inserted into their libration point orbits in 1978, 1995, 1996, 1997, and 2001, respectively [Farquhar et al., 2004]. The first mission whose orbits were completely designed using dynamical systems theory was Genesis, which made use of a Sun-Earth L_1 orbit [Lo et al., 2001].

The idea of using N-body orbits discovered through dynamical systems theory became the central idea to the development of a transfer design framework known as the interplanetary superhighway, sometimes also called interplanetary transport network [Lo et al., 2001; Howell et al., 1997]. This concept takes advantage of dynamical transportation pathways that are formed by the combination of many invariant manifolds – which we will discuss in detail in a later section – in various circular restricted three-body systems. This concept allows for a low-energy transfers throughout the Solar System. An artist's concept of this interplanetary superhighway is shown in Figure 3.34.

In the figure, the collections of curves indicate unstable and stable manifolds. Note that although the projections onto the position space are shown, these manifolds have shapes of distorted, squashed, and folded tubes in the phase space. Furthermore, the inside of these tubes are filled up only with so-called transit orbits, i.e., solution orbits transiting from one region to another through the L_1 or L_2 periodic orbit. Bundles of transit orbits inside these manifold tubes can considerably remain still as bundles and serve as transportation tunnels even with the presence of perturbing forces. In the Solar System, therefore, many celestial bodies, such as planets or natural satellites circulating around the Sun or a planet, trail these transportation tunnels, forming the interplanetary superhighway.

FIGURE 3.34 Artist's concept of the interplanetary superhighway. (Courtesy of NASA/JPL-Caltech.)

In the following sections, we will introduce some of the mathematics and methodologies used to find periodic orbits and discuss their applications in past and future missions, along with discussing the aforementioned "transportation tubes." The full application of these methods is covered in other texts such as Gómez et al. [2001a], Howell et al. [1997], Koon et al. [2001], and Lo et al. [1998].

3.7.1 Differential Correction

Several types of periodic orbits exist in the CR3BP. These differ based on their symmetric varieties, including axi-symmetric, doubly-symmetric, and planar. Generally, periodic orbits in the CR3BP are not easily found unless the appropriate methodology is used. Here, we will discuss an iterative method known as differential correction, or sometimes also called predictor-corrector method.

First, we assume that we have some guess for the initial conditions, $\mathbf{X}_0$, which generally does not lead to a periodic orbit. Our goal is to find a set of initial conditions that, when integrated using the equations of motion for the CR3BP, namely Equation (3.25), for a period of time T, we obtain a final state, $\mathbf{X}_f$, that is identical to that of the initial conditions. In other words, the orbit repeats every period, T, or, mathematically

$$\mathbf{X}_f = \mathbf{X}(T) = \mathbf{X}_0 \tag{3.111}$$

where the subscript "f" denotes the final conditions after the integration. Note that $\mathbf{X}$ is the six-element state comprised of position and velocity components. However, if searching for a planar orbit, the z-components of position and velocity can be eliminated, thus leaving $\mathbf{X}$ as a four-elements component and speeding up the numerical integration process significantly.

Expanding the equations of motion of the CR3BP to the first order about ($\mathbf{X}_0$, T), we get

$$\delta\mathbf{X}_f = \Phi(T)\delta\mathbf{X}_0 + \frac{\partial \mathbf{X}_f}{\partial t}\delta T \tag{3.112}$$

where Φ is the state-transition matrix that we defined in Section 3.2.6 and $\delta\mathbf{X}_f$ represents the incremental change that must be applied to the state $\mathbf{X}_0$ for the following iteration.

Depending on the symmetry of the periodic orbit that we seek to find, we have initial conditions of the form

$$\mathbf{X}_0 = \begin{bmatrix} x_0 & y_0 & z_0 & \dot{x}_0 & \dot{y}_0 & \dot{z}_0 \end{bmatrix}^T \tag{3.113}$$

For example, for a halo orbit, which is symmetric about the x–z plane, the above initial conditions can be further simplified to

$$\mathbf{X}_0 = \begin{bmatrix} x_0 & 0 & z_0 & 0 & \dot{y}_0 & 0 \end{bmatrix}^T \tag{3.114}$$

since a halo orbit must cross the x–z plane orthogonally at the point of intersection. So, if we want to find the initial conditions of a periodic halo orbit for a given z-amplitude, we must set the desired z_0 and iteratively correct the values of x_0 and $\dot{y}_0$ until we obtain a periodic orbit. However, we also do not generally know the period of the orbit, T. On the other hand, we know that for the symmetry to hold true, we must have that $y_0 = 0$ so $\delta y = 0$, which, expanding the third element of Equation (3.112) gives

$$\delta y = 0 = \Phi_{21}\delta x_0 + \Phi_{23}\delta z_0 + \Phi_{25}\delta \dot{y}_0 + \dot{y}_f \delta T \tag{3.115}$$

and solving for δT gives

$$\delta T = -\frac{1}{\dot{y}_f}\left(\Phi_{21}\delta x_0 + \Phi_{23}\delta z_0 + \Phi_{25}\delta \dot{y}_0\right) \tag{3.116}$$

Similarly, since z_0 is fixed, we can compute the incremental changes needed for x_0 and $\dot{y}_0$ as

$$\begin{Bmatrix} \delta x_0 \\ \delta \dot{y}_0 \end{Bmatrix} = \left[\begin{bmatrix} \Phi_{41} & \Phi_{45} \\ \Phi_{61} & \Phi_{65} \end{bmatrix} + -\frac{1}{\dot{y}_f}\begin{Bmatrix} \ddot{x}_f \\ \ddot{z}_f \end{Bmatrix}\begin{bmatrix} \Phi_{21} & \Phi_{25} \end{bmatrix}\right]^{-1}\begin{Bmatrix} \delta \dot{x}_f \\ \delta \dot{z}_f \end{Bmatrix} \tag{3.117}$$

If instead we wanted to keep x_0 fixed, we need to vary z_0 and $\dot{y}_0$

$$\begin{Bmatrix} \delta z_0 \\ \delta \dot{y}_0 \end{Bmatrix} = \left[\begin{bmatrix} \Phi_{43} & \Phi_{45} \\ \Phi_{63} & \Phi_{65} \end{bmatrix} + -\frac{1}{\dot{y}_f}\begin{Bmatrix} \ddot{x}_f \\ \ddot{z}_f \end{Bmatrix}\begin{bmatrix} \Phi_{21} & \Phi_{25} \end{bmatrix}\right]^{-1}\begin{Bmatrix} \delta \dot{x}_f \\ \delta \dot{z}_f \end{Bmatrix} \tag{3.118}$$

Lastly if we wanted to keep $\dot{y}_0$ fixed, we need to vary x_0 and z_0

$$\begin{Bmatrix} \delta z_0 \\ \delta \dot{y}_0 \end{Bmatrix} = \left[\begin{bmatrix} \Phi_{41} & \Phi_{43} \\ \Phi_{61} & \Phi_{63} \end{bmatrix} + -\frac{1}{\dot{y}_f}\begin{Bmatrix} \ddot{x}_f \\ \ddot{z}_f \end{Bmatrix}\begin{bmatrix} \Phi_{21} & \Phi_{23} \end{bmatrix}\right]^{-1}\begin{Bmatrix} \delta \dot{x}_f \\ \delta \dot{z}_f \end{Bmatrix} \tag{3.119}$$

So, if we wanted to find a halo orbit having a particular z-amplitude, A_z, we keep z_0 fixed, and integrate the equations of motion by varying the period using Equation (3.116) and updating x_0 and $\dot{y}_0$ using Equation (3.117) until the initial and final states differ at most by a prescribed tolerance, *tol* (in non-dimensional coordinates, this should be set to 10^{-10} or smaller to obtain results that lead to periodic orbits). Figure 3.35 summarizes the overall differential correction methodology for an orbit with x–z symmetry in a flowchart.

For halo orbits, for example, one does not need to integrate the equations of motion at each step for the entire period T. In fact, due to their symmetry, it is more computationally efficient to instead integrate the initial state for time T/2 and iterating until the orbit crosses the x–z plane orthogonally at T/2. This guarantees that the orbit obtained is periodic because of the time symmetry of the equations of motion.

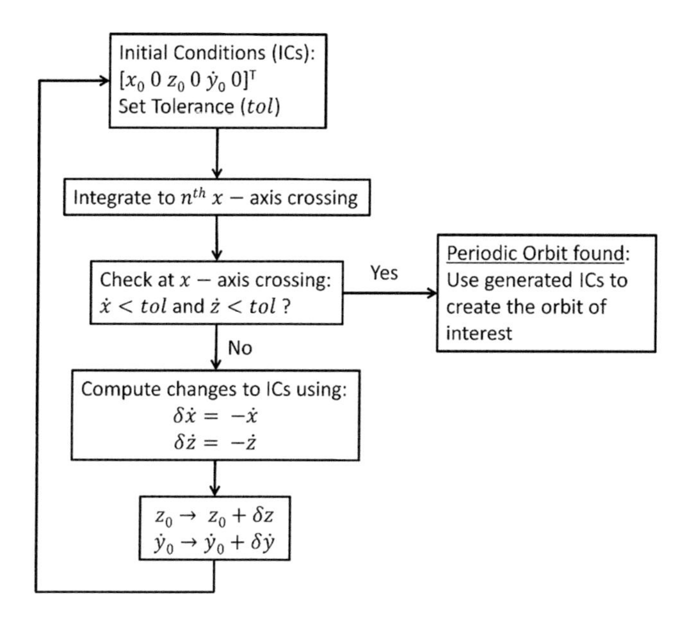

FIGURE 3.35 Differential correction flowchart for orbits that are symmetric about the x–z plane.

In the case of planar orbits, the equations needed to be simplify since the z-components are always zero. For example, if one needs to compute the initial conditions for a DRO having a given x-amplitude, then x_0 would stay fixed and the only changing parameters would be the period, T, and the orthogonal velocity, $\dot{y}_0$.

A list of initial conditions for select three-body systems for various orbits of interest is given in Appendices D and E, where we used 16 significant figures when reporting the initial conditions.

3.7.2 Poincaré Sections

As we saw in the previous section, periodic orbits must follow the condition given by Equation (3.111). However, this is true for any periodic orbit in any N-body system, not just the CR3BP. One of the classical techniques for analyzing the stability of periodic orbits in the N-body problem is called Poincaré maps, as seen in Figure 3.36. With this mapping, the stability analysis of the periodic orbit can be replaced with the stability analysis of a fixed point because of the periodicity of the orbit.

If we let $\mathbf{X}^*$ be a point on a periodic orbit, we can construct a hyperplane S (as shown in the figure) which is transverse to the orbit at $\mathbf{X}^*$. Due to the continuity of the system with respect to the initial conditions, the trajectory starting from $\mathbf{X}_1$ is going to intersects S at $P(\mathbf{X}_1)$ near $\mathbf{X}^*$, provided that $\mathbf{X}_1$ is in a sufficiently small neighborhood of $\mathbf{X}^*$. The dynamical system therefore defines a mapping onto this section S, and a periodic orbit corresponds to the fixed point of this mapping. Thus, the matrix that corresponds to this mapping is one that maps a point onto itself, making said point into a so-called fixed point. This particular matrix is known as the monodromy

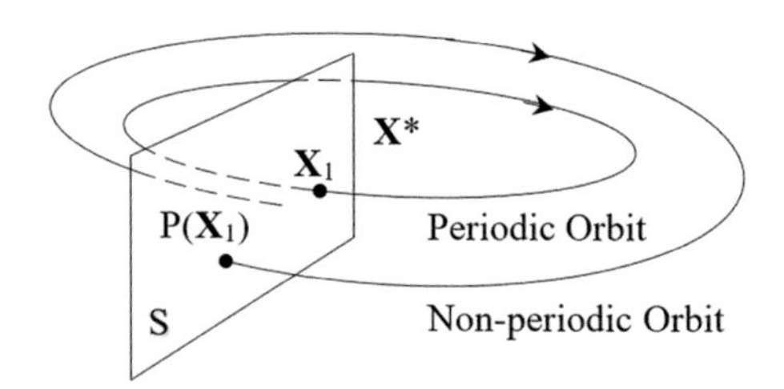

FIGURE 3.36 Poincaré map.

matrix, Π, which is a time-varying matrix (provided by the Floquet theory), and is defined as the state-transition matrix evaluated at exactly one orbital period

$$\Pi = \Phi(T) \tag{3.120}$$

This 6×6 matrix has six eigenvalues, which determine the stability of the orbit. If at least one eigenvalue has magnitude larger than 1, the orbit is considered unstable, if the largest eigenvalue has magnitude exactly equal to 1, the orbit is neutrally stable, and if all eigenvalues have magnitudes less than 1, the orbit is stable. For example, it can be shown that for any point on a generic halo orbit, the monodromy matrix has the following set of the eigenvalues

$$\begin{aligned} &\lambda_1 > 1, \quad \lambda_2 = \frac{1}{\lambda_1} \\ &\lambda_3 = \lambda_4 = 1, \quad \lambda_5 = \lambda_6, \quad |\lambda_5| = |\lambda_6| = 1 \end{aligned} \tag{3.121}$$

which generally make halo orbits unstable. However, there exist certain kinds of halo orbits, known as near-rectilinear halo orbits (NRHO, discussed in Section 3.6.5), that are neutrally stable. As previously discussed, an NRHO will be used for the Lunar Gateway thanks to its stability characteristics and relatively small perilune – facilitating landings on the Moon – and the Gateway's line-of-sight with Earth.

3.7.3 Invariant Manifolds

As we have seen, the eigenvalues of the monodromy matrix help us determine the stability of a given periodic orbit. In fact, the local behavior near periodic orbits can be characterized by these eigenspaces – eigenvalues and eigenvectors. In this section, we will use this information to obtain useful insights for the computation of so-called invariant manifolds. They exist in the CR3BP and, in general, in the N-body problem. Invariant manifolds are invariant hypersurfaces embedded in the phase space of the dynamical model of interest – such as the CR3BP or CR4BP – such that once a spacecraft is put onto a manifold, it can never leave the surface along the natural time evolution of the dynamical system (assuming that no external perturbations exist).

It should be noted that in the literature of the design method based on dynamical systems theory, however, effects of perturbations such as the other bodies and the eccentricity of primaries are still being researched. Since the principle of the orbital classification is derived from the ideal CR3BP, it is essential to investigate the effect of perturbations on the orbital classification.

Using the unstable (or stable) eigenvectors, $\boldsymbol{\lambda}$, of the monodromy operator (or matrix), an initial value of an unstable (or stable) asymptotic solution, $\mathbf{X}_{\text{perturbed}}$, can be approximated by slightly shifting an initial state $\mathbf{X}_0$ on the periodic orbit by a small value, ε, in the direction of the unstable (or stable) eigenvectors

$$\mathbf{X}_{\text{perturbed}} = \mathbf{X}_0 \pm \epsilon \boldsymbol{\lambda} \tag{3.122}$$

This perturbed state, $\mathbf{X}_{\text{perturbed}}$, can then be integrated forward or backward in time to generate trajectories that leave from or arrive at the periodic orbit whose initial condition $\mathbf{X}_0$ was used. The value of ε to be used depends on the system. For example, it suffices to set ε to a few km in the Earth-Moon system or $\varepsilon \approx 50\,\text{km}$ in the Jupiter CR3BP system (when considering any of the Galilean moons as the secondary mass). By repeating this process along a periodic orbit for various values of $\mathbf{X}_0$ that reside on the orbit, a set of initial values can be generated. The result of numerical integrations of the CR3BP from these initial values generates a family of approximated asymptotic solution trajectories that form a so-called invariant manifold tube in the phase space. Examples of stable and unstable invariant manifolds are given in Figures 3.37 and 3.38, where trajectories have been computed to the Poincaré

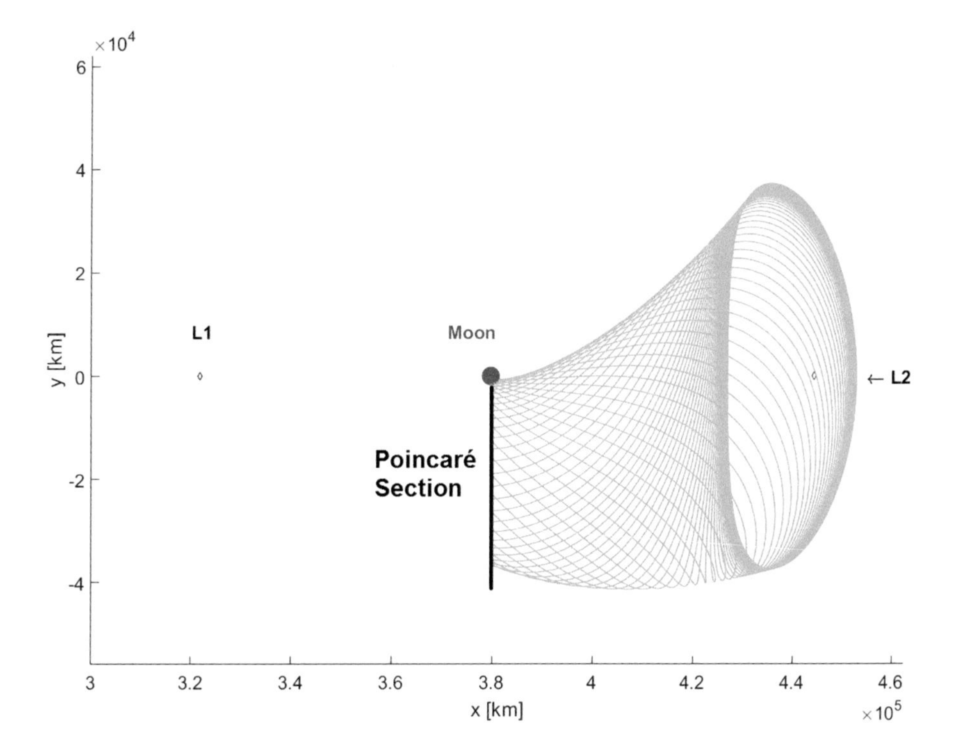

FIGURE 3.37 Stable manifolds for an Earth-Moon L_2 halo orbit.

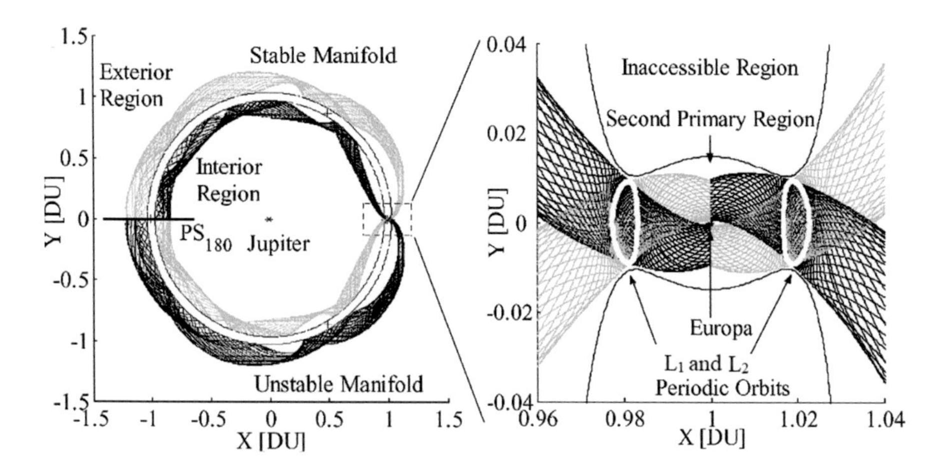

FIGURE 3.38 Stable and unstable manifolds near the Jupiter-Europa L_1 and L_2 locations [Yamato and Spencer, 2005].

sections shown. Note that these trajectories can physically intersect a celestial body, rendering them potentially useful for landing on said body, or invalid if looking for a transfer trajectory.

As shown in detail in the literature [Gómez et al., 2001b; Howell et al., 1997], many periodic orbits of interest present stable and unstable manifolds, which are regions of the phase space that asymptotically approach said orbits of interest. Therefore, they have practical significance for certain types of orbital transfers. In fact, a vehicle on a stable manifold, even at a distant location from a halo orbit, will approach the orbit gradually and can be inserted into it with a negligibly small Δv. It should be kept in mind that invariant manifold tubes are formed by an infinite number of actual solution trajectories. Since the existence and the uniqueness of solutions with respect to initial values are guaranteed except for singular points at the primaries, a spacecraft with any initial values cannot cross the surface of manifold tubes without implementing an additional propulsive maneuver. In the actual computation, numerical integrations are terminated when the spacecraft reaches the Poincaré section specified in advance. The information about the flight time is also gained from the manifold computation.

3.7.4 Orbit Classification

A significant feature in the CR3BP phase space is the ability to classify orbits based on their behavior near the L_1 and L_2 libration points. This type of orbital classification separates the solution trajectories which pass by L_1 or L_2 into four types of orbits, as shown in Figure 3.39. The first and second types are periodic orbits and trajectories asymptotic to them. The third type of trajectory is an orbit that transits from one region to another, for example, from the exterior region to the second primary realm. Finally, the last type of orbit is a non-transit orbit, which remains in the same domain (the exterior, the interior, or the second primary region).

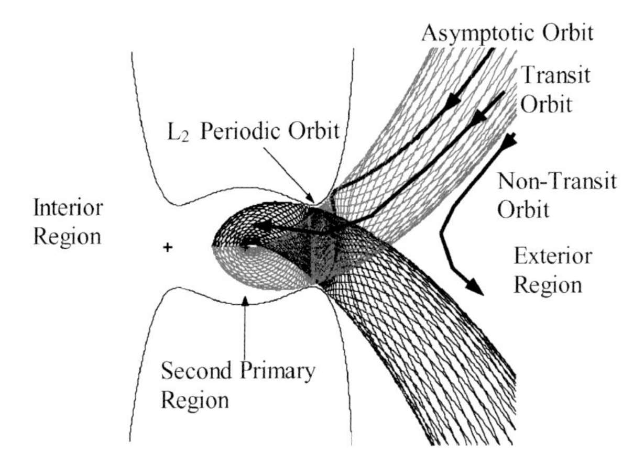

FIGURE 3.39 Four types of solution trajectories near L_2 [Yamato and Spencer, 2005].

Furthermore, it can be shown that unstable and stable manifolds partition different regimes of motion in the phase space [Howell et al., 1997; Lo et al., 2001]. While the dynamical flows lying on the surface of a manifold tube must be trajectories asymptotic to the associated halo orbit, the solution trajectories inside (or outside) the manifold tube must be transit (or non-transit) orbits. In other words, for the given value of the Jacobi integral, the inside of the unstable or stable manifold "tube" is filled up only with trajectories which are going to depart from or enter the smaller primary region, such as the Moon in the Earth-Moon system. The manifold tube can be viewed as a boundary (or separatrix) of a set of transit orbits from this viewpoint.

Orbital classification is important when searching for an appropriate orbital transfer or parking orbit during the mission design phase. For instance, if a spacecraft needs to encounter a certain celestial object that is orbiting the Sun or a moon of a planet, the feasibility of the close encounter is determined by the class of the orbit that is chosen. This, of course, means that enough energy (or a high enough Jacobi constant) is required to enter the region of the smaller body. Then, it is simply necessary to set the spacecraft on a course that encounters a stable manifold and, with a rather small amount of Δv, the spacecraft can be captured by the smaller body.

3.7.5 Transit Orbit Search

Finding transit orbits means searching for a set of initial states that propagates into the desired region (which is usually close to the secondary mass) on an arbitrarily chosen Poincaré section in the CR3BP. Often, the term Poincaré cut is used to indicate an intersection between a computed manifold (or transportation) tube and the Poincaré section of interest. The basic idea is to find transportation tubes that intersect at a given section so that a Δv can be applied to make the spacecraft "jump" from one manifold to the other.

Let's look at an example where we want to transfer from Ganymede to Europa, two of the Galilean moons. A series of perturbed transit orbits can systematically

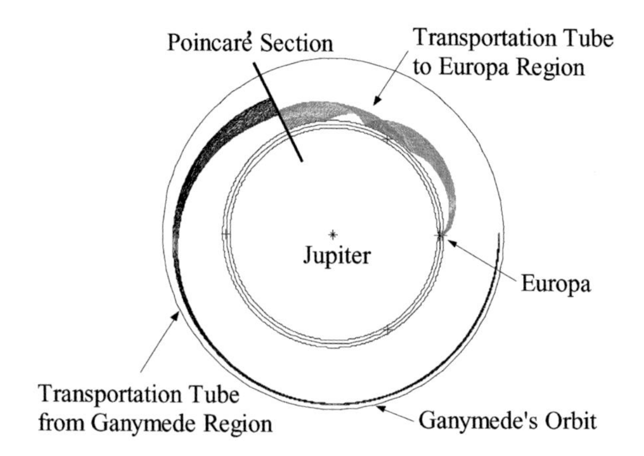

FIGURE 3.40 Departure and arrival transportation tubes between Ganymede and Europa [Yamato and Spencer, 2005].

be identified inside the transportation tube cut as presented in Figure 3.40. Once the intersection in the configuration space is located, a number of transfers – in this case from Ganymede to Europa – can be found inside the intersecting region of the transportation tunnels.

Two sets of states on the intersection are identified, such that if the Jupiter-Europa CR3BP system is numerically integrated from one of the sets forward in time the spacecraft encounters Europa, and if the Jupiter-Ganymede CR3BP system is numerically integrated from the other set backward in time the spacecraft encounters Ganymede.

The identified transit orbits correspond to states on the Poincaré section yielding transit orbits so that when they are transformed back to the inertial frame coordinates of each system – here Jupiter-Ganymede and Jupiter-Europa – the dynamical model in the inertial frame can be numerically integrated with those initial states to produce an end-to-end transfer. The velocity discontinuity between the identified states can then be used to determine the required Δv for this end-to-end transfer. In fact, Δv is generally the parameter that mission designers wish to minimize, along with accounting for transfer time. Note that the Δv's required to depart from and to arrive at the parking orbits around the targeted bodies are usually extremely small (recall the definition of the perturbed state from Equation 3.122), so generally it is sufficient to find the minimum Δv needed to "jump" between manifolds in order to find the overall minimum Δv for the entire transfer. Figure 3.41 shows the overlapping regions corresponding to the locations where the spacecraft could "jump" from one transportation tube to the other.

This method is not only limited to the CR3BP, but it is applicable to any N-body problem. In fact, going back to our Ganymede-Europa transfer example, transit orbits that connect the two celestial bodies in the CR4BP and ER4BP models exist, and Figure 3.42 shows the departing an arriving manifolds that result in a Δv of 1.387 km/s and a transfer time of 25 days [Yamato and Spencer, 2005].

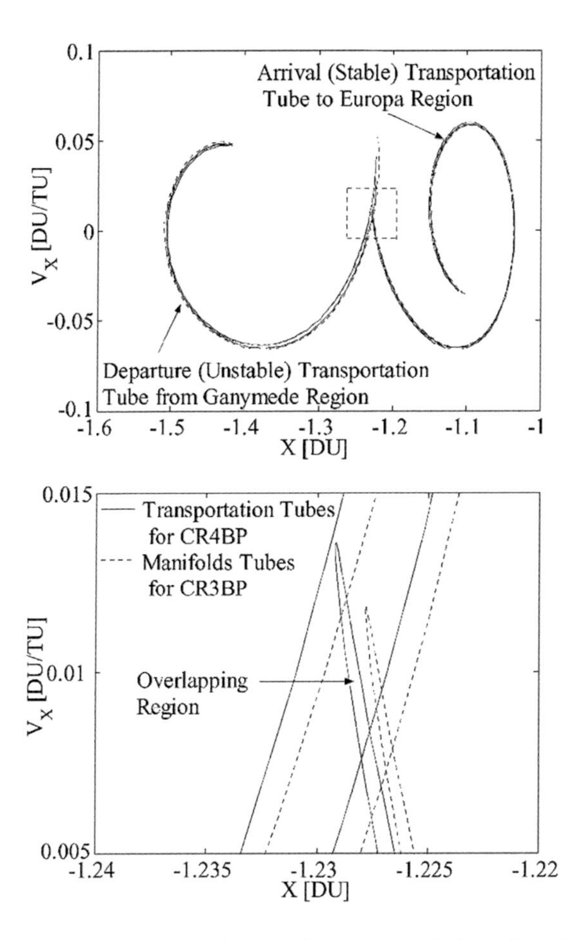

FIGURE 3.41 $x-v_x$ projection of the intersections between Poincaré section and departure and arrival transportation tubes [Yamato and Spencer, 2005].

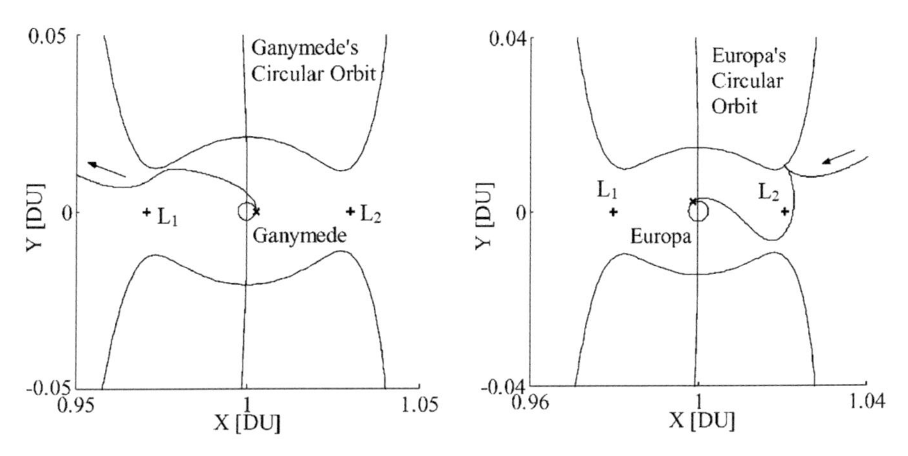

FIGURE 3.42 Ganymede departure and Europa arrival trajectories in the CR4BP [Yamato and Spencer, 2005].

In contrast, a Hohmann transfer between Ganymede and Europa requires approximately a Δv of 2.822 km/s, which is significantly more, but only a transfer time of 3 days.

It is important to note that any points inside the manifold cut can produce transit orbits, allowing the generation of various trajectories involving the transition to the second primary region. In general, the inverse-square gravity field causes the system to be highly sensitive to initial conditions, as we have discussed in previous sections. Because of this, the search of transit orbits can sometimes be extremely challenging. However, the analysis based on invariant manifold structures provides an excellent framework for the construction and, ultimately, design of spacecraft orbits in N-body systems.

3.7.6 Weak Stability Boundary

One of the most important applications of dynamical systems theory is weak stability boundary (WSB), which was first introduced by Belbruno and Miller [1990]. Trajectories that make use of this concept are generally more propellant-efficient at the cost of travel time. The WSB is used to construct low-energy transfers to secondary bodies in the N-body problem, such as orbits to the Earth's Moon that requiring little or no propellant for capture into lunar orbit. The first application of WSB in a space mission was in 1991 with the rescue of the Japanese mission Hiten, which was not designed to enter lunar orbit but managed to release its probe Hagoromo during a lunar flyby which, despite a communications failure, entered lunar orbit [Belbruno and Miller, 1990]. WSB was also used in more recent missions, such as ESA's mission BepiColombo, aimed at exploring the planet Mercury in 2013, and is planned to be used in upcoming missions.

WSB transfers follow the interplanetary superhighway (that we previously discussed) in a way that allows spacecraft to be utilize the gravitational perturbation of a far-away celestial body, such as the Sun, to be captured by a much closer targeted body, such as the Moon, using nearly no propulsive maneuver. This is the key feature of WSB trajectories: the final orbit insertion is performed almost solely ballistically, i.e., requiring virtually no Δv. While the overall maneuver requires significantly more time, the overall mission Δv is usually much smaller when compared to other transfers, such as Hohmann transfers.

For example, an Earth-Moon mission departing from LEO and directed to LLO could save up to 25% of Δv on the burn applied after leaving low Earth orbit, while a mission targeting the Martian moons, Phobos and Deimos, would result in a Δv saving of up to 12% (for Phobos) and 20% (for Deimos) [Belbruno and Carrico, 2000].

The Gravity Recovery and Interior Laboratory (GRAIL) is one of the most notable NASA missions to have used WSB trajectories for its lunar insertion. GRAIL was composed of two main probes: GRAIL-A (Ebb) and GRAIL-B (Flow) which were launched together in September of 2011 and stayed in very similar orbits after launch. They both followed WSB trajectories to the Moon and managed to perform their orbit insertion almost entirely ballistically thanks to the gravitational perturbation of the Sun over the course of the approximately 3.5 months required to arrive at the Moon. Figure 3.43 shows the trajectory of the GRAIL probes.

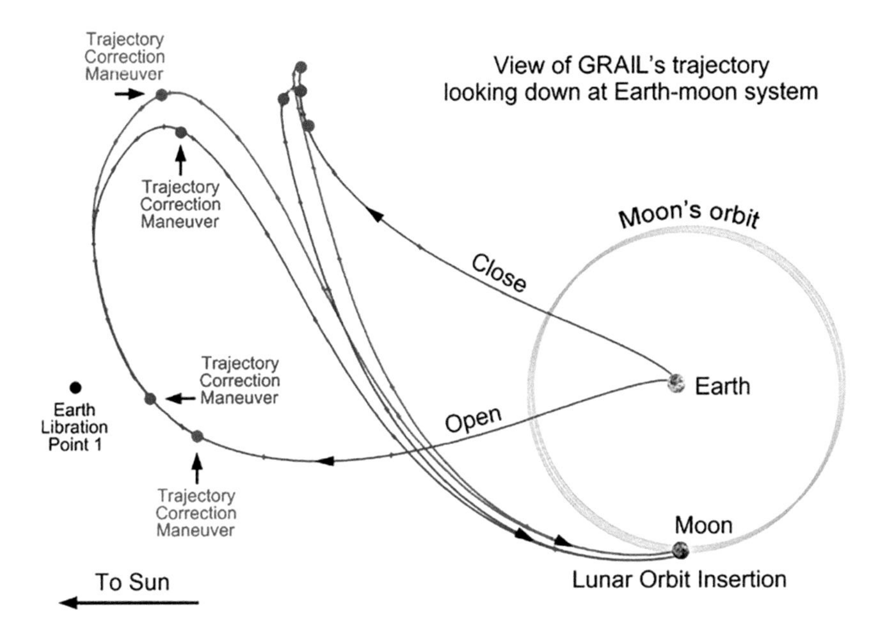

FIGURE 3.43 WSB trajectories of GRAIL-A (red) and GRAIL-B (blue). (Courtesy of NASA/JPL-Caltech.)

3.8 APPLICATIONS OF LIBRATION POINT ORBITS

As was shown in earlier sections, in the ideal restricted three-body problem, an object placed at the collinear libration points will not stay there due to the instability of the points. However, objects placed in an ideal restricted three-body system's triangular libration points, L_4 and L_5, will remain at those points. Before the space age, the only known instances of libration point objects were the Trojan and Greek asteroids in the Sun-Jupiter system. The Trojan asteroids orbit near the Sun-Jupiter L_5 point while the Greek asteroids orbit near the Sun-Jupiter L_4 point. These objects are subject to many perturbations, so they continually move relative to these points (recall that these libration points are moving with the planet's orbit around the system barycenter).

It wasn't until the space age that the concept of stationkeeping allowed mission designers to overcome the natural drifting and take advantage of the stability characteristics of objects placed near the collinear points. As was seen in the zero-velocity curves shown in Section 3.2.4, the amount of energy (which directly relates to Δv) needed to reach the triangular points is much greater than that needed to reach the collinear points. The combination of the energy requirement and stationkeeping ability, as well as the stability characteristics of halo and Lissajous orbits combine to make the collinear points the logical choice of where to send spacecraft. Two additional characteristics that make collinear libration point orbits attractive is the ability to place a spacecraft between the Sun and the Earth (at the L_1 point), or at a point outside of the Sun-Earth system to point away from the Sun and Earth (at the L_2 point). Figure 3.44 shows the location of libration points and example halo orbits in the Sun-Earth and Earth-Moon systems.

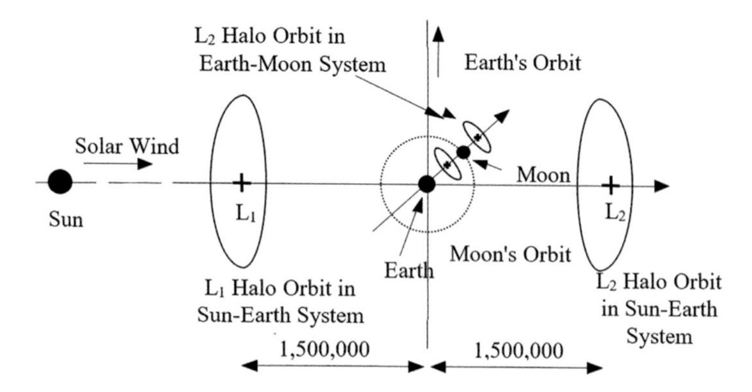

FIGURE 3.44 Libration points and example orbits in the Sun-Earth and Earth-Moon systems. Distances are in km.

3.8.1 Libration Point Missions

Several spacecraft have flown or are currently flying in orbits about the collinear libration points in the Sun-Earth/Moon system. In this section, we will discuss some of the most influential missions to such locations.

The first applications for halo orbits came with the Apollo program [Farquhar, 1970], where a potential landing on the far side of the Moon would lose line-of-sight communication with the Earth. This problem would be overcome by placing a relay satellite in a halo orbit about the Earth-Moon L_2 libration point. This satellite would have line-of-sight communications with both the crew on the far side of the Moon and the Earth. While this system has yet to be used, it is still feasible for use in future far-side lunar missions that require near-constant communication and/or line-of-sight. The first space mission to use libration point orbits was the International Sun-Earth Explorer 3 (ISEE-3) spacecraft [Farquhar, 1998; Farquhar et al., 1977]. This spacecraft was launched in September 1978, and was inserted into a halo orbit about the Sun-Earth L_1 point. A schematic of the transfer into the halo orbit is shown in Figure 3.45.

Upon completion of its original mission, the spacecraft transferred out of this orbit in 1982 and put on a trajectory to intercept the comet P/Giacobini-Zinner. The spacecraft was renamed the International Cometary Explorer (ICE) and passed through the tail of the comet in September, 1985. The spacecraft passed in the vicinity of the Earth-Moon system in August, 2014.

Later missions placed satellites into various libration point orbits. NASA's WIND,[7] SOHO, and ACE spacecraft were placed in orbits about the Sun-Earth L_1 libration point. Each mission built on knowledge of libration point orbits gained from earlier missions and was successfully more complex missions. The WIND spacecraft's path to its libration point orbit took it through many lunar flyby orbits, and the spacecraft was eventually inserted into an orbit around the libration point. This path is shown

[7] http://www-istp.gsfc.nasa.gov/istp/wind/wind.html

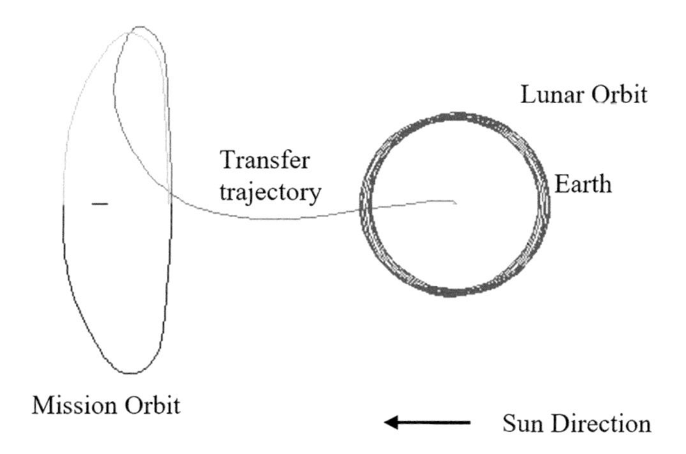

FIGURE 3.45 ISEE-3 transfer and mission trajectory in the rotating coordinate frame. (Courtesy of NASA/Goddard Space Flight Center.)

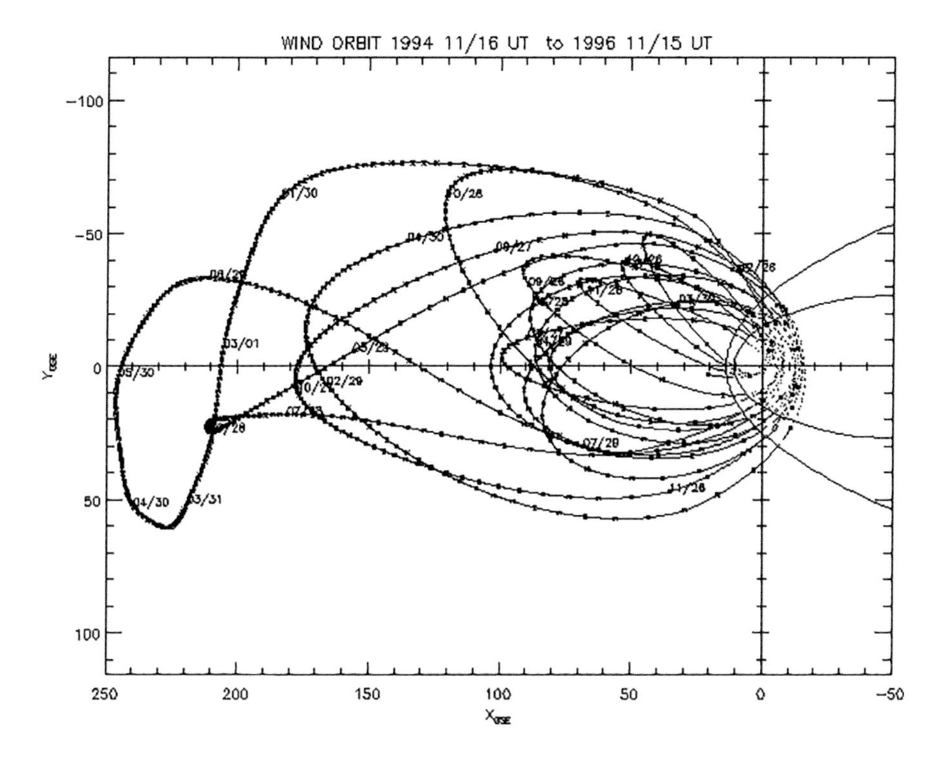

FIGURE 3.46 WIND trajectory. (Courtesy of NASA/JPL-Caltech.)

in Figure 3.46. Later, the mission was moved out of the L_1 orbit and was put on a trajectory that allowed the spacecraft to oscillate back and forth between the L_1 and L_2 libration points.

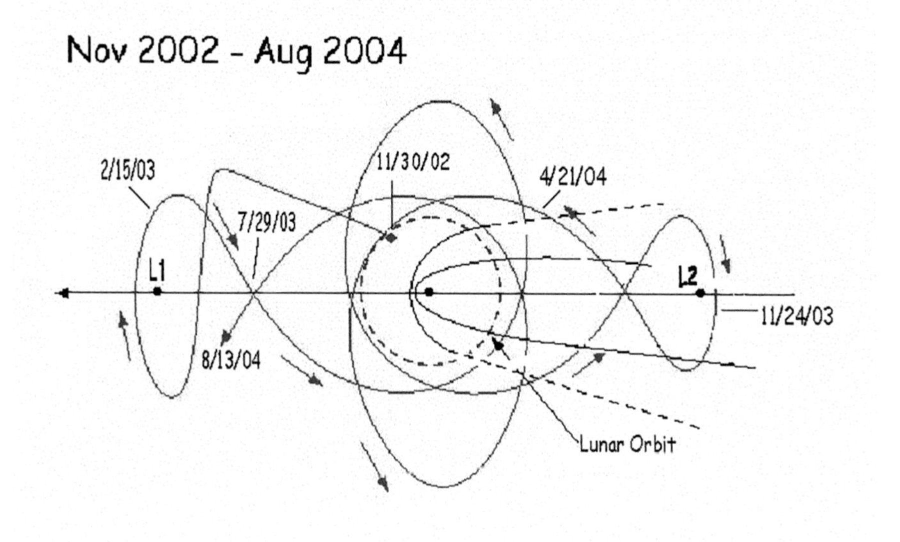

FIGURE 3.47 Post-mission WIND trajectory. (Courtesy of NASA.)

The spacecraft's mission officially ended in October, 2001, but the spacecraft continued to pass near the Moon and oscillate between the L_1 and L_2 points. The post-mission trajectory from November 2002 to August 2004 is shown in Figure 3.47.

The Solar and Heliospheric Observatory (SOHO) was launched in December, 1995 and is a joint project between ESA and NASA. Its primary mission objective is to study the Sun from the core to the solar wind. In order to efficiently accomplish this, the spacecraft was inserted into a halo orbit around the Sun-Earth L_1 libration point. The schematic of this orbit is shown in Figure 3.48. As of this writing, SOHO is still operating in its libration point orbit and is expected to continue operating until 2025.

The next mission launched into a libration point orbit was the Advanced Composition Explorer (ACE). This mission, launched in August 1997, placed the spacecraft into a Lissajous orbit. The schematic of the transfer into the operational orbit is shown in Figure 3.49. This figure also shows the location of three maneuvers (MCC-1: mid-course correction; OSM: orbit shaping maneuver; HOI: halo orbit insertion). All times are given relative to launch (L). At current propellant usage rate, it is estimated that ACE can continue to operate until 2024.

The most recent completed mission launched into a libration point orbit was NASA's Genesis spacecraft. Genesis was a unique program, as it was the first libration point mission to return to Earth. The purpose of the Genesis mission was to return samples of solar wind particles collected, while the spacecraft was in orbit around the Sun-Earth L_1 point. While the mission was only a partial success (the spacecraft's parachute failed to open after reentry), the libration point orbit trajectory phase was a tremendous success. This was the first application of dynamical systems

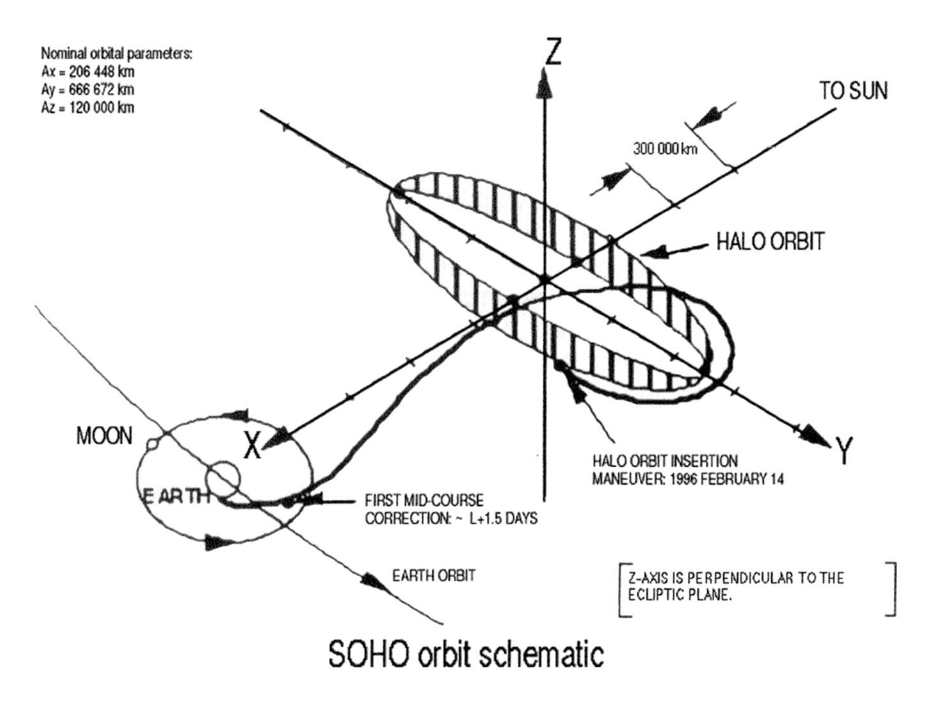

FIGURE 3.48 SOHO orbit. (Courtesy NASA/Goddard Space Flight Center.)

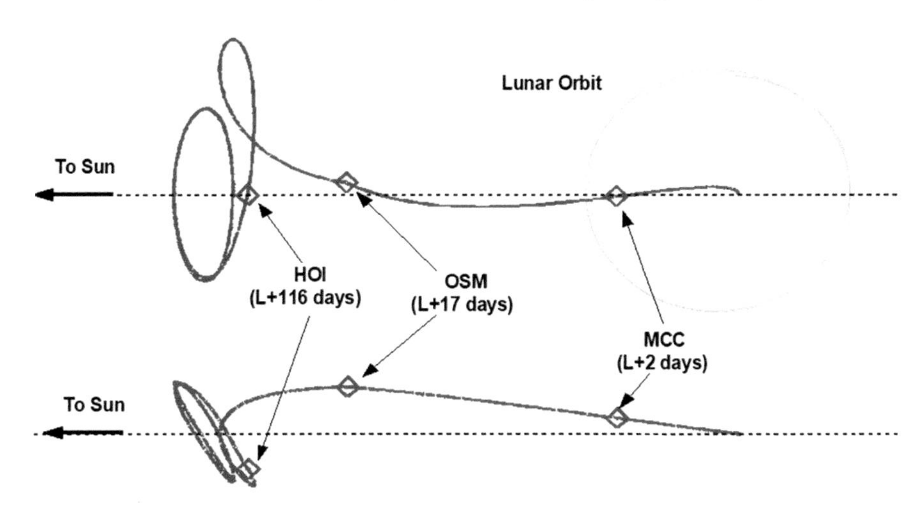

FIGURE 3.49 Orbit insertion process for ACE orbit. (Courtesy of NASA/Goddard Space Flight Center.)

theory to space mission design. Genesis took advantage of a combination of lunar flybys and invariant manifolds to both insert it into its operational orbit and after 37 months, returned to Earth and sent the sample-return capsule on an Earth reentry trajectory. The spacecraft bus that dropped off the sample-return capsule headed back to an orbit around the L_1 point and then drifted off on a trajectory that escaped

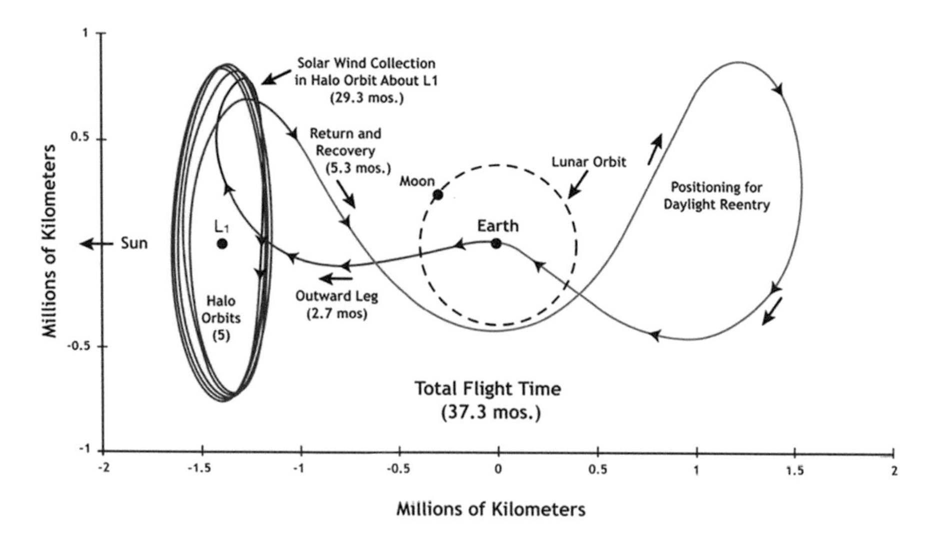

FIGURE 3.50 Genesis sample-return trajectory profile. (Courtesy NASA/JPL-Caltech.)

the Earth-Moon system and is now in orbit around the Sun. The flight profile for the sample-return portion of the mission is shown in Figure 3.50.

The Wilkinson Microwave Anisotropy Probe (WMAP) gathered data that allowed the science team to produce a full sky map of the variations of the cosmic background radiation while the spacecraft orbited the Sun-Earth L_2 point. The spacecraft was launched in June 2002 and was transferred into an orbit around L_2 during three phasing loops over a period of around 100 days. The spacecraft's transfer trajectory is shown in Figure 3.51.

Planck and Herschel are two European Space Agency spacecraft launched in May 2009. The spacecraft, launched together on an Arianne-5 launch vehicle, operated in different Lissajous orbits around the Sun-Earth L_2 point. The Planck spacecraft performed imaging of the whole sky temperature anisotropies of the cosmic microwave background at very high resolutions [Tauber and Clavel, 2001]. The Herschel spacecraft was also investigating the cosmic background radiation by observing the far-infrared at an extremely high precision [Crone et al., 2006]. The trajectories of the Planck and Herschel spacecraft are shown in Figure 3.52. Both spacecraft operated for approximately 4 years.

Gaia is an ESA mission that was launched in December 2013 to a Lissajous orbit around the Sun-Earth L_2 libration point, as shown in Figure 3.26. The objectives of this mission were to measure the positions, distances, and motions of stars with the highest accuracy to date. This would result in the largest and most precise 3D space catalog created, including ~1 billion astronomical objects such as stars, planets, comets, asteroids, and quasars. Gaia is expected to remain operational until 2025.

LISA Pathfinder was launched in December 2015 to a Sun-Earth L_1 Lissajous orbit and was operational until June 2017. The mission tested technologies needed for the Laser Interferometer Space Antenna (LISA), which is a mission planned by ESA for 2037. The goal of LISA will be to study gravitational waves utilizing a

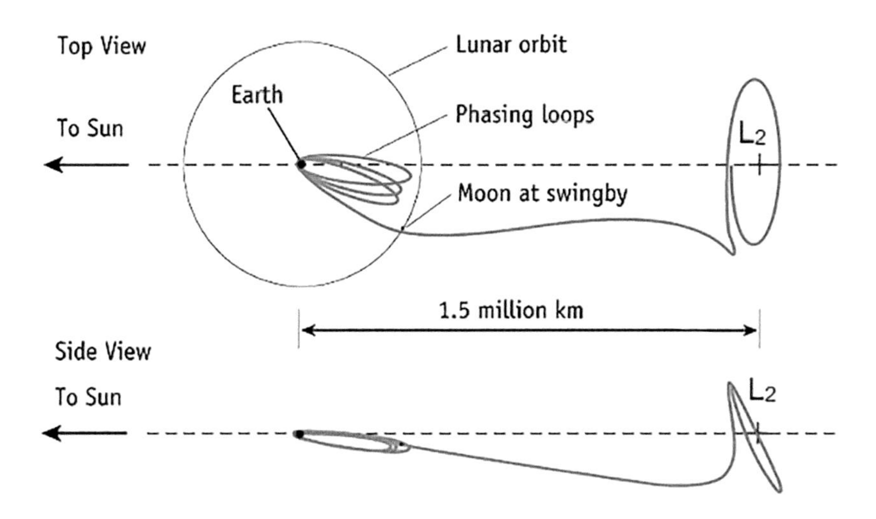

FIGURE 3.51 WMAP trajectory. (Courtesy of NASA Goddard Space Flight Center.)

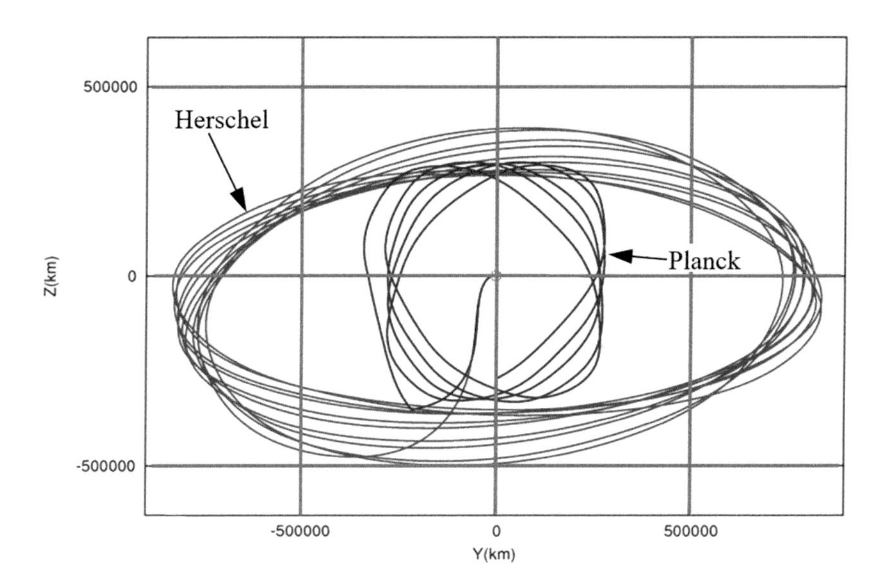

FIGURE 3.52 Planck and Herschel trajectories. (Courtesy of ESA.)

constellation of three spacecraft located in an equilateral triangle configuration, as shown in Figure 3.53.

One of the most anticipated missions of the 2020s was undoubtedly the James Webb Space Telescope (JWST). This spacecraft is the follow-on for the Hubble Space Telescope, and it was launched in December 2021. The spacecraft was inserted into its operations halo orbit around the Sun-Earth L_2 point in January 2022. Figure 3.54 shows the complex telescope's deployment phase into its final orbit. JWST is the

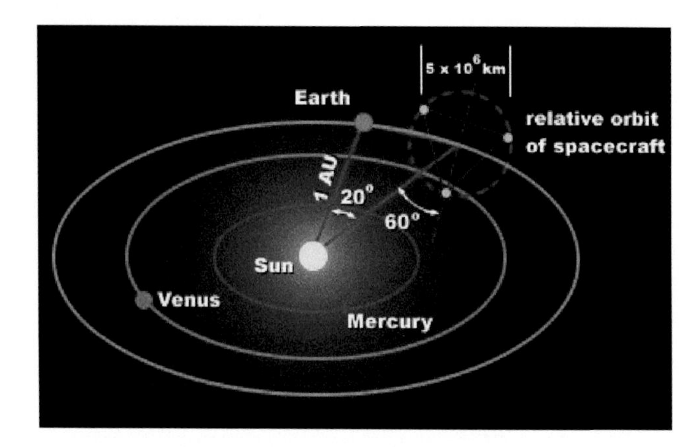

FIGURE 3.53 LISA's planned trajectory. (Courtesy of ESA.)

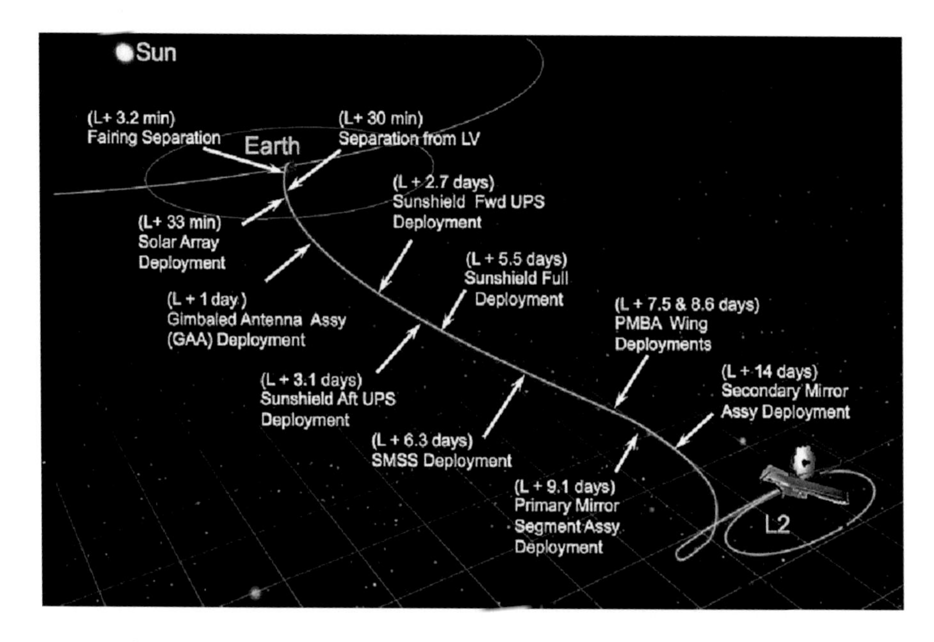

FIGURE 3.54 JWST's deployment into its operating orbit. (Courtesy of NASA.)

largest space telescope to date and is designed primarily to conduct infrared astronomy. Its objectives are to observe the early Universe, including the first stars and the formation of the first galaxies. Additionally, JWST will be searching for potentially habitable exoplanets – planets that reside outside of the Solar System – and characterize their atmospheric compositions.

Chang'e 5 (which we introduced in Chapter 1) was China's first lunar sample-return mission. However, as part of its extended mission, the orbiter was successfully captured into a Sun-Earth/Moon L_1 6-month periodic halo orbit in March 2021, becoming the first Chinese spacecraft to reach this Lagrange point.

TABLE 3.6
Libration point missions orbital characteristics

Satellite	Libration point	Orbit type	A_x (km)	A_y (km)	A_z (km)
ISEE-3	L_1	Halo	175,000	660,670	120,000
WIND	L_1	Lissajous	100,000	350,000	250,000
SOHO	L_1	Lissajous	206,448	666,672	120,000
ACE	L_1	Lissajous	81,775	264,071	157,406
Genesis	L_1	Lissajous	250,000	800,000	250,000
WMAP	L_2	Lissajous	90,000	264,000	264,000
Planck	L_2	Lissajous	400,000	400,000	400,000
Herschel	L_2	Lissajous	800,000	800,000	800,000
Gaia	L_2	Lissajous	263,000	707,000	370,000
LISA Pathfinder	L_1	Lissajous	500,000	800,000	800,000
JWST	L_2	Halo	250,000	800,000	832,000

Table 3.6 summarizes the approximate orbital characteristics of the libration point orbits used by the missions presented in this section.

New missions are constantly being proposed, so it is difficult to accurately list future missions since new libration point orbiting spacecraft are continually being proposed. Some popular topics in recent years have included libration point orbits in other planetary systems, such as Sun-Mars, Sun-Jupiter, for example.

For example, NASA's Lucy mission, named after the Lucy hominin fossils, was launched in October 2021 with the aim of flying by a series of eight asteroids, including Greek asteroids (located near the Sun-Jupiter L_4 libration point) and Trojan asteroids (located near the Sun-Jupiter L_5 libration point). Lucy is the first mission to target so many destinations in independent orbits around our Sun.

PROBLEMS

1. Compute the period, time unit (in seconds), and dimension unit (in kilometers) of the systems listed in Table 3.1. You may find it useful to consult the Appendix A for various constants you will need, such as semimajor axes and gravitational parameters.
2. Compute the x- and y-coordinates of the collinear Lagrange points for the following systems:
   a. Earth-Moon
   b. Sun-Mercury
   c. Sun-Venus
   d. Sun-Earth
   e. Sun-Mars
   f. Sun-Jupiter
   g. Sun-Saturn
   h. Mars-Phobos
   i. Mars-Deimos

j. Jupiter-Ganymede
k. Saturn-Titan
l. Pluto-Charon

3. How do the values you computed in Problem 2 compare with the approximations for L_1, L_2, and L_3 given by Equation (3.55)?
4. The non-linear equations of motion in the restricted three-body problem are:

$$\ddot{x} - 2\dot{y} - x = -\frac{(1-\mu)(x-\mu)}{r_1^3} - \frac{\mu(x+1-\mu)}{r_2^3}$$

$$\ddot{y} + 2\dot{x} - y = -\frac{(1-\mu)y}{r_1^3} - \frac{\mu y}{r_2^3}$$

$$\ddot{z} = -\frac{(1-\mu)z}{r_1^3} - \frac{\mu z}{r_2^3}$$

Where

$$r_1 = \left[(x-\mu)^2 + y^2 + z^2\right]^{1/2}$$

$$r_2 = \left[(x+1-\mu)^2 + y^2 + z^2\right]^{1/2}$$

For the restricted three-body problem, the complete set of equations of motion for the linear solutions about the Lagrange points are found as:

$$\delta\ddot{x} - 2\delta\dot{y} - \delta x = -\delta x\left\{(1-\mu)\left[\frac{1}{r_{1e}^3} - 3\frac{(x_e-\mu)^2}{r_{1e}^5}\right] + \mu\left[\frac{1}{r_{2e}^3} - 3\frac{(x_e+1-\mu)^2}{r_{2e}^5}\right]\right\}$$

$$+\delta y\left\{3(1-\mu)\frac{(x_e-\mu)y_e}{r_{1e}^5} + 3\mu\frac{(x_e+1-\mu)y_e}{r_{2e}^5}\right\}$$

$$\delta\ddot{y} + 2\delta\dot{x} - \delta y = \delta x\left\{3(1-\mu)\frac{(x_e-\mu)y_e}{r_{1e}^5} + 3\mu\frac{(x_e+1-\mu)y_e}{r_{2e}^5}\right\}$$

$$-\delta y\left\{(1-\mu)\left[\frac{1}{r_{1e}^3} - 3\frac{y_e^2}{r_{1e}^5}\right] + \mu\left[\frac{1}{r_{2e}^3} - 3\frac{y_e^2}{r_{2e}^5}\right]\right\}$$

Using these equations, assess the stability of L_1 for the systems listed in Problem 2.

5. Compute the value of the Jacobi integral at each Lagrange point computed in Problem 2 assuming zero-velocity components in the x-, y-, and z-directions.

6. Using the non-linear equations of motion in the restricted three-body problem:
   a. if the state vector is: $\vec{X} = \left(x\ y\ z\ \dot{x}\ \dot{y}\ \dot{z}\right)^T$, and using the values of $\mu = 0.16$ and the following initial conditions: $\vec{X}(0) = \left(-1.024114\ 0\ 0.528576\ 0\ 0.381137\ 0\right)^T$, numerically integrate the equations of motion for one period (you will have to experimentally determine what the period is (*hint*: think of the symmetry of the problem and where does the orbit cross a plane of symmetry again?)). Show plots of x vs. y, x vs. z, y vs. z and a three-dimensional plot of x vs. y vs. z. You should get a periodic, halo orbit. Identify the period of the orbit.
   b. Confirm the invariance of these equations of motion under the transformation $y \to -y$ and $t \to -t$.
   c. For another case, we don't have the exact initial conditions for x_0 and z_0. If $\mu = 0.2$ and the initial conditions are $\vec{X}(0) = \left(-1.01\ 0\ 0.58\ 0\ 0.413250\ 0\right)^T$, use a differential correction process to correct the initial conditions to obtain a periodic halo orbit. Show three iterations of your differential correction results with the corrected initial conditions and period (or half period) after each iteration.

7. Numerically integrate the equations of motion of the circular restricted three-body problem using the mass ratio of the Earth-Moon system ($\mu = 0.01215060$) for the following sets of initial conditions:

   a. $\vec{X}(0) = \left(-1.180732845\ 0\ 0.013007284\ 0\ 0.156830543\ 0\right)^T$

   b. $\vec{X}(0) = \left(-1.178004826\ 0\ 0.052029136\ 0\ 0.169784129\ 0\right)^T$

   c $\vec{X}(0) = \left(-1.125032004\ 0\ 0.182101977\ 0\ 0.225430661\ 0\right)^T$

   d. $\vec{X}(0) = \left(-1.014568158\ 0\ 0\ 0\ -0.9385765507\ 0\right)^T$

   e. $\vec{X}(0) = \left(-0.7500000000\ 0\ 0.6204002141\ 0\ -0.3228789168\ 0\right)^T$

   f. $\vec{X}(0) = \left(-1.156957\ 0\ 0\ 0\ 0.4800000000\ 0\right)^T$

   You will need to determine the period of these orbits to correctly plot them (*hint*: think of the symmetry of the problem). What orbit types do you get? Match and verify that the orbits you obtained are the same as the orbits plotted on Figure 3.55.

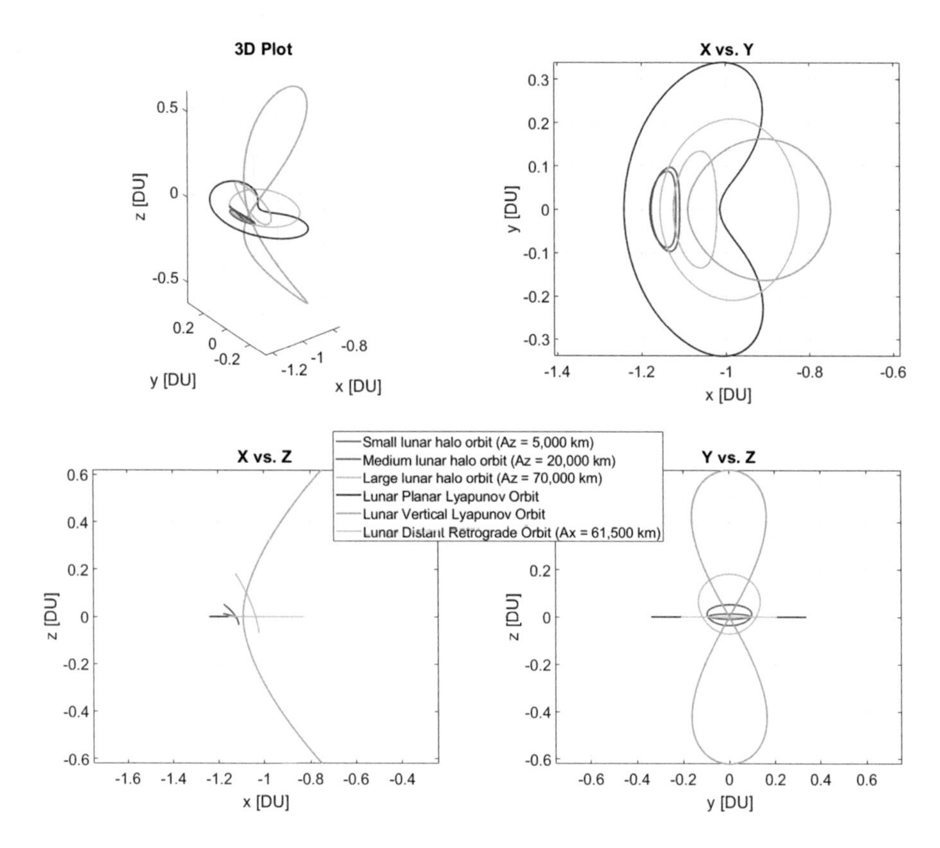

FIGURE 3.55 Various periodic orbits in the CR3BP.

8. In the planar-restricted three-body problem, the equations for the zero-velocity curve can be written as:

$$C = -\frac{(1-\mu)}{2}\left(r_1^2 + \frac{2}{r_1}\right) - \frac{\mu}{2}\left(r_2^2 + \frac{2}{r_2}\right) + \frac{\mu(1-\mu)}{2}$$

and

$$C = -\frac{1}{2}\left(x^2 + y^2\right) - \frac{1-\mu}{r_1} - \frac{\mu}{r_2}$$

a. Show that these two expressions are equivalent.
b. Using the first equation above, show that the zero-velocity curves disappear where $C = \frac{1}{2}\left[-3 + \mu(1-\mu)\right]$
c. For a system with $\mu = 0.25$, choose the value of the Jacobi constant halfway between $C(L_1)$ and $C(L_2)$, i.e., the arithmetic average between $C(L_1)$ and $C(L_2)$ (be sure to state this value of the Jacobi constant) and develop

a computer program to plot the zero-velocity curve for this value. Be sure to annotate the location of the collinear libration points. Test your code at one more location between $C(L_2)$ and $C(L_3)$ and identify the Jacobi constant value you use.

4 Coordinate Frames, Time, and Planetary Ephemerides

4.1 INTRODUCTION TO COORDINATE SYSTEMS

In order to define the state (position and velocity) of an object in space, we must be able to reference it to some sort of a coordinate frame. In the field of interplanetary astrodynamics, there are many coordinate frames that are used, depending on the application. Detailed information on fundamental coordinate frames is found in existing literature [Walter and Sovers, 2000], but here we are focusing primarily on those that are most important for interplanetary mission design.

Coordinate frames for celestial objects were first introduced for the cataloging of stars. Hipparchus (190–120 B.C.E.) built on the earlier work of Timocharis and Aristyllus to discover that the Earth's axis of rotation precesses. In order to determine with sufficient accuracy the location and motion of other celestial bodies, precise understanding of various coordinate frames and time systems is necessary.

4.2 INERTIAL COORDINATE SYSTEMS

An inertial coordinate system is a fundamental, non-moving, coordinate frame fixed at some central location. For example, in our Solar System, the Sun could be thought of as the center of an inertial coordinate system. In this inertial coordinate system, all planets move relative to this coordinate frame. This means that the Cartesian XYZ coordinate frame attached to the center of the Sun remains fixed and does not move. In reality, however, there is no truly inertial coordinate frame that is fixed in space. The Sun-centered inertial coordinate frame that was just discussed is really a frame that is moving with respect to the center of the galaxy, which is moving with respect to the rest of the universe. However, for the applications used in astrodynamics and the time frames involved, these inertial coordinate systems are generally sufficiently accurate to be able to describe the motion of the bodies orbiting it to sufficient accuracy. This is due to the fact that the rotational accelerations resulting from the Sun moving around the galaxy, and the galaxy moving around the universe are many orders of magnitudes smaller than the gravitational acceleration of the Sun acting on the planets for the time intervals we are currently considering for spaceflight.

DOI: 10.1201/9781003165071-4

4.2.1 Planet-Centered Inertial

A planet-centered (or planetcentric) inertial coordinate frame has its origin at the center of a planet. Whether the planet rotates or not, the coordinate frame remains fixed in inertial space. A common planetcentric coordinate frame includes the Earth-centered inertial, which is used primarily for Earth-orbiting spacecraft, but also for interplanetary departures. We looked at this reference frame in Chapter 2 when we defined the classical orbital elements (Section 2.5). A rotating, planetcentric coordinate frame is generally related to the planetcentric inertial coordinate frame by the planet's rotation rate. The relationship between these two coordinate frames can therefore be developed using a relatively simple direction cosine matrix. A commonly used Earth-centered rotating coordinate frame is the latitude/longitude coordinate frame. For Earth, a common plane (the equator) exists between the inertial and rotating coordinate frames. The zero line is the prime meridian (zero longitude, and has been set to the location of the Greenwich Observatory in London).

Another useful Earth-centered inertial frame is the Earth Mean Equatorial (EME or EME2000, Figure 4.1) reference frame, which has basis vectors defined as $\mathbf{Z}_{\text{EME}}$ being normal to Earth's mean equator of epoch J2000,[1] $\mathbf{X}_{\text{EME}}$ being parallel to the vernal equinox of the Earth mean orbit at J2000, and $\mathbf{Y}_{\text{EME}}$ completes the right-handed system such that $\mathbf{X}_{\text{EME}} \times \mathbf{Y}_{\text{EME}} = \mathbf{Z}_{\text{EME}}$.

Figure 4.1 shows this reference frame with respect to the ecliptic plane and Earth's obliquity, $\epsilon_{\oplus}$, which we will discuss in detail in the next section.

In the realm of interplanetary trajectories, Earth-centered coordinate frames are usually not the only coordinate frames used. It is many times more convenient to define a planet-, moon-, or asteroid-/comet-centered reference frame based on the

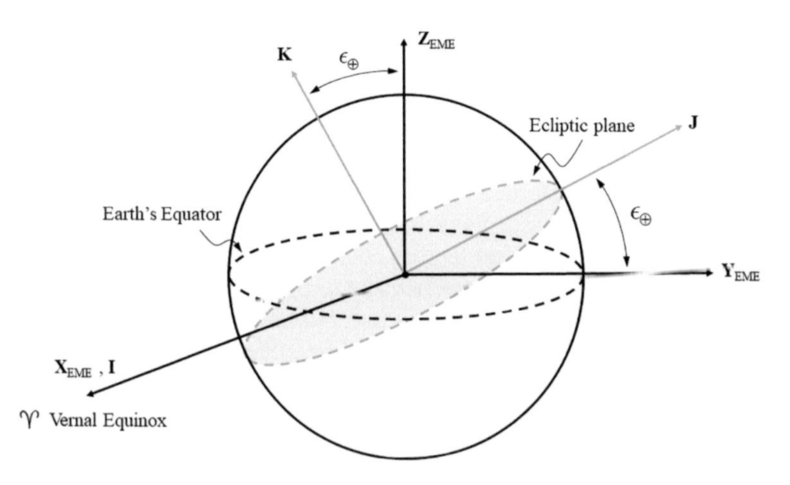

FIGURE 4.1 Definition of the Earth Mean Equatorial (EME) reference frame with respect to the ecliptic plane.

[1] J2000 is defined as the Gregorian date January 1, 2000, at 12:00 TT (Terrestrial Time). The equivalent Julian date is 2,451,545.0 TT. We discuss Julian date and TT later in the chapter.

trajectory and destination of the spacecraft. Nonetheless, Earth-centered coordinates are used for communication and telemetry purposes given that the ground stations reside on Earth's surface. For example, for a spacecraft orbiting Mars, mission operations would likely make use of Mars-centered reference frames. For convenience, it is useful to derive a Mars-centered inertial reference frame based on Mars' equator, or Mars Mean Equatorial (MME). This helps with the analysis of the arrival trajectories and B-plane targeting (which we discuss in Section 4.2.7) for Mars missions, in this case. This coordinate frame is defined such that $\mathbf{Z}_{MME}$ is normal to Mars' mean equator at epoch J2000, $\mathbf{X}_{MME}$ is parallel to the vernal equinox of Mars' mean orbit at J2000, and $\mathbf{Y}_{MME}$ completes the right-handed system such that $\mathbf{X}_{MME} \times \mathbf{Y}_{MME} = \mathbf{Z}_{MME}$. The direction of the Martian North Pole with respect to the EME reference frame is given in degrees in terms of its right ascension α and declination δ as

$$\begin{aligned} \alpha &= 317.68143^{\circ} - 0.1061\mathrm{T} \\ \delta &= 52.8865^{\circ} - 0.0609\mathrm{T} \end{aligned} \tag{4.1}$$

where T is the time in Julian centuries of 365.25 days past the reference epoch of J2000 (Julian date is discussed in Section 4.4.7). Notice that the time dependence of right ascension and declination in Equation (4.1) causes a rather slow change in these quantities and it is an approximation that is considered to be valid until the year 3000 [Standish and Williams, 2006]. For example, at epoch J2000 (T = 0 in Equation (4.1)), converting α and δ into Cartesian coordinates results in the direction of the Mars pole vector with respect to the **XYZ** basis vector of the EME reference frame

$$\begin{aligned} \mathbf{Z}_{MME} &= 0.446159\mathbf{X}_{EME} - 0.406238\mathbf{Y}_{EME} + \\ &+ 0.797442\mathbf{Z}_{EME} \end{aligned} \tag{4.2}$$

In order to define a coordinate frame, at least two perpendicular unit vectors must be specified. The Mars vernal equinox direction can be determined by finding the direction of the mean Mars orbit angular momentum vector at the same epoch, J2000. At J2000, the position and velocity directions, i.e., unit vectors pointing in the direction of position and velocity, of Mars can be found in the Sun-centered EME reference frame by applying Standish and Williams' method (discussed in detail in Section 4.7.1)

$$\begin{aligned} \hat{\mathbf{r}}_{Mars}^{EME} &= 0.999647\mathbf{X}_{EME} + 0.001007\mathbf{Y}_{EME} + \\ &- 0.026567\mathbf{Z}_{EME} \\ \hat{\mathbf{v}}_{Mars}^{EME} &= 0.044163\mathbf{X}_{EME} + 0.908515\mathbf{Y}_{EME} + \\ &+ 0.415513\mathbf{Z}_{EME} \end{aligned} \tag{4.3}$$

Therefore, knowing that the angular momentum of an orbit is $\mathbf{h} = \mathbf{r} \times \mathbf{v}$, the unit vector pointing in the direction of Mars' angular momentum is

$$\hat{\mathbf{h}}_{\text{Mars}}^{\text{EME}} = \hat{\mathbf{r}}_{\text{Mars}}^{\text{EME}} \times \hat{\mathbf{v}}_{\text{Mars}}^{\text{EME}} =$$
$$0.024569\mathbf{X}_{\text{EME}} - 0.416780\mathbf{Y}_{\text{EME}} + \tag{4.4}$$
$$+ 0.908675\mathbf{Z}_{\text{EME}}$$

Then, to find the direction of the vernal equinox of Mars ($\mathbf{X}_{\text{MME}}$), we take the cross-product between $\mathbf{Z}_{\text{MME}}$ and $\hat{\mathbf{h}}_{\text{Mars}}^{\text{EME}}$ and reduce it to the following unit vector

$$\mathbf{X}_{\text{MME}} = \mathbf{Z}_{\text{MME}} \times \hat{\mathbf{h}}_{\text{Mars}}^{\text{EME}} =$$
$$- 0.086410\mathbf{X}_{\text{EME}} - 0.906433\mathbf{Y}_{\text{EME}} + \tag{4.5}$$
$$- 0.413415\mathbf{Z}_{\text{EME}}$$

The remaining basis vector, $\mathbf{Y}_{\text{MME}}$ must complete the right-handed reference frame such that

$$\mathbf{Y}_{\text{MME}} = \mathbf{X}_{\text{MME}} \times \mathbf{Z}_{\text{MME}} =$$
$$0.890773\mathbf{X}_{\text{EME}} + 0.115542\mathbf{Y}_{\text{EME}} + \tag{4.6}$$
$$- 0.439516\mathbf{Z}_{\text{EME}}$$

Figure 4.2 shows the orientation of $\mathbf{X}_{\text{MME}}$, $\mathbf{Y}_{\text{MME}}$, and $\mathbf{Z}_{\text{MME}}$ with respect to Mars' orbital plane and its obliquity, ϵ_{Mars}. This reference frame is technically a rotating frame because of the time dependence in Equation (4.1). However, for the times of flight considered in a Mars mission, one may assume that such reference frame is "inertial enough."

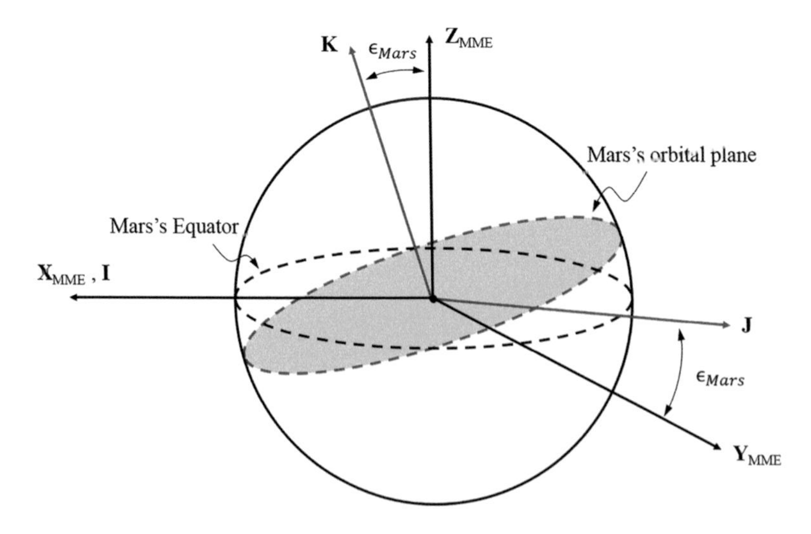

FIGURE 4.2 Definition of the Mars Mean Equatorial (MME) reference frame with respect to the orbital plane of Mars.

Combining the definitions of $\mathbf{X}_{\text{MME}}$, $\mathbf{Y}_{\text{MME}}$, and $\mathbf{Z}_{\text{MME}}$ expressed in the EME reference frame allows us to define the MME reference frame in terms of the Earth Mean Equator inertial reference frame. For example, using J2000, the direction cosine matrix to transform any vector from EME to MME can be constructed as

$$\mathbf{C}_{\text{EME}}^{\text{MME}} = \begin{bmatrix} \mathbf{X}_{\text{MME}} \\ \mathbf{Y}_{\text{MME}} \\ \mathbf{Z}_{\text{MME}} \end{bmatrix} = \begin{bmatrix} -0.086410 & -0.906433 & -0.413415 \\ 0.890773 & 0.115542 & -0.439516 \\ 0.446159 & -0.406238 & 0.797442 \end{bmatrix} \tag{4.7}$$

Here, we presented an example using Mars, but the same procedure can be repeated for other celestial bodies in our Solar System.

4.2.2 Heliocentric

For a large portion of interplanetary trajectories, heliocentric (Sun-centered) coordinates are used to reference the state vector of the satellite. This frame's fundamental XY plane is the Sun-Earth ecliptic plane. This fundamental plane is fixed in inertial space and the X-axis of this plane points in the direction of the Earth's Northern hemisphere vernal equinox (first point of Aries) which generally occurs on March 20 or 21. This axis points directly to Earth on the vernal equinox. The Z-axis points in the direction of celestial North (which does not correspond to the North Pole) and is perpendicular to the fundamental plane. The Y-axis completes the right-handed frame, such that $\mathbf{I} \times \mathbf{J} = \mathbf{K}$. When patched conics (Section 5.5) are used to model an interplanetary transfer, the coast phase is modeled using the dynamical equations of motion of the spacecraft with respect to the heliocentric inertial frame. This does not necessarily mean that the only acceleration used in the equations of motion is the gravitational acceleration of the Sun, but rather that *all* accelerations being considered must be written in terms of the heliocentric reference frame. An example of a heliocentric reference frame that uses the ecliptic plane as its fundamental reference frame is shown in Figure 4.3, where, by definition, the ecliptic plane contains the position and velocity vectors of the Earth with respect to the Sun, $\mathbf{r}_{\text{Earth/Sun}}$ and $\mathbf{v}_{\text{Earth/Sun}}$, respectively, and $\mathbf{K}$ points in the direction of Earth's angular momentum with respect to the Sun. For this reference, the fundamental plane of the ecliptic is defined at epoch J2000. Thus, this reference frame is often referred to as J2000 ecliptic.

In order to convert from the J2000 ecliptic to EME, one simple rotation is necessary. Since the ecliptic plane is defined to be the orbital plane of Earth about the Sun, the only change in orientation of the EME frame with respect to the J2000 ecliptic frame is the obliquity of the Earth, $\epsilon_{\oplus}$. This angle – sometimes referred to as the obliquity of the ecliptic – is Earth's axial tilt with respect to the ecliptic plane, and it varies between 22.1° and 24.5° over a cycle of about 41,000 years. Currently, $\epsilon_{\oplus}$ is approximately

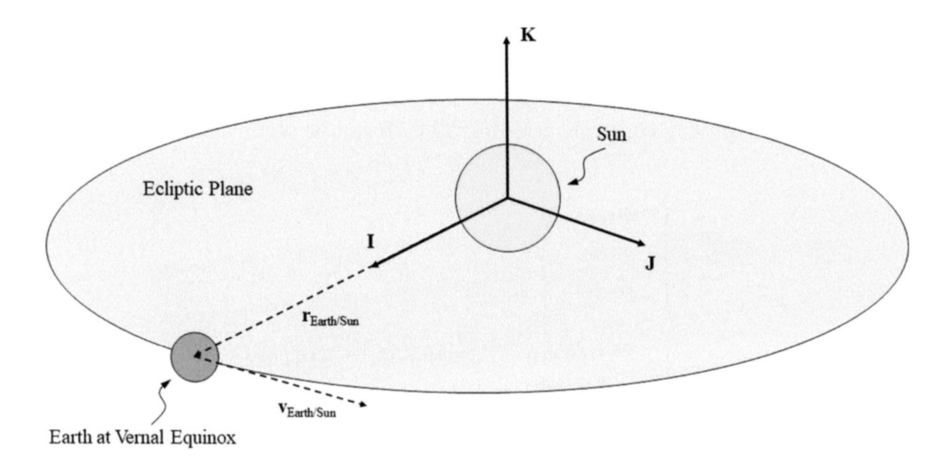

FIGURE 4.3 Definition of a heliocentric reference frame using the ecliptic plane as fundamental plane.

23.43928°, which is about halfway between its extreme values. The tilt last reached its maximum in 8700 B.C.E., and it is currently decreasing and projected to reach its minimum around the year 11800 C.E. Although the value of $\epsilon_{\oplus}$ can be estimated following a complex model known as Milankovitch cycles, it can be approximated as

$$\begin{aligned}\epsilon_{\oplus} = 23.4393^{\circ} - 0.0130^{\circ} \times T_{JD} + \\ - 2.0000^{\circ} \times 10^{-7} T_{JD}^{2} + 5.0000^{\circ} \times 10^{-7} T_{JD}^{3}\end{aligned} \tag{4.8}$$

where T_{JD} is number of Julian centuries, i.e., century composed of 36,525 days, since epoch J2000 (January 1, 2000) [Hughes et al., 1989]. Thus, knowing the value of Earth's obliquity permits to compute the single 1-axis direction cosine matrix that transforms from the ecliptic to the EME references frames as

$$\mathbf{C}_{\mathrm{Ecl2000}}^{\mathrm{EME2000}} = \mathbf{C}_1(\epsilon_{\oplus}) = \begin{bmatrix} 1 & 0 & 0 \\ 0 & \cos\epsilon_{\oplus} & \sin\epsilon_{\oplus} \\ 0 & -\sin\epsilon_{\oplus} & \cos\epsilon_{\oplus} \end{bmatrix} \tag{4.9}$$

The collection of changes in Earth's orbit, rotation, nutation, and precession are part of the so-called Milankovitch cycles. Figure 4.4 shows the estimated short-term changes in Earth's obliquity as a function of time for the past and next 10,000 years due to the Milankovitch cycles.

Understanding the phenomena related to the Milankovitch cycles is an essential part of today's science, which goes beyond the scope of this book. However, for the reader who wishes to learn more about how these phenomena affect space missions, we suggest "Theory of Satellite Geodesy – Applications of Satellites to Geodesy" by William M. Kaula [1966], which discusses geodesy (the science of accurately measuring and understanding the Earth's geometric shape, its orientation in space, and its gravity field as a function of time).

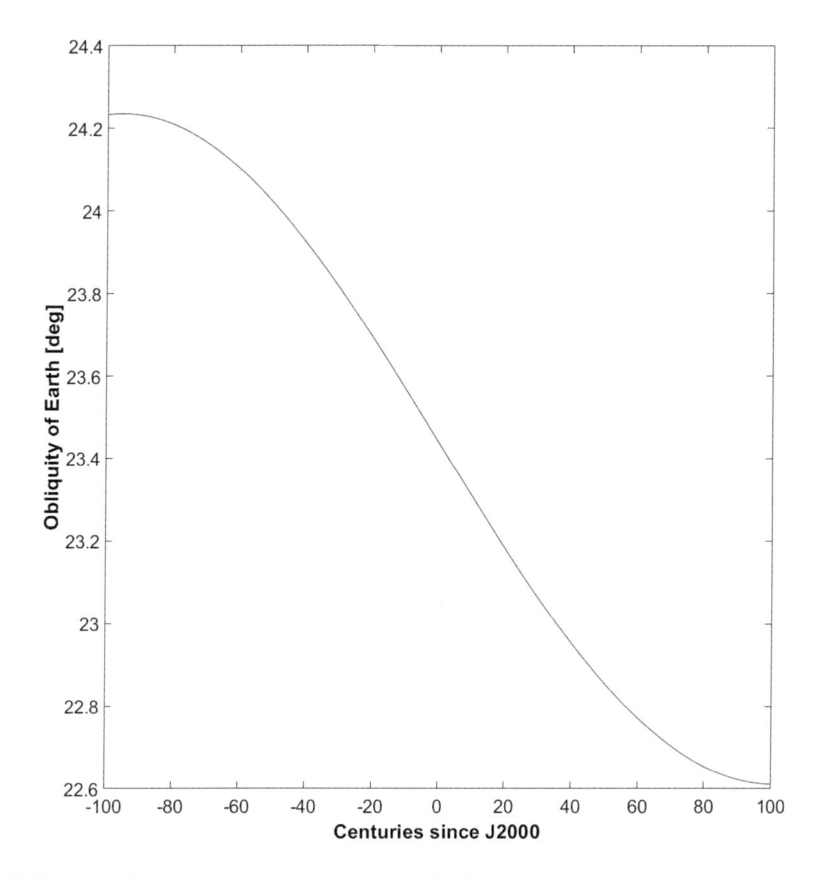

FIGURE 4.4 Short-term changes in Earth's obliquity as a function of time.

4.2.3 Catalogs of Fundamental Stars (FK)

The catalog of fundamental stars, or simply fundamental catalogs, is a series of six high-precision astrometric catalogs which include positional data for a selection of stars used to define a celestial reference frame. This is a standard coordinate system for measuring positions of stars. The fundamental catalogs were first introduced in 1879 by 19th-century German astronomer Georg Friedrich Julius Arthur von Auwers with the name *Fundamentalkatalog* (FK), which literally translates to "fundamental catalog." The second, updated version of the FK catalog, the *Neuer Fundamentalkatalog* (NFK), or "new fundamental catalog," was introduced by Peters in 1907. For simplicity, all subsequent version of the FK were numbered: FK3, FK4, and FK5. The International Astronomical Union (IAU) first recommended switching to the FK3 catalog in 1938, which was replaced in 1963 by FK4, and eventually by the FK5 in 1985. Eichhorn [1974] outlined the development of these catalogs. The FK5 catalog [Fricke et al., 1988] had positions and dynamical motion of 1,535 stars brighter than magnitude 7.5. The average position error ranged from 30 to 40 milliarcseconds. An extension of this catalog, FK5 Extension [Fricke et al., 1991], contains another

3,117 stars brighter than magnitude 11. The sixth iteration of the fundamental catalog (FK6) was introduced in 2000 and comes in two parts: FK6(I), which contains 878 stars, and FK6(II) and FK6(III), which together contain 3,272 stars. Both are updated and amended versions of FK5 using the catalog data of ESA's Hipparcos spacecraft. ESA's Gaia spacecraft, launched in 2013, was a follow-up mission that has further helped us update and refine astrometric measurements.

4.2.4 International Celestial Reference Frame

In 1991, the IAU established a new celestial reference frame based on kinematic instead of dynamic relationships. This system, known as the International Celestial Reference Frame (ICRF), replaced the FK5 system as the fundamental coordinate frame. This system is based on the positions of several hundred extragalactic radio sources observed using very long baseline interferometry. A subset of these objects is comprised of defining sources and establishes the orientation of the reference frames axes. The orientation of these axes is set relative to Earth's equator and vernal equinox at 2000.0 (00:00:00 UTC on January 1, 2000; we will discuss timing conventions further in Section 4.4). The origin of the ICRF is chosen as either the barycenter or geocenter of the Solar System. More information on this coordinate frame can be found in McCarthy [2000], Capitaine et al. [2002], and Seeber [2003], among others.

4.2.5 Barycentric and Geocentric Celestial Reference Systems

The IAU designed the barycentric and the geocentric celestial reference frames for use in astronomy with one at the origin of the Solar System barycenter and one with its origin at the geocenter. The barycentric celestial reference system describes the positions and motions of bodies outside the immediate vicinity of the Earth. It is used to describe the state vectors of objects in orbit around the Sun from the solution of the dynamical equations of motion. The geocentric celestial reference system is a local reference frame centered at the geocenter and is used extensively for Earth-based measurements of celestial objects. Both systems take into account many higher-order terms, such as relativistic effects on orbital motion. Relationships between the ICRF and these two reference frames are explained further, along with a more detailed explanation of these systems in Kaplan [2005].

4.2.6 B-Plane

A commonly used reference frame for interplanetary travel is the body-plane, or B-plane, frame. The B-plane frame is used for targeting the arrival at a planet and it is therefore primarily used when sending a spacecraft to a celestial body in the Solar System. It is especially useful to analyze B-plane parameters when considering targeting specific orbits within the sphere of influence of the arrival planet. Note that the B-plane approach can be used also when a spacecraft needs to perform an orbital transfer from an asteroid or the moon of a planet. Thus, here we use the words "arrival body" or simply "body" to refer to all of these celestial bodies being targeted for a mission.

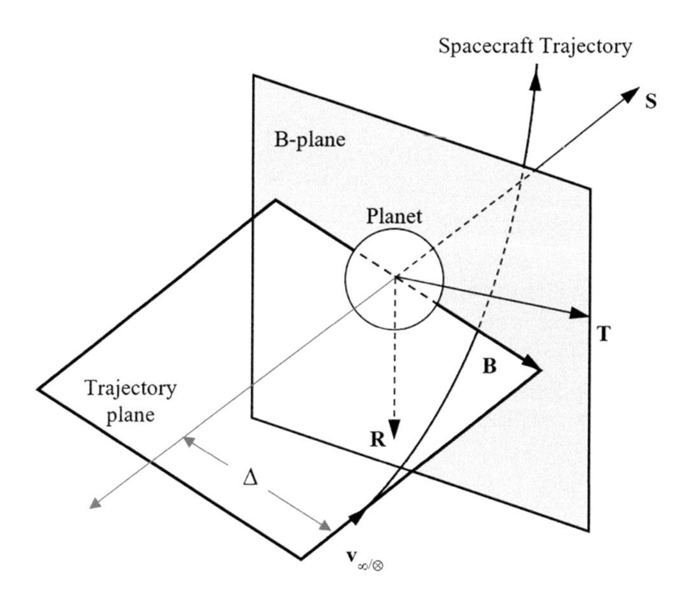

FIGURE 4.5 B-plane geometry.

The fundamental plane of the B-plane is the plane perpendicular to the plane of motion of the arrival planet around the Sun. The three axes of this reference frame, **RST**, are shown in Figure 4.2.

The B-plane is a plane which is normal to the asymptote of the incoming hyperbolic orbit and contains the target body's center of mass, as shown in Figure 4.2. The B-plane is normal to the velocity vector at "infinity." Here, infinity is defined where the spacecraft is far enough away from the body such that it essentially lies on the asymptote of the hyperbolic orbit.

S is a unit vector that has origins at the center of mass of the body being targeted and is parallel to the approach asymptote of the spacecraft's arrival orbit. Defining the hyperbolic excess speed of the spacecraft with respect to the target body as $\mathbf{v}_{\infty/\otimes}$ and using the definition of the perifocal coordinate system (Section 2.5.1), one can write

$$\mathbf{S}\cdot\mathbf{e} = e\cos\theta_\infty \tag{4.10}$$

where θ_∞ is the true anomaly corresponding to the spacecraft having velocity $\mathbf{v}_{\infty/\otimes}$ such that

$$\cos\theta_\infty = \lim_{r\to\infty}\cos\theta = \lim_{r\to\infty}\frac{h^2/\mu - r}{er} = -\frac{1}{e} \tag{4.11}$$

and the negative in the $-1/e$ term is taken care by the fact that $\theta_\infty > 180°$. Furthermore, **S** can be expressed as

$$\mathbf{S} = -\left(\cos\theta_\infty\mathbf{p} + \sin\theta_\infty\mathbf{q}\right) \tag{4.12}$$

where **p** and **q** are two of the unit vectors defining the perifocal reference frame and $\sin\theta_\infty$ may be written as

$$\sin\theta_\infty = -\sqrt{1-\frac{1}{e^2}} \tag{4.13}$$

In Equation (4.13), the square root will always result in a real value since the eccentricity e of a hyperbolic orbit is always greater than 1. Notice that θ_∞ is not defined for elliptical and circular orbits.

B is a unit vector that is defined originating from the target body's center of mass to the point where the spacecraft intersects the B-plane. **B** points in the direction of closest approach if the body were massless. This distance is also known as impact parameter Δ and, through some simple algebraic manipulations [Jah, 2002], it can be shown that

$$\Delta = \frac{r_p v_p}{v_\infty} \tag{4.14}$$

and, by using the vis viva equation

$$\Delta = r_p\sqrt{1+\frac{2\mu}{r_p v_\infty^2}} \tag{4.15}$$

For convenience, we have dropped the $/\otimes$ notation since it is implied that all orbital parameters here are measured with respect to the target body.

T is a unit vector normal to **S** that lies along the plane of reference of the target body. This plane is usually the equatorial plane, or a plane parallel to the ecliptic or Earth's mean equator. Lastly, **R** is defined as $\mathbf{R} = \mathbf{S}\times\mathbf{T}$ to complete the right-handed system of orthogonal unit vectors. It should be noted that the basis vectors for the B-plane are **R**, **S**, and **T**, although **B** is usually used for navigation purposes. The significance of this reference frame is due to the intersection of the trajectory's plane of motion of the spacecraft with respect to the B-plane, which, as described here, is defined based on the target body's orientation and fundamental plane (equatorial, or ecliptic, etc.). Spacecraft navigators target a point on the line where these two planes intersect. Further discussion of the B-plane and B-plane targeting applied to trajectory design for interplanetary orbital transfers is presented throughout the rest of this book.

4.3 ROTATING COORDINATE SYSTEMS

Rotating coordinate systems are inherently non-inertial reference frames because of the rotational acceleration they experience. We introduced the Earth-Centered Inertial (ECI) coordinate frame in Chapter 2 when we defined the classical orbital elements. While ECI is often used when describing Earth orbits, sometimes it may be more convenient to define a spacecraft position using a rotating reference frame that rotates at Earth's rotation rate, $\omega_\oplus$. The Earth-Centered Earth-Fixed (ECEF) frame is defined such that its basis vector **x** points in the direction of the prime meridian (Greenwich) and lies on the equatorial plane; the **z** unit vector points in the direction

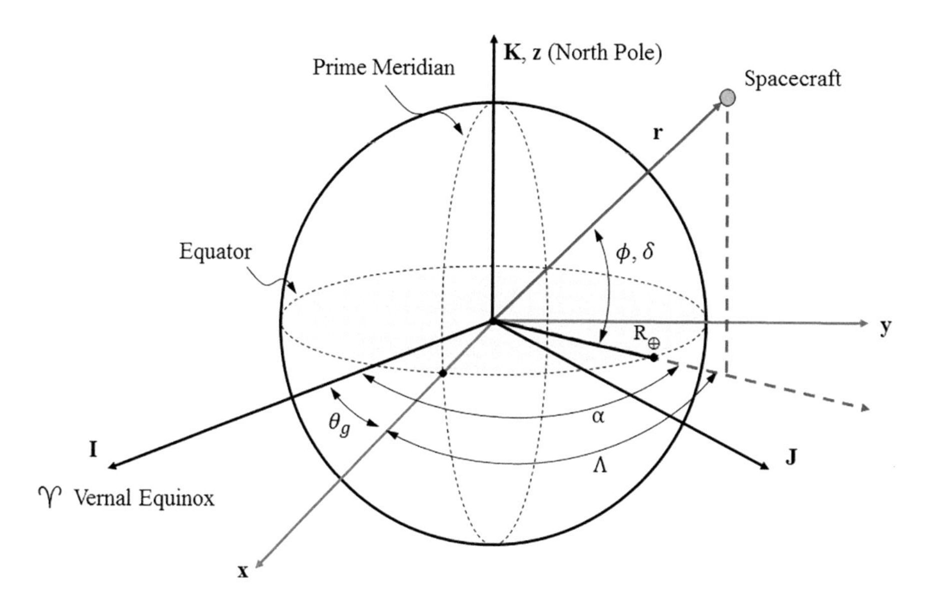

FIGURE 4.6 Relationship between ECI (rotating) and ECEF (non-rotating).

of Earth's geographical North Pole, and lastly **y** completes the right-handed system such that $\mathbf{x} \times \mathbf{y} = \mathbf{z}$. Figure 4.6 shows the relationship between ECI and ECEF.

In Figure 4.6, Λ represents the longitude, measured positively due East, ϕ is the latitude, measured positively due North from Earth's equator, δ is the declination, which, for this reference frame, is equivalent to ϕ, α is right ascension, and θ_g is the Greenwich location angle. θ_g moves over time following this approximate expression,

$$\begin{aligned}\theta_g = 2\pi\big[&0.7790572732640 + \\ &+1.00273781191135448(\mathrm{JD} - 2{,}451{,}545)\big]\end{aligned} \tag{4.16}$$

where JD is the Julian date, as explained in Section 4.4.7. Note that here $(\mathrm{JD} - 2{,}451{,}545)$ represents the number of days since January 1, 2000, noon Universal Coordinated Time (UTC; see Section 4.4.6). Thus, for a given spacecraft position vector $\mathbf{r} = r_x\mathbf{I} + r_y\mathbf{J} + r_z\mathbf{K}$, one can compute its corresponding right ascension and declination as follows

$$\delta = \sin^{-1}\left(\frac{r_x}{r}\right)$$

$$\alpha = \begin{cases} \cos^{-1}\left(\dfrac{r_x}{r\cos\delta}\right) & \text{if } m \geq 0 \\ 360^\circ - \cos^{-1}\left(\dfrac{r_x}{r\cos\delta}\right) & \text{if } m < 0 \end{cases} \tag{4.17}$$

where no quadrant check is needed for declination δ since $-90° \le \delta \le 90°$, but a quadrant correction is required for right ascension α since $0° \le \alpha \le 360°$. Converting between ECI and ECEF requires a three-axis direction cosine matrix where the angle of rotation is θ_g

$$\mathbf{C}_{\text{ECI}}^{\text{ECEF}} = \mathbf{C}_3(\theta_g) = \begin{bmatrix} \cos\theta_g & \sin\theta_g & 0 \\ -\sin\theta_g & \cos\theta_g & 0 \\ 0 & 0 & 1 \end{bmatrix} \tag{4.18}$$

such that any position vector can be transformed from ECI to ECEF with

$$\mathbf{r}^{\text{ECEF}} = \mathbf{C}_{\text{ECI}}^{\text{ECEF}} \mathbf{r}^{\text{ECI}} \tag{4.19}$$

4.3.1 Ground Tracks

Transforming right ascension and declination into latitude and longitude is rather simple

$$\begin{aligned} \Lambda &= \alpha - \theta_g \\ \phi &= \delta \end{aligned} \tag{4.20}$$

Orbital observations expressed in terms of right ascension and declination can be used to plot ground tracks. Figure 4.7 shows an example of the ground track of the International Space Station (ISS) for approximately half a day. Tick marks of the ISS trajectory are separated by 60 seconds.

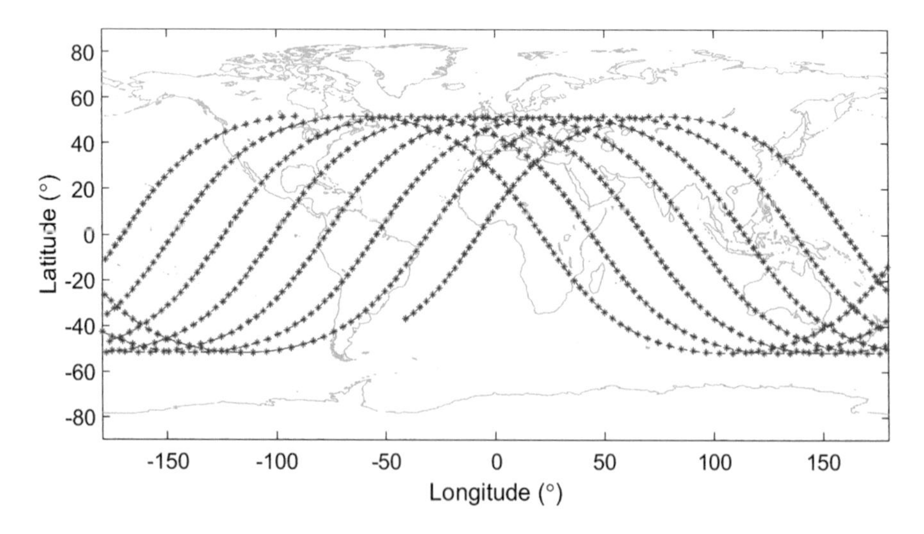

FIGURE 4.7 Example of a ground track depicting the ISS over 12 hours.

4.3.2 Topocentric Coordinate Frames

Topocentric coordinate systems are centered at the observer's location on the surface of the Earth. Because the Earth is not a perfect sphere (Figure 4.8) primarily due to its rotation, it is often useful to approximate Earth's shape using an ellipsoid with a flattening parameter f such that

$$f = \frac{R_e - R_p}{R_e} \tag{4.21}$$

where R_e and R_p are the radii of the Earth measured at the equator and the poles, respectively. Thus, approximating the shape of the Earth as an ellipsoid, R_e and R_p represent the semimajor and semiminor axes of the ellipse that results from a cross section of the Earth along its axis of rotation, as shown in Figure 4.8. This ellipse has an eccentricity e that can be related to f as

$$f = 1 - \sqrt{1 - e^2} \tag{4.22}$$

Figure 4.8 shows the difference between an ellipsoidal Earth (exaggerated in the figure) and a perfectly spherical Earth with respect to the position of an arbitrary ground station (GS). The circle representing the cross-sectional area of the spherical Earth is drawn such that the position vector $\mathbf{R}'$ is perpendicular to the surface where GS is located.

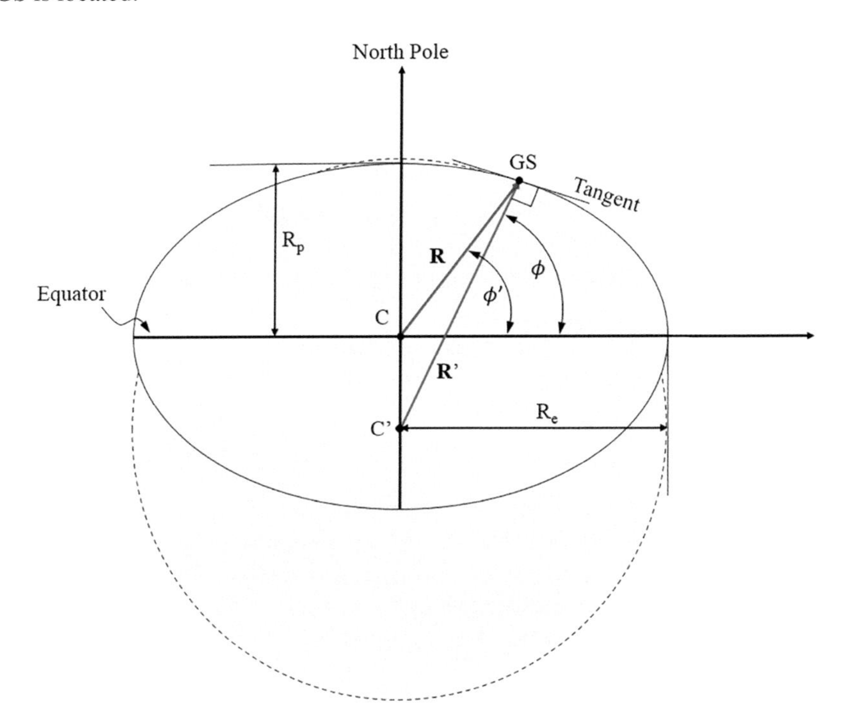

FIGURE 4.8 Difference in latitude measurements for a perfectly spherical Earth (dotted line) and an ellipsoid (solid line).

For Earth, f is approximately 0.003353. Although we present this section using the Earth as an example, the same analysis can be made with other planets. For example, Mars has a flattening of 0.00648. Generally, faster rotating objects tend to have a higher flattening, which often correlates to higher oblateness effects or, in other words, a higher value of J_2 (Section 2.7.1).

This flattening requires topocentric coordinates that are expressed in terms of latitude to be corrected. Thus, a ground station (GS) on the surface of a planet would have position

$$\begin{aligned}\mathbf{R} = & \left(\frac{R_e}{\sqrt{1-\left(2f-f^2\right)\sin^2\phi}}+H\right)\cos\phi\cos\theta\mathbf{I}+ \\ & +\left(\frac{R_e}{\sqrt{1-\left(2f-f^2\right)\sin^2\phi}}+H\right)\cos\phi\sin\theta\mathbf{J}+ \\ & +\left(\frac{R_e(1-f)^2}{\sqrt{1-\left(2f-f^2\right)\sin^2\phi}}+H\right)\sin\phi\mathbf{K}\end{aligned} \tag{4.23}$$

where H, θ, and ϕ are the altitude, longitude, and latitude of GS, respectively. ϕ is also known as the geodetic latitude, which would correspond to the latitude of GS if the planet were a perfect sphere (see Figure 4.8). Detailed derivations of Equations (4.22) and (4.23) can be found in Curtis [2019] and Vallado [2013]. It is often convenient to express **R** in terms of the geocentric latitude ϕ',

$$\mathbf{R} = R_e\cos\phi'\cos\theta\mathbf{I}+R_e\cos\phi'\sin\theta\mathbf{J}+R_e\sin\phi\mathbf{K} \tag{4.24}$$

where ϕ' includes the flattening parameter f such that, at altitude H of zero

$$\tan\phi' = (1-f)^2\tan\phi \tag{4.25}$$

Two topocentric horizon coordinate systems that are often used are the topocentric equatorial coordinate frame (inertial) and the topocentric horizon coordinate frame (rotating).

The topocentric equatorial coordinate frame (Figure 4.9) is an inertial reference frame that has origin at a ground station (GS) on the surface of Earth such that the **xyz** basis vectors that define this frame's orientation align with the **IJK** basis vectors that define the ECI coordinate frame (which has origin set at the center of the Earth).

The position of the spacecraft with respect to the center of the Earth **r** is thus

$$\mathbf{r} = \mathbf{R}+\rho \tag{4.26}$$

where ρ can be written in terms of either the **IJK** or **xyz** basis vectors as

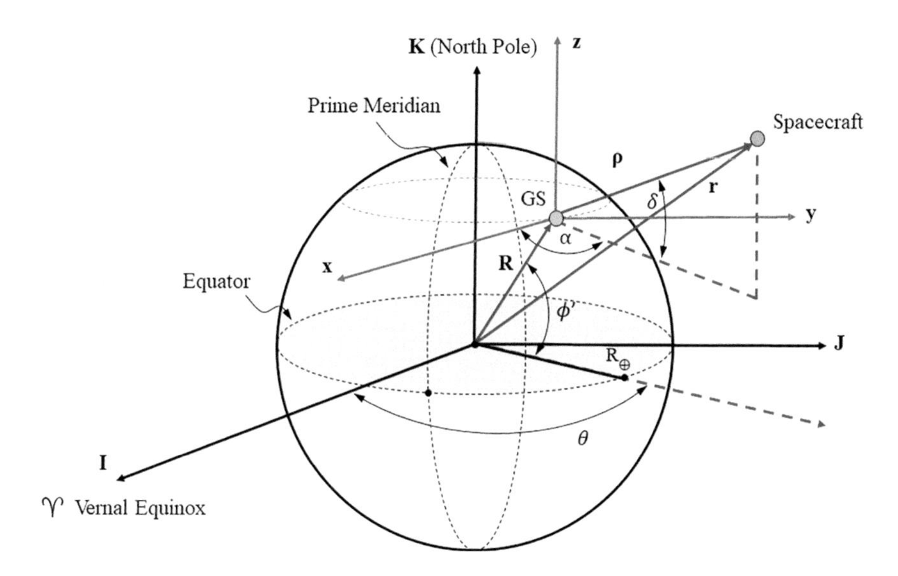

FIGURE 4.9 Topocentric equatorial coordinates **xyz**.

$$\boldsymbol{\rho} = \rho \cos\delta \cos\alpha \mathbf{I} + \rho \cos\delta \sin\alpha \mathbf{J} + \rho \sin\delta \mathbf{K} \tag{4.27}$$

or

$$\boldsymbol{\rho} = \rho \cos\delta \cos\alpha \mathbf{x} + \rho \cos\delta \sin\alpha \mathbf{y} + \rho \sin\delta \mathbf{z} \tag{4.28}$$

However, since the origins of **IJK** and **xyz** do not coincide, the direction cosines **r** and $\boldsymbol{\rho}$ will generally not be the same. In other words, topocentric right ascension and declination of a spacecraft will not generally be the same as the geocentric right ascension and declination, resulting in an effect known as parallax. For Earth-orbiting spacecraft, this effect is significant. Contrarily, when observing far away stars, this effect is negligible since $r \gg R$.

The topocentric horizon coordinate frame is a rotating reference frame that has origin at a ground station on the surface of the Earth. However, unlike the topocentric equatorial reference frame, the topocentric horizon coordinate frame has basis vectors that rotate, and it is therefore a non-inertial frame. The xy plane corresponds to the ground station's local horizon and is tangent to the ellipsoid at GS. Two main ways of defining a topocentric horizon frame are used: south-east-zenith (SEZ) and east-north-zenith (ENZ). Here, we will discuss the former, (SEZ; Figure 4.10), where the **xyz** basis vectors align with the local south (**x**), east (**y**), and zenith (**z**). Any vector can be converted between SEZ and ENZ using a simple direction cosine matrix with a rotation angle of ±90° about the z-axis

$$\begin{aligned} \mathbf{C}_{\mathrm{ENZ}}^{\mathrm{SEZ}} &= \mathbf{C}_3(-90^\circ) \\ \mathbf{C}_{\mathrm{SEZ}}^{\mathrm{ENZ}} &= \mathbf{C}_3(90^\circ) \end{aligned} \tag{4.29}$$

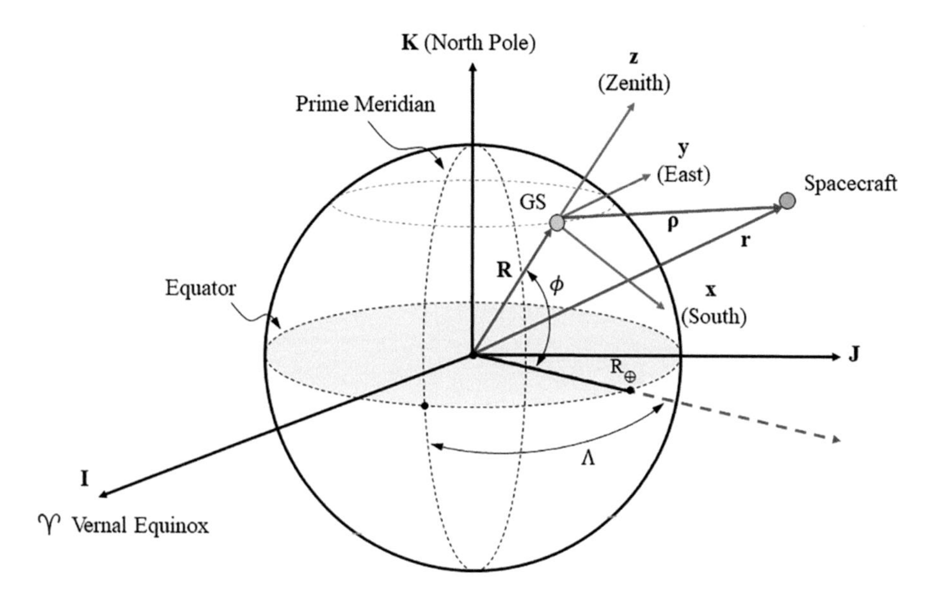

FIGURE 4.10 Topocentric horizon coordinates SEZ (**xyz**).

where C_3 is the direction cosine matrix corresponding to a three-axis rotation such that for a generic angle α, $\mathbf{C}_3(\alpha) = \begin{bmatrix} \cos\alpha & \sin\alpha & 0 \\ -\sin\alpha & \cos\alpha & 0 \\ 0 & 0 & 1 \end{bmatrix}$.

The position of the ground station can be written in the SEZ frame as

$$\mathbf{R}^{SEZ} = (R_\oplus + H)\mathbf{z} \tag{4.30}$$

where $R_\oplus + H$ is the location of the ground station on Earth's surface (H is the elevation). Since the ground station is located on the surface of the Earth, it moves along with the planet's rotation

$$\boldsymbol{\omega}_\oplus^{SEZ} = \omega_\oplus(-\cos\phi\mathbf{x} + \sin\phi\mathbf{z}) \tag{4.31}$$

such that the inertial velocity, $\dot{\mathbf{R}}$, of the ground station can be expressed in terms of the SEZ frame as

$$\dot{\mathbf{R}} = \dot{\mathbf{R}}^{SEZ} + \boldsymbol{\omega}_\oplus^{SEZ} \times \mathbf{R} \tag{4.32}$$

where $\dot{\mathbf{R}}^{SEZ}$ is zero unless the observations are being collected from a moving vehicle. Here, we assume $\dot{\mathbf{R}}^{SEZ} = \mathbf{0}$ since we are referring to ground stations on the surface of the Earth, which results in

$$\dot{\mathbf{R}} = \omega_\oplus(R_\oplus + H)\cos\phi\mathbf{y} \tag{4.33}$$

If the position of the spacecraft is needed in ECI, it is possible to convert between ECI and ECEF using Equation (4.9). Then, in order to convert between ECEF and SEZ, the following direction cosine matrix can be used

$$\mathbf{C}_{\text{ECEF}}^{\text{SEZ}} = \begin{bmatrix} \sin\phi\cos\Lambda & \sin\phi\sin\Lambda & -\cos\phi \\ -\sin\Lambda & \cos\Lambda & 0 \\ \cos\phi\cos\Lambda & \cos\phi\sin\Lambda & \sin\phi \end{bmatrix} \tag{4.34}$$

where ϕ and Λ are the latitude and longitudes of the ground station, respectively. The state of the spacecraft can then be converted between ECEF and SEZ using the following relationships

$$\begin{aligned} \mathbf{r}^{\text{SEZ}} &= \mathbf{C}_{\text{ECEF}}^{\text{SEZ}}\mathbf{r}^{\text{ECEF}} \\ \dot{\mathbf{r}}^{\text{SEZ}} &= \mathbf{C}_{\text{ECEF}}^{\text{SEZ}}\dot{\mathbf{r}}^{\text{ECEF}} \end{aligned} \tag{4.35}$$

Generally, ground stations acquire observations recording the azimuth θ_a and elevation θ_e (Figure 4.11), which are the angles measured from the local south and the local horizon, respectively.

Azimuth and elevation can be related to the spacecraft position with respect to the ground station as follows

$$\begin{aligned} \tan\theta_a &= \frac{\boldsymbol{\rho}\cdot\mathbf{y}}{\boldsymbol{\rho}\cdot\mathbf{x}} \\ \sin\theta_e &= \frac{\boldsymbol{\rho}\cdot\mathbf{z}}{\rho} \end{aligned} \tag{4.36}$$

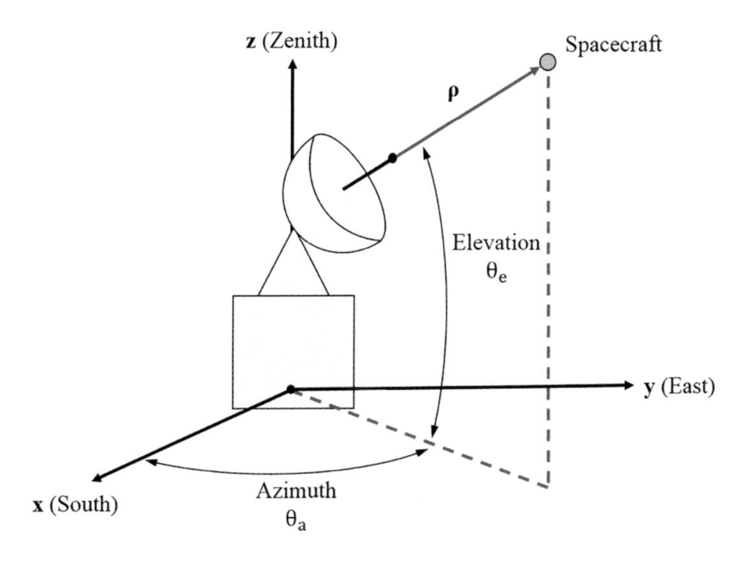

FIGURE 4.11 Azimuth and elevation angles in the topocentric horizon coordinates SEZ.

where a quadrant correction is needed for azimuth since $0° \le \theta_a \le 360°$, but not for elevation since $0° \le \theta_a \le 180°$.[2] Furthermore, the velocity of a tracked spacecraft with respect to the ground station in SEZ coordinates is

$$\dot{\rho}^{SEZ} = \dot{\mathbf{r}}^{SEZ} - \dot{\mathbf{R}}^{SEZ} + \omega_{\oplus}^{SEZ} \times \rho^{SEZ} \tag{4.37}$$

where, as stated above, $\dot{\mathbf{R}}^{SEZ} = \mathbf{0}$ for ground stations. Then, if the velocity of the spacecraft is needed in ECEF or ECI, Equations (4.18) and (4.34) can be used to make the appropriate coordinate transformations.

4.4 TIME

Another important aspect of using coordinate frames is the time at which events take place, observations are made, and data is collected. The instant at which a specific event takes place is known as the epoch of that event, which is reported as a date. While we have developed a variety of tools to measure time intervals, it was necessary to agree upon a fundamental epoch from which time would be measured.

Time measurements have changed dramatically over the course of history. Ancients used sundials and clepsydras ("water clocks," used by Greeks and Romans), or incense clocks and astrolabes, used by Chinese and Persians, respectively. The first mechanical clocks were invented in Europe only around the start of the 14th century. They remained the standard timekeeping device until the invention of the pendulum clock, in 1656, which is considered to have started the era of precision timekeeping. Electric clocks made their debut in 1815, greatly facilitating the measurement of time and increasing its accuracy. More modern and eventually digital watches were introduced in the 20th century and, since then, the accuracy of timekeeping has increased dramatically.

Even time definitions have evolved significantly throughout history. For example, while ancient Romans divided the day into 24 hours like we do today, they did not use minutes and seconds, unlike the Babylonians, who used a sexagesimal (counting in 60s) system for mathematics and astronomy, including timekeeping. Even the calendar has gone through many iterations based on civilizations and time periods. For example, the French revolution introduced new conventions, making the day the autumnal equinox occurred in Paris the first day of the year, which had 12 months of 30 days each. Each month was given a new name based on natural phenomena, such as *Brumaire* ("mist") and *Floréal* ("flower"). They also used 10-hour days, and each hour was divided into 100 decimal minutes, and each decimal minute into 100 decimal seconds. This unusual calendar was used by the French government for about 12 years from late 1793 to 1805. Another example of an unusual calendar-related event can be seen when the Gregorian calendar was adopted. In fact, the Julian calendar, used predominantly in the Christian world for the first millennium C.E. and part of the second millennium, was an improvement over the Roman republican calendar

[2] Theoretically, one could measure a negative elevation if a ground station is located, for example, at the peak of a mountain. Similarly, if a sensor is mounted on an aircraft or spacecraft, elevation could be negative too. However, no quadrant correction is needed since the arcsine function is uniquely defined between −180° and 180°.

that it replaced, but it was 11 minutes and 14 seconds longer than the tropical year (the time it takes the Sun to return to the same position, as seen from Earth). As a result, the calendar drifted about 1 day for every 314 years. To fix this problem, Pope Gregory XIII signed a papal bull in February 1582 promulgating the reformed calendar, the Gregorian calendar. However, to implement the new calendar in October 1582, 10 days were dropped from the calendar to bring the vernal equinox from March 11 back to March 21. This resulted in the days between October 5, 1582, and October 14, 1582 (included) to have never existed.

Time intervals can be measured with the familiar days, hours (h), minutes (m), and seconds (s). However, it may be more convenient to refer to them as angles at times. The conversion between times measured in seconds (s) to degrees is defined based on the fact that Earth rotates on its axis 360 degrees (°) each day (24 hours), i.e., 15° each hour. Subunits of degrees are arcminutes (′) and arcseconds (″), such that

$$\begin{aligned} 1^{h} &= 60^{m} = 3{,}600^{s} \\ 1^{\circ} &= 60' = 3{,}600'' \end{aligned} \tag{4.38}$$

from which

$$\begin{aligned} 1^{h} &= 15^{\circ} \\ 1^{m} &= 15' \\ 1^{s} &= 15'' \end{aligned} \tag{4.39}$$

From the perspective on an observer on the surface of the Earth, the Sun and Moon both have approximately the same apparent size of 30’ in the sky. We are going to discuss various time systems and report some of the most useful time conversions for interplanetary applications. We refer the reader to the literature [Vallado, 2013; Curtis, 2019] for additional time conversion algorithms not discussed here.

4.4.1 Solar and Sidereal Time

Since the Sun regulates nearly all human activities on Earth, we generally base our time systems on the apparent motion of the Sun in the sky. Solar time is defined by successive transitions of the Sun over a reference point, usually a local meridian. In 1884, Greenwich (pronounced “GREN-ich,” based on its original British English pronunciation) was assigned 0° latitude, making it the official prime meridian. Therefore, a solar day is defined as the time the Sun takes to reappear at the same local meridian, or 24^{h}. This time, however, does not account for the fact that the Earth orbits the Sun in approximately 365.24 days, or about 1°/day. Since Earth’s orbit is prograde around the Sun, but Earth rotates on its axis due East, this means that the time Earth takes to complete a full rotation on its axis with respect to far away “fixed” stars is slightly smaller than a solar day. This time is known as sidereal day, which is approximately $23^{h}\ 56^{m}\ 4.0905^{s}$. Thus, 24 *sidereal hours* make up one sidereal day. The difference between solar and sidereal day is shown in Figure 4.12.

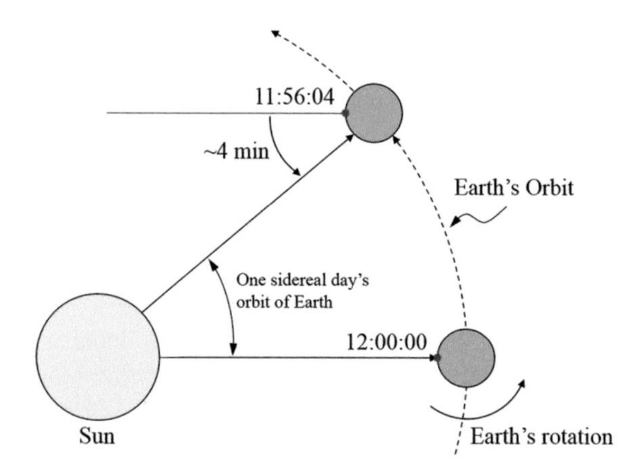

FIGURE 4.12 Geometry of sidereal day vs. solar day.

As shown in Figure 4.12, sidereal and solar time are different. Because of this, different definitions of "a day" can be used, depending on the application. For example, a mean sidereal day is 24 sidereal hours, which is the time the Earth takes to rotate one time on its axis with respect to far away stars. In order to convert between sidereal and solar time, the following conversion factors can be used, as provided in the Astronomical Almanac [1984],

$$\begin{aligned} 1 \text{ solar day} &= 1.002737909350795 \text{ sidereal day} \\ 1 \text{ sidereal day} &= 0.997269566329084 \text{ solar day} \\ 1 \text{ solar day} &= 24^{h}\, 3^{m}\, 56.5553678^{s} \text{ sidereal time} \\ 1 \text{ sidereal day} &= 23^{h}\, 56^{m}\, 4.090524^{s} \text{ solar time} \end{aligned} \tag{4.40}$$

However, due to the small yet measurable changes in the orientation of Earth's axis and its rotation, the definitions of solar day and sidereal day provided above are not constant, including the conversion factors used in Equation (4.40). In fact, as provided in the Astronomical Almanac [1984], Earth's precession rate causes variations in the conversion factors between solar and sidereal day, such that

$$\begin{aligned} 1 \text{ solar day} = 1.002737909350795 &+ 5.9006 \times 10^{-11} T_{UT1} \\ &- 5.9 \times 10^{-15} T_{UT1}^{2} \text{ sidereal day} \end{aligned} \tag{4.41}$$

where T_{UT1} is the number of Julian centuries since the J2000 epoch, resulting in a negative value for T_{UT1} for dates before January 1, 2000.

While using solar time may be convenient for everyday life on Earth, it is not sufficiently accurate for many applications, especially for interplanetary missions. Because of this, universal time (UT) was instead adopted in 1928, when UT was defined as the mean solar time of the Greenwich meridian (0° longitude). Universal time replaced the designation Greenwich Mean Time (GMT), and it is now used to denote the solar time when an accuracy of about 1 second is sufficient. Later, UTC[3] was adopted, as discussed in more details in Section 4.4.6.

Time can also be measured using the motion of certain celestial bodies as reference. This time, called dynamical time, uses the fact that time is the independent variable in the equations of motion that govern the dynamics of, for example, the Earth's motion with respect to the Sun or the Moon's motion with respect to the Earth. In order to precisely adopt this time, relativist corrections must be implemented. One time convention that is generally used for high-precision measurements is atomic time, which is based on the concept of an atomic second, as described in the next section.

The SI unit of time, the second, is a fundamental unit of time, which is defined in terms of the radiation frequency at which atoms of the element cesium change from one state to another. While the second may seem a rather obvious unit of measurement to the reader, its definition has seen significant changes over the course of history. In fact, the second was formerly defined as 1/86,400 of the mean solar day, but it proved to be too imprecise. As a result, in 1956 the International Committee on Weights and Measures redefined the second as 1/31,556,925.9747 of the length of the tropical (seasonal) year in 1900. In 1967, the 13th General Conference on Weights and Measures provisionally defined the second as 9,192,631,770 cycles of radiation associated with the transition between the two hyperfine levels of the ground state of the cesium-133 atom, which is how it is implemented in atomic time. The number of cycles of radiation was chosen to make the length of the defined second correspond as closely as possible to that of the now obsolete astronomically determined second of Ephemeris Time (the fraction of the tropical year given above) [BIPM, 2006].

4.4.2 International Atomic Time (TAI)

International Atomic Time (TAI, from the French name "*temps atomique international*") is a high-precision atomic coordinate time standard. TAI is a continuous scale of time that does not include leap seconds. This time is the basis for UTC. Atomic clocks are extremely precise, and they are predicted to be off by less than 1 second in more than 50 million years [BIPM, 2015].

4.4.3 Ephemeris Time (ET)

Ephemeris time (ET) refers to time measured in conjunction with the ephemeris[4] of a given celestial object, i.e., the object's position and velocity vectors at a given time. ET usually refers to the time standard adopted in 1952 by the IAU, which was then replaced in 1976s. ET was proposed in 1948 as a Newtonian-based time scale

[3] This abbreviation was chosen by the International Telecommunication Union and the International Astronomical Union, which agreed on wanting to use the same abbreviation regardless of language. In English CUT ("coordinated universal time") was proposed, while in French TUC ("*temps universel coordonné*") was proposed. The result is a combination of both, UTC, which conforms to the pattern for the abbreviations of the variants of Universal Time (UT0, UT1, etc.).

[4] Ephemeris comes from the Greek ἐφήμερος (*ephēmeros*), meaning "daily" or "lasting only 1 day." The plural of ephemeris is ephemerides, which is pronounced "eh-fuh-MEH-ruh-deez".

to overcome the disadvantages of irregularly fluctuating mean solar time. ET thus became the first application of the concept of a dynamical time scale, where the time and time scale are defined by and inferred from the observed ephemeris data (usually position and velocity) of an astronomical object through its motion. However, ephemeris time sometimes refers to a modern relativistic coordinate time scale that was implemented by the Jet Propulsion Laboratory (JPL). To avoid confusion, here we refer to JPL's definition of ephemeris time as T_{eph}. JPL developed a series of numerically integrated Development Ephemerides (DE) that have been updated since their first application with DE102, created in 1981. At the time of writing, the most recent is DE441, which was created in June 2020 [Park et al., 2021]. All DE versions are available in the various published articles, but for simplicity we refer the reader to a DE full list on JPL's own website where all such publications are linked.[5] We discuss JPL's planetary ephemerides in more detail in Section 4.8.

4.4.4 Barycentric Dynamical Time (TDB) and Barycentric Coordinate Time (TCB)

Barycentric Dynamical Time (TDB, from French: *Temps Dynamique Barycentrique*) is a relativistic coordinate time scale which considers the time dilation (or contraction) due to relativistic effects. This time is especially used when considering the ephemeris of celestial bodies such as planets, moons, asteroid, and comets. TDB is used in conjunction with the Solar System-barycentric reference frame, and it was first defined in 1976 to replace the non-relativistic ephemeris time. This time is therefore useful also for interplanetary spacecraft that travel really close to or far from the Sun. In 2006, the IAU defined TDB as a linear scaling of Barycentric Coordinate Time (TCB) [IAU, 2006]. From the perspective of Earth's surface, TDB has a difference from Terrestrial Time (TT; Section 4.4.5) that is less than 2 milliseconds for several millennia. For reference, TCB progresses faster than TDB (approximately 0.5 second/year), while TDB and TT remain close. As of the beginning of 2011, the difference between TDB and TCB has amounted to about 16.6 seconds. TDB and TCB are related to each other as follows

$$\mathrm{TDB} = \mathrm{TCB} - 1.550519768 \times 10^{-8} \times \left(\mathrm{JD}_{\mathrm{TCB}} - \mathrm{T}_0\right) \times 86{,}400 + \mathrm{TDB}_0 \quad (4.42)$$

where $\mathrm{TDB}_0 = -6.55 \times 10^{-5}\,\mathrm{s}$, $\mathrm{T}_0 = 2{,}443{,}144.5003725$, and $\mathrm{JD}_{\mathrm{TCB}}$ is the TCB Julian date, i.e., a quantity which was equal to T_0 on January 1, 1977, 00:00:00 TAI which increases by one every 86,400 seconds of TCB [IAU, 2006].

4.4.5 Terrestrial Time (TT) and Terrestrial Dynamical Time (TDT)

Terrestrial time (TT) – formerly terrestrial dynamical time (TDT), which replaced ephemeris time (ET) – is used primarily for time measurements of astronomical observations made from the surface of Earth. TT is a timescale of apparent geocentric ephemerides of other celestial bodies in the Solar System. Thus, TT is used when developing the ephemerides of the Sun, Moon and planets as seen from Earth.

[5] https://ssd.jpl.nasa.gov/planets/eph_export.html

Because of this, TT is useful for interplanetary applications, especially for observations made from Earth and for communication with interplanetary spacecraft. TT is independent of the rotation of the Earth and is measured using the SI second adopted by atomic clocks. However, TT is not defined by atomic clocks, but it is a theoretical time that real clocks can only approximate. TT is related to other time systems discussed in this chapter by the relationships

$$\begin{aligned} \mathrm{UTC} &= \mathrm{UT1} - \Delta\mathrm{UT1} \\ \mathrm{TAI} &= \mathrm{UTC} + \Delta\mathrm{AT} \\ \mathrm{TT} &= \mathrm{TAI} + 32.184^{\mathrm{s}} \end{aligned} \tag{4.43}$$

where ΔAT and ΔUT1 are accumulated differences in atomic time and universal time, respectively. The 32.184 s offset in Equation (4.43) was chosen so as to provide continuity with ET. TT is defined to be exactly 32.184 seconds offset from TAI at the start of January 1, 1977.

4.4.6 Universal Time and Its Variations

Universal time (UT) was adopted in 1928, replacing Greenwich Mean Time (GMT), as we saw in Section 4.4.1. Variations of UT exist, and the most used and known are UT0, UT1, and UT2. UT0 is related to the local hour angle as

$$\mathrm{UT0} = 12^{\mathrm{h}} + \mathrm{GHA}_{\mathrm{Sun}} = 12^{\mathrm{h}} + \mathrm{LHA}_{\mathrm{Sun}} - \lambda \tag{4.44}$$

where $\mathrm{GHA}_{\mathrm{Sun}}$ is the Greenwich Hour Angle of the Sun, $\mathrm{LHA}_{\mathrm{Sun}}$ is the local hour angle and λ is the longitude. The 12^{h} is added to ensure that 12:00 corresponds to noon (when the Sun can be seen) and 00:00 corresponds to midnight. According to Equation (4.44), UT0 is noon at Greenwich when the Sun is at an hour angle of zero.

UT1 accounts for polar motion, and it is therefore independent of station location, while UT2 takes into account seasonal variations too. However, UT2 is considered to be obsolete, while the difference between UT0 and UT1 is on the order of tens of milliseconds.

4.4.7 Julian Date (JD) and Modified Julian Date (MJD)

The Julian day is the continuous count of days since the beginning of the Julian period. The first Julian day, which is assigned number 0, starts at noon on Monday, January 1, 4713 B.CE. This date may seem arbitrary and while it does not correspond to any known historical event, it is the result of combining three multiyear cycles (Roman indiction, solar, and lunar cycles[6]) suggested by Joseph Scaliger[7] in 1582

---

[6] Roman indiction was a periodic reassessment of taxation in the Roman Empire which took place every 15 years. The solar cycle is a 28-year cycle of the Julian calendar, and 400-year cycle of the Gregorian calendar with respect to the week. It occurs because leap years occur every 4 years and there are 7 possible days to start a leap year, making a 28-year sequence. The Metonic or lunar cycle is a period of approximately 19 years after which the phases of the moon recur at the same time of the year.

[7] The adjective "Julian" comes from Scaliger's father's name.

such that Julian day 0 falls on the occurrence of all such cycles. The Julian date (JD) is the number of days between Monday, January 1, 4713 B.C.E. and the day of interest. For example, the Julian day number for the day starting at 12:00 UT (noon) on January 1, 2000, was 2,451,545. The Julian date of any instant is the Julian day number plus the fraction of a day since the preceding noon in UT. For example, the Julian date for 17:05:00.0 UT January 22, 1990, is 2,447,914.21181. The Julian period is a chronological interval of 7,980 years; year 1 of the Julian period was 4713 B.C.E. (−4712). This means that the next Julian period begins in the year 3268 C.E. In order to convert from a date given in years (yr), months (mo), days (d), hours (hr), minutes (min), and seconds (s) to Julian date, the following equation can be used

$$\begin{aligned} JD = 367\,yr - INT\left[\frac{7}{4}\left(yr + INT\frac{mo+9}{12}\right)\right] + \\ + INT\left(\frac{275\,mo}{9}\right) + d + 1{,}721{,}013.5 + \\ + \frac{1}{24}\left[\frac{\frac{s}{60^*} + min}{60} + hr\right] \end{aligned} \tag{4.45}$$

where 60^* should be substituted with 61 seconds for days with leap seconds. When using Equation (4.45), the results in JD are in the order of millions. However, since these represent days, it is recommended that at least 8 decimal places (~10^{-4}s accuracy) be kept in these calculations. From a computational perspective, it maybe be more efficient to separate the integer part from the decimal part in order to maintain an appropriate level of significant figures.

In 1957, the Smithsonian Astrophysical Observatory (SAO) introduced the modified Julian date (MJD). The purpose of MJD is to simplify the lengthy Julian date values so that computers using a small word length (bit counts) could still operate correctly without the danger of overflowing. MJD is simply defined as

$$MJD = JD - 2{,}400{,}000.5 \tag{4.46}$$

which corresponds to MJD starting at midnight on November 17, 1858. Since JD and MJD are both time intervals represented as real numbers, they find much more applications than the Gregorian date format.

4.4.8 Leap Seconds

Since the rate of rotation of the Earth constantly changes, it is necessary to occasionally add a second during the year to ensure the atomic timescale UTC stays in synchronization with nature. In fact, UTC deviates from TAI by a number of whole seconds, known as leap seconds. As of January 1, 2017, when the last leap second to date was put into effect, UTC is currently exactly 37 seconds behind TAI. The 37 seconds result from the initial difference of 10 seconds at the start of 1972, plus 27 leap seconds in UTC since 1972.

4.4.9 One-Way Light Time (OWLT)

Because signals used to communicate between ground stations on Earth and spacecraft travel at the finite speed of light, there is a delay between the time a transmission (usually in the form of a radio signal) is sent from a spacecraft and arrives at a ground station, or vice versa. The time it takes for light to travel between two objects is called one-way light time (OWLT). For interplanetary applications, this is the time that a radio signal takes to travel between Earth and a spacecraft or other body in space. OWLT is generally measured in milliseconds, and it is a value that changes continuously because of the motion of a given spacecraft through space, therefore changing the relative motion between it and the ground stations on Earth. While OWLT for Earth-orbiting missions can be on the order of a few milliseconds, interplanetary missions can experience OWLT on the order of minutes or hours. For example, Mars missions experience a OWLT of 5–20 minutes, depending on the relative positions of Earth and Mars in their orbits around the Sun, while far away spacecraft such as Voyager 1 and Voyager 2 experience a OWLT of approximately 21.5 and 17.8 hours, respectively, as of the end of 2022. These values are increasing since both Voyager spacecraft are traveling away from us. Sometimes, OWLT is referred to as light time. For example, saying that the New Horizon spacecraft is 7 light-hours away corresponds to its OWLT being 7 hours. OWLT added to transmission time, which is the time it takes for a given instrument to compute and prepare data to be transmitted or received, is equal to spacecraft event time. This means that, even for simple instructions that require near-instantaneous computations once received by the spacecraft command and data handling system, mission controllers must still account for OWLT in order to accurately assign instructions to a spacecraft.

4.4.10 Spacecraft Event Time (SCET)

Spacecraft Event Time (SCET) is the local time that the spacecraft uses to measure events that occur at the spacecraft. SCET is used by mission controllers to give commands that control the timing of spacecraft operations and to identify when specific events occur on the spacecraft relative to Earth time. Most spacecraft operations need to occur at specific times, such as orbit insertion maneuvers. In fact, signals are sent from ground stations to the spacecraft with instructions given in SCET, such that a spacecraft can autonomously perform a set of instructions without needing immediate intervention from mission control, which would be impossible due to the minutes or hours that separate the spacecraft from Earth. For example, assuming that signals are received and acted upon instantaneously, a signal from a spacecraft orbiting Mars 20 light-minutes away would be received by a ground station on Earth at exactly 15:00 UTC from a spacecraft showing that it had just completed an orbit insertion maneuver, the SCET time of the maneuver would be 20 minutes earlier (14:40 UTC). Conversely, if an orbit insertion maneuver is required by a spacecraft arriving at Mars 20 light-minutes away from Earth, the ground station would need to send instructions to the spacecraft taking into account the 20 minutes that said signal takes to reach the spacecraft. Spacecraft Event Time in UTC is also known as orbiter UTC and Earth-received time as ground UTC.

4.4.11 Mission Elapsed Time (MET)

Mission Elapsed Time (MET) is the amount of time recorded from the instant a spacecraft is launched. Following this definition, MET is zero until the moment the spacecraft launches, and the elapsed time of a flight is then measured in days, hours, minutes, and seconds. For example, if MET for a given mission is recorded as 5/07:23:36, this means that the spacecraft launched 5 days, 7 hours, 23 minutes, and 36 seconds ago. MET is recorded by clocks on board a spacecraft, and it was first adopted by NASA during their crewed space missions, most notably during their Space Shuttle missions. If certain mission operations depend on the time of launch, such events can be easily scheduled using MET, instead of UTC, for example. This avoids constant rescheduling of events in case a launch needs to be rescheduled. The International Space Station (ISS) does not use an MET clock since it is considered a permanent and international mission. Instead, the ISS uses UTC. When the Space Shuttle visited ISS, the ISS-crew would adjust their workday to the MET clock of the shuttle to facilitate scheduling, similar to how one adjusts to jet-lag after a long flight. The shuttles also carried UTC clocks so that astronauts could convert between ISS and shuttle times prior to arrival and docking. A notable MET-related event that caused the launch of a spacecraft to be postponed was when in 2019, a test flight of the Boeing CST-100 Starliner spacecraft suffered a mission anomaly through an incorrectly set MET on their vehicle.[8]

4.5 LUNAR EPHEMERIDES RELATIVE TO THE EARTH

Lunar ephemerides are used for lunar missions, and they are often described with respect to Earth-Centered Inertial (ECI). Accurate lunar ephemerides are needed for most applications, and thus JPL's Development Ephemerides (DE) are used. We will discuss these in detail in Section 4.8. However, for preliminary mission design, an approximate form of lunar ephemerides can be used, as provided by Simpson [1999]. This is because the motion of the Moon with respect to the Earth is quasi-periodic, with a period of approximately 18.6 years. These simplified ephemerides, which we will refer to as Simpson's lunar ephemerides, were developed based on DE200 using Fourier transforms, resulting in sine and cosine wave approximations as

$$X_n = \sum_{m=1}^{N_n} a_{nm} \sin\left(\omega_{nm} T_{JD} + \delta_{nm}\right) \tag{4.47}$$

where X_n corresponds to the Cartesian coordinates of the Moon's center of mass position in ECI, such that $X_1 = x$, $X_2 = y$, and $X_3 = z$; N_n is the order of the series, such that higher N_n results in more accurate ephemerides; a_{nm}, ω_{nm}, and δ_{nm} represent the amplitudes, frequencies, and phase constants of the series, respectively; and T_{JD} is time measured in Julian centuries since epoch J2000. The values of the matrices a_{nm}, ω_{nm}, and δ_{nm} for a seventh-order approximation are given in Simpson [1999] and reported here for convenience

[8] https://blogs.nasa.gov/commercialcrew/2020/02/07/nasa-shares-initial-findings-from-boeing-starliner-orbital-flight-test-investigation/

$$a_{nm} = \begin{bmatrix} 383.0 & 31.5 & 10.6 & 6.2 & 3.2 & 2.3 & 0.8 \\ 351.0 & 28.9 & 13.7 & 9.7 & 5.7 & 2.9 & 2.1 \\ 153.2 & 31.5 & 12.5 & 4.2 & 2.5 & 3.0 & 1.8 \end{bmatrix} \times 10^3 \text{ km} \quad (4.48)$$

$$\omega_{nm} = \left[\begin{matrix} 8399.685 & 70.990 & 16728.377 & 1185.622 \\ 8399.687 & 70.997 & 8433.466 & 16728.380 \\ 8399.672 & 8433.464 & 70.996 & 16728.364 \end{matrix} \right. \quad \ldots$$

$$\ldots \left. \begin{matrix} 7143.070 & 15613.745 & 8467.263 \\ 1185.667 & 7143.058 & 15613.755 \\ 1185.645 & 104.881 & 8399.116 \end{matrix} \right] \text{rad/century} \quad (4.49)$$

$$\delta_{nm} = \left[\begin{matrix} 5.381 & 6.169 & 1.453 & 0.481 \\ 3.811 & 4.596 & 4.766 & 6.165 \\ 3.807 & 1.629 & 4.595 & 6.162 \end{matrix} \right. \quad \ldots$$

$$\ldots \left. \begin{matrix} 5.017 & 0.857 & 1.010 \\ 5.164 & 0.300 & 5.565 \\ 5.167 & 2.555 & 6.248 \end{matrix} \right] \text{rad} \quad (4.50)$$

Note that because the frequencies are measured in rad/century, we have added more significant figures to Equation (4.49). Taking a time derivative of Equation (4.47) results in the Cartesian velocity components of the Moon with respect to ECI,

$$\dot{X}_n = \frac{1}{T_C} \sum_{m=1}^{N_n} a_{nm} \omega_{nm} \cos\left(\omega_{nm} T_{JD} + \delta_{nm}\right) \quad (4.51)$$

where $\dot{X}_n$ corresponds to the Cartesian velocity coordinates, such that $\dot{X}_1 = \dot{x}$, $\dot{X}_2 = \dot{y}$, and $\dot{X}_3 = \dot{z}$; T_C is a conversion factor used to convert from velocities in km/century to km/s, such that $T_C = 3.1557600 \times 10^9$ s/centuries; and the values of a_{nm}, ω_{nm}, and δ_{nm} used for a seventh-order approximation are the same as given in Equations (4.48) through (4.50). Using the ephemerides provided by JPL – refer to Section 4.8 – we obtain the inclination of the Moon's orbit with respect to the ecliptic for the years 2000–2100, as shown in Figure 4.13. These changes are primarily attributed to the gravitational perturbations of the Sun on the Moon.

In Figure 4.13, the periodicity of the inclination of the Moon's orbit can be seen, with values ranging between 18.4° and 28.5° – shown as dotted lines on the figure – over a period of 18.6 years. Computing the difference between the high-fidelity JPL model and Simpson's approximation presented in this section, we get a relatively small difference, as seen in Figure 4.14.

In comparison, Earth's orbital plane has a periodicity of approximately 26,000 years. The Moon's orbital plane is inclined with respect to Earth's equatorial plane by approximately 5.14°, as shown in Figure 4.15. This means that lunar

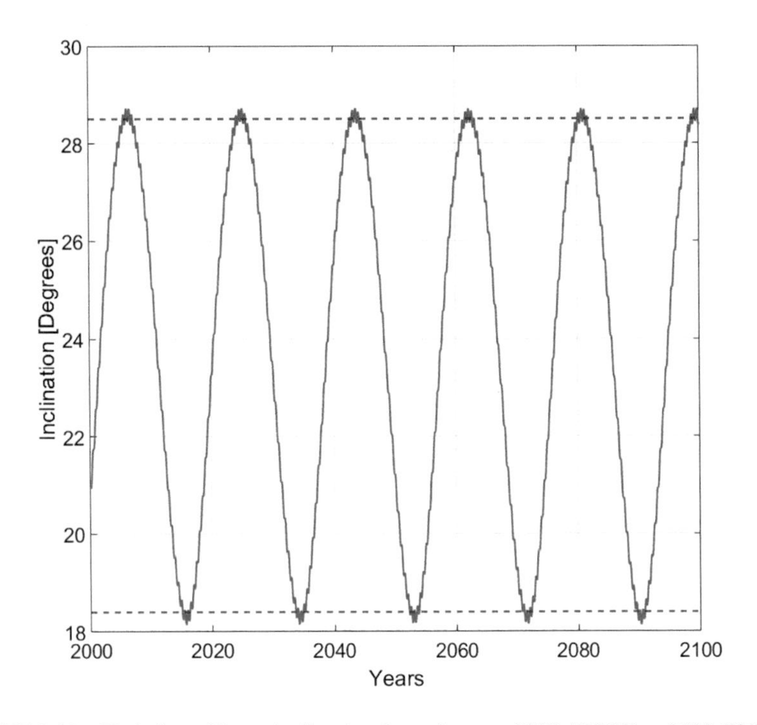

FIGURE 4.13 Variation of lunar inclination from the year 2000 (J2000) to 2100 (J2100).

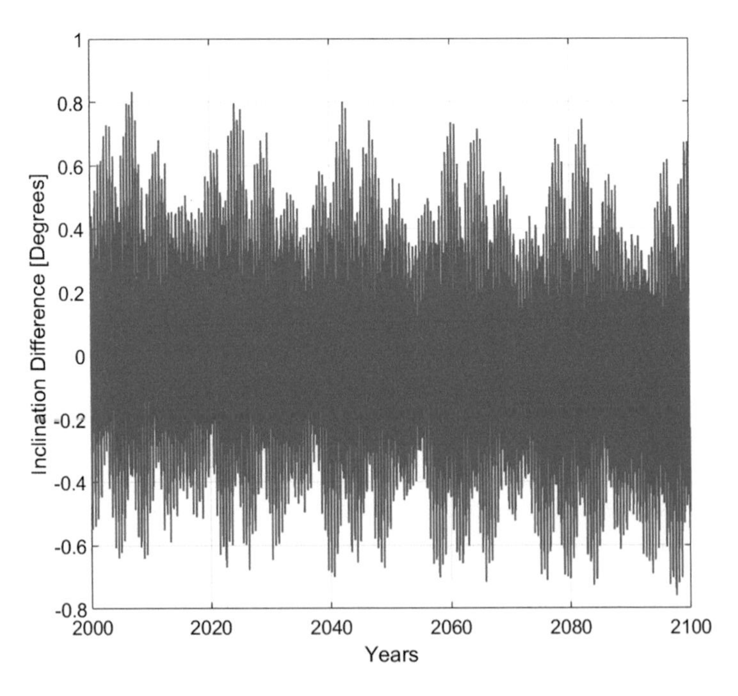

FIGURE 4.14 Difference in variation of lunar inclination (JPL minus Simpson's approximation) from the year 2000 (J2000) to 2100 (J2100).

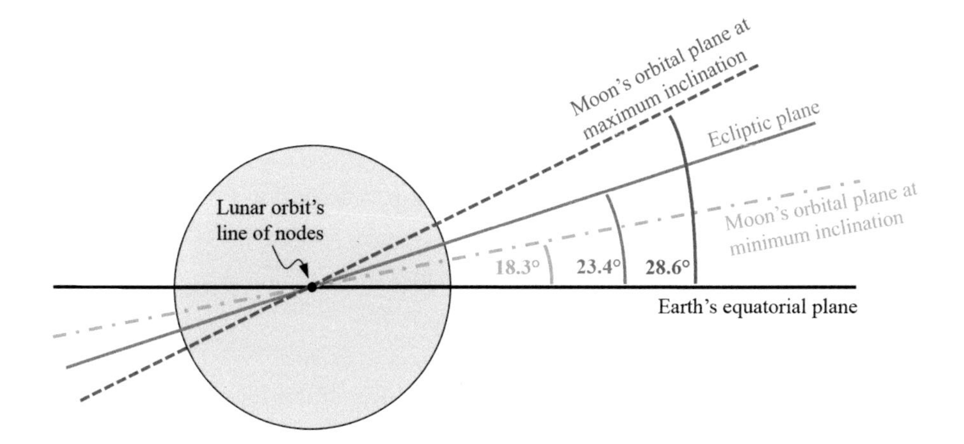

FIGURE 4.15 Lunar orbital minimum and maximum inclinations with respect to the ecliptic plane and Earth's equatorial plane.

missions that target the plane of the Moon's orbit can be scheduled in a way that tries to minimize the Δv required from any inclination changes during the 18.6-year cycle. For example, missions departing from Kennedy Space Center, which has a latitude of approximately 28.57°, can launch almost directly in the plane of the Moon's orbit only when the Moon's inclination is near its maximum. The next available opportunities when such conditions are met will be in 2025.

4.6 SOLAR EPHEMERIDES RELATIVE TO THE EARTH

Solar ephemerides are needed to accurately determine the position of the Sun with respect to a spacecraft. This helps determine eclipse conditions, sizing solar arrays appropriately, and ensuring that remote sensing equipment points in the appropriate direction. Usually, solar ephemerides are computed with respect to the Earth in the J2000 ecliptic coordinate system (Section 4.2.2), with the Earth at the center of the reference frame. This results in computing the apparent motion of the Sun with respect to the Earth. A first-order approximation of determining the position of the Sun involves calculating its mean longitude as provided in the Astronomical Almanac [2006]

$$\lambda_{M_\odot} = 280.460^\circ + 36{,}000.771 T_{JD} \tag{4.52}$$

from which the Sun's mean anomaly can be computed

$$M_\odot = 357.5291092^\circ + 35{,}999.050 T_{JD} \tag{4.53}$$

where T_{JD} is the number of Julian centuries since the epoch J2000. We suggest reducing the results of Equations (4.52) and (4.53) to a range between 0° and 360° for easier numerical computations. Assuming that the Earth's eccentricity does not change significantly and remains small, we can then set the latitude of the Sun with respect to the ecliptic to be exactly zero. Furthermore, the longitude of the Sun with respect to the ecliptic plane can be approximated as a function of the Sun's mean anomaly, resulting in

$$\lambda_{ecliptic} = \lambda_{M_\odot} + 1.914666471^\circ \sin M_\odot + \\ + 0.019994643 \sin(2M_\odot) \tag{4.54}$$

$$\phi_{ecliptic} = 0^\circ$$

The magnitude of the position vector of the Sun with respect to Earth can then be determined as

$$r_\odot = 1.000140612 - 0.016708617 \cos M_\odot + \\ - 0.000139589 \cos(2M_\odot) \tag{4.55}$$

where $r_\odot$ is measured in astronomical units (AU[9]). Recall that the inclination of the ecliptic plane $\epsilon_\oplus$ can be approximated using Equation (4.8) so that the position vector of the Sun can thus be determined as

$$\mathbf{r}_\odot = r_\odot \cos \lambda_{ecliptic} \mathbf{I} + r_\odot \cos \epsilon_\oplus \sin \lambda_{ecliptic} \mathbf{J} + \\ + r_\odot \sin \epsilon_\oplus \sin \lambda_{ecliptic} \mathbf{K} \tag{4.56}$$

From Equation (4.56), right ascension and declination can be determined using Equation (4.17) if needed.

Although we used Earth as the center of our coordinate system in this analysis, it is convenient to utilize a coordinate system centered at the body being orbited. For example, a Mars mission would use a Mars-centered coordinate system, such as MME (Section 4.2.1).

4.7 PLANETARY MOTION RELATIVE TO THE SUN

Planetary motion, or planetary ephemerides, with respect to the Sun need to be determined in order to design and plan an interplanetary mission. Although high accuracy is eventually needed to mission planning, low-fidelity planetary motion can be used in preliminary mission and trajectory design. However, certain kinds of operations can be achieved using high-fidelity planetary motion, such as landing on a celestial body.

4.7.1 Low-Fidelity Motion

Lower-fidelity models have been developed to provide the approximate location of planets at various times. Using higher fidelity models, a simple linear regression is adopted to approximate high-precision ephemerides, thus providing a lower-fidelity planetary model. These lower-fidelity models can be used for observation scheduling, telescope pointing, and preliminary mission planning. For applications requiring high accuracy, such as high-precision targeting, a lower-fidelity model may be used for preliminary results, but should then be refined using a higher fidelity model.

Standish and Williams [2006] developed linear models that show an initial state vector and a secular change in elements. These models have various time lengths of validity,

[9] 1 AU = 149,597,870.7 km exactly, as defined in 2012 by the IAU.

and fitted data has been computed for various time periods. Using the planetary elements (semimajor axis, eccentricity, inclination, mean longitude, longitude of perihelion, and longitude of the ascending node), ephemeris data can be approximated as

$$\begin{aligned}
a &= a_0 + \frac{da}{dt} T_{JD} \\
e &= e_0 + \frac{de}{dt} T_{JD} \\
i &= i_0 + \frac{di}{dt} T_{JD} \\
L &= L_0 + \frac{dL}{dt} T_{JD} \\
\varpi &= \varpi_0 + \frac{d\varpi}{dt} T_{JD} \\
\Omega &= \Omega_0 + \frac{d\Omega}{dt} T
\end{aligned} \tag{4.57}$$

where T_{JD} is the number of Julian centuries since the epoch J2000 such that $T_{JD} = \left(T_{eph} - 2{,}451{,}545.0\right)/36{,}525$, and T_{eph} = Julian Ephemeris Date. The values for the initial conditions and rates are found in Appendix C. These tables are subject to being updated as additional observation data is collected and processed, so the reader is encouraged to use the most current data as found at the Jet Propulsion Laboratory's Solar System Dynamics Group's website (http://ssd.jpl.nasa.gov). Tables C.1a and C1.b are valid during the time interval 1800 AD–2050 AD while Tables C.2a and C.2b are valid from 3000 BC to 3000 AD. For this longer time, the mean anomalies of Jupiter through Pluto are augmented with additional times. After choosing a time in which to compute the state vector of the planet, we compute the argument of perihelion, ω, as

$$\omega = \varpi - \Omega \tag{4.58}$$

Then, mean anomaly, M, is computed as

$$M = L - \varpi + bT_{JD}^{2} + c\cos\left(fT_{JD}\right) + s\sin\left(fT_{JD}\right) \tag{4.59}$$

where the terms b, c, f, and s are found in Table C.2c and are used for determining the location of the planets from 3000 BC to 3000 AD. The mean anomaly in Equation (4.59) is modulated such that $-180° \le M \le 180°$. Using Kepler's equation, (Equation 2.97) and the associated algorithm discussed in Section 2.6, we can use the mean anomaly to find the eccentric anomaly. It should be noted that the values given in Appendix C use the ecliptic plane as reference frame for computing i, ϖ, and Ω.

4.7.2 High-Fidelity Motion

While the lower-fidelity model developed by Standish and Williams provides a very good estimate of the location of a planetary body, and generally provides sufficient accuracy for mission design, higher fidelity models are needed for applications requiring higher accuracy. Two such examples include spacecraft mission execution and prediction of planetary impacts. In the next section, we discuss how to retrieve high-fidelity ephemerides from JPL Horizons, which is a free-access platform that represents the current state-of-the-art in terms of ephemeris data. Here, we provide a brief introduction to JPL Horizons and its capabilities, including access methods, and settings. Also, a short example of a JPL Horizons request for ephemeris data for NASA's New Horizons spacecraft, which became the first spacecraft to explore Pluto and its moons up close.

4.8 JPL HORIZONS

NASA's Jet Propulsion Laboratory has developed an online software that provides highly accurate Solar System object ephemerides. This software, known as JPL Horizons, can produce information regarding position and velocity (or equivalent data, e.g., orbital elements) for 1,224,841 asteroids, 3,818 comets, 211 planetary satellites, 8 planets, the Sun, Lagrange points, select spacecraft, and system barycenters.[10] Over time, the amount of data increases, and more objects are included. Also, observation data is continuously being taken, and this data is used to update object orbital characteristics. Solar System ephemeris data can be produced for objects in the past, present, and future. Future ephemerides are computed via numerically integrating the current state to the desired future time using the model stated in the currently used Development Ephemerides (DE) as discussed in Section 4.4.3. At the time of writing the most recent DE corresponds to DE 441. It should be noted that all DEs have a given timeframe of validity, during which they can be used to predict the ephemerides of various celestial objects.

The simplest JPL Horizons access method is the online web-interface. Though it is missing some functionality of the telnet and email techniques that are described in the following section, it is still robust and the best choice for new users. When accessing the JPL Horizons' website, the user can click on "App" to access the system that will generate ephemerides. There are many options for each of the five main settings present on this page:

- Ephemeris Type
- Target Body
- Coordinate Center
- Time Specification
- Table Settings

Table 4.1 summarizes these settings and their options with a brief explanation.

Once these settings are specified, it is possible to generate the ephemeris data needed by pressing the green button "Generate Ephemeris," as shown in Figure 4.16. Details regarding "Table Settings" are given in Figure 4.17.

The result of a JPL Horizons request is given in the form of text or a spreadsheet with data regarding position, velocity, or orbital elements of a specified body

[10] https://ssd.jpl.nasa.gov/horizons/, accessed in September 2022.

TABLE 4.1
Summary and explanation of JPL Horizons web-interface user settings, https://ssd.jpl.nasa.gov/horizons/app.html#/

Setting	Explanation
Ephemeris type	Desired ephemeris data type. "Observer Table" will return quantities observable from a position on Earth such as right ascension and declination. "Vector Table" will return Cartesian state vector table with respect to a major body. "Osculating Orbital elements" will return the osculating (i.e., instantaneous) orbital elements with respect to an appropriate major body. "Small-Body SPK File" is used for small bodies, such as asteroids, and will return an SPK file,[12] a binary file which, using SPICE Toolkit APIs, may be smoothly interpolated to retrieve an object's position and velocity at any instant within the file time span.
Target Body	Body user wishes to track. Available targets are small and large bodies and dynamic points including planets, asteroids, comets, spacecraft, and Lagrange points.
Coordinate center	Observer location or coordinate center. Origins can be looked up, specified via city, observatory, or precise latitude/longitude/altitude.
Time specification	Discrete or continuous time spans of which ephemeris data is desired. Data step size can also be specified in years, days, hours, seconds, etc.
Table settings	Changeable output quantities and preferences such as astrometric or apparent right ascension and declination, illuminated fraction, orbit angle, constellation ID, units, data format, reference system, precision, etc. This setting also includes the ability to choose the format of the output.

[12] SPK stands for "Spacecraft and Planet Kernel". Such files can be used as input to SPICE-enabled visualization and mission design programs, allowing them to quickly retrieve accurate target body observation and data analysis ephemerides without having to integrate equations of motion. An SPK file could be considered a "recording" of the integrator.

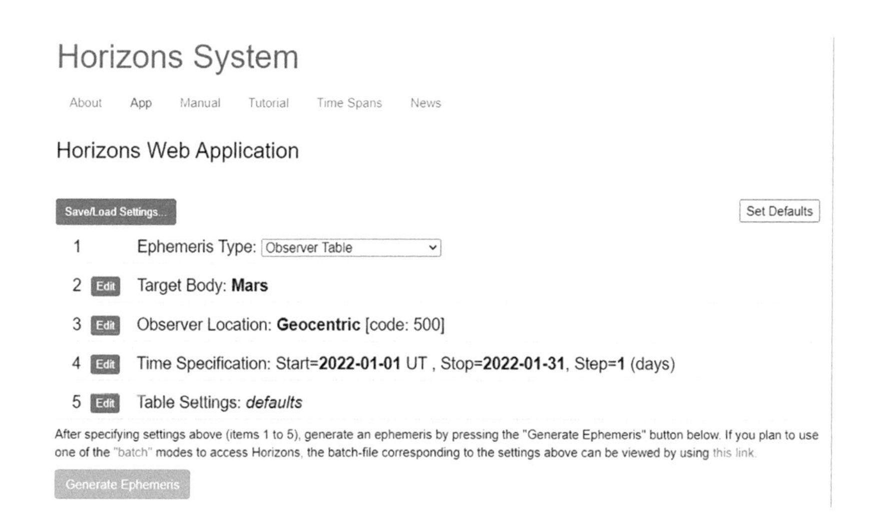

FIGURE 4.16 Screenshot of JPL Horizons web-interface, https://ssd.jpl.nasa.gov/horizons/app.html#/.

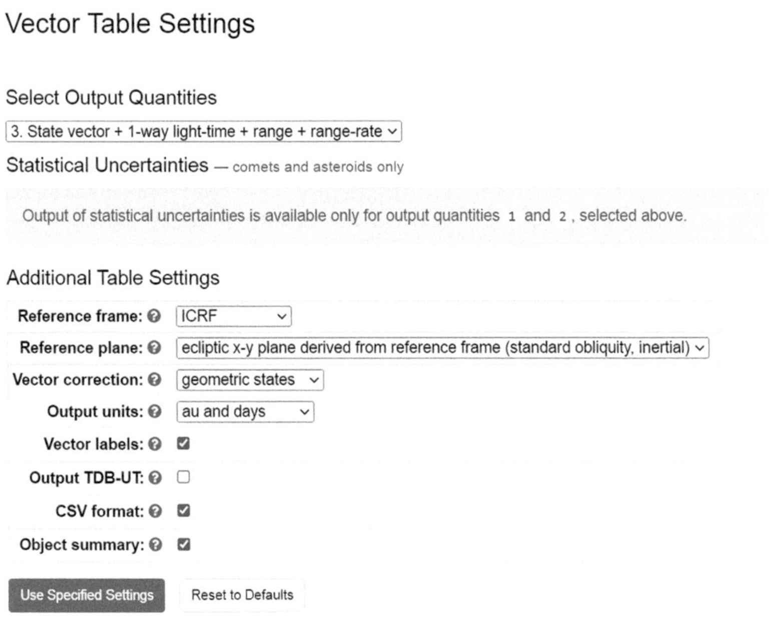

Vector Table Settings

Select Output Quantities

3. State vector + 1-way light-time + range + range-rate

Statistical Uncertainties — comets and asteroids only

Output of statistical uncertainties is available only for output quantities 1 and 2 , selected above.

Additional Table Settings

Reference frame: ICRF

Reference plane: ecliptic x-y plane derived from reference frame (standard obliquity, inertial)

Vector correction: geometric states

Output units: au and days

Vector labels: ☑

Output TDB-UT: ☐

CSV format: ☑

Object summary: ☑

Use Specified Settings   Reset to Defaults

FIGURE 4.17 Screenshot of additional table settings, https://ssd.jpl.nasa.gov/horizons/app.html#/.

with respect to a chosen reference frame. The JPL Horizons software is a robust ephemeris generation software system. The full extent of the software's capabilities is beyond the scope of this book. However, the reader who is interested in learning the specifics of how to use the software is referred to their website. While the goal of this section of this book is not to sound like an advertisement for this software, the sophistication of this software is beyond anything else that is available.

4.8.1 Alternative Access Methods

The software can be accessed via telnet through a telnet console. Most browsers support telnet as well, and the software can be used in "terminal" mode by typing telnet://ssd.jpl.nasa.gov:6775 into the browser address bar. A JPL Horizons telnet terminal is shown in Figure 4.18.

The JPL Horizons software can also be accessed via email at horizons@ssd.jpl.nasa.gov. This technique is more suitable for experienced users who wish to execute batch jobs without a user-interface or command window. An email sent to this address with the subject "BATCH-LONG" indicates the sender's request for an example batch job. The reply to this email is a fully commented example job request. To request a job, the user edits this example file and sends it back to horizons@ssd.jpl.nasa.gov with the subject "JOB." The specific request's corresponding ephemeris data is sent back to the user.

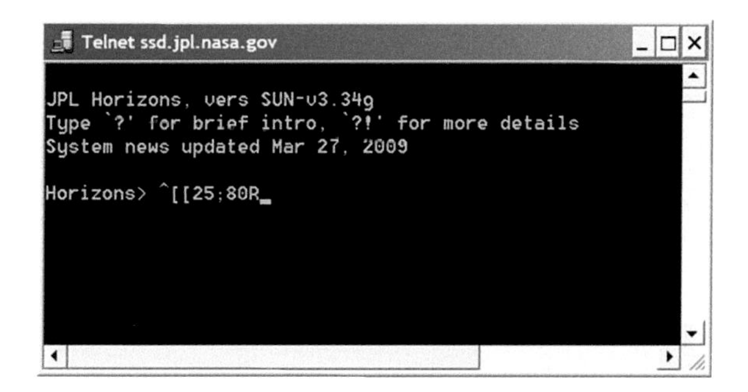

FIGURE 4.18 Screenshot of JPL Horizons telnet console.

4.8.2 Tracking the New Horizons Spacecraft – A JPL Horizons Example

An example application of the Horizons software's capabilities is shown for the New Horizons spacecraft trajectory from its Earth launch to its Pluto flyby in 2015. Here, a step-by-step walkthrough of this example by using the JPL Horizons web-interface is shown.

The ephemeris type is specified as vector so the position and velocity vectors are returned in Cartesian coordinates. The target body is selected as New Horizons spacecraft [NH New_Horizons (Spacecraft)] [–98] from the Spacecraft category. The coordinate origin is specified as the center of the Sun. The time span is specified between the launch on February 1st, 2006 and arrival at Pluto on July 14, 2015. The step size is chosen to be 1 day. No corrections to the state vector, such as light-time and stellar aberrations, are selected. Figure 4.19 shows the settings used for this example.

Horizons System

About App Manual Tutorial Time Spans News

Horizons Web Application

Save/Load Settings... Set Defaults

1 Ephemeris Type: Vector Table

2 Edit Target Body: **New Horizons (spacecraft) [NH New_Horizons]**

3 Edit Coordinate Center: **Sun (body center)** [500@10]

4 Edit Time Specification: Start=**2006-02-01** TDB , Stop=**2015-07-14**, Step=**1** (days)

5 Edit Table Settings: *custom*

After specifying settings above (items 1 to 5), generate an ephemeris by pressing the "Generate Ephemeris" button below. If you plan to use one of the "batch" modes to access Horizons, the batch-file corresponding to the settings above can be viewed by using this link.

Generate Ephemeris

FIGURE 4.19 Screenshot of JPL Horizons web-interface for the example discussed in this section, https://ssd.jpl.nasa.gov/horizons/app.html#/.

The output is provided in the CSV format option to make it easier to import the data into a spreadsheet program. The resulting ephemeris data is shown in Figures 4.20 and 4.21 for the first 15 time steps.

In addition to the data shown, the file also contains mission schedules, objectives, and parameters. The Cartesian states for position (X, Y, Z) and velocity (VX, VY, VZ) are given for each time step, along with one-way down-leg Newtonian light-time (LT), range (RG), and range rate (RR). Time is given in JDTDB, which corresponds to Julian day number, Barycentric Dynamical Time. Lastly, $$SOE means "Start of Ephemeris." For convenience, the data can be imported and saved as a comma separated values (CSV) file or Excel file, which can be easily imported into various programs (Figure 4.22).

The resulting ephemeris data, state vectors as a function of time, can be useful for many trajectory analyses. An example application of this ephemeris data for the New Horizons spacecraft is shown in Figures 4.23–4.26, where the position vectors of the spacecraft are plotted along with the orbits of Earth, Mars, Jupiter, Saturn, Uranus, Neptune, and Pluto for various points of view. Figure 4.23 shows the overall trajectory of New Horizons along with Earth launch (January 19, 2006; white dot), Jupiter flyby (February 28, 2007; light gray dot), Saturn flyby (June 8, 2008; dark gray dot), and Pluto flyby (July 14, 2015; black dot). Figure 4.24 shows how the planets are mostly located near the ecliptic plane, while Pluto has a much higher inclination with respect to the ecliptic (approximately 17.16°). From this figure, it can be seen that Pluto was targeted at a location which passes very close to the ecliptic plane. Currently, Pluto is moving

```
Target body name: New Horizons (spacecraft) (-98) {source: NH_merged}
Center body name: Sun (10)                         {source: NH_merged}
Center-site name: BODY CENTER
*******************************************************************************
Start time      : A.D. 2006-Feb-01 00:00:00.0000 TDB
Stop  time      : A.D. 2015-Jul-14 00:00:00.0000 TDB
Step-size       : 1440 minutes
*******************************************************************************
Center geodetic : 0.00000000,0.00000000,0.0000000 {E-lon(deg),Lat(deg),Alt(km)}
Center cylindric: 0.00000000,0.00000000,0.0000000 {E-lon(deg),Dxy(km),Dz(km)}
Center radii    : 696000.0 x 696000.0 x 696000.0 k{Equator, meridian, pole}
Output units    : AU-D
Output type     : GEOMETRIC cartesian states
Output format   : 3 (position, velocity, LT, range, range-rate)
Reference frame : Ecliptic of J2000.0
*******************************************************************************
            JDTDB,            Calendar Date (TDB),                      X,                      Y,                      Z,
**************************************************************************************************************************
$$SOE
2453767.500000000, A.D. 2006-Feb-01 00:00:00.0000, -7.337886697076316E-01,  6.893823703121267E-01,  4.439152605700138E-03,
2453768.500000000, A.D. 2006-Feb-02 00:00:00.0000, -7.529209013583591E-01,  6.742107629024322E-01,  4.805976573009162E-03,
2453769.500000000, A.D. 2006-Feb-03 00:00:00.0000, -7.718372589575184E-01,  6.588459557358844E-01,  5.171418944773972E-03,
2453770.500000000, A.D. 2006-Feb-04 00:00:00.0000, -7.905350212340824E-01,  6.432946402377342E-01,  5.535393274468265E-03,
2453771.500000000, A.D. 2006-Feb-05 00:00:00.0000, -8.090117959031080E-01,  6.275635661465764E-01,  5.897817774676325E-03,
2453772.500000000, A.D. 2006-Feb-06 00:00:00.0000, -8.272655141980926E-01,  6.116595193698969E-01,  6.258614084740139E-03,
2453773.500000000, A.D. 2006-Feb-07 00:00:00.0000, -8.452944264526822E-01,  5.955893014323026E-01,  6.617707885928126E-03,
2453774.500000000, A.D. 2006-Feb-08 00:00:00.0000, -8.630970940113096E-01,  5.793597081696704E-01,  6.975029086400034E-03,
2453775.500000000, A.D. 2006-Feb-09 00:00:00.0000, -8.806723883547134E-01,  5.629775174813614E-01,  7.330511597071225E-03,
2453776.500000000, A.D. 2006-Feb-10 00:00:00.0000, -8.980194743422173E-01,  5.464494644552182E-01,  7.684093909606301E-03,
2453777.500000000, A.D. 2006-Feb-11 00:00:00.0000, -9.151378026770294E-01,  5.297822280725842E-01,  8.035718809184493E-03,
2453778.500000000, A.D. 2006-Feb-12 00:00:00.0000, -9.320271026319664E-01,  5.129824173849667E-01,  8.385332049613536E-03,
2453779.500000000, A.D. 2006-Feb-13 00:00:00.0000, -9.486873668344268E-01,  4.960565553892171E-01,  8.732884018234381E-03,
2453780.500000000, A.D. 2006-Feb-14 00:00:00.0000, -9.651188398190388E-01,  4.790110668274132E-01,  9.078329329581761E-03,
2453781.500000000, A.D. 2006-Feb-15 00:00:00.0000, -9.813220059795777E-01,  4.618522669474494E-01,  9.421626575949252E-03,
```

FIGURE 4.20 Screenshot of JPL Horizons' output showing the New Horizons spacecraft ephemeris data (part 1 of 2). (Courtesy of NASA/JPL-Caltech.)

```
                 VX,                      VY,                      VZ,                      LT,                      RG,                      RR,
**************************************************************************************************************************************************

-1.923917944388471E-02, -1.507279590317120E-02, 3.674851012014410E-04, 5.814984973540145E-03, 1.006833437249836E+00, 3.702899911691087E-03,
-1.902477557067774E-02, 1.526931660669707E-02, 3.661478836684454E-04, 5.837198214341226E-03, 1.010679540668843E+00, 3.988577099224355E-03,
-1.880748560305221E-02, -1.545918228173491E-02, 3.647221363363197E-04, 5.861048508270356E-03, 1.014809091050375E+00, 4.269753374911642E-03,
-1.858764015667920E-02, -1.564232344417086E-02, 3.632128063090369E-04, 5.086509191287153E-03, 1.019217471658954E+00, 4.546201696910684E-03,
-1.836556446919285E-02, -1.581869341044573E-02, 3.616231322259405E-04, 5.913552344457770E-03, 1.023099848481034E+00, 4.817712826320141E-03,
-1.814158066867046E-02, -1.598826618845279E-02, 3.599570986199468E-04, 5.942148901718854E-03, 1.028851188883752E+00, 5.084099369209335E-03,
-1.791600356449925E-02, -1.615103715496877E-02, 3.582188289218535E-04, 5.972268772808824E-03, 1.034066282899820E+00, 5.345192759705433E-03,
-1.768914442310129E-02, -1.630701849980823E-02, 3.564125186983247E-04, 6.003880950002320E-03, 1.039539761708021E+00, 5.600849404814573E-03,
-1.746130533282020E-02, -1.645623998290573E-02, 3.545421886243626E-04, 6.036953680333731E-03, 1.045266127452787E+00, 5.850945662154804E-03,
-1.723277524459839E-02, -1.659875253024948E-02, 3.526128898415963E-04, 6.071454520536018E-03, 1.051239762756565E+00, 6.095372410180985E-03,
-1.700383686801981E-02, -1.673461833140777E-02, 3.506277378030313E-04, 6.107350471790063E-03, 1.057454954050939E+00, 6.334045760691408E-03,
-1.677476194178723E-02, -1.686391570377098E-02, 3.485905042716549E-04, 6.144608110686283E-03, 1.063905914251934E+00, 6.566899835018138E-03,
-1.654580742721324E-02, -1.698673730630726E-02, 3.465058699880384E-04, 6.183193670087974E-03, 1.070586796761070E+00, 6.793883599520655E-03,
-1.631721948799032E-02, -1.710318710530576E-02, 3.443778529159353E-04, 6.223073144681759E-03, 1.077491713740853E+00, 7.014964773307712E-03,
-1.608923162494044E-02, -1.721337991900055E-02, 3.422103882959758E-04, 6.264212389899944E-03, 1.084614753242651E+00, 7.230127259916630E-03,
```

FIGURE 4.21 Screenshot of JPL Horizons' output showing the New Horizons spacecraft ephemeris data (part 2 of 2). (Courtesy of NASA/JPL-Caltech.)

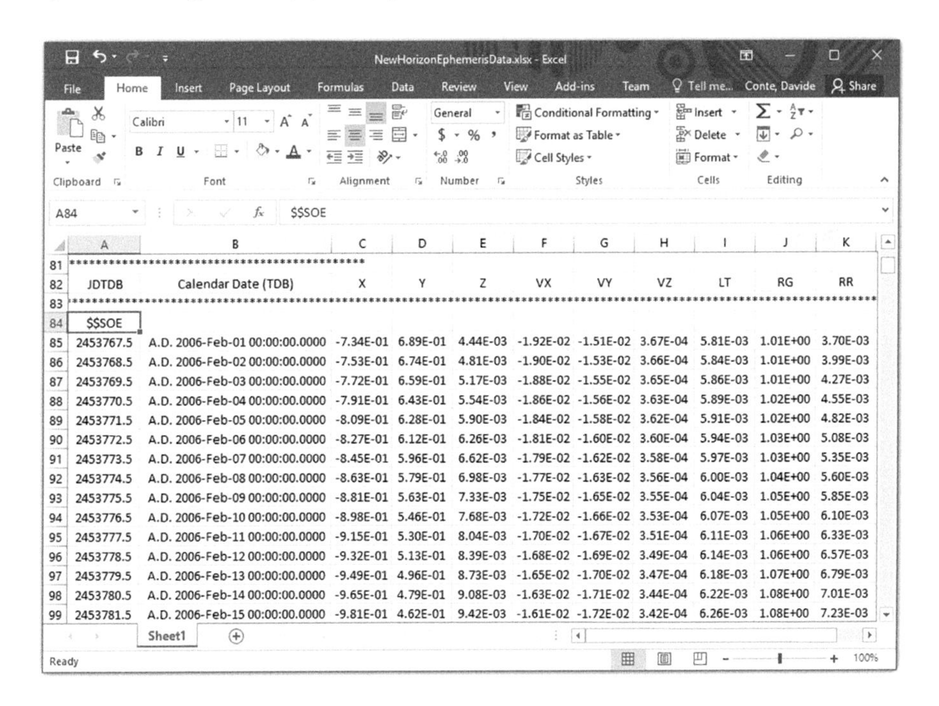

	A	B	C	D	E	F	G	H	I	J	K
81	***										
82	JDTDB	Calendar Date (TDB)	X	Y	Z	VX	VY	VZ	LT	RG	RR
83	**										
84	$$SOE										
85	2453767.5	A.D. 2006-Feb-01 00:00:00.0000	-7.34E-01	6.89E-01	4.44E-03	-1.92E-02	-1.51E-02	3.67E-04	5.81E-03	1.01E+00	3.70E-03
86	2453768.5	A.D. 2006-Feb-02 00:00:00.0000	-7.53E-01	6.74E-01	4.81E-03	-1.90E-02	-1.53E-02	3.66E-04	5.84E-03	1.01E+00	3.99E-03
87	2453769.5	A.D. 2006-Feb-03 00:00:00.0000	-7.72E-01	6.59E-01	5.17E-03	-1.88E-02	-1.55E-02	3.65E-04	5.86E-03	1.01E+00	4.27E-03
88	2453770.5	A.D. 2006-Feb-04 00:00:00.0000	-7.91E-01	6.43E-01	5.54E-03	-1.86E-02	-1.56E-02	3.63E-04	5.89E-03	1.02E+00	4.55E-03
89	2453771.5	A.D. 2006-Feb-05 00:00:00.0000	-8.09E-01	6.28E-01	5.90E-03	-1.84E-02	-1.58E-02	3.62E-04	5.91E-03	1.02E+00	4.82E-03
90	2453772.5	A.D. 2006-Feb-06 00:00:00.0000	-8.27E-01	6.12E-01	6.26E-03	-1.81E-02	-1.60E-02	3.60E-04	5.94E-03	1.03E+00	5.08E-03
91	2453773.5	A.D. 2006-Feb-07 00:00:00.0000	-8.45E-01	5.96E-01	6.62E-03	-1.79E-02	-1.62E-02	3.58E-04	5.97E-03	1.03E+00	5.35E-03
92	2453774.5	A.D. 2006-Feb-08 00:00:00.0000	-8.63E-01	5.79E-01	6.98E-03	-1.77E-02	-1.63E-02	3.56E-04	6.00E-03	1.04E+00	5.60E-03
93	2453775.5	A.D. 2006-Feb-09 00:00:00.0000	-8.81E-01	5.63E-01	7.33E-03	-1.75E-02	-1.65E-02	3.55E-04	6.04E-03	1.05E+00	5.85E-03
94	2453776.5	A.D. 2006-Feb-10 00:00:00.0000	-8.98E-01	5.46E-01	7.68E-03	-1.72E-02	-1.66E-02	3.53E-04	6.07E-03	1.05E+00	6.10E-03
95	2453777.5	A.D. 2006-Feb-11 00:00:00.0000	-9.15E-01	5.30E-01	8.04E-03	-1.70E-02	-1.67E-02	3.51E-04	6.11E-03	1.06E+00	6.33E-03
96	2453778.5	A.D. 2006-Feb-12 00:00:00.0000	-9.32E-01	5.13E-01	8.39E-03	-1.68E-02	-1.69E-02	3.49E-04	6.14E-03	1.06E+00	6.57E-03
97	2453779.5	A.D. 2006-Feb-13 00:00:00.0000	-9.49E-01	4.96E-01	8.73E-03	-1.65E-02	-1.70E-02	3.47E-04	6.18E-03	1.07E+00	6.79E-03
98	2453780.5	A.D. 2006-Feb-14 00:00:00.0000	-9.65E-01	4.79E-01	9.08E-03	-1.63E-02	-1.71E-02	3.44E-04	6.22E-03	1.08E+00	7.01E-03
99	2453781.5	A.D. 2006-Feb-15 00:00:00.0000	-9.81E-01	4.62E-01	9.42E-03	-1.61E-02	-1.72E-02	3.42E-04	6.26E-03	1.08E+00	7.23E-03

FIGURE 4.22 Screenshot of JPL Horizons' output imported into excel.

away from the ecliptic, heading toward the celestial North. Figures 4.25 and 4.26 are projections of the previous figures on the ecliptic plane.

Figure 4.27 shows the distance of New Horizons from the Sun and its orbital speed, as a function of time. The effects of flying by Jupiter (indicated by an arrow in the figure) are noticeable since they correspond to an abrupt increase in orbit speed with respect to the Sun. In comparison, the Saturn flyby has a significantly smaller effect.

For any type of problem where an object moves from one point to another, it is vital to know at minimum where these two points are relative to each other, but, even better, where these points are relative to some common reference frame. In order to

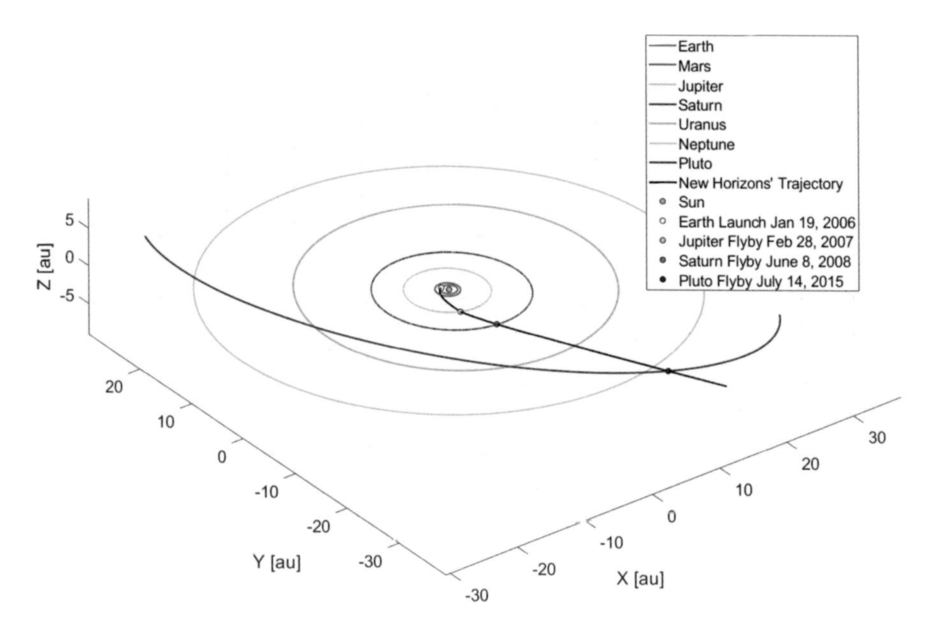

FIGURE 4.23 New Horizons' trajectory (3D view).

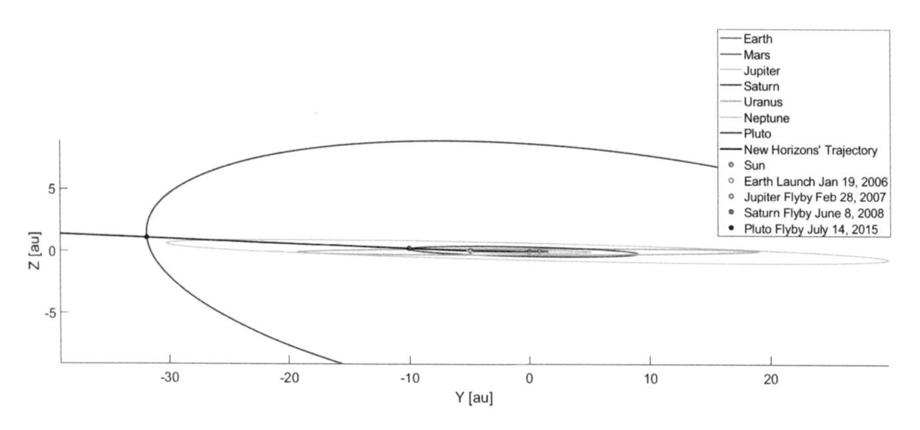

FIGURE 4.24 New Horizons' trajectory, YZ plane.

move from orbit around one body to orbit around another, an accurate description of the time-tagged location of these bodies is necessary.

Determining and describing the location of Solar System bodies is a branch of astronomy known as dynamical astronomy. Dynamical astronomy is the study of the natural motion of astronomical objects, subject to various gravitational forces. Dynamical astronomy has its origins in celestial mechanics and differs from astrodynamics simply because dynamical astronomy focuses solely on the motion of natural objects (planets, moons, asteroids, comets – all the way down to dust particles). Dynamical astronomy also differs from astrodynamics in that many of the larger objects have their own non-negligible gravitational forces that affect other bodies in the same system.

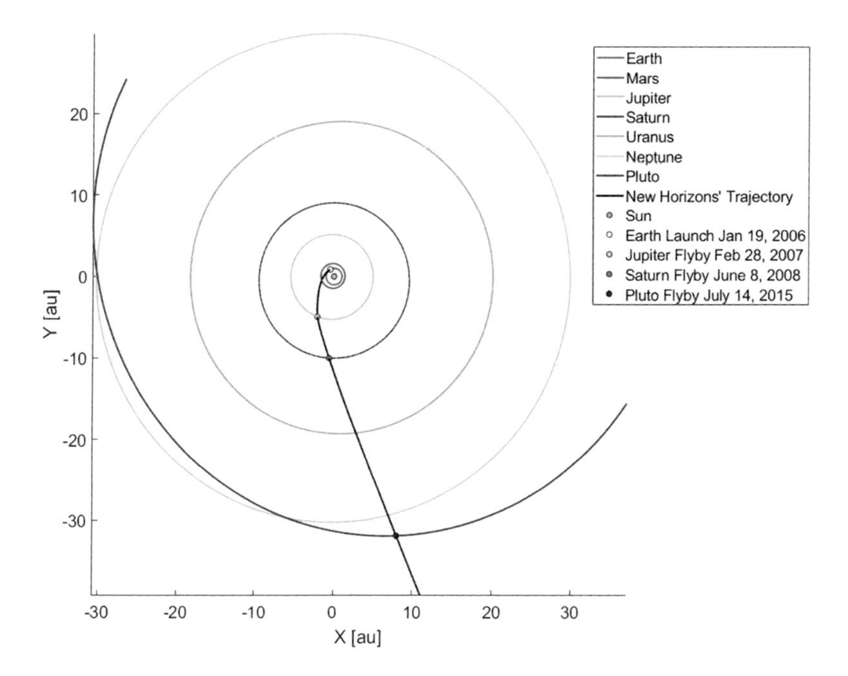

FIGURE 4.25 New Horizons' trajectory, XY plane (ecliptic).

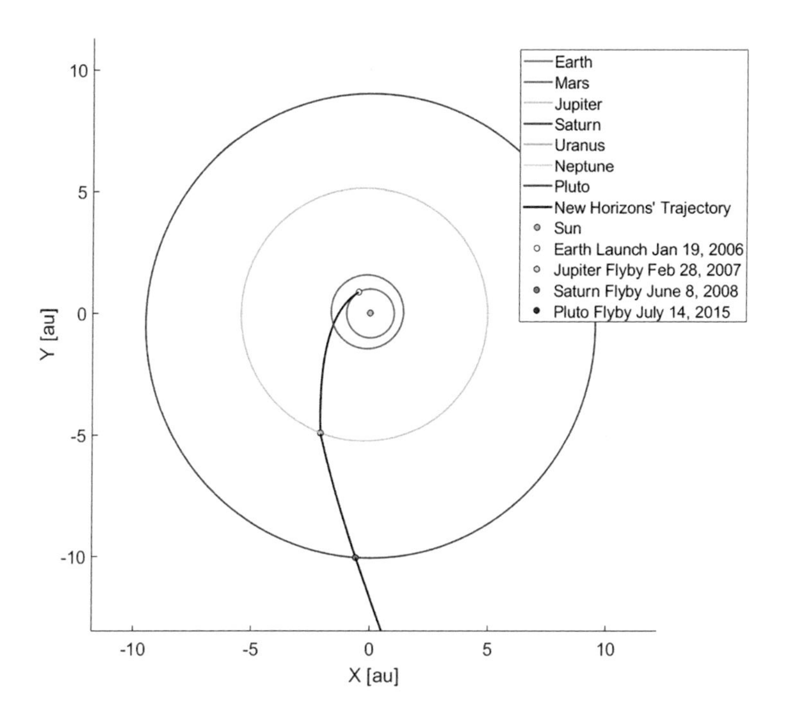

FIGURE 4.26 New Horizons' trajectory, XY plane (ecliptic), detail.

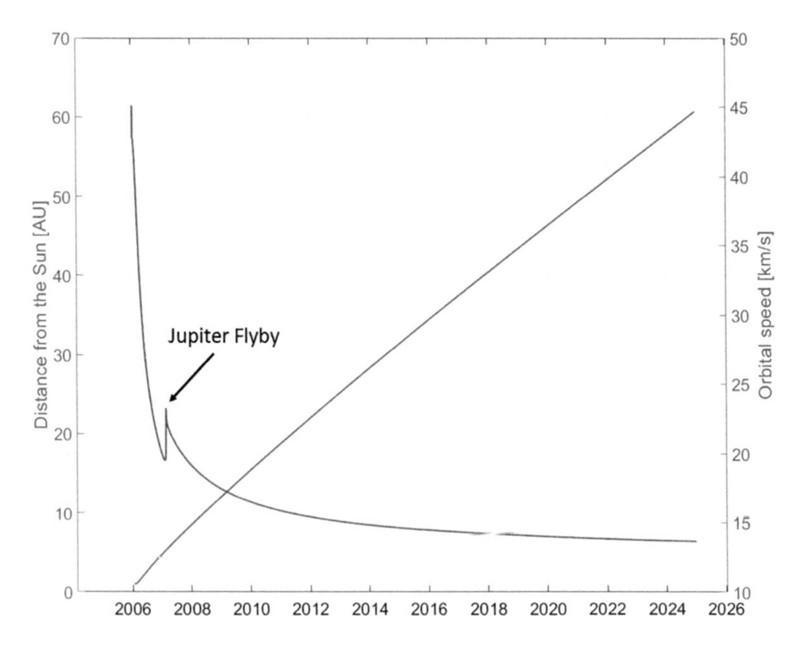

FIGURE 4.27 New Horizons' distance (on the left, in blue) and orbital speed (on the right, in red).

Much of the foundation of modern orbital mechanics owes its origins to the field of dynamical astronomy. Many of the historical mathematicians who developed many of the methods that became the foundation of modern astrodynamics were also astronomers and mathematicians. Dynamical astronomy is important in the field of interplanetary astrodynamics because we must be able to predict where planetary objects are at a given time, whether it be to land, orbit, or simply fly by that body. Planetary locations also have implications on tracking of spacecraft in a planetary system. Considerations such as eclipsing, conjunctions, and other geometrical constraints affect operations and future planning for mission execution.

4.8.3 Comparing Low-Fidelity Motion with High-Fidelity Motion

In this section, we present two brief examples meant to compare the results obtained between using low-fidelity motion and using high-fidelity motion. The first example we present shows the difference between the position, velocity, and orbital elements of Earth and Mars computed using the approximations derived by Standish and Williams, and using JPL Horizons. Computing the Earth's and Mars' ephemerides between January 1, 2030 and January 1, 2050 results in Figure 4.28, where we plotted the percent errors of Standish and Williams' approximation (low-fidelity) with respect to JPL Horizons' ephemerides (high-fidelity) for the distance of the planets from the Sun and their orbital speeds. In Figure 4.29, we plotted the absolute difference between the semimajor axis, eccentricity, and inclination (with respect to the J2000 ecliptic plane) of both Earth and Mars computed using low-fidelity vs. high-fidelity. From Figure 4.28, we can see that position is always accurate to within 0.8%, while orbital

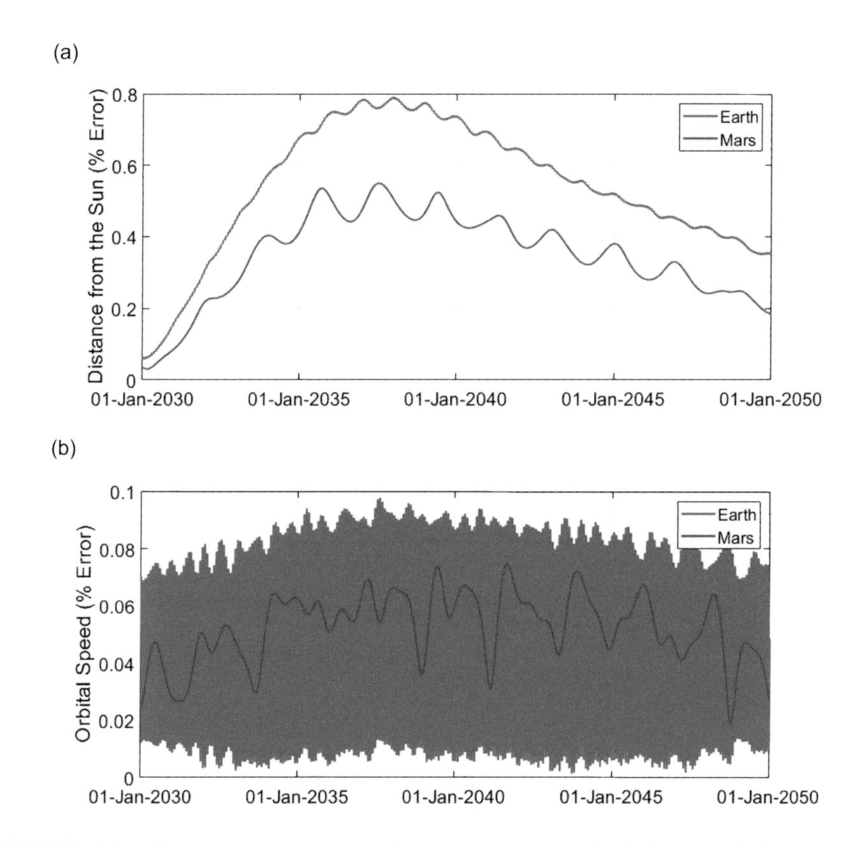

FIGURE 4.28 Percent error in position (a) and orbital speed (b) for Earth and Mars.

speed never exceeds 0.1% error, thus indicating that the low-fidelity model is still able to produce significantly accurate ephemerides. Additionally, from Figure 4.29, we can see that the error in semimajor axis is within 0.02 AU difference (2% error for Earth and ~1.3% error for Mars). However, eccentricity has an error of almost 0.01 in the mid-late 2030s for Earth, corresponding to nearly 50% error for Earth, but only about 10% for Mars. Lastly, Earth's inclination deviates by at most approximately 0.02° in the 2040s, while Mars' only by 0.015° in the mid-late 2030s (or about 8% error).

For the second example, we computed Earth-Mars porkchop plots using the motions of Earth and Mars computed from the same two methods to compare how the resulting Δv changes. As discussed in Chapter 5, porkchop plots are preliminary mission design tools used by engineers to understand how Δv changes as a function of departure and arrival times. Computing Δv for Earth-Mars trajectories for the time frame between the years 2049 and 2051 results in Figure 4.30. In this figure, the high-fidelity model was used to determine the positions and velocities of Earth and Mars. For all transfers, we assumed an Earth departure altitude of 300 km and a Mars arrival altitude of 300 km.

Replicating Figure 4.30 using a low-fidelity ephemeris model results in an extremely similar figure. On the other hand, we instead took the percent difference between the resulting Δv considering the result obtained from the high-fidelity case

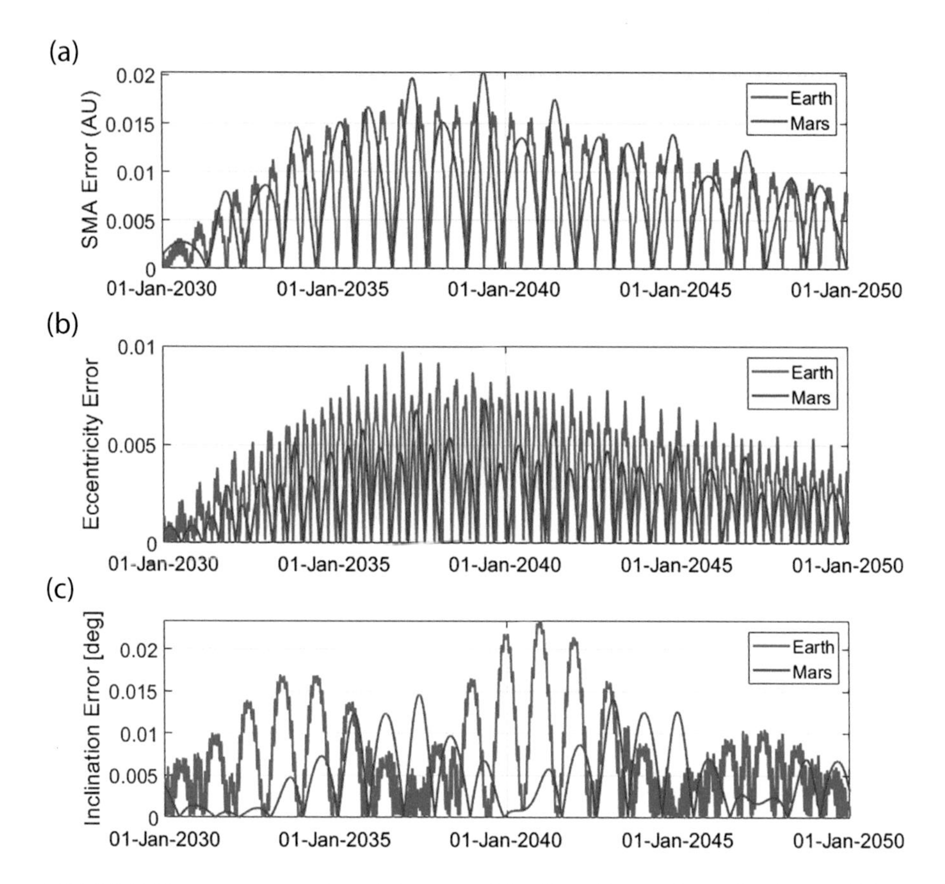

FIGURE 4.29 Errors semimajor axis (a), eccentricity (b), and inclination (c) for Earth and Mars.

as "true values." Figure 4.31 shows such percent difference between the minimum computed for this case (approximately $1.6 \times 10^{-6\%}$) to 1%. As shown in Figure 4.31, the low-fidelity model can still result in extremely accurate Δv values that can be used for the preliminary stage of mission design.

Table 4.2 shows the numerical comparison between the two methods in terms of the lowest total Δv, split into the Earth departure maneuver (Δv_1), the Mars arrival maneuver (Δv_2), and their sum (Δv_{total}). It additionally gives the departure and arrival dates, and the time of flight (TOF) required for the maneuver. The figures in this section were created using MATLAB. It should be noted that in MATLAB one can add the publicly available function planetEphemeris,[11] which can calculate the ephemerides of all eight planets, the Earth's Moon, and Pluto with respect to the Sun or any of the aforementioned celestial body. This function can use various DEs, but at the time of writing it only supports up to DE430 and DE432t. The function uses the Chebyshev coefficients provided by NASA JPL to estimate the ephemeris of a desired body. While a convenient implementation in a widely used engineering program such as MATLAB, this function presents severe limitations when compared to JPL Horizons.

[11] https://www.mathworks.com/help/aerotbx/ug/planetephemeris.html

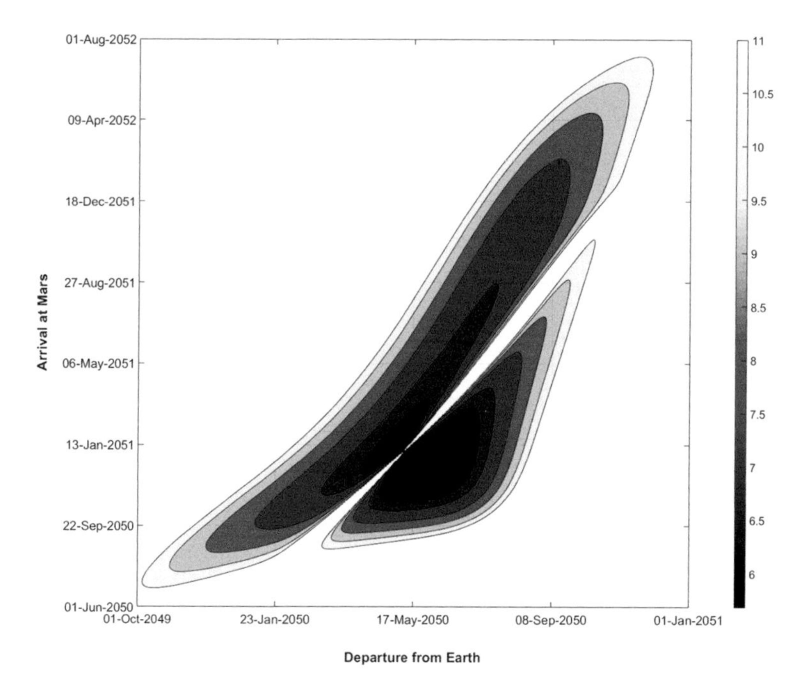

FIGURE 4.30 Porkchop plot showing the total Δv for an interplanetary transfer between Earth and Mars using high-fidelity ephemeris data for Earth and Mars.

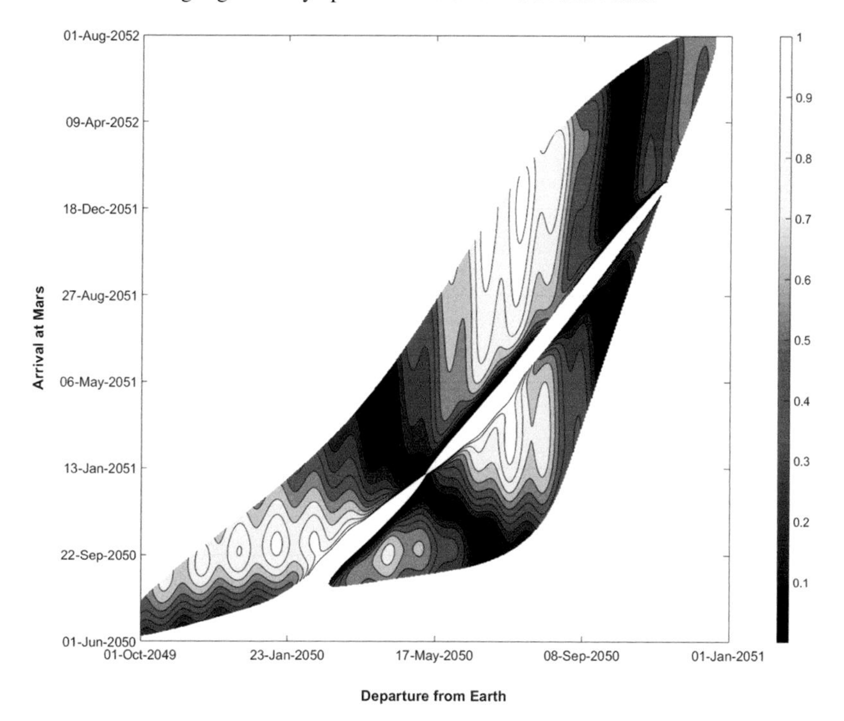

FIGURE 4.31 Percent difference in Δv for an interplanetary transfer between Earth and Mars between using low-fidelity vs. high-fidelity ephemeris data.

TABLE 4.2
Minimum total Δv solutions using the high-fidelity ephemeris and the low-fidelity ephemeris models

Model Used	Δv_1 (km/s)	Δv_2 (km/s)	Δv_{total} (km/s)	Launch Date	Arrival Date	TOF (Days)
High-fidelity	3.5292	2.1514	5.6805	26/05/2050	19/12/2050	207
Low-fidelity	3.5424	2.1525	5.6949	26/05/2050	18/12/2050	206
Difference	0.0132	0.0011	0.0144	None	1 day	1

PROBLEMS

1. Using Equation (4.8), plot a graph of Earth's obliquity vs. time between January 1, 2000 and January 1, 3000. How do the initial and final values of obliquity compare? What is their difference? Create another plot using the same approximation, but this time from January 1, 2000 for 10,000 years. When does the approximation start diverging significantly from the prediction given in Figure 4.4?
2. Consider an interplanetary Hohmann transfer from Earth to Mars (refer to Chapter 5 if needed). Using the resultant Mars arrival v_∞, compute the impact parameter and eccentricity of the arrival trajectory assuming that your targeted periapsis with respect to Mars has an altitude of 300 km.
3. Compute the Julian date (JD) and Modified Julian Date (MJD) for a given time (e.g., when your homework is due). Make sure to use enough decimal places so that your accuracy is within 1 ms (10^{-3} s). Then, compute the Greenwich location angle corresponding to this time.
4. Create a ground track similar to that of Figure 4.7 for an Earth-orbiting spacecraft that has classical orbital elements with respect to ECI as follows:
   a. semimajor axis: 6,671 km
   b. eccentricity: 0.3
   c. inclination: 45°
   d. RAAN: 20°
   e. argument of periapsis: 95°

   Assume that the spacecraft starts at periapsis (true anomaly = 0) on January 1, 2030, and propagate its orbit using the unperturbed two-body model for five full orbits.
5. Compute the coordinates of your school/institution using the topocentric equatorial coordinates (**IJK**) discussed in Section 4.3.2 for a given time (e.g., when your homework is due). Compute these coordinates assuming that the Earth is a perfect sphere (flattening parameter = 0) and using a flattening parameter of 0.003353. How do the results differ?
6. Compute the coordinates of the NASA Perseverance landing location using the topocentric equatorial coordinates (**IJK**) discussed in Section 4.3.2 for a given time (e.g., when your homework is due). Compute these coordinates

assuming that Mars is a perfect sphere (flattening parameter = 0) and using a flattening parameter of 0.00648. How do the results differ?

7. Compute the JD and MJD for your next birthday. If you know it, include the hours and minutes correspond to your time of birth, otherwise, use midnight UTC.
8. Using Equation (4.42), compute the date on which TDB and TCB will start differing by more than a minute.
9. Compute the ephemeris of the Moon using Simpson's approximation (discussed in Section 4.5) for a given time, e.g., from when your homework is due for the following 10 years. Plot the orbital elements of the Moon with respect to ECI for this time frame.
10. Using JPL Horizons, compute the position of Earth and Mars for a given time (e.g., when your homework is due). Compute the OWLT for this time. Then, obtain the positions of Earth and Mars from JPL Horizons for one synodic period and plot the OWLT for this time.
11. Repeat Problem 10, but use the low-fidelity ephemeris model discussed in this chapter (Standish and Williams' method).
12. Using the method explained in Section 4.6, compute the position of the Sun with respect to the Earth for a given time (e.g., when your homework is due).
13. Using JPL Horizons and the low-fidelity model discussed in Section 4.7.1, generate the ephemeris data for Venus from a given time (e.g., when your homework is due) for 10 years. How do the two methods compare? Plots the difference in distance of the planet with respect to the Sun and the planet's orbital speed.
14. Repeat Problem 13 for Mercury.
15. Repeat Problem 13 for Jupiter.
16. Repeat Problem 13 for Saturn.
17. Repeat Problem 13 for Uranus.
18. Repeat Problem 13 for Neptune.
19. Repeat Problem 13 for Pluto.
20. Use Standish and Williams' method to *estimate* the date on which the next conjunction between Earth and Mars occurs, i.e., when Earth and Mars are on the opposite side of the Sun.

5 Trajectory Design

5.1 INTRODUCTION TO TRAJECTORY DESIGN

In order to design a mission, it is necessary 1) to understand the behavior of the spacecraft's trajectory, taking into account the forces acting on the spacecraft, and 2) to possibly change the orbit of a spacecraft. The natural behavior of a spacecraft moving under the influence of gravitational and nongravitational forces was presented in Chapters 2 and 3. We can also apply a force on the spacecraft from a maneuver. This part of the trajectory design process involves estimation of the amount of velocity change maneuver (Δv) that is needed to change the orbit from one state to another. This chapter discusses the fundamentals of various orbital transfers. We begin by introducing minimum Delta v transfers for two-burn, coplanar transfers. This is expanded to include three-burn coplanar transfers and three-dimensional transfers. Continuous-thrust maneuvers are then introduced, and a brief discussion of optimal transfers is presented. Examples of continuous-thrust systems (including electric propulsion and solar sail propulsion) are discussed. The chapter then introduces the concepts needed for orbit design, including patched conic approximations, and develops tools and techniques used by orbit designers to plan trajectories between planets, moons, and asteroids, including launch parameters, maneuvers used to correct errors, gravity assist maneuvers, and the process of orbit insertion, entry, descent and landing on destination bodies.

5.1.1 Useful Orbits

Orbital transfers are needed to change orbits. Here, we briefly discuss some of the "useful" orbits that are often used to as initial or final orbits, depending on the mission. Table 5.1 shows a list of such orbits along with their defining characteristics. Note that the listed orbits are not necessarily mutually exclusive. For example, a polar orbit – which is defined by its inclination – can be a low Earth orbit – which is defined by its altitude. Additionally, while the examples provided in Table 5.1 relate to the Earth, they can easily be applied to other celestial bodies. For example, low Martian orbit (LMO) and low lunar orbit (LLO) are also orbits which have rather low orbital altitudes with respect to Mars and the Moon, respectively. However, the actual altitude values depend on the body in question.

Table 5.1 does not list all existing orbits, but rather those that are commonly used for planetary departure and final arrival orbits for interplanetary missions. For example, synchronous orbits exist for planets other than the Earth – although they are not called *geo*synchronous orbits. Also, although we focused on two-body orbits here, we have also discussed several three-body orbits and some of their uses in Chapter 3.

Many orbits are also used as so-called staging orbits. These are orbits that are not used to accomplish any major mission objective, but rather as intermediate orbits due to the existence of other mission requirements. These can include the need to assemble a spacecraft in orbit before it is launched into an interplanetary trajectory or the need to wait for a certain orbital alignment to occur in order to complete an orbital transfer.

DOI: 10.1201/9781003165071-5

TABLE 5.1
List of useful orbits along with their characteristics

Orbit	Defining Characteristics	Applications and Comments
Low Earth orbit (LEO)	Altitude < 2,000 km	Several
Medium Earth orbit (MEO)	Altitude between 2,000 and 35,786 km	Global Navigation Satellite System (GNSS) spacecraft, such as GPS
High Earth orbit	Altitude > 35,786 km	Scientific observations, staging
Geosynchronous orbit (GSO)	Altitude of 35,786 km and orbital period of one sidereal day (23 h 56 m 4 s for Earth)	Communications
Geostationary orbit (GEO)	GSO with zero inclination	Communications
Polar orbit	Inclination close to 90° (prograde) or −90° (retrograde)	Scientific observations, mapping
Frozen orbit	Drifts due to perturbations (J_2, etc.) are minimized	Lunar missions
Sun synchronous orbit (SSO)	Spacecraft passes over any point of the planet's surface at the same local solar time	Scientific observations
High elliptical orbit (HEO)	Eccentricity > 0.3	Scientific observations, staging

5.2 ORBITAL TRANSFERS

An orbital transfer, or simply orbit transfer, is the act of going from one orbit to another. There are an infinite number of transfers possible – however, in practical operations, an orbit transfer is chosen because it meets requirements and constraints on the mission. For example, many times the desire is to minimize the amount of Δv required to transfer, ensure that the transfer does not exceed a prescribed amount of time, or a combination of both. The amount of Δv is directly proportional to the amount of mass used (Δm) to perform this transfer, and it is found using Tsiolkovsky's rocket equation (also known simply as the rocket equation),

$$\frac{\Delta m}{m_0} = 1 - e^{-\Delta v / I_{sp} g_0} \tag{5.1}$$

where m_0 is the initial mass of the vehicle, I_{sp} is the specific impulse of the engine (which is a function of the type of propulsion system used), and g_0 is the gravitational acceleration on the Earth's surface. Note that I_{sp} is measured in seconds, and it is often referred to as the efficiency of a propulsion system. I_{sp} measures the amount of time for which a given amount of propellant can deliver thrust equal to its weight. Thus, higher I_{sp} is desirable. More details on the derivation of this equation can be found in Wiesel [1997] or Prussing and Conway [2012], among many other texts.

Other types of transfers that are commonly performed are those that are chosen based on the amount of propellant available. These transfers are more costly in terms of Δv than the minimum Δv transfer, but they meet other mission requirements or satisfy other constraints on various design considerations.

5.2.1 Optimal, Two-Burn Coplanar Transfers

The simplest of all transfers is a simple circle-to-circle transfer in which we assume that both orbits are located in the same plane and there is only one gravitational source acting on the spacecraft. These assumptions allow us to use the two-body approximation that we discussed in detail in Chapter 2.

In a circular orbit, the velocity vector is always perpendicular to the radius vector. The velocity vector is changed by adding (or subtracting) a velocity increment to the tangential velocity. This change in velocity then changes the energy of the orbit (think about the *vis viva* equation, Equation 2.50). This energy change then results in a different shaped orbit. A second similar maneuver is then performed at a different location in the new orbit, thus changing the orbit further. Proper choice of the velocity increment, direction of maneuver, and location in the orbit where the change is made is used to shape the orbit change.

In a two-maneuver (two-burn) coplanar circle-to-circle transfer, the first maneuver changes the initial circular orbit to an elliptical orbit. The spacecraft travels in this elliptical orbit for half period of the elliptical orbit. A second tangential maneuver is applied when the spacecraft reaches the point one half an orbit after the first maneuver was performed. This second maneuver then circularizes this orbit into the final circular orbit.

A Hohmann transfer [Hohmann, 1960] is an optimal, impulsive, two-burn, circle-to-circle coplanar transfer [Lawden, 1963; Barrar, 1963; Hazelrigg, 1968; Marec, 1979; Battin, 1987; Prussing, 1992]. A schematic of this orbit transfer is shown in Figure 5.1, where the final circular orbit is shown to be larger than the starting circular orbit. The first burn is performed in the direction parallel to the velocity vector. This maneuver sends the spacecraft into an elliptical transfer orbit, with the apoapse at the altitude of the final orbit. As the spacecraft reaches the apoapse, another impulse is applied in the direction parallel to the velocity vector. The added velocity is calculated to circularize the orbit. In reality no orbit is exactly circular and no transfer is done from an exactly

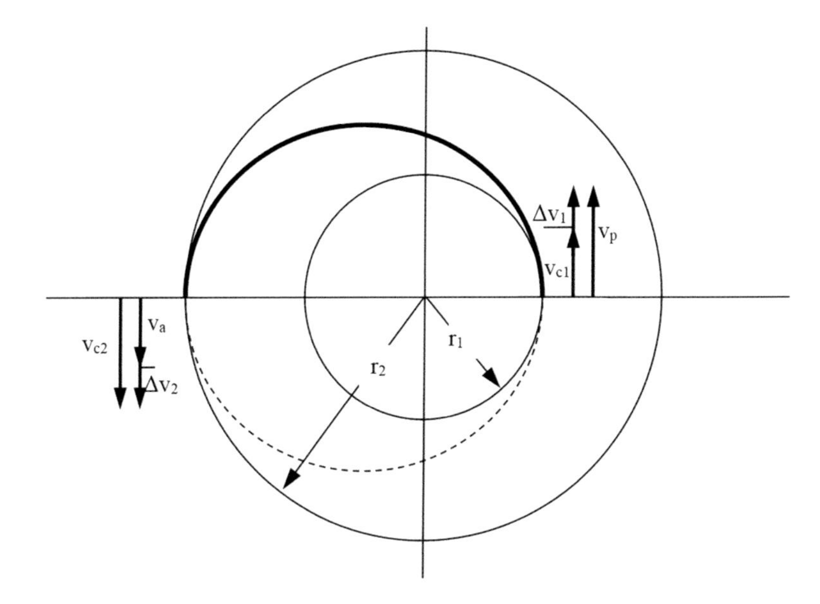

FIGURE 5.1 Schematic of the Hohmann transfer.

circular orbit to another exactly circular orbit in the same plane of motion. However, although an approximation, the Hohmann transfer nevertheless gives a reasonable result of what can be done with a high-thrust spacecraft. It should be noted that the mathematics of the Hohmann transfer from a larger circular orbit to a smaller circular orbit is identical to what is shown here, although the direction of the Δv maneuvers is inverted.

Initially, a spacecraft is in a circular orbit of radius r_1, with all of its velocity in the tangential direction, v_{c1}. A maneuver (Δv_1) is applied in the tangential direction, which changes this orbit from a circular orbit to an elliptical orbit. The periapse radius of this elliptical orbit is the same as the radius of the initial circular orbit. The velocity of the spacecraft is now v_p. The spacecraft then travels to the apoapse of the transfer orbit. At apoapse, its velocity is v_a, and a second tangential maneuver (Δv_2) is applied – one that is large enough to make the spacecraft's velocity (v_{c2}) equal to the velocity of the final circular orbit of radius r_2.

The semimajor axis of the transfer orbit, also known as the Hohmann ellipse, is

$$a_T = \frac{1}{2}(r_1 + r_2) \tag{5.2}$$

Initially, the velocity in the first circular orbit is

$$v_{c1} = \sqrt{\frac{\mu}{r_1}} \tag{5.3}$$

Inserting Equation (5.2) into the *vis viva* equation (Equation 2.50), the velocity of the spacecraft at the periapse becomes

$$v_p = \sqrt{\frac{2\mu r_2}{r_1(r_1 + r_2)}} \tag{5.4}$$

Subtracting Equation (5.3) from Equation (5.4) gives the first change in velocity,

$$\Delta v_1 = v_p - v_{c1} \tag{5.5}$$

At the apoapse of the transfer orbit, the *vis viva* equation is also used to determine the velocity at that point,

$$v_a - \sqrt{\frac{2\mu r_1}{r_2(r_1 + r_2)}} \tag{5.6}$$

The velocity in the final circular orbit is

$$v_{c2} = \sqrt{\frac{\mu}{r_2}} \tag{5.7}$$

The difference between what the spacecraft's velocity is in the transfer orbit and its final velocity is the second maneuver,

$$\Delta v_2 = v_{c2} - v_a \tag{5.8}$$

The total Δv needed for the entire transfer is given by the sum of the results of Equations (5.5) and (5.8)

$$\Delta v_{total} = \Delta v_1 + \Delta v_2 \quad (5.9)$$

To further simplify Equation (5.9), we divide it by v_{c1} and introduce the parameter $R = r_2/r_1$, resulting in

$$\frac{\Delta v_{total}}{v_{c1}} = \left(1 - \frac{1}{R}\right)\sqrt{\frac{2R}{R+1}} + \frac{1}{\sqrt{R}} - 1 \quad (5.10)$$

Note that the right-hand side of Equation (5.10) is effectively non-dimensionalized, which is useful for calculations and for further analysis, as we will see in the next section.

The last thing to compute for the Hohmann transfer is how long the transfer takes. To move from the periapse to the apoapse of the Hohmann ellipse equates to traveling for half the period of the transfer orbit, or

$$t_{Hohmann} = \pi\sqrt{\frac{1}{\mu}\left(\frac{r_1 + r_2}{2}\right)^3} \quad (5.11)$$

5.2.2 Optimal Three-Burn Coplanar Orbit Transfers

There are other ways to perform circle-to-circle transfers, such as using multiple elliptical orbits to go from the initial to the final circular orbit, also known as a bi-elliptic transfer. A bi-elliptic transfer (depicted in Figure 5.2) uses two elliptical transfer orbits whose apoapsides extend beyond the larger target orbit. Each of these elliptical orbits has periapsis tangent to one of the circular orbits, and they are tangent to each other at their apoapsides. While the starting and target orbital radii are assumed to be known, the intermediate shared apoapsis of the elliptical orbits r_i is placed sufficiently far from the focus such that Δv_2 is small when compared to Δv_1 and Δv_3. In fact, as r_i approaches infinity, the bi-elliptic transfer becomes a bi-parabolic transfer, where Δv_2 approaches zero, but the time of flight of the overall transfer approaches infinity, making the bi-parabolic transfer impractical, but a useful mathematical tool to compare various orbital transfers. The geometry of this transfer is shown in Figure 5.2.

Initially, a spacecraft is in a circular orbit of radius r_1 with velocity v_{c1}. A maneuver of the amount Δv_1 is made tangentially. This places the spacecraft on an elliptical orbit with periapse radius of r_1 and an apoapse radius of r_i, where r_i is greater than the final circular orbit radius, r_2. At apoapse of this first transfer orbit, a second tangential maneuver, Δv_2 is performed which places the spacecraft on a second transfer ellipse. This transfer orbit now has an apoapse radius of r_i and a periapse radius of r_2. After the spacecraft travels to periapse of the second transfer ellipse, a third tangential maneuver, Δv_3, is applied in a direction opposite of the velocity vector. This reduces the apoapse of the second transfer orbit to r_2, making the orbit circular of radius r_2.

The three Δv's are found the same way as was done for the Hohmann transfers (using the *vis viva* equation) and are

$$\Delta v_1 = \sqrt{\frac{\mu}{r_1}}\left[\sqrt{\frac{2(r_i/r_1)}{1+(r_i/r_1)}} - 1\right] \quad (5.12)$$

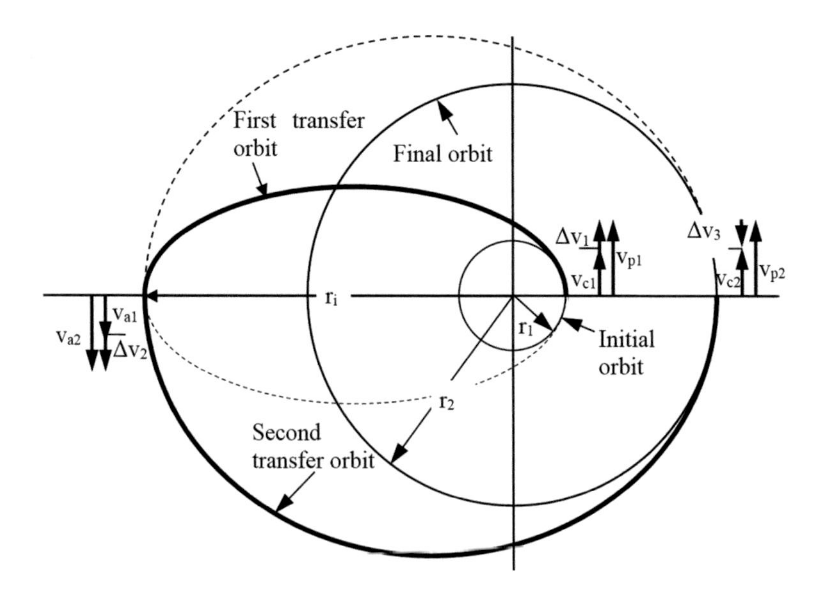

FIGURE 5.2 Bi-elliptic transfer geometry.

$$\Delta v_2 = \sqrt{\frac{\mu}{r_1}}\left[\sqrt{\frac{2(r_2/r_1)}{(r_i/r_1)\left[(r_i/r_1)+(r_2/r_1)\right]}} - \sqrt{\frac{2}{(r_i/r_1)\left[1+(r_2/r_1)\right]}}\right] \tag{5.13}$$

$$\Delta v_3 = \sqrt{\frac{\mu}{r_1}}\left\{\sqrt{\frac{2(r_i/r_1)}{(r_2/r_1)\left[(r_2/r_1)+(r_i/r_1)\right]}} - \frac{1}{r_2/r_1}\right\} \tag{5.14}$$

and the total Δv is

$$\Delta v_{total} = \Delta v_1 + \Delta v_2 + \Delta v_3 \tag{5.15}$$

Similarly to the analysis for the Hohmann transfer, the parameter $R = r_2/r_1$ is used alongside a new parameter $S = r_i/r_2$, which includes the addition of the intermediate connection point between transfer ellipses. The use of these parameters can further simplify Equation (5.15), resulting in

$$\frac{\Delta v_{total}}{v_{c1}} = \sqrt{\frac{2}{(RS+1)RS}}(RS-1) + \sqrt{\frac{2(S+1)}{RS}} - \frac{1}{\sqrt{R}} - 1 \tag{5.16}$$

Note that, like Equation (5.10), the right-hand side of Equation (5.16) is non-dimensional. However, in Equation (5.16), while R is a fixed value for a given transfer, S tends to be a design parameter which changes based on various mission requirements. So, when does the bi-elliptic transfer result in an overall less Δv_{total} than the Hohmann transfer? By inspection of Equation (5.16), for a fixed value of R, as S approaches infinity, $\Delta v_{total}/v_{c1}$ is minimized, which is the bi-parabolic case, i.e.

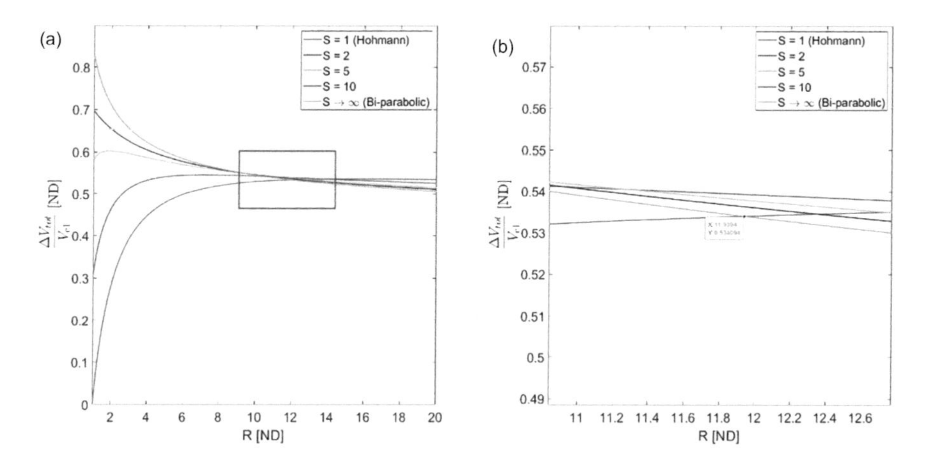

FIGURE 5.3 Non-dimensionalized Δv_{total} comparison between the Hohmann transfer (S = 1) and various bi-elliptic transfers.

$$\frac{\Delta v_{total}}{v_{c1}} = \left(\sqrt{2}-1\right)\left(1+\frac{1}{\sqrt{R}}\right) \tag{5.17}$$

Thus, equating Equations (5.10) and (5.17), and solving for R, yields a third-degree polynomial in R whose only real solution is the so-called critical R value, or $R^* \approx 11.93876$. Thus, the Hohmann transfer is better than the bi-elliptic transfer for $R < R^*$. Above this ratio, a three-impulse, bi-elliptic transfer can be found with a lower cost than the two impulse, Hohmann transfer. When R is between approximately 11.94 and 15.58, the bi-elliptic transfer is optimal for certain values of S, namely between the extreme cases for $S \to \infty$ (bi-parabolic transfer) when $R \approx 11.94$ and for $S \to 1$ (Hohmann transfer) when $R \approx 15.58$. However, after R reaches approximately 15.58 (such as a transfer from Earth to Uranus, or further), the bi-elliptic transfer is always the most efficient transfer in terms of Δv_{total}. Figure 5.3a shows this relationship, while Figure 5.3b is a zoomed-in version of the same graph with a focus on R^*.

Even though Δv_{total} is smaller for a high value of R, the total bi-elliptic transfer time is generally prohibitively high. This total transfer time is the sum of half of the periods of the two transfer orbits, i.e.,

$$t_{bi-elliptic} = \pi\left[\sqrt{\frac{(r_1+r_i)^3}{8\mu}} + \sqrt{\frac{(r_i+r_2)^3}{8\mu}}\right] \tag{5.18}$$

5.2.3 Three-Dimensional Transfers

The next simplest maneuver is changing the spacecraft's orbital plane by changing the inclination and/or the right ascension of the ascending node. Doing a purely out-of-plane maneuver can change these angles without affecting the semimajor axis or the eccentricity. The effects on the angles depend on where the maneuver occurs. Changing a spacecraft's orbital plane requires a lot of propellant,

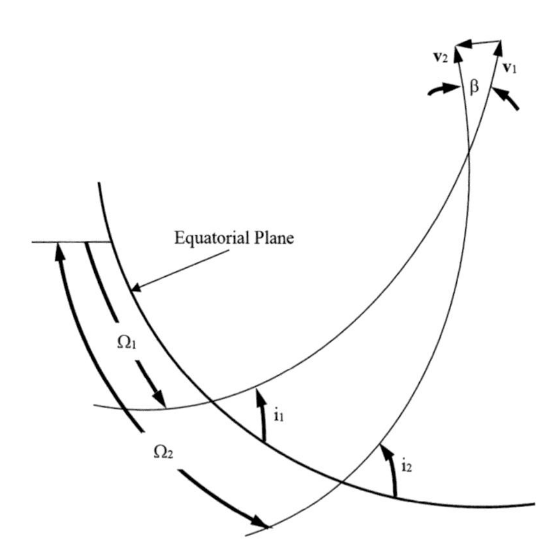

FIGURE 5.4 Schematic of plane change geometry.

so it is done sparingly. This type of maneuver is completed by a change in the angular momentum vector.

A schematic of a plane change maneuver is presented in Figure 5.4. Here, the initial orbit has an inclination and right ascension of the ascending node of i_1 and Ω_1, respectively. The spacecraft is transferred to a new orbit by applying an instantaneous out-of-plane maneuver, which changes the inclination and RAAN to new values, i_2 and Ω_2, respectively. The velocity vector is changed from $\mathbf{v}_1$ and $\mathbf{v}_2$ by applying a maneuver,

$$\mathbf{v}_1 + \Delta\mathbf{v} = \mathbf{v}_2 \tag{5.19}$$

Assuming that the entire velocity change is applied to change the plane, then the magnitude of $\mathbf{v}_1$ and $\mathbf{v}_2$ are the same. The turning angle of the maneuver is therefore β, as shown in Figure 5.4.

Let's geometrically examine the point where the two orbits cross. A larger view of this triangle is seen in Figure 5.5.

Using the law of sines,

$$\Delta v = 2v\sin\frac{\beta}{2} \tag{5.20}$$

These are only approximated as right triangles – in fact, they are what is known as spherical triangles, as discussed in Appendix B. An analogous law of cosines yields,

$$\cos\beta = \cos i_1 \cos i_2 + \sin i_1 \sin i_2 \cos(\Omega_2 - \Omega_1) \tag{5.21}$$

If it is desired to change only the inclination ($i_2 = i_1 + \Delta i$, $\Omega_2 - \Omega_1 = 0$), using some trigonometry identities we find that $\beta = \Delta i$. For this to happen, the velocity vector must be turned as the orbit crosses the equatorial plane.

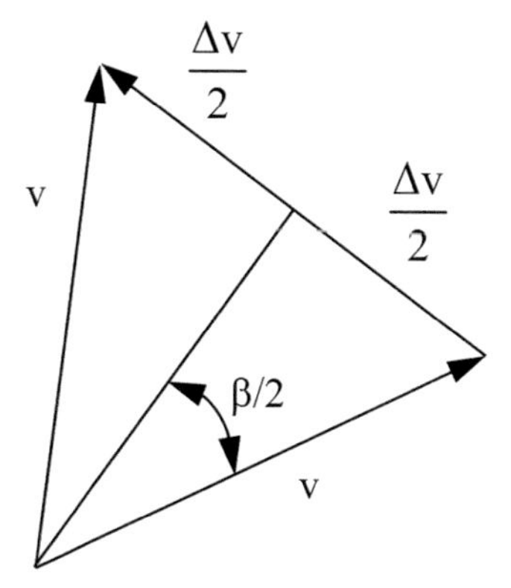

FIGURE 5.5 Geometry of pure out-of-plane maneuver.

It is possible to change both the inclination and RAAN at the same time by choosing a location other than an equatorial crossing. However, there are easier ways to change the RAAN. We have already discussed the precession of the RAAN due to the J_2 term which is a slow process (if you are changing from one RAAN to a target RAAN, recall that that RAAN is precessing as well). One reason to want to change the RAAN is so that a launch vehicle and a target vehicle can rendezvous in the same orbit plane. An excellent example of this is when a spacecraft wants to dock with the International Space Station (ISS). Instead of expending a large amount of Δv to change the orbit plane, a launch window is computed such that the plane the spacecraft is being launched into aligns with the plane that the orbit of the ISS. With the fixed ground launch site, the Earth is rotating under every orbit plane around twice per day (once on an ascending pass, once on a descending pass – there is also a slight difference due to the J_2 effects on the orbit plane).

The most common of these maneuvers is a three-dimensional orbit change. A common example is a geosynchronous transfer orbit. Most geosynchronous spacecraft launch from locations that are not exactly on the equator. NASA and the U.S. Air Force launch their geosynchronous spacecraft from Cape Canaveral in Florida (~28.5°N), and the European Space Agency launches theirs from French Guyana (~5.3°N). To achieve a nearly zero inclination, these spacecraft need a three-dimensional maneuver that reduces inclination at the same time it increases the semimajor axis. This combination is more efficient than doing these maneuvers separately. To know when and where to do the maneuvers, an ad-hoc simulation that optimizes the trajectory needs to be done, which is beyond the scope of this chapter. For more information, the reader is referred to Vallado [2013] and Lawden [1963], among others.

5.3 LAMBERT'S PROBLEM

An important aspect of interplanetary mission design is the need to design orbits that depart from a specific location at a given time and arrive at another place at another time. Given these physical starting and ending points and the time of flight between them, this two-point boundary value problem (2PBVP) has been a focal aspect of mission design and was first formally formulated by the Swiss-French astronomer, physicist, and mathematician Johann Heinrich Lambert (1728–1777), who originally developed a geometric method for solving it. The underlying theorem which leads into solutions of the so-called Lambert's problem is based on Lambert's 1761 realization: the time of flight spent traversing an elliptic arc between two points P_1 and P_2, as seen in Figure 5.6, depends only on the semimajor axis of the ellipse and on two dimensions taken from a triangle drawn between P_1, P_2 and the focus, the chord, c, connecting P_1 and P_2 and not the eccentricity of the transfer ellipse.

Mathematically, this can be expressed as

$$\sqrt{\mu}\left(t_2 - t_1\right) = F\left(a, r_1 + r_2, c\right) \tag{5.22}$$

Equation (5.22) is often referred to as Lambert's theorem, which was proven in 1778 by Lagrange, 1 year after Lambert's death.

The triangle formed by FP_1P_2 from Figure 5.6 is often referred to as the space triangle for a given orbital transfer, and r_1 and r_2 represent the orbital radii at P_1 and P_2, respectively. Examining Figure 5.6 further, we find that the difference in true anomalies between points P_1 and P_2 is

$$\cos\Delta\theta = \frac{r_1 \cdot r_2}{r_1 r_2} \tag{5.23}$$

Even though this gives us the transfer angle, it does not resolve the quadrant ambiguity – there are two solutions, $\Delta\theta$ or $-\Delta\theta$. To resolve this quadrant ambiguity, we need to develop a relationship involving the sine of the angular change. Let's start by examining the **Z** component of the vector product of $r_1 \times r_2$,

$$\begin{aligned}\left(r_1 \times r_2\right)_Z &= \mathbf{K} \cdot \left(\mathbf{r}_1 \times \mathbf{r}_2\right) = \mathbf{K} \cdot \left(r_1 r_2 \sin\Delta\theta \, \mathbf{w}\right) \\ &= r_1 r_2 \sin\Delta\theta \left(\mathbf{K} \cdot \mathbf{w}\right)\end{aligned} \tag{5.24}$$

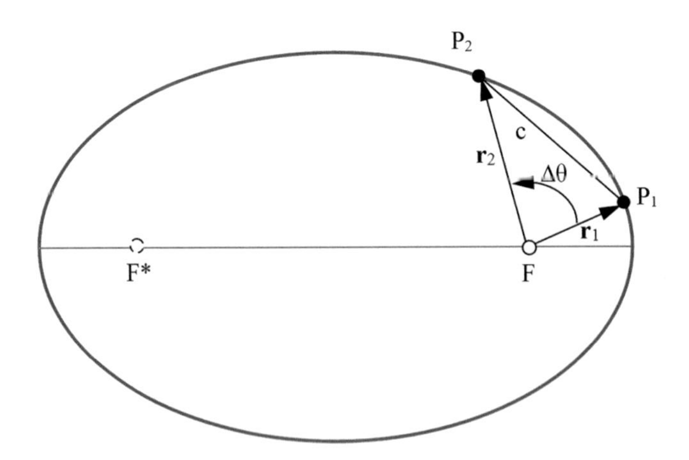

FIGURE 5.6 Lambert's problem geometry.

The dot product term is related to the inclination,

$$\cos i = \mathbf{K} \cdot \mathbf{w} \tag{5.25}$$

which results in two equations and two unknowns (angle and quadrant),

$$\cos \Delta\theta = \frac{\mathbf{r}_1 \cdot \mathbf{r}_2}{r_1 r_2} \quad \text{and} \quad \sin \Delta\theta = \frac{\mathbf{K} \cdot (\mathbf{r}_1 \times \mathbf{r}_2)}{r_1 r_2 \sin i} \tag{5.26}$$

There are an infinite number of trajectories between any two points P_1 and P_2 (this is shown in Figure 5.7). What distinguishes these various orbits is the time of flight between these two points. Additionally, there are two possible routes for each of these trajectories. Commonly referred to as the "short way" (Figure 5.8) and the "long way" (Figure 5.9), these two routes have different times of flight. Additionally, the "short way" does not go near the vacant focus, while the "long way" does. Typically, the most efficient way (in terms of Δv needed to transfer) is to move in the same direction (prograde or retrograde) as the original trajectory.

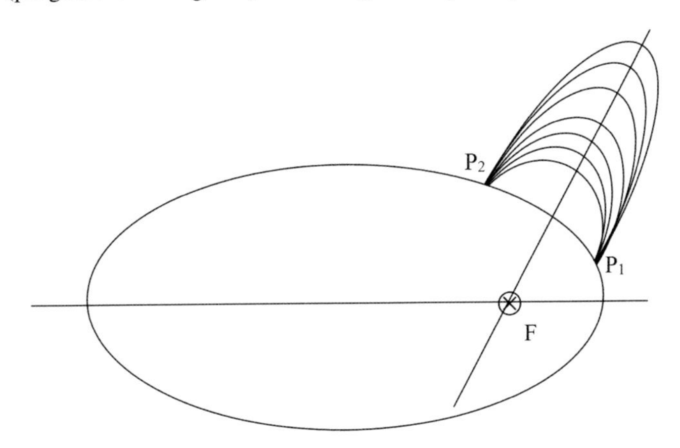

FIGURE 5.7 Some of the infinite number of paths connecting two points.

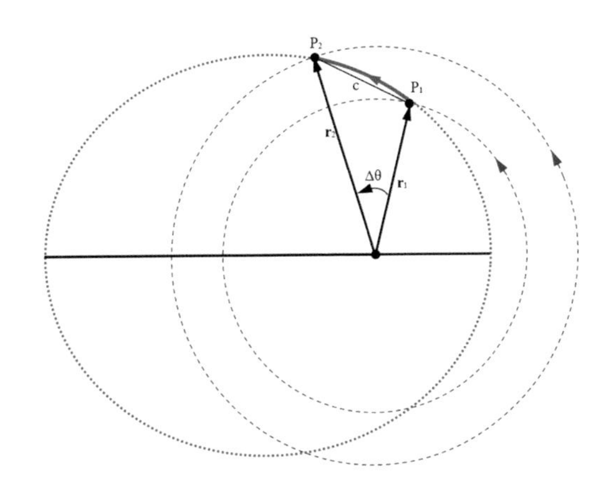

FIGURE 5.8 Lambert's problem – the "short way."

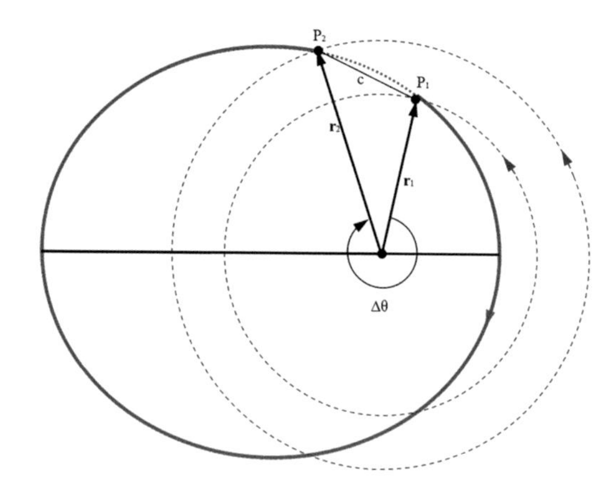

FIGURE 5.9 Lambert's problem – the "long way."

A transfer orbit using any of the various conic sections (ellipse, parabola, and hyperbola) can be used. The type of conic section depends on a number of things, but primarily on the time of flight.

Lambert's problem states that given two position vectors and the time of flight between these two vectors, we can find an orbit that connects these two points.

Kepler's problem (which we first saw in Chapter 2) shows that the time between two points can be related to one of the anomalies. We will show relationships with eccentric anomaly, hyperbolic anomaly, and universal anomaly (we'll skip parabolic anomaly, as this is a single point between eccentric and hyperbolic, and can be encompassed in the universal anomaly approach).

Using eccentric anomaly, Kepler's equation can be written as

$$\sqrt{\mu}\,(t_2 - t_1) = a^{3/2}\left[E_2 - E_1 - e(\sin E_2 - \sin E_1)\right] \tag{5.27}$$

Define two variables,

$$E_{mean} = \frac{1}{2}(E_2 + E_1) \tag{5.28}$$

and

$$E_{middle} = \frac{1}{2}(E_2 - E_1) \tag{5.29}$$

We also know that the orbit equation in terms of eccentric anomaly can be written as

$$r = a(1 - e\cos E) \tag{5.30}$$

Applying this to the two positions, r_1 and r_2, and adding them, we get

$$r_1 + r_2 = a\left[2 - e(\cos E_2 + \cos E_1)\right] \tag{5.31}$$

Using the trigonometric identity

$$\cos E_2 + \cos E_1 = 2\cos\left(\frac{E_2 + E_1}{2}\right)\cos\left(\frac{E_2 - E_1}{2}\right) \\ = 2\cos E_{mean}\cos E_{middle} \tag{5.32}$$

We can simplify Equation (5.31) into

$$r_1 + r_2 = 2a\left(1 - e\cos E_{mean}\cos E_{middle}\right) \tag{5.33}$$

Examining the chord connecting P_1 and P_2, we can write the Cartesian coordinates (x- and y-components) relative to the center of the ellipse as

$$\begin{aligned} x_1 &= a\cos E_1 \qquad y_1 = b\sin E_1 = a\sqrt{1-e^2}\sin E_1 \\ x_2 &= a\cos E_2 \qquad y_2 = b\sin E_2 = a\sqrt{1-e^2}\sin E_2 \end{aligned} \tag{5.34}$$

the chord is

$$c = \sqrt{(x_2 - x_1)^2 + (y_2 - y_1)^2} \tag{5.35}$$

Inserting Equation (5.34) into Equation (5.35), and simplifying,

$$c^2 = 4a^2\sin^2 E_{middle}\left(1 - e^2\cos^2 E_{mean}\right) \tag{5.36}$$

If we define another auxiliary equation as $\cos\xi = e\cos E_{mean}$, then the chord length becomes

$$c = 2a\sin E_{middle}\sin\xi \tag{5.37}$$

In terms of these anomaly functions,

$$r_1 + r_2 = 2a\left(1 - \cos E_{middle}\cos\xi\right) \tag{5.38}$$

Next, we define another set of auxiliary variables, α and β as

$$\alpha = \xi + E_{middle} \tag{5.39}$$

and

$$\beta = \xi - E_{middle} \tag{5.40}$$

Developing relationships for r_1, r_2 and c as functions of semimajor axis, α and β, we find these as

$$r_1 + r_2 - c = 2a\left(1 - \cos\alpha\right) = 4a\sin^2\left(\frac{\alpha}{2}\right) \tag{5.41}$$

and

$$r_1 + r_2 - c = 2a(1-\cos\beta) = 4a\sin^2\left(\frac{\beta}{2}\right) \tag{5.42}$$

Plugging these auxiliary anomaly variables into Kepler's equation give us

$$\sqrt{\mu}(t_2 - t_1) = 2a^{3/2}(E_{middle} - \cos\xi \sin E_{middle}) \tag{5.43}$$

In terms of α and β, Kepler's equation becomes

$$\sqrt{\mu}(t_2 - t_1) = a^{3/2}\left[\alpha - \beta - (\sin\alpha - \sin\beta)\right] \tag{5.44}$$

Equation (5.44) is also known as Lambert's equation. Noting that $r_1 + r_2 + c$ is the perimeter s of the triangle formed by the origin, P_1 and P_2, we find that

$$\sin\left(\frac{\alpha}{2}\right) = \sqrt{\frac{s}{2a}} \tag{5.45}$$

and

$$\sin\left(\frac{\beta}{2}\right) = \sqrt{\frac{s-c}{2a}} \tag{5.46}$$

Angles α and β[1] are now only a function of a, $r_1 + r_2$ and c, so we have shown that Kepler's equation can be transformed into Lambert's equation, as presented in Equation (5.44), and we have proven Equation (5.22). In fact, Lambert's problem is only one of a few important instances in astrodynamics where a property of an orbit – here given by Equation (5.22) – is not dependent on eccentricity. Other examples of such kind include the orbital period and energy of an orbit, which are both dependent on semimajor axis, but not on eccentricity.

Lambert's problem is considered one of the fundamental cornerstones of trajectory design, and we will often refer to it (and its solutions) throughout this and other chapters. Lambert's problem is so important that it is even featured in the cover art of Prussing and Conway's "Orbital Mechanics" textbook. There are several ways of solving Lambert's problem. Here, we will look at two main solutions methods: the so-called classical solution method and a solutions method that makes use of the universal variable. Solving Lambert's problem over and over provides information to construct what is known as a "porkchop" plot. Porkchop plots are used by mission designers and mission planners to design a mission from a departure location to an arrival location. Thus, Lambert's problem generally represents the first step toward designing an interplanetary trajectory. We briefly introduced porkchop plots in the example shown in Section 4.8.3, but this chapter expands on their utility further in Section 5.8.

[1] The geometric interpretation of the angles α and β can be found in Prussing and Conway [2012].

5.3.1 Classical Solution to Lambert's Problem

The classical solution to Lambert's problem makes use of the problem's geometry for the minimum energy transfer – refer to Figure 5.10 – and the given time of flight to choose which conic section the transfer must be in order to solve Lambert's problem. Finding the type of conic section before starting to solve Lambert's equation is not required in all solution methods, such as the solution to Lambert's problem using the universal variable, which we will see in the next section. The basic geometry of the minimum energy transfer is constructed such that P_1 and P_2 are the centers of circles (in red and blue) whose intersection is the vacant focus, F^*, that results from the minimum energy ellipse, whose transfer arc is shown in green in the figure along with its line of apsides (LOA, dotted).

From Figure 5.10, using the properties of ellipses, we can write

$$\begin{aligned} \overline{P_1F^*} &= 2a - r_1 \\ \overline{P_2F^*} &= 2a - r_2 \end{aligned} \tag{5.47}$$

and

$$c = \overline{P_1F^*} + \overline{P_2F^*} \tag{5.48}$$

Plugging Equation (5.47) into (5.48) and letting $a = a_m$, i.e., the smallest semimajor axis that connects P_1 and P_2 as shown in Figure 5.10, yields

$$c = (2a_m - r_1) + (2a_m - r_2) \tag{5.49}$$

and if we define with s the semiperimeter of the space triangle, then

$$s = \frac{1}{2}(r_1 + r_2 + c) \tag{5.50}$$

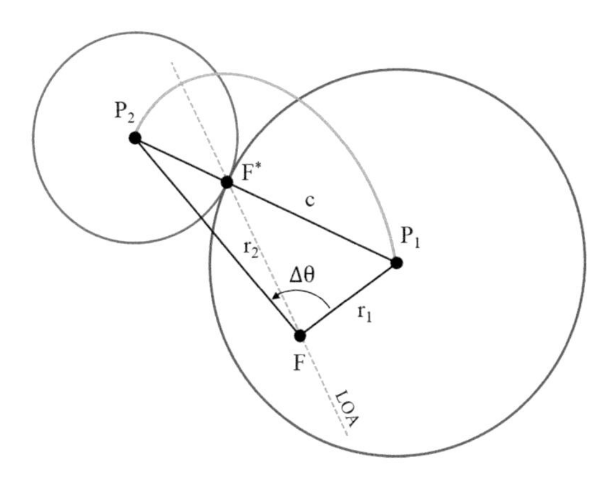

FIGURE 5.10 Geometry of the minimum energy solution.

so that

$$a_m = \frac{1}{2}s \tag{5.51}$$

is the minimum possible semimajor axis for a given set of P_1, P_2, and $\Delta\theta$. Note that the chord c can be computed using the law of cosines on the space triangle. Therefore, the minimum energy corresponding to a_m is

$$E_m = -\frac{\mu}{2a_m} \tag{5.52}$$

The ellipse having semimajor axis a_m and energy E_m is called the minimum energy ellipse. We can also compute the α and β values corresponding to the minimum energy ellipse by substituting Equation (5.51) into Equations (5.45) and (5.46) to get

$$\begin{aligned} \alpha_m &= \pi \\ \beta_m &= 2\,\mathrm{sgn}(\sin\Delta\theta)\sin^{-1}\sqrt{\frac{s-c}{s}} \end{aligned} \tag{5.53}$$

where sgn is the sign function – not to be confused with the trigonometric sine function – and is defined as

$$\mathrm{sgn}(x) = \begin{cases} +1 & \text{for } x > 0 \\ -1 & \text{for } x < 0 \end{cases} \tag{5.54}$$

keeping in mind that no quadrant correction for the arcsine function is needed for Equation (5.53). Additionally, the time of flight corresponding with the minimum energy ellipse, t_m, is computed as follows

$$t_m = \sqrt{\frac{s^3}{8\mu}}\left(\pi - \beta_m + \sin\beta_m\right) \tag{5.55}$$

Another important parameter is the parabolic time of flight, t_p, which is computed by taking the limit as semimajor axis a approaches infinity using Equation (5.44), i.e.

$$t_p = \lim_{a\to\infty}(t_2 - t_1) = \frac{a^{3/2}}{\sqrt{\mu}}\left[\alpha - \beta - (\sin\alpha - \sin\beta)\right] \tag{5.56}$$

To solve this limit, we substitute semimajor axis with energy (which is exactly zero for a parabola), we use the definitions of α and β as provided by Equations (5.45) and (5.46), and simplify to get

$$t_p = \frac{\sqrt{2}}{3\sqrt{\mu}}\left[s^{3/2} - \mathrm{sgn}(\sin\Delta\theta)(s-c)^{3/2}\right] \tag{5.57}$$

Thus, in order to determine the orbit type (the conic section), it is simply necessary to compare the given time of flight, $TOF = t_2 - t_1$ to the computed t_p from Equation (5.57) such that

$$\begin{aligned} TOF > t_p &\rightarrow \text{orbit type} = \text{ellipse or circle} \\ TOF = t_p &\rightarrow \text{orbit type} = \text{parabola} \\ TOF < t_p &\rightarrow \text{orbit type} = \text{hyperbola} \end{aligned} \tag{5.58}$$

A parabolic orbit has known values of semimajor axis and eccentricity, which are infinity and 1, respectively. Therefore, the case for which the orbit type is found to be a parabola – which is generally a rare occurrence – does not necessitate that we solve Lambert's equation numerically. However, if we determine that the orbit type is an ellipse (or a circle, which is a special case of an ellipse), we need to solve Lambert's equation as provided by Equation (5.44). In order to proceed, we need to determine the values of α and β that we need to use, which are dependent on the values of TOF and $\Delta\theta$ as provided by the following

$$\alpha = \begin{cases} \alpha_0 & \text{if } TOF \le t_m \\ 2\pi - \alpha_0 & \text{if } TOF > t_m \end{cases} \tag{5.59}$$

and

$$\beta = \begin{cases} +\beta_0 & \text{if } 0 \le \Delta\theta < \pi \\ -\beta_0 & \text{if } \pi \le \Delta\theta < 2\pi \end{cases} \tag{5.60}$$

where

$$\begin{aligned} \alpha_0 &= 2\sin^{-1}\sqrt{\frac{s}{2a}} \\ \beta_0 &= 2\sin^{-1}\sqrt{\frac{s-c}{2a}} \end{aligned} \tag{5.61}$$

keeping in mind that no quadrant correction is necessary for these angles.

Combining the definitions of α and β given in this section allows us to solve Lambert's equation with the guarantee that one and only one solution exists. In order to solve Equation (5.44), several numerical methods exist. Here, we suggest a simple yet very effective method known as the secant method. This method is an iterative root-finding method that we can use here to solve the following equation

$$f(a) = a^{3/2}\left[\alpha - \beta - (\sin\alpha - \sin\beta)\right] - \sqrt{\mu}\,TOF \tag{5.62}$$

which is simply a rewriting of Equation (5.44) where we used the definition of $TOF = t_2 - t_1$.

Our goal is to find the value of semimajor axis a that makes $f(a)=0$. Using the secant method, we need to choose two initial values of a, initial guesses a_1 and a_2, and compute their corresponding function evaluations using Equation (5.62). a_1 and a_2 need to be chosen such that they are located on opposite sides of the solution a^*, i.e.,

$$a_1 < a^* < a_2 \tag{5.63}$$

and since the value of a^* is unknown – otherwise we wouldn't need to solve this problem – we can ensure the condition in Equation (5.63) by checking that

$$f(a_1)f(a_2) < 0 \tag{5.64}$$

since $f(a^*)$ is the only root of f(a). Furthermore, the domain of f(a) is $a \geq a_m$, so initial guesses must be chosen within this range of values.

Once we know that our initial guesses satisfy Equation (5.64), we need to iterate to find subsequent values of a using the secant line that is formed between the points $(a_1, f(a_1))$ and $(a_2, f(a_2))$, or more generally

$$a_n = a_{n-1} - f(a_{n-1})\frac{a_{n-1} - a_{n-2}}{f(a_{n-1}) - f(a_{n-2})} \tag{5.65}$$

and new computed values of a_n are used to replace either a_{n-1} or a_{n-2} in order to satisfy Equation (5.64). Iterations continue using the update step provided in Equation (5.65) until a given absolute and/or relative tolerance are met.

In contrast with the Newton-Raphson method that we discussed in Chapter 2 to solve Kepler's time equation, the secant method does not require that we compute and evaluate the derivative of the function f(a). However, the secant method generally converges more slowly than the Newton-Raphson method. Furthermore, Newton-Raphson requires that we divide by the derivative of f(a), which can take on small values, or even be zero for certain iterations while solving Lambert's problem, and thus would result in the algorithm not converging.

Above we discussed the solution to Lambert's problem for the elliptical case, which is also valid for the special case of an orbit being circular. When, however, $TOF < t_p$, the orbit connecting P_1 and P_2 is hyperbolic, as provided by Equation (5.58). Lambert's equation can thus be rederived for the hyperbolic case, and it can be shown to be

$$\sqrt{\mu}(t_2 - t_1) = a^{3/2}[\sinh\alpha_H - \alpha_H + - \operatorname{sgn}(\sin\Delta\theta)(\sin\beta_H + \beta_H) \tag{5.66}$$

where, in summary, the hyperbolic sine function replaces the sine function used in Equation (5.44). Equation (5.66) is also known as Lambert's equation for hyperbolic orbits, and α_H and β_H are given by

$$\sinh\left(\frac{\alpha_H}{2}\right) = \sqrt{\frac{s}{-2a}} \tag{5.67}$$

and

$$\sinh\left(\frac{\beta_H}{2}\right) = \sqrt{\frac{s-c}{-2a}} \tag{5.68}$$

where $a < 0$ for hyperbolic orbits. Equation (5.66) can also be solved using the secant method described in this section.

Once Lambert's equation is solved – whether it be for the elliptical/circular case or the hyperbolic case – additional orbital parameters can be found. Eccentricity can be computed through the use of semilatus rectum

$$p = \frac{4a(s-r_1)(s-r_2)}{c^2}\sin^2\left(\frac{\alpha+\beta}{2}\right) \tag{5.69}$$

for elliptical orbits and

$$p = \frac{-4a(s-r_1)(s-r_2)}{c^2}\sinh^2\left(\frac{\alpha_H+\beta_H}{2}\right) \tag{5.70}$$

for hyperbolic orbits, such that

$$e = \sqrt{1-\frac{p}{a}} \tag{5.71}$$

Geometrically, it can be shown that the minimum possible eccentricity for a transfer orbit is

$$e_s = \frac{r_2 - r_1}{c} \tag{5.72}$$

as was proven in Prussing and Conway [2012].

Lambert's problem is usually used to compute the change in velocity necessary to insert a spacecraft from a given orbit to an orbital transfer between P_1 and P_2. Thus, we need to compute the velocity vectors at P_1 and P_2 that the spacecraft would need to be on the computed transfer orbit. Figure 5.11 shows a set of unit vectors based on the orientation of the space triangle, such that

$$\hat{\mathbf{u}}_1 = \frac{\mathbf{r}_1}{r_1}, \quad \hat{\mathbf{u}}_2 = \frac{\mathbf{r}_2}{r_2}, \quad \hat{\mathbf{u}}_c = \frac{\mathbf{r}_2 - \mathbf{r}_1}{c} \tag{5.73}$$

Therefore, the velocities at P_1 and P_2 can be computed as

$$\begin{aligned} \mathbf{v}_1 &= (B+A)\hat{\mathbf{u}}_c + (B-A)\hat{\mathbf{u}}_1 \\ \mathbf{v}_2 &= (B+A)\hat{\mathbf{u}}_c - (B-A)\hat{\mathbf{u}}_2 \end{aligned} \tag{5.74}$$

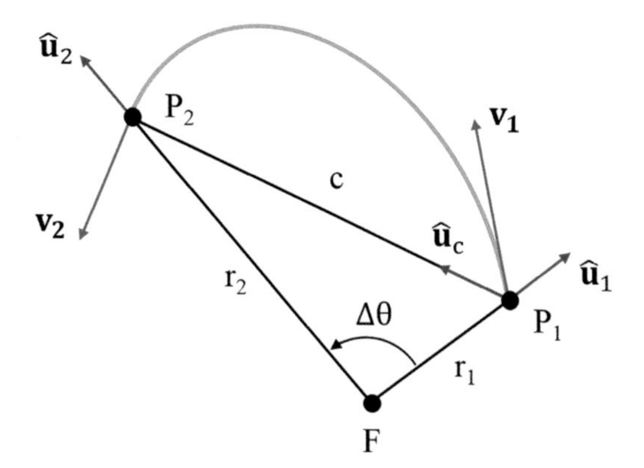

FIGURE 5.11 Unit vectors based on the space triangle.

where

$$A = \sqrt{\frac{\mu}{4a}} \cot\left(\frac{\alpha}{2}\right)$$
$$B = \sqrt{\frac{\mu}{4a}} \cot\left(\frac{\beta}{2}\right) \tag{5.75}$$

for elliptical orbits and

$$A = \sqrt{\frac{\mu}{-4a}} \coth\left(\frac{\alpha_H}{2}\right)$$
$$B = \sqrt{\frac{\mu}{-4a}} \coth\left(\frac{\beta_H}{2}\right) \tag{5.76}$$

for hyperbolic orbits.

Equations (5.44) and (5.66) can be also solved using alternative methods, such as those developed by Battin [1999] and Gooding [1990]. In the next section, we discuss an alternative formulation of Lambert's problem which makes use of the universal variable, similar to that discussed in Section 2.6.4.

5.3.2 Universal Variable Solution to Lambert's Problem

A universal variable can also be applied to solve Lambert's problem – refer to Section 2.6 for the definition of the universal variable. Since the value of semimajor axis of Lambert's equation (Equation 5.44) is unknown, so is the type of conic section that the transfer orbit represents. Similarly to how we solved Kepler's time equation with the use of the universal variable, we will make use of the Lagrange coefficients as defined by Equation (2.153) which, rewritten using the variables we introduced in the last two sections, become

$$f = 1 - \frac{\mu r_2}{h^2}(1 - \cos\Delta\theta)$$

$$g = \frac{r_1 r_2}{h}\sin\Delta\theta$$

$$\dot{f} = \frac{\mu}{h}\left(\frac{1 - \cos\Delta\theta}{\sin\Delta\theta}\right)\left[\frac{\mu}{h^2}(1 - \cos\Delta\theta) - \frac{1}{r_1} - \frac{1}{r_2}\right] \tag{5.77}$$

$$\dot{g} = 1 - \frac{\mu r_1}{h^2}(1 - \cos\Delta\theta)$$

so that we can write the velocity vectors at P_1 and P_2 as a function of Lagrange's coefficients as

$$\mathbf{v}_1 = \frac{1}{g}(\mathbf{r}_2 - f\mathbf{r}_1)$$

$$\mathbf{v}_2 = \frac{1}{g}(\dot{g}\mathbf{r}_2 - \mathbf{r}_1) \tag{5.78}$$

In terms of the universal anomaly χ, Lagrange's coefficients become

$$f = 1 - \frac{\chi^2}{r_1}C_2(z)$$

$$g = TOF - \frac{1}{\sqrt{\mu}}\chi^3 C_3(z)$$

$$\dot{f} = \frac{\sqrt{\mu}}{r_1 r_2}\chi(zC_3(z) - 1) \tag{5.79}$$

$$\dot{g} = 1 - \frac{\chi^2}{r_2}C_2(z)$$

where

$$z = \alpha\chi^2 \tag{5.80}$$

and where C_2 and C_3 are the Stumpff functions that we defined in Equations (2.142) and (2.143). Note that the expressions for f and g in Equation (5.79) do not depend on eccentricity. Additionally, the only unknowns in Equations (5.77) and (5.79) are h, χ, and z since TOF, $\Delta\theta$, r_1 and r_2 are known. Also, although we have a total of four equations and only three unknowns, one equation is redundant since through the conservation of angular momentum and using Equation (2.151), it can be shown that

$$f\dot{g} - \dot{f}g = 1 \tag{5.81}$$

Equating the expressions for g in Equations (5.77) and (5.79) gives

$$\frac{r_1 r_2}{h}\sin\Delta\theta = \text{TOF} - \frac{1}{\mu}\chi^3 C_3(z) \tag{5.82}$$

and doing the same for the expressions for f gives

$$1 - \frac{\mu r_2}{h^2}(1-\cos\Delta\theta) = 1 - \frac{\chi^2}{r_1}C_2(z) \tag{5.83}$$

Solving for the angular momentum in Equation (5.83) yields

$$h = \sqrt{\frac{\mu r_1 r_2 (1-\cos\Delta\theta)}{\chi^2 C(z)}} \tag{5.84}$$

which, when substituted into Equation (5.82), gives a preliminary alternative formulation of Lambert's equation

$$\sqrt{\mu}\,\text{TOF} = \chi^3 C_3(z) + \chi\sqrt{C_2(z)}\left(\sin\Delta\theta\sqrt{\frac{r_1 r_2}{1-\cos\Delta\theta}}\right) \tag{5.85}$$

Letting the terms in parentheses – which are all known and are constant – equal to

$$A = \sin\Delta\theta\sqrt{\frac{r_1 r_2}{1-\cos\Delta\theta}} \tag{5.86}$$

we can simplify Equation (5.85) as

$$\sqrt{\mu}\,\text{TOF} = \chi^3 C_3(z) + A\chi\sqrt{C_2(z)} \tag{5.87}$$

where we still have two unknowns, χ and z. As shown in Bate et al. [1971] and Curtis [2019], equating the expressions for $\dot{f}$ in Equations (5.77) and (5.79), and simplifying gives the following relationship between χ and z

$$\chi = \sqrt{\frac{y(z)}{C_2(z)}} \tag{5.88}$$

where

$$y(z) = r_1 + r_2 + A\frac{zC_3(z)-1}{\sqrt{C_2(z)}} \tag{5.89}$$

Substituting Equation (5.88) into Equation (5.87) gives

$$\sqrt{\mu}\,\text{TOF} = \left(\frac{y(z)}{C_2(z)}\right)^{3/2} C_3(z) + A\sqrt{y(z)} \tag{5.90}$$

which is known as the universal variable formulation of Lambert's equation. Similarly to how we solved Equations (5.44) and (5.66), we can solve Equation (5.90) for z using the secant method or, more efficiently, Newton-Raphson method as originally discussed in Section 2.6. If we decide to use Newton-Raphson, we need to rewrite Equation (5.90) as

$$F(z)=\left(\frac{y(z)}{C_2(z)}\right)^{3/2}C_3(z)+A\sqrt{y(z)}-\sqrt{\mu}\,TOF \tag{5.91}$$

and compute its first derivative which, after several algebraic manipulations and by using the definitions of the Stumpff functions, becomes

$$\begin{aligned}F'(z)=&\left(\frac{y(z)}{C_2(z)}\right)^{3/2}\left[\frac{1}{2z}\left(C_2(z)-\frac{3C_3(z)}{2C_2(z)}\right)+\frac{3C_3^2(z)}{4C_2(z)}\right]\\&+\frac{A}{8}\left(\frac{3C_3(z)}{C_2(z)}\sqrt{y(z)}+A\sqrt{\frac{C_2(z)}{y(z)}}\right)\end{aligned} \tag{5.92}$$

for $z\neq0$ and

$$F'(z)=\frac{\sqrt{2}}{40}y(0)^{3/2}+\frac{A}{8}\left(\sqrt{y(0)}+A\sqrt{\frac{1}{2y(0)}}\right) \tag{5.93}$$

for $z=0$. Here, we chose to use capital F instead of lower-case f to denote the function resulting from Lambert's equation to avoid confusing lower-case f with the Lagrange coefficient f.

Although we do not know the orbit type a priori, we can determine it based on the sign of z. Positive z values correspond to elliptical (and circular) orbits, negative z values to hyperbolic orbits, and $z=0$ corresponds to the parabolic case. Starting the Newton-Raphson iterations by letting $z=0$ as the initial guess is therefore considered a reasonable choice. Subsequent values of z can be found with the following update step

$$z_n=z_{n-1}-\frac{F(z_{n-1})}{F'(z_{n-1})} \tag{5.94}$$

and iterations are done until a certain tolerance is met.

The Lagrange coefficients, which are then used to compute $\mathbf{v}_1$ and $\mathbf{v}_2$, can be computed using the following expressions once the value of z is found

$$f=1-\frac{y(z)}{r_1}$$

$$g=A\sqrt{\frac{y(z)}{\mu}}$$

$$\dot{f} = \frac{\sqrt{\mu}}{r_1 r_2}\sqrt{\frac{y(z)}{C_2(z)}}\left(zC_3(z)-1\right)$$
$$\dot{g} = 1-\frac{y(z)}{r_2} \tag{5.95}$$

5.3.3 Multi-rev Lambert's Problem

In the previous sections, we discussed Lambert's problem in terms of two of its most famous formulations along with their solutions constraining our assumptions to $\Delta\theta < 2\pi$. However, for long-duration flights – high values of TOF – it is often more beneficial to perform multiple revolutions around the gravitational source instead of an orbital transfer with a large value of a (and therefore requiring a large amount of energy, and thus Δv). In other words, if TOF is large enough, one or more revolutions can be done around the gravitational source, with the last portion of the last revolution resulting in the desired intercept maneuver at P_2. This means that the transfer angle is now bound to the number of revolutions N such that

$$2N\pi \leq \Delta\theta < 2(N+1)\pi \tag{5.96}$$

where $N=0$ results in the formulations described in the previous sections. However, for $N>1$, Equations (5.44) and (5.66) become

$$\sqrt{\mu}\,\text{TOF} = a^{3/2}\left[2N\pi+\alpha-\beta-(\sin\alpha-\sin\beta)\right] \tag{5.97}$$

for the elliptical case. No parabolic or hyperbolic multi-rev Lambert's problem solutions exist since parabolic and hyperbolic orbits are not periodic.

5.3.4 Examples of Lambert's Problem

Here we want to present a numerical example of Lambert's problem by solving it using both solutions methods that we presented in Sections 5.3.1 and 5.3.2.

Let's consider an interplanetary Earth-Venus transfer such that r_1 and r_2 are 1 AU and 0.723 AU, respectively, the transfer angle $\Delta\theta$ is 75°, and the time of flight TOF is 180 days. In vector form, let's define the Earth to be located on the x-axis such that $\mathbf{r}_1=[1\ 0\ 0]$ AU and Venus is located at $\mathbf{r}_2=[0.723\cos\Delta\theta\ 0.723\sin\Delta\theta\ 0]=[0.187\ 0.698\ 0]$ AU – here we assumed that Earth and Venus are located on the same plane. We want to find the orbit that allows us to perform this transfer and the resultant required semimajor axis, eccentricity, and the velocities required at departure and arrival.

Following the classical solution described in Section 5.3.1, we compute the parameters summarized in Table 5.2. Since TOF = 0.4928 years > t_p = 0.1200 years, the type of orbit is elliptical, and thus the elliptical formulation of Lambert's problem is used. Additionally, since TOF > t_m, $\alpha=2\pi-\alpha_0$ and since $0<\Delta\theta\leq 180°$ then $\beta=\beta_0$.

TABLE 5.2
Computed parameters used to solve Lambert's problem following the classical formulation

Parameter	Value	Equation Used
c	1.072 AU	Law of cosines
s	1.397 AU	(5.50)
a_m	0.6987 AU	(5.51)
β_m	57.73°	(5.53)
t_m	0.2769 years	(5.55)
t_p	0.1200 years	(5.56)

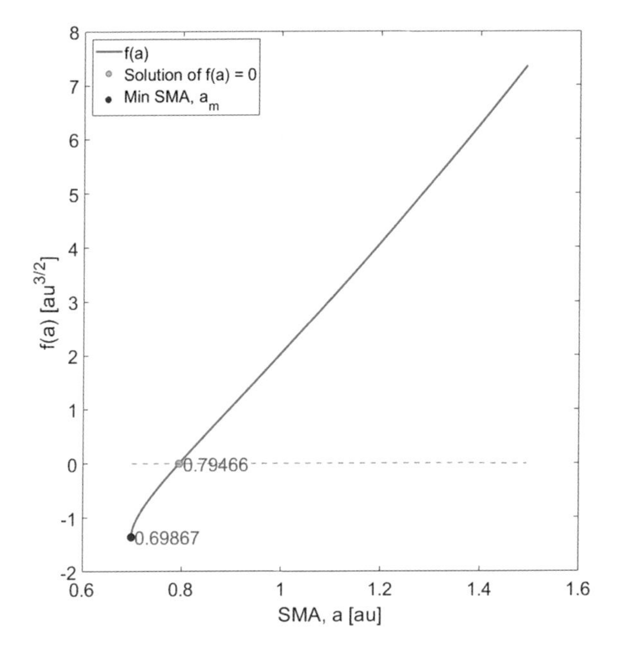

FIGURE 5.12 Plot of f(a) for an Earth-Venus transfer example.

Solving Lambert's equation for the elliptical case (Equation 5.44) results in a semimajor axis a of 0.7947 AU. Figure 5.12 shows a graph of f(a) as provided by Equation (5.62) plotted for some values of a.

The secant method converges to the solution in five steps with a relative tolerance of 10^{-8}. A summary of each iteration is given in Table 5.3, where we kept eight significant figures for a to see how it converges. Note how a converges to its final result while the values of f(a) converge to zero.

Once semimajor axis is known, semilatus rectum can be computed with Equation (5.69) and eccentricity with Equation (5.71), resulting in 0.3416 AU and 0.7550, respectively. Lastly, the velocities can be computed using Equations (5.74) and (5.75),

TABLE 5.3
Values of a and f(a) for each secant method iteration

Iteration	a (au)	f(a) ($au^{3/2}$)
1 (initial guess, a_m)	0.69866751	−1.3564145
2 (initial guess, $2a_m$)	1.3973350	6.2482643
3	0.82328589	0.28941246
4	0.79540518	7.6216228×10^{-3}
5	0.79465109	$-1.1114626 \times 10^{-4}$
6	0.79466193	5.1809561×10^{-8}
7	0.79466192	$<10^{-12}$

and are listed in Table 5.4, resulting in magnitudes of approximately $v_1 = 25.67$ km/s and $v_2 = 36.60$ km/s.

Let's now solve this same problem, but using the universal variable formulation. In order to do so, we need to compute the parameter A using Equation (5.86), which gives 0.9540 AU for this example. Now, we need to iterate to solve for the root of the function F in Equation (5.91). Using Newton-Raphson and an initial z value of 0, z converges to 8.4799 within three iterations using the same tolerance as before (10^{-8}). While iterating, we get the z and F values that are summarized in Table 5.5. The fast-decreasing values of F(z) are an indication that the algorithm is able to converge very quickly.

The resulting Lagrange coefficients are given in Table 5.6.

TABLE 5.4
Resulting velocity vectors from solving Lambert's problem for an Earth-Venus example

Velocity	Value
$\mathbf{v}_1$	[+18.849585 +17.421723 0] km/s
$\mathbf{v}_2$	[−30.405379 −20.372940 0] km/s

TABLE 5.5
Values of z and F(z) for each secant method iteration

Iteration	z	F(z)
1 (initial guess)	0	−2.4053915
2	8.4798991184	1.1747988×10^{-5}
3	8.4798722163	$2.0992541 \times 10^{-11}$
4	8.4798722162	$<10^{-15}$

TABLE 5.6
Lagrange coefficients resulting from solving Lambert's problem

Coefficient	Value
f	−0.56847525
g	0.19015588 yrs
$\dot{f}$	−1.7628999 yrs^{-1}
$\dot{g}$	−1.1693987

Once the Lagrange coefficients are known, it is possible to compute the velocities at P_1 and P_2 using Equation (5.78), resulting in the same velocity vectors shown in Table 5.4, as expected.

In Section 5.7, we will use the classical formulation of Lambert's problem to solve for an Earth-Mars transfer using planetary ephemeris data. In fact, throughout this chapter, we will make use of Lambert's problem several times to derive and discuss extremely useful tools that mission designers use, such as porkchop plots, as previously stated.

5.4 RELATIVE MOTION

An operational rendezvous between two objects in space is a complex process, but a simplified trajectory analysis can be used as a valuable approximation. Originally developed by Hill [1878] to describe the relative motion between two astronomical objects, Clohessy and Wiltshire [1960] applied Hill's development to describe the relative motion between two artificial objects, and to show how one object can be maneuvered so that the two objects can be at the same place at the same time, i.e., they can rendezvous and eventually dock to each other. By linearizing the non-linear equations of motion of the two-body problem, analytical equations of relative motion can be developed. Here, we will develop such equations, which are also known as the Hill-Clohessy-Wiltshire (HCW) equations, named after their main contributors. Then, we will use the HCW equations to determine how one can estimate the Δv to initiate and complete a rendezvous maneuver for two spacecraft orbiting the same body.

5.4.1 The Hill-Clohessy-Wiltshire Equations

Let's assume that two objects are in orbit around a common central body, with one object in a circular orbit and the other object in a nearly circular orbit, as seen in Figure 5.13. The relative motion of the second object, often called the chaser or the deputy, relative to the first object, often called the target or the chief, is desired. Using a rotating coordinate frame, shown in Figure 5.13, allows us to develop the linearized radial, along-track, and cross-track $\left(\mathbf{e}_r, \mathbf{e}_\theta, \mathbf{e}_h\right)$ equations of motion, i.e., the position

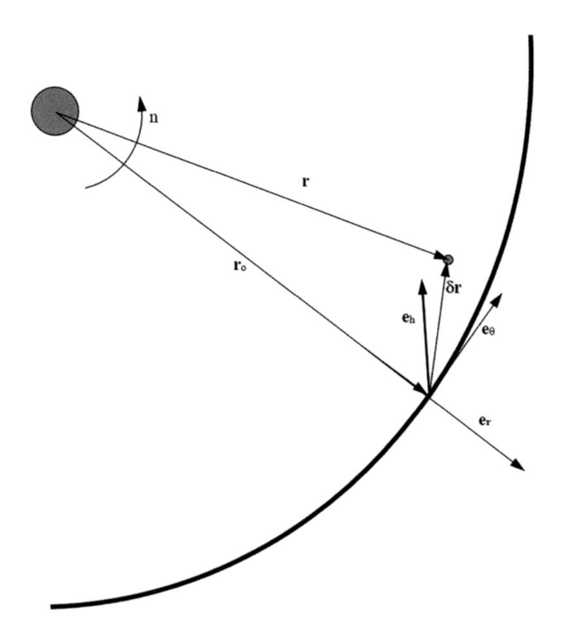

FIGURE 5.13 Rotating Hill-Clohessy-Wiltshire reference frame.

and velocity of the chaser with respect to the target – in other words, the target is always at the origin of this coordinate frame. The solution of these linearized equations is known as the Hill-Clohessy-Wiltshire (or HCW) equations.

In the two-body problem, the differential equations of motion in the inertial reference frame (I) are

$$\frac{d^2\mathbf{r}^{(I)}}{dt^2} + \frac{\mu \mathbf{r}}{r^3} = 0 \tag{5.99}$$

The position vector, **r**, can be written as the vector sum of the vector from the origin of the inertial frame to the origin of the rotating frame plus the distance from the origin of the rotating frame to the object,

$$\mathbf{r} = \mathbf{r}_0 + \delta\mathbf{r} = (r_0 + \delta r)\mathbf{e}_r + r_0\delta\theta\mathbf{e}_\theta + \delta z\mathbf{e}_h \tag{5.100}$$

In this equation, $\mathbf{r}_0 = r_0\mathbf{e}_r$ and $\delta\mathbf{r} = \delta r\mathbf{e}_r + r_0\delta\theta\mathbf{e}_\theta + \delta z\mathbf{e}_h$. From Equation (2.13), recall that the acceleration of an object in an inertial frame (I) and rotating reference frame (R) is

$$\frac{d^2\mathbf{r}^{(I)}}{dt^2} = \frac{d^2\mathbf{r}^{(R)}}{dt^2} + 2\boldsymbol{\omega} \times \frac{d\mathbf{r}^{(R)}}{dt} + \boldsymbol{\omega} \times (\boldsymbol{\omega} \times \mathbf{r}) \tag{5.101}$$

The rotation rate of the rotating frame, **ω**, is constant for a circular orbit and is

$$\boldsymbol{\omega} = \omega\mathbf{e}_h = n\mathbf{e}_h \tag{5.102}$$

In fact, here n is both the mean motion and the instantaneous angular rate of the target spacecraft around the focus. The velocity and acceleration (first and second derivatives of Equation (5.100)) in the rotating frame are

$$\frac{d\mathbf{r}^{(R)}}{dt} = \delta\dot{r}\mathbf{e}_r + r_0\delta\dot{\theta}\mathbf{e}_\theta + \delta\dot{z}\mathbf{e}_z \tag{5.103}$$

and

$$\frac{d^2\mathbf{r}^{(R)}}{dt^2} = \delta\ddot{r}\mathbf{e}_r + r_0\delta\ddot{\theta}\mathbf{e}_\theta + \delta\ddot{z}\mathbf{e}_z \tag{5.104}$$

Inserting Equations (5.102) – into Equation (5.101) results in

$$\begin{aligned} \frac{d^2\mathbf{r}^{(I)}}{dt^2} &= \left[\delta\ddot{r} - 2nr_0\delta\dot{\theta} - n^2\left(r_0 + \delta r\right)\right]\mathbf{e}_r + \left(r_0\delta\ddot{\theta} + 2n\delta\dot{r} - n^2 r_0\delta\theta\right)\mathbf{e}_\theta + \delta\ddot{z}\mathbf{e}_h \\ &= -\frac{\mu\left[\left(r_0 + \delta r\right)\mathbf{e}_r + r_0\delta\theta\mathbf{e}_\theta + \delta z\mathbf{e}_h\right]}{r^3} \end{aligned} \tag{5.105}$$

Let's look at the right-hand side of Equation (5.105). Our goal is to simplify the r^3 term in the denominator. To do so, let's look at the following

$$r^2 = \mathbf{r}\cdot\mathbf{r} = \left[\left(r_0 + \delta r\right)\mathbf{e}_r + r_0\delta\theta\mathbf{e}_\theta + \delta z\mathbf{e}_h\right]\cdot\left[\left(r_0 + \delta r\right)\mathbf{e}_r + r_0\delta\theta\mathbf{e}_\theta + \delta z\mathbf{e}_h\right] \tag{5.106}$$

Expanding this out,

$$\begin{aligned} r^2 = \mathbf{r}\cdot\mathbf{r} &= \left(r_0 + \delta r\right)^2 + \left(r_0\delta\theta\right)^2 + \left(\delta z\right)^2 \\ &= r_0^2 + 2r_0\delta r + \left(\delta r\right)^2 + \left(r_0\delta\theta\right)^2 + \left(\delta z\right)^2 \end{aligned} \tag{5.107}$$

Factoring out r_0^2 from Equation (5.107),

$$r^2 = r_0^2\left[1 + \frac{2\delta r}{r_0} + \left(\frac{\delta r}{r_0}\right)^2 + \left(\delta\theta\right)^2 + \left(\frac{\delta z}{r_0}\right)^2\right] \tag{5.108}$$

The terms $\delta r, \delta\theta$ and δz are small when compared to r_0, so we can neglect these higher-order terms,

$$r^2 \cong r_0^2\left(1 + \frac{2\delta r}{r_0}\right) \tag{5.109}$$

This can also be written as

$$r^{-3} \cong r_0^{-3}\left(1 + \frac{2\delta r}{r_0}\right)^{-3/2} \tag{5.110}$$

Applying the binomial theorem,

$$r^{-3} \cong r_0^{-3}\left(1-\frac{3\delta r}{r_0}\right) \tag{5.111}$$

Thus, the right-hand side of Equation (5.63) becomes,

$$-\frac{\mu \mathbf{r}}{r^3} = -\frac{\mu\left[\left(r_0+\delta r\right)\mathbf{e}_r + r_0\delta\theta\mathbf{e}_\theta + \delta z\mathbf{e}_h\right]\left(1-\frac{3\delta r}{r_0}\right)}{r_0^3} \tag{5.112}$$

where here we substituted the acceleration term with the gravitational acceleration, which is the only acceleration present in the unperturbed two-body problem. Expanding Equation (5.112), dropping higher-order terms, and recalling that $n^2 = \mu/r_0^3$ (definition of the mean motion),

$$\ddot{\mathbf{r}} = -n^2\left[\left(r_0 - 2\delta r\right)\mathbf{e}_r + r_0\delta\theta\mathbf{e}_\theta + \delta z\mathbf{e}_h\right] \tag{5.113}$$

Equating Equation (5.104) to Equation (5.113), and grouping terms, we obtain the following equations of motion

$$\begin{aligned}
&\delta\ddot{r} - 2nr_0\delta\dot{\theta} - 3n^2\delta r = 0\\
&r_0\delta\ddot{\theta} + 2n\delta\dot{r} = 0\\
&\delta\ddot{z} + n^2\delta z = 0
\end{aligned} \tag{5.114}$$

Equation (5.114) are the Hill-Clohessy-Wiltshire (HCW) equations, which are linear coupled differential equations that describe the motion of the chaser spacecraft with respect to the target spacecraft when they are "close" to each other. The initial conditions for Equation (5.114) are given by

$$\delta \mathbf{r}_0 = \begin{Bmatrix} \delta r_0 \\ r_0\delta\theta_0 \\ \delta z_0 \end{Bmatrix} \qquad \delta\dot{\mathbf{r}}_0 = \begin{Bmatrix} \delta\dot{r}_0 \\ r_0\delta\dot{\theta}_0 \\ \delta\dot{z}_0 \end{Bmatrix} \tag{5.115}$$

Note that the out-of-plane motion ($\mathbf{e}_h$) of Equation (5.114) is uncoupled from the in-plane motion ($\mathbf{e}_r$_$\mathbf{e}_\theta$). In fact, the solution of the out-of-plane motion is a simple harmonic oscillator. When applying initial conditions δz_0 for position and $\delta\dot{z}_0$ for velocity, the solution to the third equation in Equation (5.114) becomes

$$\delta z(t) = \delta z_0 \cos nt + \frac{\delta\dot{z}_0}{n}\sin nt \tag{5.116}$$

The in-plane motion is coupled. Integrating the second equation in Equation (5.114) and solving for the constant of integration as a function of the initial conditions yield

$$r_0\delta\dot{\theta} = r_0\delta\dot{\theta}_0 + 2n\left(\delta r_0 - \delta r\right) \tag{5.117}$$

Inserting this into the first equation of Equation (5.114) and grouping terms, it takes the form of an oscillator with a constant forcing function,

$$\delta\ddot{r} + n^2\delta r = 2n\left(r_0\delta\dot{\theta}_0 + 2n\delta r_0\right) \tag{5.118}$$

This equation has homogeneous and particular solutions. The homogeneous solution is a simple harmonic oscillator,

$$\delta r_h = C_1 \cos nt + C_2 \sin nt \tag{5.119}$$

and the particular solution is

$$\delta r_p = C_3 \tag{5.120}$$

Solving for the constants of integration, C_1, C_2, and C_3, the solution to Equation (5.118) becomes

$$\delta r = -\left(\frac{2}{n} r_0\delta\dot{\theta}_0 + 3\delta r_0\right)\cos nt + \frac{\delta\dot{r}_0}{n}\sin nt + \frac{2}{n} r_0\delta\dot{\theta}_0 + 4\delta r_0 \tag{5.121}$$

Taking the derivative of Equation (5.121), we find an expression to describe the relative radial velocity,

$$\delta\dot{r} = \left(2r_0\delta\dot{\theta}_0 + 3n\delta r_0\right)\sin nt + \delta\dot{r}_0 \cos nt \tag{5.122}$$

The time history of the tangential velocity can be found by inserting Equation (5.122) into Equation (5.117), so that

$$r_0\delta\dot{\theta} = \left(-3r_0\delta\dot{\theta}_0 - 6n\delta r_0\right) + \left(6n\delta r_0 + 4r_0\delta\dot{\theta}_0\right)\cos nt + 2\delta\dot{r}_0 \sin nt \tag{5.123}$$

Integrating Equation (5.123), we get

$$r_0\delta\theta = r_0\delta\theta - \left(3r_0\delta\dot{\theta}_0 + 6n\delta r_0\right)t + \left(\frac{4r_0\delta\dot{\theta}_0}{n} + 6\delta r_0\right)\sin nt + \frac{2\delta\dot{r}_0}{n}\cos nt - \frac{2\delta\dot{r}_0}{n} \tag{5.124}$$

We can write the solutions to the HCW equations in matrix form,

$$\left\{\begin{array}{c} \delta r(t) \\ \hline \delta\dot{r}(t) \end{array}\right\} = \left[\begin{array}{c|c} \phi_{rr} & \phi_{rv} \\ \hline \phi_{vr} & \phi_{vv} \end{array}\right] \left\{\begin{array}{c} \delta r_0 \\ \hline \delta\dot{r}_0 \end{array}\right\} \tag{5.125}$$

where each of these time-varying elements, ϕ_{ij} (i, j = r or v) is a 3×3 matrix, and together the 6×6 matrix in the right-hand side of Equation (5.125) is known as a state transition matrix. The state vector and its derivative are

$$\delta \mathbf{r}(t) = \begin{Bmatrix} \delta r \\ r_0 \delta\theta \\ \delta z \end{Bmatrix} \qquad \delta \dot{\mathbf{r}}(t) = \begin{Bmatrix} \delta \dot{r} \\ r_0 \delta\dot{\theta} \\ \delta \dot{z} \end{Bmatrix} \tag{5.126}$$

where the initial conditions are given by Equation (5.115).

The elements of the state transition matrix are

$$[\phi_{rr}] = \begin{bmatrix} 4 - 3\cos nt & 0 & 0 \\ 6(\sin nt - nt) & 1 & 0 \\ 0 & 0 & \cos nt \end{bmatrix} \tag{5.127}$$

$$[\phi_{rv}] = \frac{1}{n}\begin{bmatrix} \sin nt & 2(1 - \cos nt) & 0 \\ 2(\cos nt - 1) & 4\sin nt - 3nt & 0 \\ 0 & 0 & \sin nt \end{bmatrix} \tag{5.128}$$

$$[\phi_{vr}] = \begin{bmatrix} 3n\sin nt & 0 & 0 \\ 6n(\cos nt - 1) & 0 & 0 \\ 0 & 0 & -n\sin nt \end{bmatrix} \tag{5.129}$$

and

$$[\phi_{vv}] = \begin{bmatrix} \cos nt & 2\sin nt & 0 \\ -2\sin nt & -3 + 4\cos nt & 0 \\ 0 & 0 & \cos nt \end{bmatrix} \tag{5.130}$$

Thus, for a given target spacecraft orbit, which is a circular orbit with angular rate n, and for a given set of initial conditions of the relative position and velocity of the chaser with respect to the target, it is possible to compute the time history of the position and velocity vectors of the chaser, keeping in mind that this approximation, i.e., the HCW equations, is only valid for small values of δr, $\delta\theta$, and δz.

5.4.2 Rendezvous between Spacecraft

In addition to obtaining the time history of the relative position and velocity, the HCW equations can also be used for rendezvous analysis.

In rendezvous, the goal is to have two objects at the same place at the same time. When the two objects reach the same location, a propulsive maneuver is executed to cancel out their relative velocity, thus putting them in the same place with the same velocity, i.e., performing a rendezvous. At the time of rendezvous, t_f, the relative position vector, $\delta \mathbf{r}$, must be equal to zero. Thus, expanding out Equation (5.125), in block matrix form, yields,

$$\delta \mathbf{r}(t_f) = \mathbf{0} = \phi_{rr}(t_f)\delta \mathbf{r}_0 + \phi_{rv}(t_f)\delta \mathbf{v}_0 \tag{5.131}$$

$$\delta \mathbf{v}(t_f) = \phi_{vr}(t_f)\delta \mathbf{r}_0 + \phi_{vv}(t_f)\delta \mathbf{v}_0 \tag{5.132}$$

Given the initial relative position and solving for the initial relative velocity needed to rendezvous at time t_f gives

$$\delta \mathbf{v}_0 = -\phi_{rv}^{-1}(t_f)\phi_{rr}(t_f)\delta \mathbf{r}_0 \tag{5.133}$$

If there is a relative velocity vector that already exists which is generally the case the first Δv maneuver needed to initiate the rendezvous is

$$\Delta \mathbf{v}_1 = \delta \mathbf{v}_{actual} - \delta \mathbf{v}_0 \tag{5.134}$$

and the magnitude of the Δv maneuver is simply the magnitude of Equation (5.134). Once the chaser spacecraft reaches the point where the relative position is zero, it is necessary to apply another maneuver to ensure that the chaser does not crash into the target unless that is the desired outcome (e.g., impacting an asteroid such as in NASA's Deep Impact mission[2]). Inserting the initial relative velocity (Equation 5.133) into Equation (5.132) the relative velocity at the rendezvous point becomes

$$\delta \mathbf{v}(t_f) = \left(\phi_{vr} - \phi_{vv}\phi_{rv}^{-1}\phi_{rr}\right)\delta \mathbf{r}_0 \tag{5.135}$$

Thus, the velocity needed to complete the rendezvous maneuver and ensure that the final relative velocity is zero is given by

$$\Delta \mathbf{v}_2 = -\delta \mathbf{v}(t_f) \tag{5.136}$$

5.4.3 Rendezvous around a Distant Body

The Hill-Clohessy-Wiltshire relative equations of motion have been primarily applied to relative motion of two objects orbiting Earth. However, there is nothing exclusive to Earth – this solution can be used for any gravitational source by simply using the appropriate value of μ.

Objects rendezvousing around a distant body are nothing new. During the Apollo missions, lunar orbit rendezvous was chosen as the mission architecture. In this plan, the lunar module detached from the command module, landed on the Moon, and then returned and rendezvoused with the command module. All of this was done while in orbit around the Moon. The Apollo program's techniques that were used in lunar orbit were developed and refined during the Gemini program in orbit around the Earth. The rendezvous system for the Apollo (and Gemini) programs relied on active sensors and onboard human participation. The lunar module system included a rendezvous radar with a 740-km range. It also relied on a landing radar,

[2] https://solarsystem.nasa.gov/missions/deep-impact-epoxi/in-depth/

optical instruments, and an inertial measurement unit. The command module was able to perform very high-frequency (VHF) ranging, with an operational range of 370 km. The crew also used optical systems (sextant and scanning telescopes), along with an inertial measurement unit.[3]

Several different rendezvous conditions were planned while in orbit around the Moon. The sequence performed for all lunar missions ended up being a nominal rendezvous following a nominal lunar module ascent. Fortunately, the emergency sequences were never needed, although the astronauts and ground controllers were prepared to execute them if required. Had the astronauts had some failure after leaving the command module, many mitigating sequences involved the command module come after the lunar module, i.e., the opposite maneuver of the nominal plan. This could have occurred after the undocking and separation, after the lunar module had performed its descent burn, after an abort following a lunar module abort from the lunar surface after landing, rescue after a nominal lunar module ascent, or rescue following an unplanned lunar module lift-off from the lunar surface. In the nominal case, the HCW equations were defined such that the command module (the target spacecraft) was the origin of the rotating coordinate frame, while the lunar module (the chaser spacecraft) was the object maneuvering relative to the command module. In many of the emergency contingencies (especially when the lunar module was unable to maneuver), the location of the lunar module would become the origin of the rotating reference frame and the command module would become the object maneuvering relative to the rotating reference frame.

Additionally, it should be noted that the Hill-Clohessy-Wiltshire equations are only an approximation of the relative motion in the two-body problem. However, formulations of the relative motion equations for the three-body problem exist and are documented in the literature [Conte, 2019], mostly dealing with periodic three-body orbits, such as halo orbits or distant retrograde orbits, as seen in Chapter 3.

5.5 BASICS OF ORBIT DESIGN

When performing a ballistic transfer from one body to another (short burn with a long coast), the transfer orbit can be classified by what is known as *types*. This classification system is used when describing the transfers – type I is a transfer of less than 180° (e.g., all of the Apollo lunar trajectories were type I transfers), a type II transfer is one where the transfer angle is between 180° and 360° (e.g., some Mars missions have used these), a type III transfer has a transfer angle of between 360° and 540°, and so forth. A Hohmann transfer is the boundary between a type I and type II transfer since its transfer angle is exactly 180°. However, this classification is not used for low-thrust transfers or for gravity assist transfers. For a gravity assist, each phase of the transfer is categorized as a transfer type. For example, the Galileo spacecraft, which used a single gravity assist from Venus and two gravity assists from Earth, performed a type I Earth-Venus transfer, a type II Venus-Earth transfer, and a resonant Earth-Earth transfer (this was a special case, where the spacecraft performed one revolution around the Sun while the Earth performed two orbits around the Sun,

[3] NASA's Apollo Mission Reports: https://www.hq.nasa.gov/alsj/alsj-mrs.html

putting Galileo and Earth into a 2:1 resonant orbit configuration). Like the Hohmann transfer, this transfer was the boundary between a type II and type III transfer. Finally, the spacecraft performed a type I Earth-Jupiter transfer. This entire transfer was shown in Figure 1.7. To date, no interplanetary or lunar trajectory greater than a type II transfer has been flown.

Once a spacecraft leaves from its orbit around a planet, it is on an orbit around the Sun. Although the orbital dynamics remain similar, to continue the analysis, one needs to use the Sun's gravitational parameter instead of the planet's gravitational parameter, and the origin of the coordinate frame moves from the planet's center to the Sun's center. To escape from orbit around a given planet, a spacecraft must add energy (velocity) and reach at least the so-called escape velocity. As the spacecraft adds velocity, its orbit changes from an ellipse to a parabola, and eventually to a hyperbola. As it transitions to the hyperbolic orbit, it reaches escape velocity when it enters a parabolic orbit (eccentricity = 1, semimajor axis approaches ∞). Using the *vis viva* equation (Equation 2.50), the escape velocity can be found by setting letting a approach ∞ $\left(1/a \to 0\right)$ and solving for the escape velocity, v_{escape}. Thus,

$$v_{escape} = \sqrt{\frac{2\mu}{r}} \tag{5.137}$$

For a spacecraft initially in a circular orbit $\left(v_{circular} = \sqrt{\mu/a_0}\right)$ the amount of velocity a spacecraft must acquire to escape is the difference between the initial and escape velocities

$$\Delta v = \left(\sqrt{2} - 1\right)\sqrt{\frac{\mu}{a_0}} \tag{5.138}$$

This equates to increasing the spacecraft's velocity by approximately 41% $\left(\sqrt{2} - 1\right)$. A larger Δv maneuver than what is given by Equation (5.138) will transfer the spacecraft onto a hyperbolic orbit. Once the spacecraft has escaped the planet's orbit, it is in orbit around the Sun, generally in a heliocentric (Sun-centered) elliptical orbit. To enter an orbit around another planet, the spacecraft must slow down near that planet, or it would simply fly by. The process is similar to what occurs when the spacecraft escapes, only in reverse.

5.5.1 Patched Conics

Combining the solutions of the problems of departure from a planet, heliocentric cruise, and planetary arrival is known as the method of patched conics, in which multiple conic sections are patched together in various phases. This is an approximate solution in which several simplifications are made to allow for an analytical solution. Patched conic approximations lead to a solution that is relatively close to the actual solution. Other more accurate methods can include the integration of the full-force dynamical equations of motion (as we talked about in Chapters 3 and 4), but tend to be more computationally intensive and many times lack the necessary insight to find adequate solutions for space missions. As a result, patched conic approaches are usually used to obtain an approximate initial solution to the problem that can then be used to obtain and refine the results in a full-force model. In real-world missions, other maneuvers such as mid-course maneuvers or trajectory correction maneuvers (TCMs) are added to correct for mismodeling of trajectories, thrust-vector pointing errors and over or under performance of the propulsion system or other spacecraft components.

While there are at least three phases presented here, a space mission can have multiple planetary flybys, deep space maneuvers, TCMs, etc. However, here we will focus on the simplest case in which a spacecraft leaves a planet, orbits the Sun, and arrives at a different planet. In order to compute the necessary Δv's for all of these phases, the heliocentric cruise phase must be assessed first. In this phase, the transfer orbit between where the departure planet is at the time of departure (relative to the Sun) and where the arrival planet is at the time of arrival (relative to the Sun) must be determined. Note that this can be determined in various ways, as we discussed in Section 4.7. Once these locations are determined, a transfer orbit connecting these points can be constructed. At the departure and arrival locations, the velocity vectors of the spacecraft are then found. These velocity vectors are patched to the Phase 1 (planetary departure) and Phase 3 (planetary arrival) orbits to ultimately determine the Δv needed to jump on or jump off the heliocentric transfer orbit.

Two main scenarios exist. The first, described as a transfer to an inner planet, is a transfer from a planet (such as Earth) to a planet that is closer to the Sun (such as Venus or Mercury). The second, described as a transfer to an outer planet, is a transfer from a planet to one that is farther from the Sun (such as a transfer from Earth to Mars, Jupiter, Saturn, or beyond).

5.5.2 Sphere of Influence

Although a spacecraft has moved to an escape trajectory, the question of when it has exited the gravitational influence of the body it is departing from and when it is entering the gravitational influence of the Sun arises. This "boundary" is defined by the sphere of influence (SOI). We will first (incorrectly – and purposely) assume that the boundary of the sphere of influence is the point where the gravitational influence of the Sun is greater than the gravitational influence of the planet the spacecraft is orbiting. We will then show that this is not the correct definition of a planet's SOI with respect to the Sun, which also applies to moons with respect to their parent planet.

In an inverse square gravitational field, the magnitude of the gravitational acceleration due to the Sun is

$$a_{\odot} = \frac{\mu_{\odot}}{r_{\odot \rightarrow s/c}^2} \tag{5.139}$$

where $r_{\odot \rightarrow s/c}$ is the distance between the Sun and the spacecraft (s/c). The magnitude of the gravitational acceleration due to the planet is

$$a_{\otimes} = \frac{\mu_{\otimes}}{r_{\otimes \rightarrow s/c}^2} \tag{5.140}$$

where $r_{\otimes \to s/c}$ is the distance between the planet and the spacecraft. Equating Equations (5.139) and (5.140) to determine where these two accelerations are equal, and recognizing that the sum of the distances between the spacecraft and the planet and the spacecraft and the Sun is equal to the distance between the planet and the Sun, a little algebra shows that the distance between the spacecraft and the planet is

$$r_{\otimes \to s/c} = \left(\frac{\sqrt{\mu_{\otimes}/\mu_{\odot}}}{1 + \sqrt{\mu_{\otimes}/\mu_{\odot}}} \right) r_{\otimes \to \odot} \tag{5.141}$$

Using the gravitational constants for the Sun and the Earth, their ratio is approximately 3×10^{-6}, and the resulting (incorrect) SOI is approximately 1.73×10^{-3} AU, which is roughly 259,500 km. However, the distance between the Earth and the Moon is about 384,400 km. Using this formulation, then, means that the Earth's Moon would be outside of the boundary of the sphere of influence of the Earth and it will be orbiting the Sun not the Earth.

Obviously, the Moon orbits the Earth, so a different approach is needed. The correct formulation is based on work done by Laplace and is presented in detail in Battin [1987, pp. 395–397], among others. Summarizing this formulation is as follows.

Laplace's method starts by assuming that the spacecraft is in orbit around one body and its orbit is perturbed by the other body. We then switch the equations by assuming that the spacecraft is in orbit around the other body and its orbit is perturbed by first body. Expanding the law of gravitational acceleration for three bodies, i.e., by setting $N = 3$, the accelerations of two bodies on the spacecraft become

$$\frac{d^2 \mathbf{R}_1}{dt^2} = \frac{Gm_2}{r_{12}^3} \mathbf{r}_{12} + \frac{Gm_3}{r_{13}^3} \mathbf{r}_{13} \tag{5.142}$$

and

$$\frac{d^2 \mathbf{R}_2}{dt^2} = \frac{Gm_1}{r_{21}^3} \mathbf{r}_{21} + \frac{Gm_3}{r_{23}^3} \mathbf{r}_{23} \tag{5.143}$$

Subtracting Equation (5.142) from Equation (5.143), and recognizing that $\mathbf{r} = \mathbf{r}_{12} = -\mathbf{r}_{21}$, the equations of motion for the spacecraft become

$$\frac{d^2 \mathbf{r}}{dt^2} + \frac{\mu}{r^3} \mathbf{r} = -Gm_3 \left(\frac{\mathbf{r}_{13}}{r_{13}^3} - \frac{\mathbf{r}_{23}}{r_{23}^3} \right) \tag{5.144}$$

If we arbitrarily set body 1 as the larger massive body (e.g., the Sun), and body 3 as the perturbing body, which we assume to be the smaller body (e.g., the Earth), the equations of motion of the infinitesimal body orbiting the central body is

$$\frac{d^2 \mathbf{r}_s}{dt^2} - \mathbf{A}_s = \mathbf{P}_p \tag{5.145}$$

Should the central body be the smaller of the two bodies, and the perturbing acceleration come from the larger body, the equations of motion can be written as

$$\frac{d^2\mathbf{r}_p}{dt^2} - \mathbf{A}_p = \mathbf{P}_s \tag{5.146}$$

The vector terms $\mathbf{A}_s$ and $\mathbf{A}_p$ represent the gravitational acceleration of the central body, and $\mathbf{P}_s$ and $\mathbf{P}_s$ represent the perturbing acceleration due to the third body. If we compare the ratio of the perturbation acceleration to the central body acceleration of both systems, we get

$$\frac{P_p}{A_s} = \frac{P_s}{A_p} \tag{5.147}$$

If the left-hand side of Equation (5.147) is greater than the right-hand side, the spacecraft is inside the sphere of influence of the smaller body (e.g., the Earth). Conversely, if the left-hand side is larger, the spacecraft is inside the sphere of influence of the larger body (e.g., the Sun). After some algebra, and some truncation of higher-order terms, we find an approximate location of the boundary of the sphere of influence (as developed by Laplace and presented in Battin [1987]), and we can therefore approximate the radius of the sphere of influence as

$$r_{SOI} \approx \left(\frac{\mu_\otimes}{\mu_\odot}\right)^{2/5} r_{\odot \to \otimes} \tag{5.148}$$

For the Sun-Earth system, the radius of the sphere of influence is approximately 924,000 km. With this formulation, Earth's Moon is well inside the sphere of influence of the Earth. Note that these distances are approximations and are used for a macroscopic evaluation of the overall gravitational boundaries.

Based on constants found in Appendix A, Table 5.7 shows the values of the radii of the sphere of influence for various planets (as well as Pluto) in our Solar System. Note that the planets are in nearly circular orbits, and the distance between the planet and the Sun is assumed to be the semimajor axis. However, Pluto is in a relatively high eccentric orbit about the Sun, so both aphelion and perihelion radii are shown to demonstrate the changing boundary as a function of where Pluto is in its orbit. Additionally, the radius of the sphere of influence of the Moon is also shown.

Table A.2 in Appendix A shows the synodic period of each planet with respect to each other planet of the Solar System. The concept of the sphere of influence is used to break the entire transfer, from departure from one planet to arrival at another planet, into three separate problems, and patch these three solutions together into a complete transfer.

5.5.3 Phases

As discussed in the last section, the patchwork of trajectories can be broken into three phases. We assume that a spacecraft starts in orbit around the departing planet. A Δv maneuver is applied to send the spacecraft onto an escape trajectory that will

TABLE 5.7
Sphere of influence information ($\mu_\odot = 1.327 \times 10^{11}\,km^3/s^2$)

Gravitational Source	μ (km^3/s^2)	Distance from Sun to Planet (km)	Radius of Sphere of Influence (km)	Notes
Mercury	2.203×10^4	5.791×10^7	1.124×10^5	
Venus	3.248×10^5	1.082×10^8	6.162×10^5	
Earth	3.986×10^5	1.496×10^8	9.247×10^5	
Earth's Moon	4.905×10^3	3.844×10^5	6.619×10^4	Radius around the Earth
Mars	4.283×10^4	2.279×10^8	5.772×10^5	
Jupiter	1.267×10^8	7.785×10^8	4.822×10^7	
Saturn	3.793×10^7	1.433×10^9	5.479×10^7	
Uranus	5.794×10^6	2.871×10^9	5.177×10^7	
Neptune	6.836×10^6	4.498×10^9	8.665×10^7	
Pluto	8.710×10^2	7.376×10^9	3.933×10^6	Aphelion
		4.437×10^9	2.366×10^6	Perihelion

bring it beyond the sphere of influence of the departing planet and into interplanetary space, where the gravity of the Sun is the dominant gravitational force. Thus, the second phase starts when the spacecraft exits said sphere of influence. The third phase occurs when the spacecraft enters the sphere of influence of the arrival planet. Each of these systems is treated as separate two-body problems – each using the gravitational parameter of the body that exerts the most gravitational force on the spacecraft – although it is necessary to make sure that where one phase ends and the next one begins, the position and velocity of the spacecraft is continuous. This section examines each of these phases and develops the mathematical method to assure continuity between phases is achieved.

5.5.3.1 Phase 1: Planetary Departure

The first maneuver in an interplanetary transfer occurs while the spacecraft is in orbit about the originating, or departing, planet. The location of this maneuver and its magnitude depend upon several factors, including the location of the arrival planet relative to the departure planet, and the desired time of flight. More details on this aspect of this phase are provided in Section 5.10.

During the planetary departure phase, the spacecraft is inside the sphere of influence of the departing planet. The geometry of the departure that targets an inner planet (e.g., an Earth-Venus transfer) is shown in Figure 5.14. The geometry of the departure targeting an outer planet (e.g., an Earth-Mars transfer) is shown in Figure 5.15. In both of these figures, it is assumed that the spacecraft is moving in a clockwise direction (when looking directly at the figures) around the departure planet, and that the planet is moving in a counterclockwise motion with respect to the Sun.

At the boundary of the sphere of influence (assuming that the radius at the exit is very large), an energy balance (which means using the *vis viva* equation) between the planetary departure and SOI exit yields,

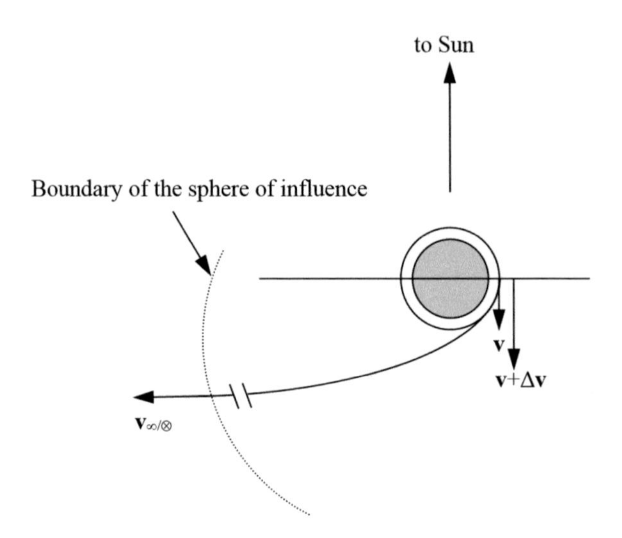

FIGURE 5.14 Geometry of departure to inner planet.

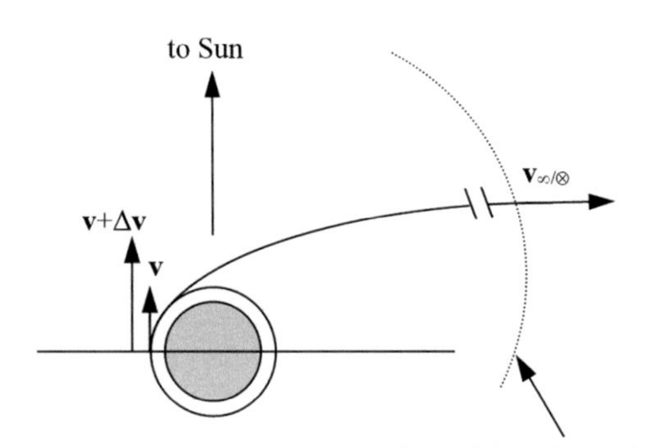

FIGURE 5.15 Geometry of departure to outer planet.

$$\frac{(\mathbf{v}+\Delta\mathbf{v})\cdot(\mathbf{v}+\Delta\mathbf{v})}{2}-\left(\frac{\mu_{\otimes}}{r_{\text{parking orbit}}}\right)=\frac{\mathbf{V}_{\infty/\otimes}\cdot\mathbf{V}_{\infty/\otimes}}{2}-\left(\frac{\mu_{\otimes}}{r_{SOI}\approx\infty}-0\right) \quad (5.149)$$

This velocity vector at the boundary of the sphere of influence must be patched to the velocity vector in the heliocentric cruise phase. The vector geometry of this patching process is shown for an inner planet transfer in Figure 5.16a and for an outer planet transfer in Figure 5.16b.

For an inner planet transfer, the velocity vector of the spacecraft with respect to the Sun is now the apoapse velocity vector of the transfer orbit. For a transfer to an outer planet, the velocity vector of the spacecraft with respect to the Sun is the periapse velocity vector of the transfer orbit.

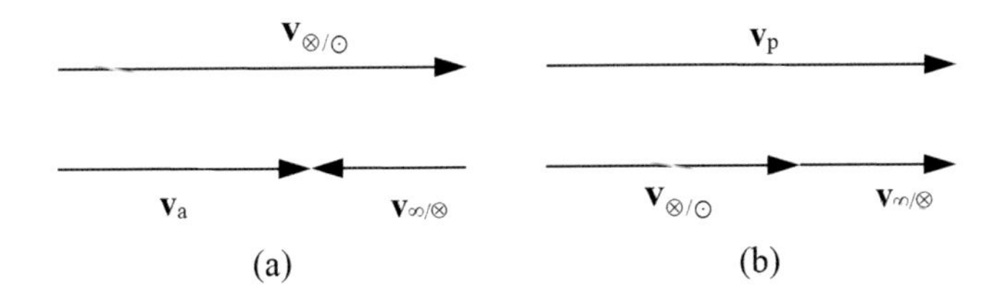

FIGURE 5.16 Vector geometry of velocity vectors to (a) inner planet and (b) outer planet.

5.5.3.2 Phase 2: Heliocentric Cruise

During the heliocentric cruise phase, the velocity vectors determined at the departure and arrival locations are used to connect the three phases. While it is common to use a Hohmann transfer for this phase, the transfer angle can be any value other than 180° (recall that since the Hohmann transfer is the minimum two-burn transfer for orbit transfers – other transfers, such as the bi-elliptic transfer and the non-Hohmann transfer, have disadvantages such as very long transfer times or higher Δv costs). In fact, note that in the previous section, we assumed that all vectors are parallel. This is true for a Hohmann transfer, bi-elliptic transfers, and other common-apse transfers. An example of an interplanetary Hohmann transfer is given in Section 5.6. However, for a general interplanetary transfer, these vectors are not necessarily parallel. This means that the orbital transfer must be solved using Lambert's problem. An example of such transfer is provided in Section 5.7.

One item that is commonly overlooked is that during the heliocentric cruise, the gravitational constant (μ) that is used in all equations is the gravitational constant of the Sun. Additionally, the spacecraft spends most of its time during this phase, meaning that the time of flight of the entire interplanetary transfer is primarily decided by the orbit chosen in this phase.

5.5.3.3 Phase 3: Planetary Arrival

Upon arrival at the destination planet, the reverse of the Phase 1 process occurs. At arrival, the velocity vector of the spacecraft with respect to the Sun is known. In order to obtain the velocity vector of the spacecraft with respect to the destination planet, the velocity vector of the destination planet with respect to the Sun is subtracted,

$$\mathbf{v}_{\infty/\otimes} = \mathbf{v}_{\infty/\odot} - \mathbf{v}_{\otimes/\odot} \tag{5.150}$$

where $\mathbf{v}_{\infty/\odot}$ is the velocity vector at arrival (for a Hohmann transfer, this is the velocity vector at the periapse or apoapse point for an inner or outer planet transfer, respectively). As the spacecraft passes through the boundary of the sphere of influence, it is then on a hyperbolic trajectory with respect to the destination planet. When it reaches the periapse location of this orbit, a braking Δv is applied, subtracting off the velocity difference between the hyperbolic periapse velocity and the velocity of the final orbit. An approximation is frequently made that the orbit that the spacecraft enters around the destination planet is a circle (this simplifies the analysis), although insertion orbits can just as easily be elliptical.

Lastly, when considering the time of flight for the overall transfer, Kepler's time equation for the elliptical case (phase 2) or hyperbolic case (phases 1 and 3) can be easily solved for the time component, as we discussed in Chapter 2.

5.6 EXAMPLE 1: INTERPLANETARY HOHMANN TRANSFER BETWEEN EARTH AND MARS

Here, we provide an example of a full Earth-Mars transfer using the analysis tools developed in the previous sections. For this example, we assume that a spacecraft is initially in a 400 km altitude circular orbit about Earth and is injected into an interplanetary Hohmann transfer through a heliocentric ellipse to Mars. At arrival, the vehicle is to pass within 600 km of the surface of Mars on the sunlit side. A heliocentric view of this transfer is shown in Figure 5.17.

The first part of the problem is to compute the required Δv to reach Mars. First, we assume that both Earth and Mars lie in the same orbital plane and that both Earth and Mars are in circular orbits. The initial Δv is determined by examining the heliocentric transfer requirements. Therefore, as the spacecraft escapes Earth, it must be at perihelion upon entering its heliocentric cruise. This fact helps us determine the velocity that the spacecraft must achieve in order to target Mars. This is computed using Equation (5.4) for the spacecraft's perihelion velocity and Equation (5.6) for the spacecraft's aphelion velocity.

Several physical quantities must be computed before we can uniquely determine the various velocities of the spacecraft throughout its journey to Mars. The velocity of Earth around the Sun is approximately 29.78 km/s, while the velocity of Mars around the Sun is approximately 24.13 km/s. The semimajor axis of the Hohmann transfer orbit is the average of Earth's and Mars's, resulting in approximately 1.888×10^8 km, or about 1.262 AU. The velocity at the periapse of the heliocentric elliptical transfer

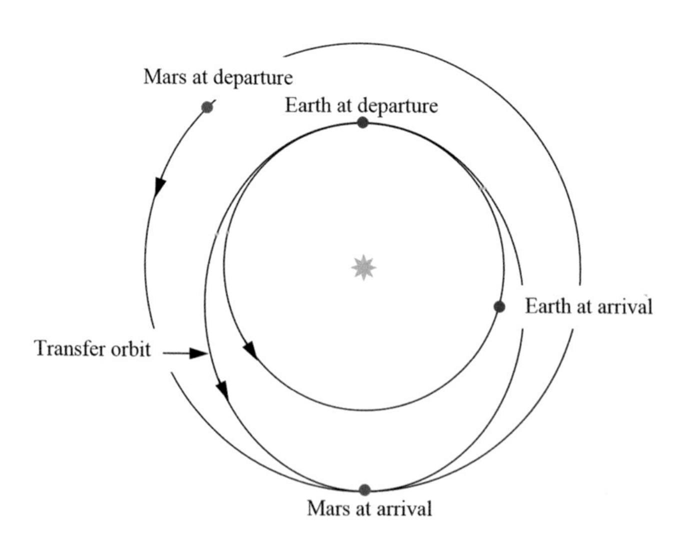

FIGURE 5.17 Heliocentric view of Earth to Mars transfer.

orbit is computed as 21.48 km/s with Equation (5.4) and the velocity at the apoapse of the elliptical transfer orbit is computed as 32.73 km/s with Equation (5.6). Using the given altitude, the gravitational parameter and the physical equatorial radius of the Earth, we can compute the circular velocity of the spacecraft around the Earth as 7.668 km/s, while the circular velocity that the spacecraft must achieve around Mars upon arrival is 3.273 km/s.

A vector diagram of the departure geometry is shown in Figure 5.18.

We can drop the vector notation, since all velocities are collinear for an interplanetary Hohmann transfer. This will not be true in general, as we will see in the next section. The velocity of the spacecraft at the boundary of the sphere of influence of Earth is

$$v_{\infty/\oplus} = v_p - v_\oplus \tag{5.151}$$

and is computed to be 2.948 km/s.

Balancing energy between the departing geocentric hyperbolic orbit and the energy of the spacecraft at the boundary of Earth's sphere of influence results in the following

$$E = -\frac{\mu_\oplus}{2a} = \frac{v_{\infty/\oplus}^2}{2} = \frac{v_{\text{Departure}}^2}{2} - \frac{\mu_\oplus}{r_p} \tag{5.152}$$

Solving for the velocity at the departure point, we get 11.24 km/s, and the Δv needed to depart Earth is therefore $v_{\text{Departure}} - v_{\text{Circular}} = 11.24 - 7.668 = 3.571$ km/s.

The spacecraft will then coast from Earth to Mars over a period of

$$T_{12} = \pi\sqrt{\frac{a_H^3}{\mu_e}} = 258.9 \text{ days} \tag{5.153}$$

After this period of time, the spacecraft approaches the sphere of influence of Mars. The velocity vectors at Mars arrival are shown in Figure 5.19.

At the entry into the sphere of influence of Mars, the velocity of the spacecraft relative to Mars is

$$v_{\infty/♂} = v_{♂} - v_a \tag{5.154}$$

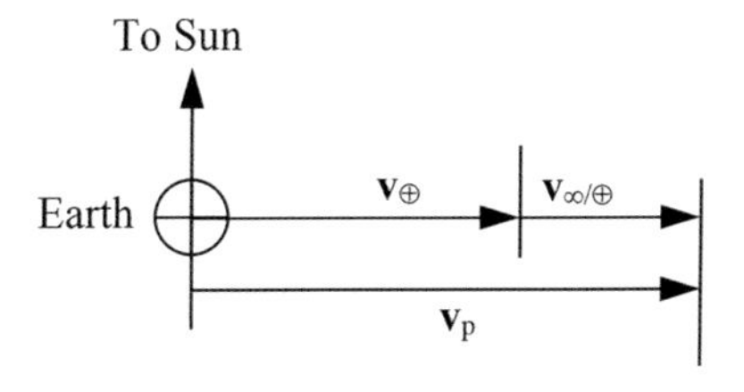

FIGURE 5.18 Velocity vectors at Earth departure.

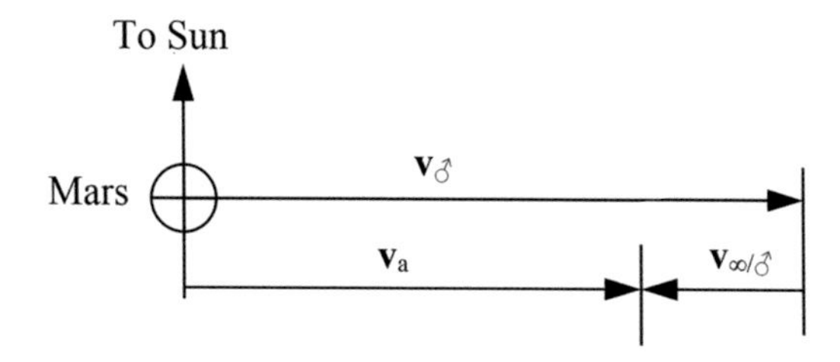

FIGURE 5.19 Velocity vectors at Mars arrival.

Computing the velocity at the boundary of the sphere of influence of Mars is 24.13−21.48 = 2.65 km/s. Performing an energy balance between the inbound hyperbolic orbit and the energy of the spacecraft at the boundary of Mars's sphere of influence gives

$$E = -\frac{\mu_{♂}}{2a} = \frac{v_{\infty/♂}^2}{2} = \frac{v_{Arrival}^2}{2} - \frac{\mu_{♂}}{r_p} \tag{5.155}$$

Solving for the arrival velocity, we find that $v_{Arrival}$ = 5.334 km/s. A second Δv of 5.334−3.274 = 2.060 km/s is needed to complete the transfer. Therefore, the total mission Δv is 3.571 + 2.060 = 5.631 km/s.

Now, let's consider the opposite transfer, from Mars to Earth. For example, this could be a return trajectory for a crewed mission or a sample-return mission. Here, we will assume that the spacecraft is initially in a 400 km circular orbit about Mars and is injected into a Hohmann heliocentric ellipse targeting Earth. The vehicle is to pass within 600 km of the surface of Earth on the sunlit side. A heliocentric view of this transfer is shown in Figure 5.20.

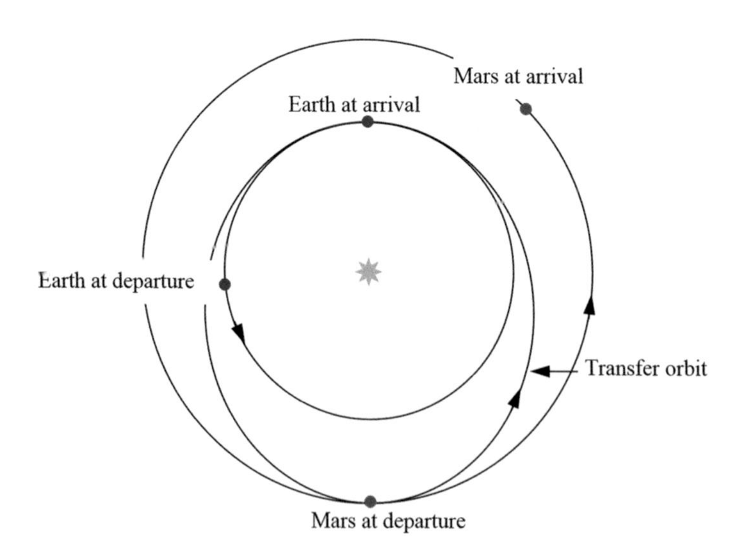

FIGURE 5.20 Heliocentric view of Mars to Earth transfer.

The first part of the problem is to compute the required Δv to escape Mars. As before, we assume that both Earth and Mars lie in the same orbital plane and that both Earth and Mars are in circular orbits. The initial Δv is determined by examining the heliocentric transfer requirements, as we did before. Therefore, as it escapes Mars's orbit, the spacecraft must be at aphelion when leaving Mars's sphere of influence. Note that the mathematics is exactly identical to the Earth-Mars transfer, although it will result into different numerical values, and slightly different diagrams.

The velocities of the Earth and Mars with respect to the Sun remain unchanged from before. Similarly, the Hohmann ellipse also has the same value as before, except that now the spacecraft is traveling from Mars to Earth. The velocities of the spacecraft at the apoapse and periapse of the heliocentric elliptical transfer orbit are also unchanged from before. The circular velocity that the spacecraft is targeting around the Earth at arrival is 7.558 km/s and the circular velocity of the spacecraft around Mars at departure (prior to the first Δv) is 3.360 km/s. The vector geometry of the Mars departure is identical to that shown in Figure 5.19.

The velocity of the spacecraft at the boundary of the sphere of influence of Mars is

$$v_{\infty/\male} = v_{\male} - v_a \tag{5.156}$$

and is computed to be 2.650 km/s.

Balancing the energy of the departing hyperbolic orbit with the energy of the spacecraft at the boundary of the sphere of influence of Mars gives

$$E = -\frac{\mu_{\male}}{2a} = \frac{v_{\infty/\male}^2}{2} = \frac{v_{\text{Departure}}^2}{2} - \frac{\mu_{\male}}{r_p} \tag{5.157}$$

Thus, computing the velocity at the departure point gives 4.988 km/s, and the Δv needed to depart Mars is $v_{\text{Departure}} - v_{\text{Circular}} = 4.988 - 3.359 = 1.629$ km/s.

The spacecraft will then coast from Mars to Earth over a period of

$$T_{12} = \pi\sqrt{\frac{a_H^3}{\mu_\odot}} = 258.9 \text{ days} \tag{5.158}$$

After that period of time – which is the same time computed for the inverse transfer – the spacecraft approaches the sphere of influence of Earth. The velocity vectors at Earth arrival are identical to those already shown in Figure 5.18. At the entry into the sphere of influence of Earth, the velocity of the spacecraft relative to Earth is

$$v_{\infty/\oplus} = v_p - v_{\oplus} \tag{5.159}$$

Computing this value becomes 32.73–29.78 = 2.948 km/s. Performing an energy balance of the inbound hyperbolic orbit,

$$E = -\frac{\mu_{\oplus}}{2a} = \frac{v_{\infty/\oplus}^2}{2} = \frac{v_{\text{Arrival}}^2}{2} - \frac{\mu_{\oplus}}{r_p} \tag{5.160}$$

Solving for the arrival velocity, $v_{Arrival} = 11.09$ km/s. Thus, a second $\Delta v = 11.09 - 7.558 = 3.532$ km/s. Therefore, the total mission Δv for this Mars-Earth transfer is $1.629 + 3.532 = 5.161$ km/s.

5.7 EXAMPLE 2: REALISTIC INTERPLANETARY TRANSFER BETWEEN EARTH AND MARS

In this section, we want to present an example similar to that of the last section (Earth-Mars transfer) using actual ephemeris data and solving the heliocentric cruise through the use of Lambert's problem. Like for the last section, we will use the same physical constants, such as gravitational parameters, planetary radii, and the same departing and arriving altitudes.

In order to make use of true planetary ephemeris data, we need to specify departure and arrival dates. We decided to choose the Earth departure and Mars arrival of the NASA 2020 Mars Rover "Perseverance," i.e., July 30, 2020 and February 18, 2021, respectively. Using JPL Horizons as described in Sections 4.7 and 4.8 (or a similar approach, e.g., MATLAB's *planetEphemeris*[4] function), we can compute the positions and velocities of Earth and Mars with respect to the Sun as provided in Table 5.8. We are using the astronomical unit[5] (AU) as the preferred length unit (LU), and the year[6] as the preferred time unit (TU) for this problem.

From the ephemeris data, we can compute the speeds of the Earth (29.3134 km/s) and Mars (23.4137 km/s) which are similar to what we computed in the previous section. However, the velocity vectors of the planets and spacecraft will no longer be collinear. In fact, in order to proceed to computing the required Δv's for Earth departure and Mars arrival, we need to first solve for the orbital parameters of the heliocentric cruise. To do so, we must solve Lambert's problem, where the boundary conditions P_1 and P_2 are the positions of the Earth and Mars with respect to the Sun, as given in Table 5.8.

TABLE 5.8
Planetary ephemeris data of Earth on July 30, 2020 and Mars on February 18, 2021 in ECI

Planetary Ephemeris	X-Coordinate	Y-Coordinate	Z-Coordinate
Earth's position [AU]	+0.605775	−0.737464	+0.319614
Mars's position [AU]	−0.0130041	+1.43238	+0.657156
Earth's velocity [AU/year]	+4.90940	+3.44919	+1.49554
Mars's velocity [AU/year]	−4.92010	+0.326355	+0.282519

[4] Note that this function uses slightly different approximations than the full-force model provided by JPL Horizons, so the planetary ephemeris data obtained from *planetEphemeris* would be slightly different than that of JPL Horizons.

[5] By definition, 1 AU = 149,597,870.700 km exactly.

[6] We use the astronomical definition of year here, i.e., 1 year = 365.24219 days as opposed to its calendar definition.

Solving Lambert's problem as described in Section 5.3 yields the results provided in Table 5.9. Note that because of the choice of units, we can also non-dimensionalize the gravitational parameter of the Sun as $\mu_{\odot} = 4\pi^2\ AU^3/year^2 = 4\pi^2\ LU^3/TU^2$. For readability and for comparison with the example from the previous section, we will convert the Δv's into km/s.

A graphical solution of Lambert's equation can be seen in Figure 5.21, where we solved Equation (5.62) using the method described in Section 5.3.1. Thus, the solution for a (semimajor axis) such that $f(a) = 0$ was found using the secant method with initial guesses of $a_1 = a_m$ and $a_2 = 1.001 \times a_m$ and converged to the solution in eight

TABLE 5.9
Results from Lambert's problem using the boundary conditions in Table 5.8

Parameter Name, Symbol	Value	Units
Time of flight, TOF	203	Days
Parabolic TOF, t_p	109	Days
Transfer angle, $\Delta\theta$	143.451	Degrees
Minimum semimajor axis, a_m	1.26028	AU
Semimajor axis, a	1.31751	AU
Semilatus rectum, p	1.24290	AU
Eccentricity, e	0.237969	–

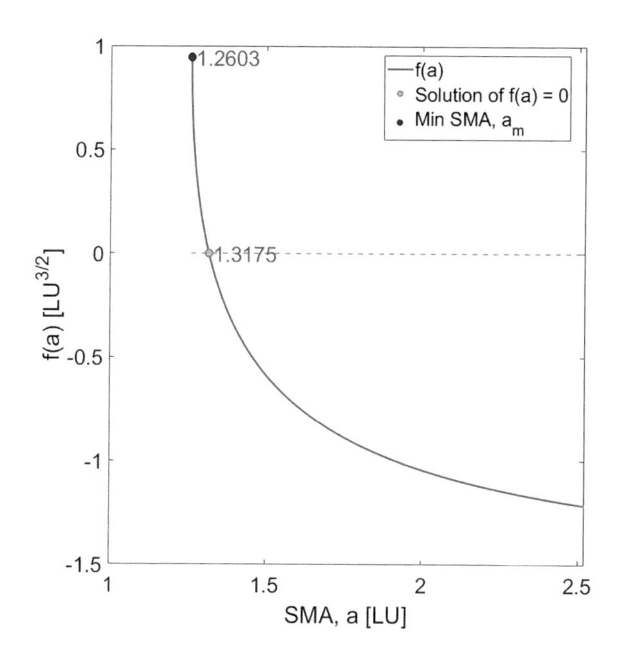

FIGURE 5.21 Graphical solution of Lambert's equation for this Earth-Mars transfer example.

steps with a tolerance of 10^{-10}. As a comparison for the reader, a generic solution to Lambert's problem should converge within a few steps if using the secant method, and should take no more than a few milliseconds on modern computers, depending on the specifications and programming language used.

And thus the velocities required at the boundaries of the spheres of influence of Earth and Mars are computed and are shown in Table 5.10. These correspond to the $\mathbf{v}_\infty$'s of the spacecraft with respect to Earth and Mars.

Computing the magnitudes of the $\mathbf{v}_\infty$'s reported in Table 5.10 gives 4.03785 and 2.64479 km/s at Earth's and Mars's boundaries of their spheres of influence, respectively. At this point, we can compute the required $v_{Departure}$ and $v_{Arrival}$ as done in Section 5.6 with energy balances, which, for an Earth departing altitude of 400 km results in $v_{Departure} = 11.5724$ km/s and for a Mars arrival altitude of 600 km results in $v_{Arrival} = 5.33529$ km/s. Thus, subtracting the values of the departing and arriving circular orbits at Earth and Mars gives us the required Δv's as $\Delta v_1 = 3.90375$ km/s, $\Delta v_2 = 2.05883$ km/s, and $\Delta v_{total} = 5.96258$ km/s. Compared to the Δv_{total} of the previous section (5.631 km/s), where we assumed that Earth's and Mars's orbits are circular and coplanar, there is a difference of only approximately 5.6%. Furthermore, note that while the geometry of two circular and coplanar orbits repeats exactly for every synodic period, the actual geometry of the orbits of Earth and Mars does not repeat exactly. However, one can always find a minimum Δv_{total} for an Earth-Mars transfer in a given time-frame, and for every synodic period the minimum Δv_{total} can be a type I or type II transfer, depending on the timespan considered.

Also, note that while we used the launch and arrival dates for the Perseverance rover, the computed Δv's are not the same since we used different departure and arrival orbits. Additionally, here we assumed that the Sun is the only gravitational source acting on the spacecraft during its heliocentric cruise, while Perseverance was under the influence of all the existing forces in the Solar System. Because of this, and other reasons, multiple trajectory correction maneuvers (TCMs) were planned during its flight, namely 15 days after launch (TCM-1), 62 days after launch (TCM-2), 62 days before landing (TCM-3), 8.6 days before landing (TCM-4), 2.6 days before landing (TCM-5), and 9 hours before landing (TCM-6). The last three were especially important in order to ensure the correct landing zone for the spacecraft [Jesick et al., 2022].

Table 5.11 summarizes the Δv values and times of flight for the examples presented in this section and the previous section.

TABLE 5.10
Required spacecraft v_∞'s for Earth departure and Mars arrival

Required v_∞	X-Coordinate	Y-Coordinate	Z-Coordinate
At Earth's SOI [AU/year]	+5.68342	+3.58578	+1.82377
At Mars's SOI [AU/year]	−4.44788	+0.579501	+0.127004
At Earth's SOI [km/s]	+26.9427	+16.9986	+8.64573
At Mars's SOI [km/s]	−21.0855	+2.74717	+0.602074

TABLE 5.11
Comparison between Hohmann transfer and realistic transfer examples

	Hohmann	Realistic
Δv_1 (km/s)	3.571	3.904
Δv_2 (km/s)	2.060	2.059
Δv_{total} (km/s)	5.631	5.963
TOF (days)	259	203

5.8 PORKCHOP PLOTS

One of the most valuable tools to interplanetary mission designers is what is known as a "porkchop" plot. This plot is a contour plot of launch dates versus arrival dates for specific body to body transfers. This plot gets its name from the porkchop shape of many of these contours. This section describes how to create such plots, how to interpret the data shown in these plots, and its uses in interplanetary astrodynamics.

Porkchop plots are used by mission designers to evaluate the necessary characteristic energy (C3 or C_3) the spacecraft needs to achieve to leave the departure body's SOI and target the destination body, the arrival velocity the spacecraft has upon entering the destination body's SOI, and the TOF required for the provided set of departure and arrival dates, the abscissa and the ordinate of the plot, respectively. An example of a porkchop plot is provided in Figure 5.22.

Porkchop plots are one of the tools that mission designers use to evaluate and budget mission Δv for the major maneuvers required during the duration of the mission. In fact, while other – and generally smaller – maneuvers are usually performed throughout a mission (e.g., trajectory correction maneuvers), these are usually not part of porkchop plots. This is because porkchop plots are used for mission design as opposed to mission planning. Mission design relates to major maneuvers and overall "big picture" mission events, while mission planning is concerned with day-to-day events and scheduling.

These mission design plots are created by solving Lambert's problem repeatedly. Recall that Lambert's problem is solved once the locations of the departure and arrival points are known, and given a time of flight between these two points. The three-dimensional locations of the departure and arrival points are the location of the two bodies at the departure and arrival times. These locations are generally found using one of the many ways to estimate ephemeris data, as discussed in detail in Chapter 4. Furthermore, because of the three-dimensional nature of this problem, the transfer orbit is generally not in the orbit plane of the departure body or the arrival body. More specifically, the plane of the orbital transfer is the plane that intersects the departure body, the arrival body, and the gravitational source (i.e., the Sun). Except for very few special cases in which these three points are close to being on a line, the plane of the orbital transfer is generally between those of the departure body and arrival body. We will look at these special cases in the next section.

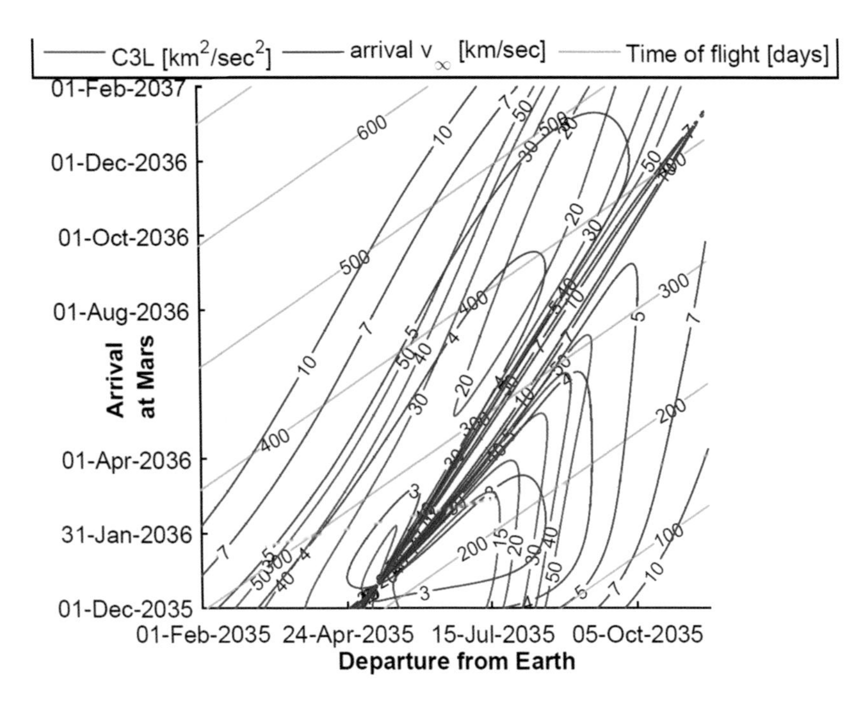

FIGURE 5.22 Porkchop plot illustrating Earth-Mars transfer for the 2035–2036 timeframe. C3L represents the required C3 a spacecraft must achieve at launch (L) [Conte, 2019].

The process of creating a porkchop plot begins by choosing a departure date. Next, either the time of flight or arrival date is chosen – only one is needed to find the other. The position vectors of the departure body at departure and the arrival body at arrival are found (see Sections 4.5–4.8 for more details). Along with the choice of a retrograde or prograde transfer orbit, we now have enough information to solve Lambert's problem. Generally, the prograde option is chosen for interplanetary transfers since nearly all known bodies of Solar System orbit the Sun in a prograde motion, including most asteroids and comets. Then, for each combination of departure and arrival dates, a solution of Lambert's problem is computed (using the methods shown in Section 5.3, or alternative methods). Note that, of course, only date combinations that give a positive time of flight (TOF) can lead to a valid solution – you can't arrive before you've departed! – although mathematically it is possible to insert a negative TOF in Lambert's equation. However, this will usually lead to an imaginary value of semimajor axis. Once Lambert's problem is solved, the required departing and arrival velocities at the departure and arrival bodies' SOIs can be computed, as details in Section 5.3. The departure v_∞ can then be used to compute the required characteristic energy at launch (C_3), as

$$C_3 = v_\infty^2 \tag{5.161}$$

C_3 is used to gauge the performance needed from the launch vehicle, which is used in most interplanetary missions to provide the initial departure Δv. The arrival v_∞

is instead used to determine the arrival Δv, which is a maneuver that is generally performed by the spacecraft's onboard propulsion system. This is why launch C_3 and arrival v_∞ are displayed separately on a porkchop plot. Based on the launch vehicle and onboard propulsion system capabilities, it is possible to then determine the largest departure and arrival windows for a mission, which correspond to a portion of the entire plot. Time of flight is also displayed to give the mission designers information regarding the timing of the mission.

While launch C_3 and arrival v_∞ are two maneuvers that are generally done by different systems, the overall total Δv is still one of the most important parameters for space missions. This is why variants of porkchop plots where only total Δv is shown exist, such as the example displayed in Figure 5.23. In this figure, the same Earth-Mars transfer as before is considered, but this time only Δv is plotted as a surface plot, where darker colors indicate a lower Δv, and the minimum overall total Δv is indicated by a yellow asterisk. In Figure 5.23, the minimum overall total Δv corresponds to 5.2120 km/s, and occurs for departure on 27 June 2035 and Mars's SOI arrival on 15 January 2036, for a total interplanetary TOF of 202 days. Note that in order to compute Δv, the initial and final orbits around the departure and arrival bodies need to be known. This is why porkchop plots generally display departure C_3 and arrival v_∞ instead in order to be more generic and usable even in the case that the departure and/or arrival orbits are changed. Lastly, note that in Figure 5.23, two local minimum Δv values exist. This is true of most porkchop plots, and they correspond to type I and type II transfers, as discussed in Section 5.5.

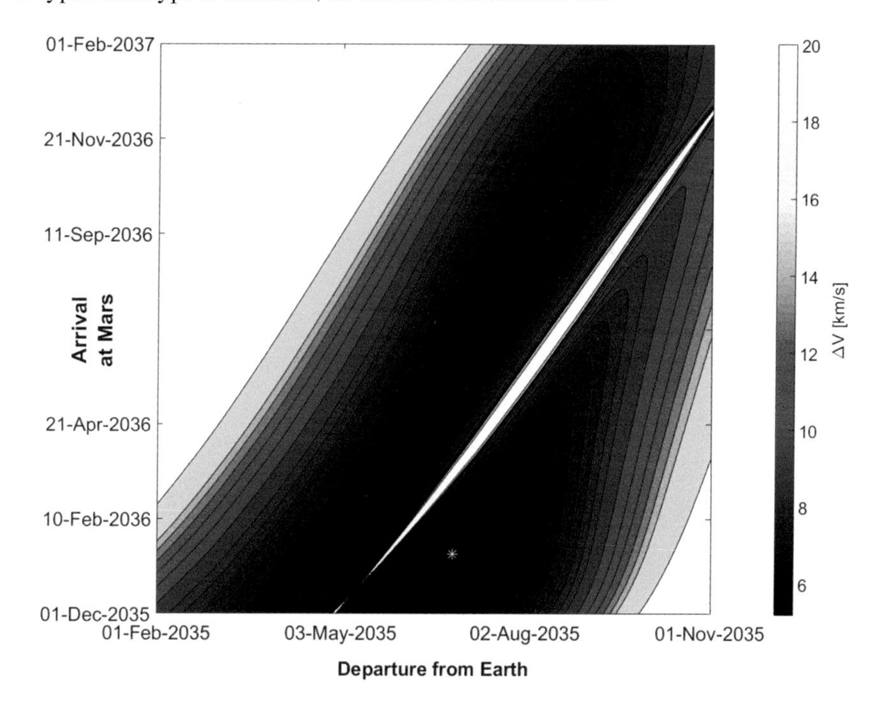

FIGURE 5.23 Porkchop plot illustrating Earth-Mars total Δv transfer for the 2035–2036 time-frame [Conte et al., 2017].

Another orbital parameter of interest during this stage of mission design is the orbital inclination of the orbit arriving at the targeted body – not to be confused with the orbital inclination of the plane of the interplanetary transfer. This is an important parameter whether the spacecraft needs to target a certain landing location or a prescribed operating orbit, and is strictly dependent on where on the B-plane the spacecraft arrives, as shown in more detail in Section 4.2.7. Figure 5.24 shows an example of the inclination of the arrival trajectory for the Earth-Mars example we have discussed. Note that, extremely high inclinations would result in near-polar orbits after the orbit insertion maneuver is completed (assuming that inclination remains unchanged). As discussed in Section 5.1.1, polar orbits have a variety of scientific applications.

Although we have only shown a few examples, various orbital parameters of interest can be computed and plotted depending on the mission at hand. For example, if we consider an orbital transfer that includes an additional intermediate stop, we could modify our porkchop plots to consider only transfers that are physically feasible between three bodies – now the orbital transfer no longer has a plane that is always defined since three bodies plus the Sun not always lie on the same plane. For example, if including the Moon as an intermediate staging location from Earth to Mars, only orbital transfers for a couple of days per month are feasible. Figure 5.25 shows such an example where a 2D lunar DRO (see Section 3.6.2) is used as intermediate stage between Earth and Mars, and only the Δv for the lunar DRO to Mars is plotted. Such a plot is called a *striped* porkchop plot since the graph appears to have stripes in place of a more homogenous surface plot.

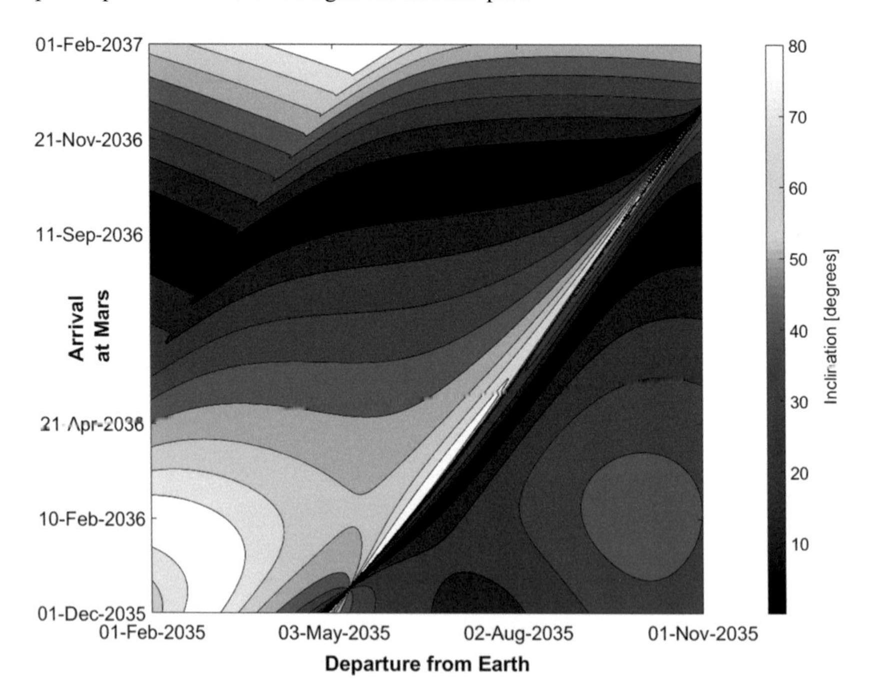

FIGURE 5.24 Surface plot illustrating the arrival inclination of the spacecraft trajectory with respect to Mars's equator [Conte et al., 2017].

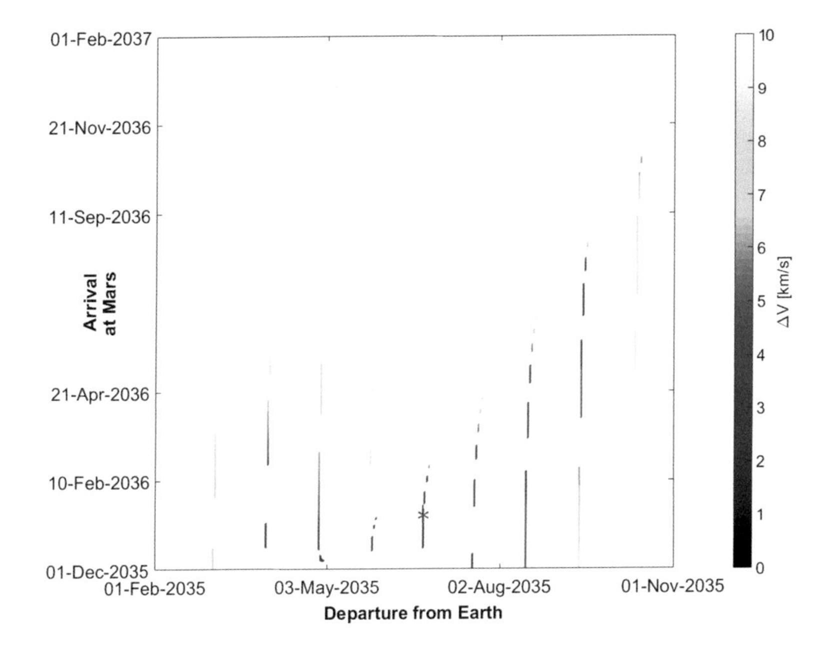

FIGURE 5.25 Striped porkchop plot for Earth-Mars transfer using a lunar DRO as intermediate staging location [Conte et al., 2017].

Other varieties of plots similar to porkchop plots exist. For example, for low-thrust maneuvers (which as discussed in more detail in Section 5.17), generally bacon plots are used, as can be seen in Woolley and Nicholas [2014].

Yet another variety of porkchop-like plots is given by a first-order Δv budget estimation of proximity operation maneuvers. While it is generally difficult to estimate the Δv required for such operations during the early stages of mission design, depending on the final targeted orbits of interest, it is possible to estimate a worst-case scenario for proximity operations [Conte, 2019]. Additionally, since proximity operations require much lower Δv than major interplanetary propulsive burns and take place in much smaller amounts of times, the resulting graphs were named *prosciutto* plots. This follows the "meat-eating" scheme of porkchop and bacon plots, albeit using the thinner counterpart to porkchops, namely the very well-known and fine Italian *prosciutto*. An example of such *prosciutto* plots is given in Figure 5.26, where a prosciutto plot for rendezvous in Martian orbit upon performing an orbit insertion is given. Note that, like for bacon plots, using an "adequate" color scheme results in a shape and color resembling that of prosciutto.

5.9 BROKEN-PLANE TRANSFERS

When talking about solving Lambert's problem and creating porkchop plots, we also discussed that two-burn maneuvers require the transfer orbit to have an inclination such that the Sun, departure body (e.g., Earth), and arrival body (e.g., Mars) all lie on such orbital plane. However, when these three points nearly lie on a straight line, the inclination of the transfer orbit can take on extremely large values, which then

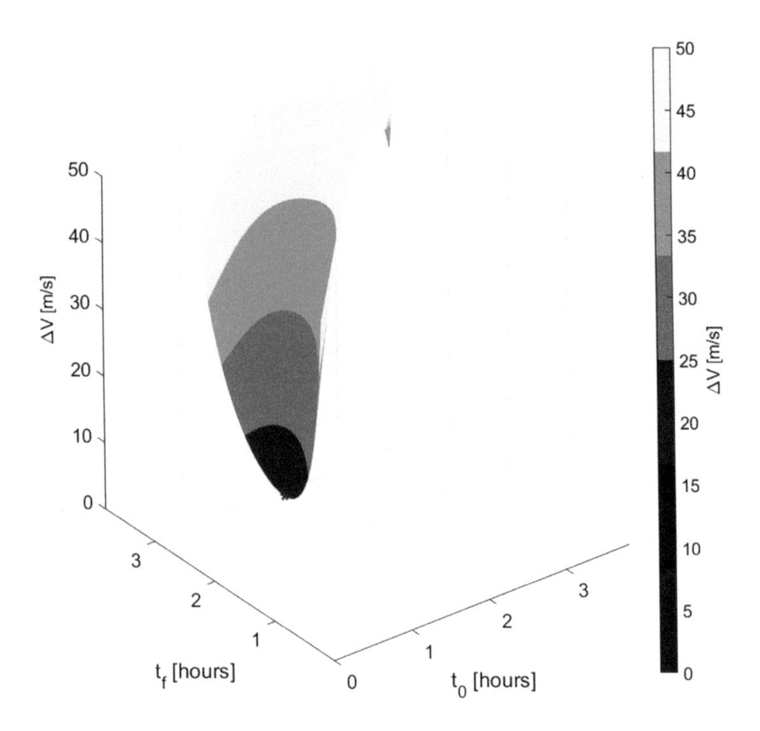

FIGURE 5.26 Example of a prosciutto plot where rendezvous is needed in Martian orbit. The lowest Δv is indicated by a blue asterisk [Conte, 2019].

correspond to very large Δv requirements. In order to avoid such large Δv's, it is possible to instead launch the spacecraft into the orbital plane of the departure body and then perform an inclination change via a deep space maneuver (DSM) after launch while in heliocentric space such that the resulting trajectory intersects the arrival body. Then, at arrival a final orbit insertion maneuver is required like we have discussed before, bringing the number of maneuvers to three instead of two. These kinds of three-burn maneuvers are called broken-plane maneuvers, and an example of the geometry associated with these maneuvers is shown in Figure 5.27. In this figure, the inclination change is done at the line of nodes, where the planes of the two planets intersect. Furthermore, note that in this figure the relative inclination between planets is exaggerated.

As shown in Figure 5.27, the name of these transfers comes from the fact that the orbital planes of the two transfer orbits connecting the departure and destination bodies appear to be broken at the line that lies between the Sun and the location of the inclination change.

Broken-plane maneuvers are especially useful in reducing total mission Δv for near 180° transfer angles and, in general, for transfers whose transfer angles are close to 180° × n, where n is an integer. Additionally, in order to minimize Δv, the inclination change should occur at the ascending or descending node (as shown in Figure 5.27), resulting in a pure inclination change as described in Section 5.2.3. However, this is not required, and it is sometimes not possible due to other mission

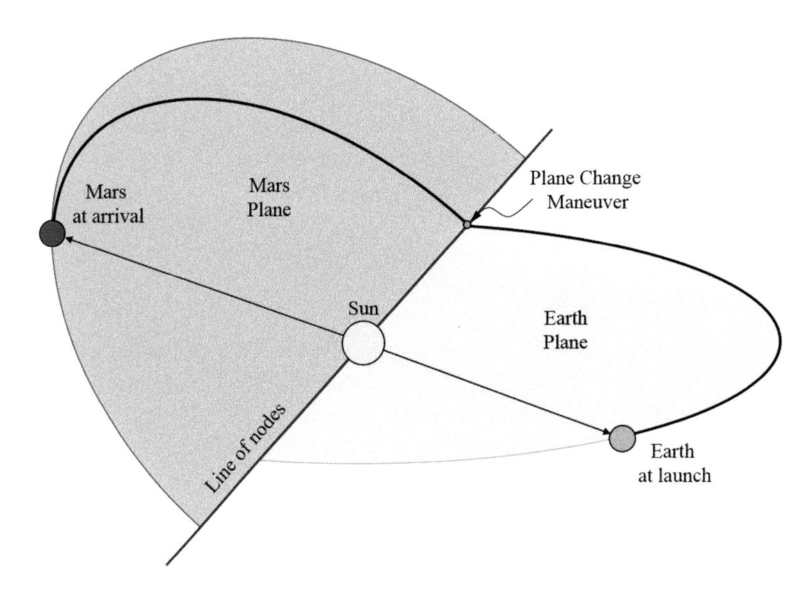

FIGURE 5.27 Geometry of broken-plane maneuvers for an Earth-Mars transfer example. The transfer orbit is the solid black curve.

requirements, such as required departure and arrival times. Mathematically, broken-plane maneuvers are a combination of two Lambert's problems – one from the departure body to the inclination change location and one from the inclination change location to the arrival body – and inclination changes. Because of this, it is possible to create porkchop plots for broken-plane maneuvers too, as the literature describes [Abilleira, 2007].

Lastly, it should be noted that appropriately adding more intermediate maneuvers, such as deep space maneuvers (DSMs), generally decreases the overall Δv of any given orbital transfers, even for same-plane transfers. In fact, DSMs can improve trajectory performance (from a C_3 perspective), but they can also open up new families of trajectories that would satisfy specific mission requirements not achievable with ballistic trajectories. One common mistake is to think that the Hohmann transfer, which is a ballistic trajectory, is *always* the best possible transfer between two orbits, but this not true. While near-Hohmann[7] transfers result in the lowest two-burn maneuvers between two circular orbits whose orbital radii have a ratio smaller than that of the critical ratio (recall Section 5.2), three-burn, four-burn, etc. maneuvers can result in smaller overall Δv. This comes, of course, at the cost of additional complexity – for example, the onboard propulsion system needs to be able to stop and restart multiple times – and the limiting case, where infinitely many infinitesimal maneuvers are performed continuously leads to low-thrust maneuvers, as we will discuss in Section 5.17.

[7] Since two circular coplanar orbits do not exist in nature, we call "near-Hohmann transfers" those transfers that are performed between two orbits that are nearly circular and nearly coplanar, such as those of the Earth and Mars.

5.10 LAUNCH GEOMETRY

In this section, we want to focus on the geometry for Earth launches, how launch windows are computed, and how launch periods – and thus how often a launch window occurs – play a role in interplanetary mission design. While here we focus on launch geometry for Earth launches, a similar analysis can be done for launches that take place elsewhere – for example, a launch from Mars for a sample-return mission.

5.10.1 LAUNCH WINDOW

A launch window is a time period during which a given mission must depart from Earth. Thus, a spacecraft targeting a specific body in the Solar System must be launched such that its trajectory and that of the target will intercept at some future time. Launch windows impose a schedule requirement on mission designers. In fact, missions can only launch during their prescribed launch windows and, if missed, an alternative launch window must be used. Depending on the mission, the time between launch windows could vary from as little as a day to as long as several years. This is why, as discussed in Section 5.8, porkchop plots are used to determine a range of departure dates for which a mission can be launched. Otherwise, a mission may be designed with alternative objectives in mind in case the mission suffers unexpected delays. The term "launch window" has also been adapted in other fields to refer to optimal times for an event to take place, like the release date of a movie or a videogame. However, its origins are rooted in the field of astrodynamics.

Objects orbiting the Earth appear in the same relative location fairly frequently. For missions requiring to rendezvous with the ISS, for example, launch windows open for only a few minutes, but they repeat daily. On the other hand, interplanetary launch windows open for a few days or weeks – depending on the launch capabilities of a given mission – and repeat much less frequently. For example, let's assume that Earth and Mars are in circular coplanar orbits for simplicity. The orbital alignment between the two planets for a Hohmann transfer to occur is shown in Figure 5.28. The angle α between Earth and Mars at departure can be computed based on the fact that Mars at arrival must be at 180° from Earth at departure. Thus, backtracking from the arrival location to where Mars must be located to where it needs to be when the launch occurs, α is computed to be 44.3° (the full computations are left as an exercise).

This Earth-Mars alignment defines the conditions required for a Hohmann transfer to take place. In reality, since Earth and Mars are not in circular coplanar orbits, this angle is not always the same for every occurrence, and it does not sweep an area that is located on the plane of Earth's or Mars's orbits. This and other orbital alignments are what define the launch window for a specific event to occur in order for a mission to launch. In the next section, we will discuss how to compute how often these launch windows take place and their importance in mission design.

5.10.2 LAUNCH PERIODS AND SYNODIC PERIOD

As discussed in Section 5.8, porkchop plots can be used to determine the timing of a given mission provided that performance of the launch vehicle (associated to C_3) and of the spacecraft propulsion system (associated with v_∞) are known. However,

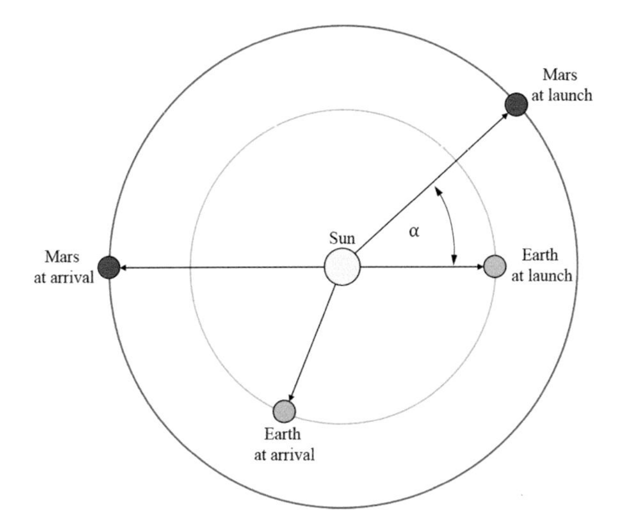

FIGURE 5.28 Orbital alignment required between Earth and Mars to perform a Hohmann transfer.

it should be noted that launch windows repeat every time that the same orbital alignment occurs. In fact, the location of any two (or more) Solar System bodies – not just of the Earth – at departure and arrival repeat periodically. The frequency of this repeat cycle is known as the synodic period, which is calculated as follows

$$\frac{1}{T_{syn}} = \left| \frac{1}{T_1} - \frac{1}{T_2} \right| \rightarrow T_{syn} = \frac{T_1 T_2}{|T_1 - T_2|} \tag{5.162}$$

where T_1 and T_2 are the orbital periods of the two bodies of interests. For example, if one object has a 500-day period and a second one has a period of 700 days, then this geometry repeats every 1,250 days, or 4.794 years. For the Earth, the synodic periods with the other planets are given in Table 5.12. Appendix A gives a more comprehensive table listing the synodic periods of all planets with respect to each other.

TABLE 5.12
Synodic period between the Earth and the other planets in the Solar System

Planet	Orbital Period (Years)	Synodic Period (Years)
Mercury	0.2408	0.3172
Venus	0.6152	1.5988
Mars	1.881	2.1351
Jupiter	11.86	1.0921
Saturn	29.46	1.0351
Uranus	84.01	1.0120
Neptune	164.8	1.0061

If using a near-Hohmann transfer between Earth and any of the planets, then Table 5.12 tells us how much time it takes for any launch window to repeat. For example, NASA's lander InSight, was originally planned for launch in March 2016. However, a persistent vacuum leak in its seismometer caused a launch delay beyond the 2016 launch window. As a result, the launch was rescheduled to May 2018 – one synodic period later – and during the wait the instrument was repaired. While the trajectory used was similar to the one that was originally planned, this delay caused other problems, such as budget overruns and the lander arriving on Mars during the Martian winter time – and thus needing to put the lander in low-energy mode.

From the perspective of a "stationary" Earth observer, the synodic period is the time required for a body within the Solar System, such as a planet, the Moon, or an artificial Earth satellite, to return to the same position in the sky.

Note also that Equation (5.162) tells us that the synodic period of two objects that have the same orbital period is undefined (because of the denominator becoming zero). In fact, looking at Table 5.12 we can see that as the distance of the planets from the Sun increases, their synodic period with respect to the Earth approaches the value of 1 year. In other words, a faraway object is essentially "stationary" when compared to the orbital period of the Earth. In fact, when considering two objects that have drastically different orbital periods, their synodic period will tend to be much closer to the period of the faster object. For example, Jupiter's orbital period is 11.862 years, which means that the synodic period of Earth and Jupiter is 1.0921, or approximately 399 days. This means that approximately every 1 year and 1 month, the orbital alignment between Earth and Jupiter repeats. This also means that a mission launching from Earth directly to Jupiter can be rescheduled 1 year and 1 month later and use an extremely similar trajectory to the originally scheduled mission. However, we saw in Section 5.2 that near-Hohmann transfers take a large amount of time, thus making them impractical for missions to far away planets such as Uranus and Neptune. On the other hand, a closer object to Earth, such as Mars, has larger synodic period – the largest of all planets, in fact – and thus requires the longest time if launch reschedules are needed, as we previously discussed. However, a ballistic trajectory from Earth to Mars is much more practical in terms of time of flight.

It should be noted that since no two objects in the Solar System are in perfectly circular orbits, the synodic period tells us that the orbital alignment of any two objects will repeat, but it won't be exactly identical. For example, while every pork-chop plot for Earth-Mars transfers has a local minimum Δv for type I transfers and one for type II transfers, the absolute minimum is not always guaranteed to be a type I (or type II) transfer. Table 5.13 gives Δv-optimal results for ballistic Earth-Mars transfers assuming planetary departing and arriving circular orbits of 400 km in altitude for both planets for every Earth-Mars synodic period from the year 2022 to the year 2040 – note that years overlap since the Earth-Mars synodic period is not exactly 2 years.

As shown in this table, sometimes type I transfers are more Δv-optimal, while other times type II transfers are more desirable.

The synodic period is a useful tool of interplanetary astrodynamics that allows mission designers to compute how often a given orbital alignment occurs. However, as we will see in more detail in later sections in this chapter, many missions utilize

TABLE 5.13
Total Δv, departure and arrival dates, and transfer orbit type (I or II) for ballistic Earth-Mars transfers

Synodic Period	Δv (km/s)	Departure Date	Arrival Date	TOF (Days)	Orbit Type
2022–2023	6.0571	29/08/2022	09/08/2023	345	II
2024–2025	5.7858	30/09/2024	28/08/2025	332	II
2026–2027	5.6995	31/10/2026	06/09/2027	310	II
2028–2029	5.8351	23/11/2028	19/09/2029	300	II
2030–2031	6.1498	26/10/2030	05/10/2031	283	II
2031–2033	5.9373	18/04/2033	04/11/2033	200	I
2034–2036	5.6169	27/06/2035	15/01/2036	202	I
2037–2038	5.9571	13/08/2037	23/07/2038	344	II
2039–2040	5.6301	18/09/2039	22/08/2040	339	II

gravity assists, and thus a spacecraft would flyby multiple planets – sometimes multiple passes of the same planet too – before arriving at its destination. When considering more than two objects, the reoccurrence of a given orbital alignment is less frequent, and thus mission rescheduling is usually not permitted without major changes to the flyby sequence and/or deep space maneuvers. For example, a trajectory as complex as that of Voyager 2 – the so-called Grand Tour, which we discussed in Chapter 1 – would repeat approximately only every 177 years. In fact, computing the synodic period between Jupiter-Saturn, Saturn-Uranus, and Uranus-Neptune results in 19.9, 45.4, and 171.4 years, respectively. Multiplying the first two by integer values[8] of 9 and 4, respectively, we get Jupiter-Saturn and Saturn-Uranus synodic periods of 178.7 and 181.5 years, respectively. Taking an average of all of these values – 178.7, 181.5, and 171.4 years – we get 177.2 years. Although only an approximation, we see that a very similar alignment to that required by Voyager 2 for its Grand Tour will occur approximately 177[9] years after its original launch, or in the year 2154.

5.11 PROPULSIVE MANEUVERS

Spacecraft maneuver for many reasons that usually are part of planning during a mission's design phase. Maneuvers can be used to move a spacecraft from an initial insertion orbit to a final orbit, to correct for orbital drifts caused by small perturbations whose effects build up over time, or dispose of a spacecraft at the end of its life.

A spacecraft is not generally inserted directly into its final orbit when launched from Earth. Many times, it is placed in a lower orbit by the launch vehicle and an additional propulsion system that is attached to (or integrated into) the spacecraft is used to insert it into its operational orbit. Additionally, repositioning maneuvers are common for satellite constellations because constellations depend on other member

[8] We can do this because any *integer* multiple of a synodic period will result in the same orbital alignment due to the periodicity of the event.

[9] For simplicity, sometimes this number is quoted as 175 years.

satellites to perform a mission. Constellations execute two main types of repositioning maneuvers, called absolute station-keeping, used to maintain the overall orbital parameters of the entire constellation of spacecraft, and relative station-keeping, used to maintain certain orbital parameters or other physical constants with respect to the other spacecraft – such as relative distance between spacecraft.

Once in its operational orbit, a spacecraft's orbit may be modeled using Newtonian dynamics, as we described so far in this and previous chapters. However, imperfections in mismodeling of the parent body's gravitational field as well as perturbations, such as atmospheric drag and solar radiation pressure, cause a spacecraft to deviate from our mathematical predictions, and drift away from its original orbit. Small maneuvers, called station-keeping maneuvers, correct for this movement. Orbits of spacecraft at low altitudes tend to get lower as drag interactions occur between the spacecraft and the non-negligible atmosphere. For example, the International Space Station's altitude decays a few kilometers every week because of the Earth's atmosphere. Periodic boosts in altitude (maneuvers) correct for this drag effect. These maneuvers are usually performed using the help of spacecraft that dock with the ISS.

At the end of the spacecraft's mission, it is common to move spacecraft out of their mission orbits for various reasons. A geosynchronous satellite can be boosted to a higher graveyard orbit to make room for a replacement satellite, for example. A low-altitude satellite can be deorbited, too, to make room for its replacement and to avoid it from accidentally colliding with other spacecraft. One well-known example of deorbiting a satellite using a propulsion system was the reentry of the Russian Mir space station in 2001. At the time of writing, the ISS is too scheduled to be deorbited eventually, according to the plans NASA and its partners have for it. A common reason for deorbiting interplanetary spacecraft is to adhere to planetary protection policies and avoid accidental contamination. In fact, even if assembled and integrated in appropriate clean rooms, spacecraft are not perfectly sterile and could thus interfere with any form of extraterrestrial life that might exist in the Solar System.

For any mission, in the absence of perturbations, spacecraft must maneuver to change their orbits. A spacecraft may need a single large or small maneuver – or a series of maneuvers – to complete the mission. These maneuvers can be represented mathematically by examining two physical parameters: orbital energy and angular momentum.

Recall that the orbital energy is found using the *vis viva* equation (Equation 2.50). Noting that the term v^2 is the dot product of the inertial velocity, this equation becomes

$$\frac{(\mathbf{v}\cdot\mathbf{v})}{2}-\frac{\mu}{r}=-\frac{\mu}{2a} \tag{5.163}$$

Adding a vector velocity change, $\Delta\mathbf{v}$, instantaneously changes the semimajor axis (and keeps the instantaneous radius constant), such that

$$\frac{(v+\Delta\mathbf{v})\cdot(v+\Delta\mathbf{v})}{2}-\frac{\mu}{r}=-\frac{\mu}{2(a+\Delta a)} \tag{5.164}$$

Adding a maneuver also changes the angular momentum vector,

$$\mathbf{h} + \Delta\mathbf{h} = \mathbf{r} \times (\mathbf{v} + \Delta\mathbf{v}) \tag{5.165}$$

Subtracting the definition for angular momentum from Equation (2.41), the change in angular momentum is

$$\Delta\mathbf{h} = \mathbf{r} \times \Delta\mathbf{v} \tag{5.166}$$

which shows that the angular momentum changes only when the $\Delta\mathbf{v}$ added to the orbit has a component in a direction other than the radial direction. We will use these two relationships to examine various simple impulsive maneuvers.

5.11.1 Impulsive Maneuvers

One common assumption that is made in approximating an orbit transfer is that the maneuver is assumed to be impulsive. The linear momentum, **H**, can be written as the product of the mass and velocity,

$$\mathrm{H} = \mathrm{m}\mathbf{v} \tag{5.167}$$

and we can write Newton's law of motion in the form

$$\mathbf{F} = \frac{\mathrm{dH}}{\mathrm{dt}} \tag{5.168}$$

If we integrate Equation (5.168) over the time interval, t_1 to t_2,

$$\int_{t_1}^{t_2} \mathrm{F}\,\mathrm{dt} = \int_{t_1}^{t_2} \frac{\mathrm{dH}}{\mathrm{dt}}\,\mathrm{dt} = \mathrm{H}_2 - \mathrm{H}_1 \tag{5.169}$$

where $\mathbf{H}_1$ and $\mathbf{H}_2$ are the values of the linear momentum at times t_1 and t_2, respectively. The time integral of the force, **F**, is the impulse $\hat{\mathrm{F}}$ of the force,

$$\hat{\mathbf{F}} = \int_{t_1}^{t_2} \mathbf{F}\,\mathrm{dt} = \mathrm{m}\mathbf{v}_2 - \mathrm{m}\mathbf{v}_1 \tag{5.170}$$

The total impulse, $\hat{\mathrm{F}}$, occurring at $t = \tau$, can be expressed as

$$\mathbf{F} = \hat{\mathbf{F}}\,\delta(t - \tau) \tag{5.171}$$

where $\delta(t - \tau)$ is the Dirac delta and is defined as

$$\delta(t) = \begin{cases} 0 & t \neq 0 \\ \infty & t = 0 \end{cases} \tag{5.172}$$

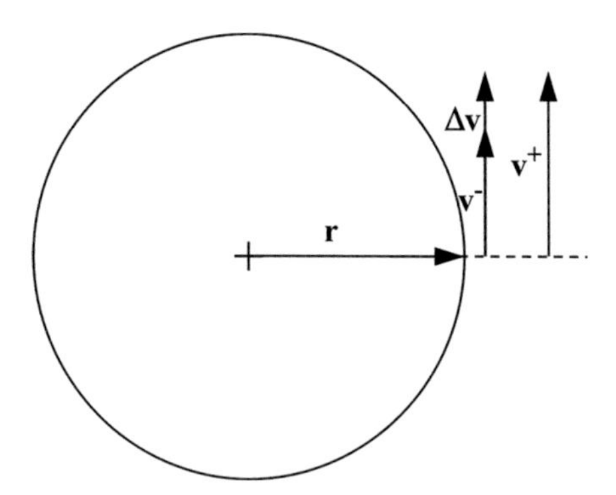

FIGURE 5.29 Tangential velocity – first maneuver in circle-to-circle transfer.

with

$$\int_{-\infty}^{\infty} \delta(t)dt = 1 \tag{5.173}$$

Applying this impulsive approximation, the maneuver instantaneously changes the velocity vector (direction and/or magnitude) from some initial value to some other final value. This type of maneuver changes only the velocity vector instantaneously while keeping the position constant. Mathematically, this can be represented as

$$\begin{aligned} \mathbf{r}^- &= \mathbf{r}^+ \\ \mathbf{v}^- + \Delta\mathbf{v} &= \mathbf{v}^+ \end{aligned} \tag{5.174}$$

where the – superscript represents the position or velocity vector before the maneuver is performed, and the + superscript is the position or velocity vector after the maneuver is performed. This geometry is shown in Figure 5.29.

In practice, a maneuver is made over a finite duration, although the approximation can be fairly accurate when considering that the burn duration is generally very small when compared to the entire transfer duration. When planning for an operational mission, the impulsive approximation can provide simplifications necessary to use analytical approximations. Further phases of the mission design implement finite-time maneuvers, which generally result in an increase in the Δv budget.

5.11.2 Trajectory Correction Maneuvers

When placing a spacecraft on an interplanetary trajectory, the transfer is modeled making assumptions, including perfect knowledge of the gravity field and perfect performance of the propulsion system.[10] Because there is no perfect knowledge

[10] This is true of the first phases of mission design. However, in later phases, statistical orbit determination – which is a topic that we will not cover here – is used to predict how a spacecraft will deviate from its nominal trajectory. Nonetheless, the statical parameters used for this analysis are still based on values, such as average and standard deviation, that could still be incorrect.

of either, the spacecraft will ultimately deviate from its modeled – or nominal – trajectory. While these uncertainties are within the design tolerances, when the trajectory is modeled over the long duration of an interplanetary coast, large deviations from the desired final state can occur. In order to compensate for this mismodeling, small mid-course corrections are performed. These corrections are calculated in similar ways to a rendezvous or station-keeping maneuvers are computed.

In determining a mid-course correction, thrusting segments are generally assumed to be impulsive. Since these maneuvers are on the order of a few meters per second and their duration can be relatively short, the impulsive approximation is reasonable (by its very nature, there could always be a mid-course correction to a mid-course correction). Mid-course corrections, often referred to as trajectory correction maneuvers (TCMs), can also be a maneuver in any of three spatial directions.

In order to plan for a mid-course correction maneuver, a mission planner can perform a sensitivity analysis on the effects of mismodeling either the gravity field or the impulsive approximation assumption. By performing a Monte Carlo analysis on these deviations, the effects of these deviations on some future state vector can be assessed. Based on the result of where the spacecraft would end up when a future mid-course correction could be performed, a new trajectory can be computed that would put the spacecraft back on course. By performing a large number of these simulations, statistical methods can then be applied to determine what the size of correction maneuver should be. This process could also be repeated for an unknown time of maneuver. In practice, the location of the TCM can be found by examination of the errors in the tracking solution (more information on navigation can be found in Chapter 6).

An example of Monte Carlo simulations for orbit insertion is provided in Figure 5.30, where an orbit insertion at the lunar halo orbit example shown in Section 3.6.5 is done at the highest z-coordinate location after a transfer from LEO was performed.

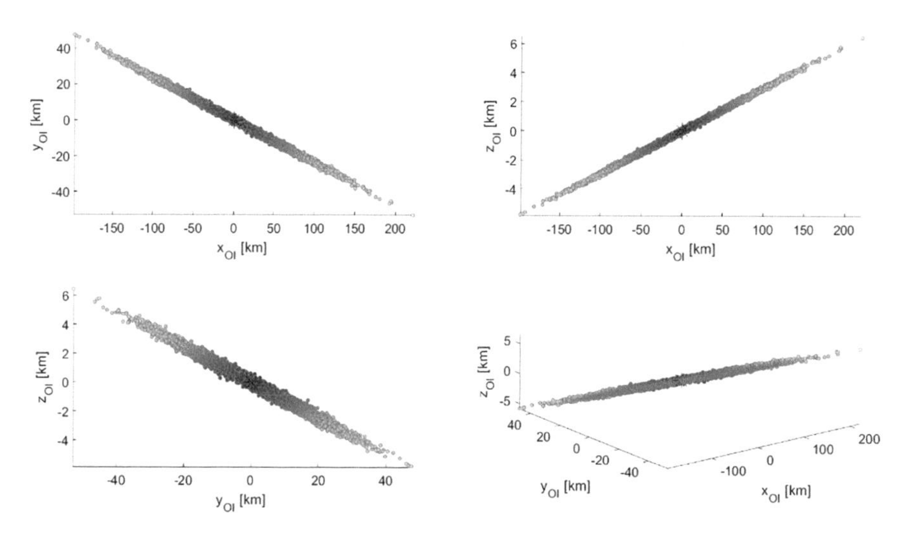

FIGURE 5.30 Monte Carlo simulation for a lunar halo orbit rendezvous with $A_z = 20{,}000$ km [Conte, 2019].

TABLE 5.14
Mars Pathfinder trajectory correction maneuver schedule

TCM	Time	Calendar Date	Mean Velocity Magnitude (m/s)	Comments
TCM 1	Launch + 37 days	10/01/1997	33.3	Remove injection bias, correct injection errors (delayed from January 4, 1997)
TCM 2	Launch + 60 days	04/02/1997	2.08	Correct TCM 1 errors
TCM 3	Arrival–60 days	07/05/1997	0.432	Target final Mars atmospheric entry point
TCM 4	Arrival–10 days	24/06/1997	0.138	Correct TCM 3 errors
TCM 5	Arrival–5 hours	04/07/1997	2.0	Correct any remaining errors

In the figure, the large red asterisk at the origin represents the nominal orbit insertion location at the halo orbit while the shade of each point is darker for closer points and brighter for further points. A standard deviation of 0.1 m/s was used to simulate 10,000 maneuvers departing from LEO, resulting in an average drift with respect to the nominal insertion point of approximately 43 km – and as high as 230 km. In order to put the spacecraft back on course, Δv and time are needed – here on the order of tens of m/s of Δ and a up to about a day, depending on the situation.

An example of how vital a mid-course correction can be illustrated by examining the basic Earth-Mars Hohmann transfer from the Mars Pathfinder mission.[11] This mission planned for five TCMs, as shown in detail in Table 5.14. The first trajectory correction maneuver was made 37 days after launch to correct launch injection errors. The second maneuver was performed 23 days later and was designed as a correction on the first TCM. The remaining TCMs were conducted to correct the inbound trajectory as the spacecraft was approaching Mars [Vaughan et al., 2004].

5.11.3 Orbit Trim Maneuver

After a spacecraft has entered orbit, the resulting orbit is unlikely to be exactly the desired orbit. To bring the orbit closer to the desired operational orbit, one or more orbit trim maneuver is executed. These maneuvers are essentially station-keeping maneuvers. The mathematics of these maneuvers is similar (although generally smaller Δv magnitudes) than the orbital correction maneuvers used to improve targeting. However, the methods used to plan these are dependent on the desired final orbit. Sometimes, it is possible to use other natural phenomena to help a spacecraft achieve its final desired orbit, such as changing the orientation of an orbit by making use of the destination body's J_2 effects (depending on the body's oblateness) or decrease orbital energy using the targeted body's atmosphere via aerobraking – we will discuss this last one in more detail in Section 5.14.

[11] https://mars.nasa.gov/MPF/mpf/mpfnavpr.html

A pristine example of orbit trim maneuvers and accuracy achieved in rocket firings is given by the Cassini-Huygens mission. During its 7-year journey to Saturn, Cassini performed 17 orbit trim trajectory correction maneuvers. The duration of a firing was executed within about 0.1% of the planned duration, and the pointing direction was executed within about 7 milliradians (0.4°).[12]

5.12 GRAVITY ASSISTS

Another consideration for interplanetary orbits is using gravity assists (GAs) or flybys – using a planet's gravity to change its orbit by transferring energy from the planet's orbit around the Sun to the spacecraft's orbit around the Sun. Interplanetary spacecraft have used gravity assists to reach Jupiter, Saturn, and beyond. While a gravity assist reduces the need for propellant and, in many cases, makes a mission possible, it also adds travel time to the mission and limits the launch window until the planets align properly. The mechanics of how gravity assists work is similar to Phase 3 of the patched conic analysis (Section 5.5.3.3) with the exception that a maneuver is not applied at the periapse flyby point – the spacecraft continues on its hyperbolic orbit. Upon exiting the flyby planet's sphere of influence, the direction of the velocity vector has been changed. It is also possible to combine flyby maneuvers with a maneuver to add energy to the spacecraft's post-flyby orbit in case the gravity assist alone cannot provide sufficient change. This type of maneuver is called a powered gravity assist – also known as synergic maneuver. An interesting historical review of gravity assists can be found in Broucke [2001].

In addition to single gravity assists, several gravity assists can be pieced together to form a multiple gravity assist transfer. Many of the more recent spacecraft that went to outer planets took advantage of this multiple gravity assist sequence, such as Galileo and Cassini, for example. Multiple gravity assists have also been used for transits to inner planets, like in the case of MESSENGER and BepiColombo, for example. Putting together such a mission is a complex undertaking.

5.12.1 Geometry

As already explained, the geometry of a gravity assist is similar to that of Phase 3 of a heliocentric transfer. However, since the spacecraft does not perform a maneuver, it will leave the sphere of influence of the arrival planet with the same magnitude of v_∞, but different exit velocity vector, as shown in Figure 5.31. In other words,

$$\begin{gathered} v_\infty^- = v_\infty^+ \\ \mathbf{v}_\infty^- \neq \mathbf{v}_\infty^+ \end{gathered} \tag{5.175}$$

where the superscript – indicates the incoming parameters and the + indicates the post-flyby parameters. Here, we dropped the subscripts relating to the "with respect to..."

[12] "Basics of Spaceflight, Section II, Chapter 13, Spacecraft Navigation", http://www2.jpl.nasa.gov/basics/bsf13-1.html

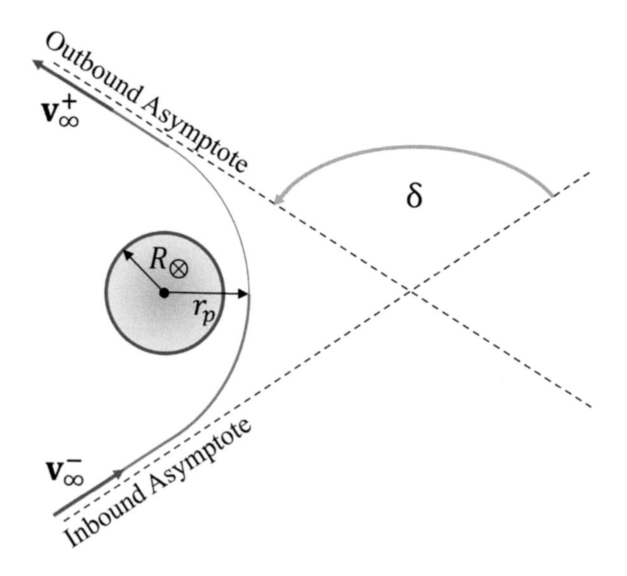

FIGURE 5.31 General geometry of a planetary gravity assist.

for simplicity since all v_∞ values are understood to be with respect to the planet the spacecraft performs the flyby.

The magnitude of the inbound and outbound velocities is the same, but their direction is different, which is what causes the spacecraft's orbit to change after the gravity assist. The amount of directional change provided by the gravity assist while in the sphere of influence of the flyby planet is known as turn angle, and is denoted by δ, and is the angle between the pre-flyby and post-flyby $\mathbf{v}_\infty$'s, which can be computed as

$$\delta = 2\theta_\infty - \pi \tag{5.176}$$

where all angles must be in radians and θ_∞ is the asymptotic true anomaly of the flyby trajectory, which is computed as

$$\theta_\infty = \cos^{-1}\left(-\frac{1}{e}\right) \tag{5.177}$$

where e is the eccentricity of the hyperbolic flyby orbit.

An alternative formulation that is used to compute turn angle is

$$\sin\left(\frac{\delta}{2}\right) = \frac{1}{1+\Psi} \tag{5.178}$$

where Ψ is a non-dimensional parameter given by

$$\Psi = \left(\frac{v_\infty}{v_s}\right)^2\left(\frac{r_p}{R_\otimes}\right) = \frac{v_\infty^2}{\mu_\otimes} r_p \tag{5.179}$$

where $R_{\otimes}$ is the physical radius of the planet and v_s is the speed an object would require to be in a circular orbit around said planet at its surface, i.e.

$$v_s = \sqrt{\frac{\mu_{\otimes}}{R_{\otimes}}} \tag{5.180}$$

The value of v_s is only used as a mathematical variable for these calculations since it is virtually impossible to safely orbit any celestial body in the Solar System in a perfectly circular orbit at their surface – also because no celestial body is perfectly spherical, and some possess atmospheres.

If we consider the spacecraft's heliocentric trajectory before and after the flyby, we then must also consider the side on which the spacecraft performs the flyby: a trailing edge gravity assist (Figure 5.32) or a leading edge gravity assist (Figure 5.33). Depending on the type of gravity assist, the resulting post-flyby heliocentric orbit can have rather different characteristics. These figures show a similar geometry of a gravity assist of Figure 5.31 including the velocity of the targeted planet ⊗ with respect to the Sun, $\mathbf{v}_{\otimes/\odot}$, and the velocity of the spacecraft with respect to the Sun before ($\mathbf{v}^{-}_{sc/\odot}$) and after ($\mathbf{v}^{+}_{sc/\odot}$) the flyby. In these figures, the velocity vectors associated with the planet with respect to the Sun, the spacecraft's heliocentric orbits, the spacecraft's planetary flyby, and the Δv resulting from the gravity assist are in green, blue, red, black, respectively.

Thus, the gravity assist provides a "free" change in pre- and post-flyby velocity given by

$$\Delta\mathbf{v}_{GA} = \mathbf{v}^{+}_{sc/\odot} - \mathbf{v}^{-}_{sc/\odot} \tag{5.181}$$

or alternatively

$$\Delta\mathbf{v}_{GA} = \mathbf{v}^{+}_{\infty/\otimes} - \mathbf{v}^{-}_{\infty/\otimes} \tag{5.182}$$

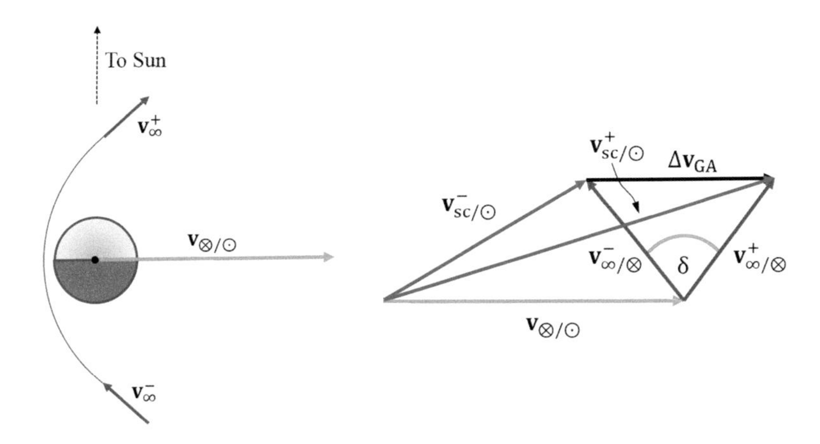

FIGURE 5.32 Geometry of a trailing edge planetary gravity assist including pre- and post-flyby velocity vectors.

It should be noted that Δv_{GA} may appear to be a "free" Δv but the reason why a spacecraft can do a gravity assist and increase (or decrease) its orbital energy with respect to the Sun is explained by conservation of linear momentum. While the speeds of a spacecraft and a planet may have comparable values, a spacecraft has insignificantly small mass when compared to a planet. Therefore when performing a gravity assist the spacecraft is receiving (or giving) some energy from (or to) the planet which results in an extremely drastic change in the spacecraft's velocity and an insignificantly small velocity change to a planet.

For any planetary body the value of Δv_{GA} can be maximized and interestingly the resulting δ and eccentricity of the flyby trajectory are independent of the physical characteristics of the flyby planet (this proof is left as an exercise). It should be noted that spacecraft can use gravity assists to both increase or decrease their heliocentric orbital energy depending on their destination. Nonetheless, the energy of the spacecraft with respect to the planet remains unchanged while it is inside its sphere of influence.

While we have been referring to gravity assists done on planets, the same analysis can be done for moons of planets. In fact, missions such as Cassini have performed a variety of gravity assists of the moons they explored, thus continuously changing their orbits with respect to their parent planet.

5.12.1.1 Two-Dimensional Gravity Assists

In the previous section, we presented the general geometry of gravity assists. Here, we want to focus on how a heliocentric trajectory changes after a gravity assist is performed. For simplicity, here we will assume that all maneuvers take place in two dimensions. Following the geometry presented in the previous section (see Figures 5.31–5.33), we see that the spacecraft's $\mathbf{v}_\infty$'s are related to each other by a simple rotation

$$\mathbf{v}^{+}_{\infty/\otimes} = \begin{bmatrix} \cos\delta & -\sin\delta \\ \sin\delta & \cos\delta \end{bmatrix} \mathbf{v}^{-}_{\infty/\otimes} \tag{5.183}$$

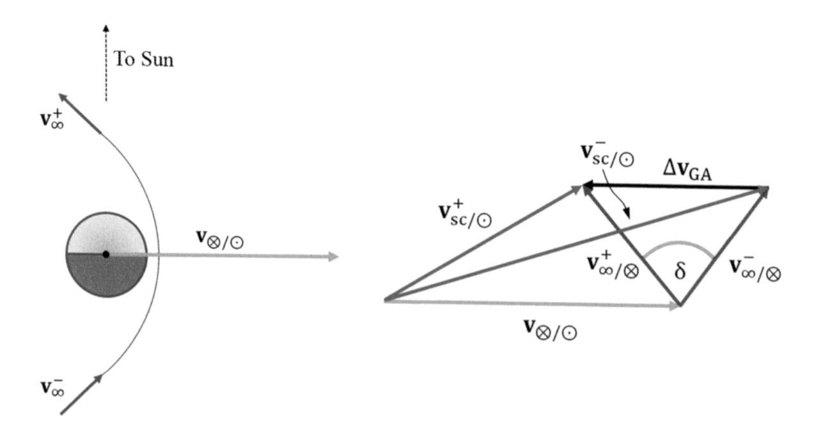

FIGURE 5.33 Geometry of a leading edge planetary gravity assist including pre- and post-flyby velocity vectors.

We also know that

$$\mathbf{v}^{-}_{\infty/\otimes} = \mathbf{v}^{-}_{sc/\odot} - \mathbf{v}_{\otimes/\odot}$$
$$\mathbf{v}^{+}_{\infty/\otimes} = \mathbf{v}^{+}_{sc/\odot} - \mathbf{v}_{\otimes/\odot} \tag{5.184}$$

Thus, inserting Equation (5.183) into the second formula of Equation (5.184), we obtain

$$\mathbf{v}^{+}_{sc/\odot} = \begin{bmatrix} \cos\delta & -\sin\delta \\ \sin\delta & \cos\delta \end{bmatrix} \mathbf{v}^{-}_{\infty/\otimes} + \mathbf{v}_{\otimes/\odot} \tag{5.185}$$

In the right-hand side of Equation (5.185), all quantities are known provided that information regarding the flyby is known, namely the targeted periapse radius, r_p and $\mathbf{v}_\infty$. Then, turn angle δ can be computed with Equations (5.178) and (5.179) to obtain the matrix in Equation (5.185). The velocity of the planet with respect to the Sun is also assumed to be known, and can be found through the use of various methods, as explained in Sections 4.7 and 4.8.

If needed, the eccentricity of the flyby trajectory can also be computed by combining Equations (5.176) and (5.177), which results in

$$e = 1 + \Psi \tag{5.186}$$

Additionally, the impact parameter Δ (which we introduced in Section 4.2.7) can be computed as

$$\Delta = r_p \sqrt{1 + \frac{2}{\Psi}} \tag{5.187}$$

Let's consider an example in which we perform a Hohmann transfer from Earth to Venus (assuming that their orbits are circular and coplanar). However, instead of performing an orbit insertion at arrival at Venus, we want to see how the heliocentric trajectory changes due to a trailing edge Venus gravity assist and how much Δv_{GA} the spacecraft receives. Earth's and Venus' velocities with respect to the Sun can be easily computed as 29.78 and 35.02 km/s since we are assuming circular orbits. Solving for the Earth departure and Venus arrival v_∞'s gives values of 2.48 and 2.71 km/s, respectively. If the spacecraft targets a flyby trajectory having periapse altitude of 600 km, we can then compute its eccentricity using Equation (5.186), resulting in e = 1.15. Turn angle is computed using Equation (5.178) and is approximately 120.6°.

If broken down into radial and tangential components and accounting for the fact that we are performing a Hohmann transfer to Venus, Equation (5.185) can be simplified as

$$v^{+}_{\theta} = v_{\otimes/\odot} + v^{+}_{\infty/\otimes} \cos\delta$$
$$v^{+}_{r} = v^{+}_{\infty/\otimes} \sin\delta \tag{5.188}$$

which results in v_r^+ and v_θ^+ of 2.29 and 33.56 km/s, respectively. Note that Equation (5.188) assumes the use of a Hohmann transfer, i.e., that the spacecraft's inbound transverse velocity component is colinear with the planet's velocity vector with respect to the Sun. While this is generally approximately true for most planetary gravity assists, it is not *always* the case.

Therefore, the spacecraft's post-flyby heliocentric trajectory has a speed of $v^+ = \sqrt{\left(v_\theta^+\right)^2 + \left(v_r^+\right)^2}$, or 33.64 km/s, a flight-path angle of $\tan^{-1}\left(v_r^+/v_\theta^+\right)$, or approximately 3.9°. Additionally, using the orbit equation, it can be computed that this new orbit has a true anomaly of approximately 142° and an orbital period of 201 days – compared to the 180° and 293 days of the pre-flyby trajectory, respectively. Figure 5.34 shows the pre- and post-gravity assist trajectories on the same heliocentric plot.

Mission designers can alter the spacecraft's heliocentric orbit quite significantly by simply targeting a different periapsis of the flyby trajectory. In fact, Figure 5.35 shows the effects of increasing the Venusian flyby periapsis on the heliocentric orbital period, amount of Δv_{GA} provided, and the flyby eccentricity and turn angle. Note that, for example, it is possible to choose a flyby periapse radius that results in a heliocentric orbital period equal to that of the pre-flyby heliocentric trajectory or equal (or close to) that of the flyby planet – here, this value is 2.84 Venus's radii or approximately 17,200 km. For the latter, this means that, exactly one orbital period later, both Venus and the spacecraft will rendezvous again, allowing the spacecraft

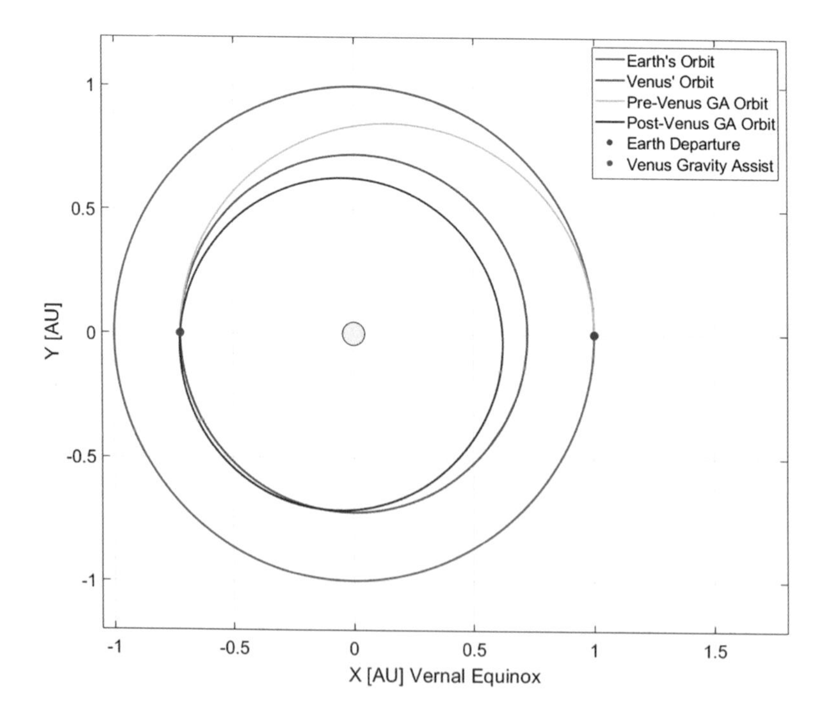

FIGURE 5.34 Venus gravity assist example – trailing edge flyby.

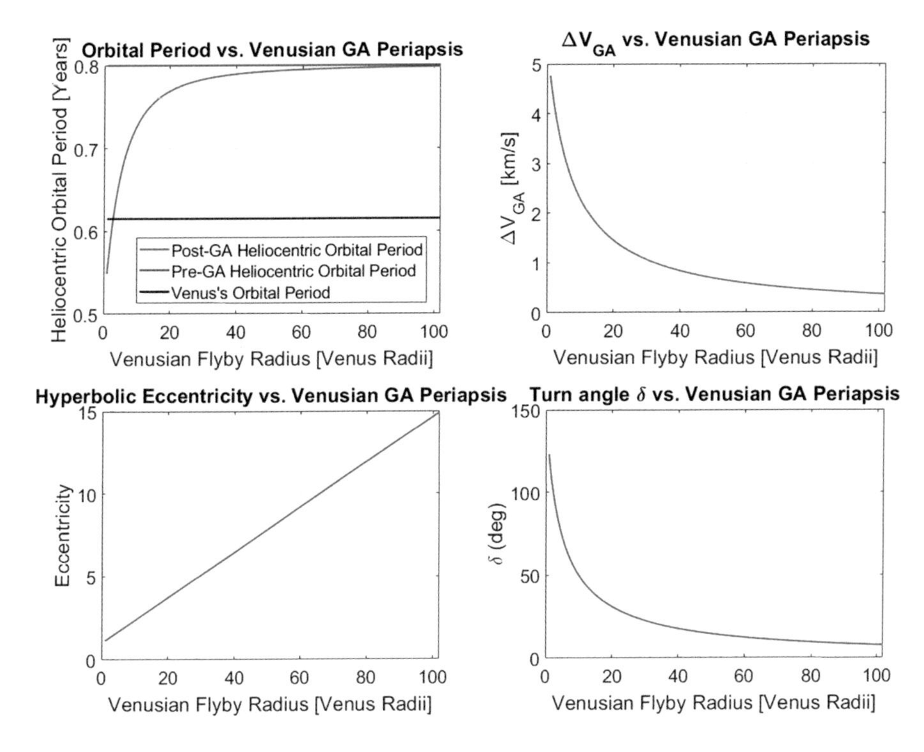

FIGURE 5.35 Effects of a Venusian gravity assist on various orbital parameters as a function of flyby radius.

to perform another gravity assist of the planet. Re-running this Earth-Venus example with a flyby altitude of approximately 17,200 km results in a post-gravity assist orbital period equal to that of Venus. However, the orbit is *not* equal to Venus's orbit, although it has the same semimajor axis, as shown in Figure 5.36.

If the mission designers wish to complete multiple gravity assists of the same planet at the same location, the spacecraft and the planet must enter a so-called orbital resonance that makes the timeline practical. For example, an orbital resonance of 2:3 means that every two orbits of one body, the other completes three orbits. In fact, this type of maneuver has been used several times in past and current missions. For example, ESA's BepiColombo was designed to make use of multiple flybys of Earth, Venus, and Mercury to lower its heliocentric energy. The spacecraft is scheduled to perform multiple flybys of Mercury at various levels of orbital resonance, such as 3:2, 4:3, 5:4, and eventually 1:1. In practice, space missions also make use of DSMs or TCMs in order to appropriate adjust their targeting.

Usually, considering gravity assists maneuvers implies that a feasible sequence of flybys must be found. A sequence of flybys is provided as a sequence of capital letters each of which denotes a planet (or moon) at which the spacecraft performs a flyby. For example, BepiColombo's gravity assist sequence is written as EVVMMMM-M, where "-M" denotes the arrival at Mercury. While "M" could be confused for Mercury, Mars, or the Moon, the sequence is usually understood among mission

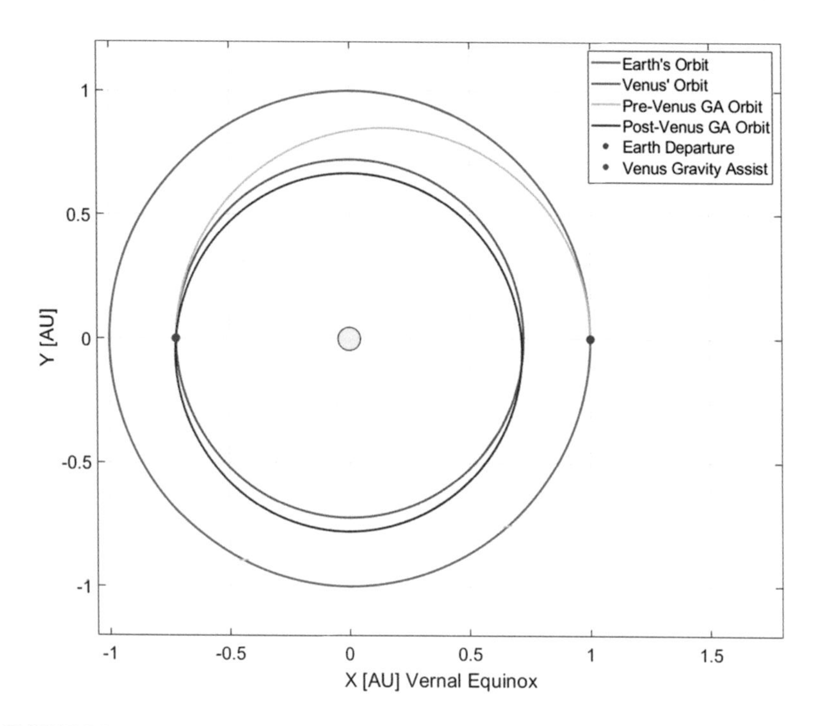

FIGURE 5.36 Venus gravity assist example with post-flyby heliocentric orbital period equal to that of Venus.

designers based on context. Additionally, the Earth departure is not considered, so the "E" in the aforementioned sequence refers to the first flyby of the spacecraft, which happens to be at Earth.

5.12.1.2 Three-Dimensional Gravity Assists

Although in the previous sections we discussed and analyzed gravity assists as two-dimensional maneuvers, they happen in three dimensions due to the fact that generally the heliocentric and planet-centric trajectories are not on the same plane, both before and after the flyby. Generally, since the planets are located on similar planes, sequences of flybys are performed in a way that each mission segment's orbital plane stays within a few degrees of each other. An example of such a sequence can be seen in Figure 1.9, where Cassini's flyby sequence was shown, including the flyby dates. Figure 1.9 shows the projection of Cassini's trajectory onto the Ecliptic plane because the out-of-plane components of each orbital segment are insignificant when compared to the other components. This, however, is not always the case, as shown in Voyager 1's interplanetary (and later interstellar) trajectory – shown in Figure 5.37 along with that of other interstellar probes, including Voyager 2, Pioneer 10, and Pioneer 11.[13] In fact, Voyager 1 performed a flyby of Saturn passing near its south pole in a way that its post-flyby heliocentric orbit had a significantly increased inclination, making Voyager 1 essentially go "out" of the Solar System's plane.

[13] New Horizons is also headed for interstellar space, although it's not shown in the figure.

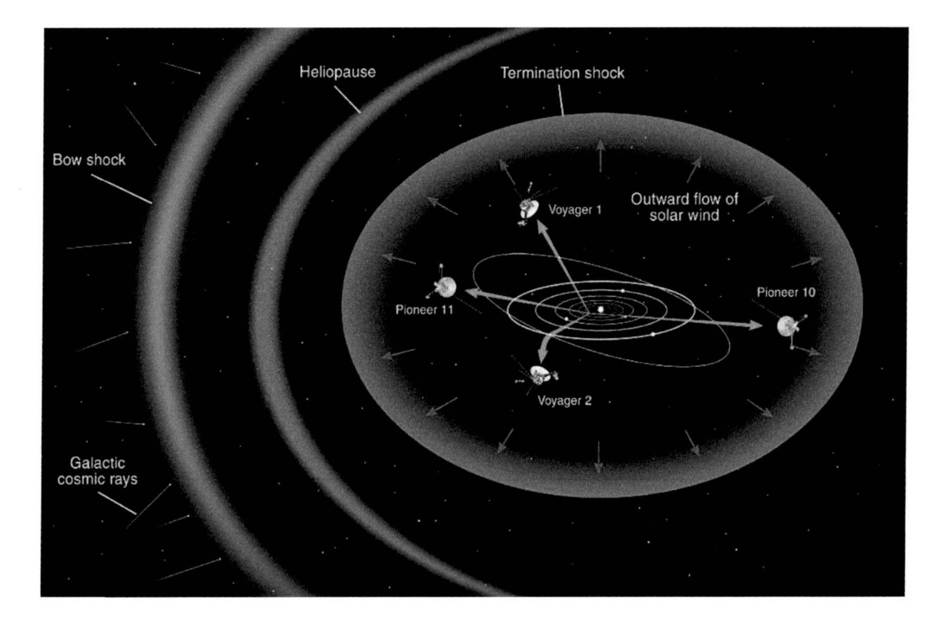

FIGURE 5.37 Trajectories of Pioneer 10, Pioneer 11, Voyager 1, and Voyager 2. (Courtesy of NASA-JPL/Caltech.)

5.13 FREE-RETURN TRAJECTORIES

A free-return trajectory is an orbital transfer from a primary celestial body, such as the Earth, to a secondary celestial body, such as the Moon, that allows the spacecraft to return to the primary body without the use of any propulsive maneuvers. This means that the spacecraft performs a flyby of the secondary body and then returns to the primary body. Figure 5.38 shows an example of the free-return trajectory that Apollo 8 would have used if the spacecraft had suffered complications on its way to the Moon.

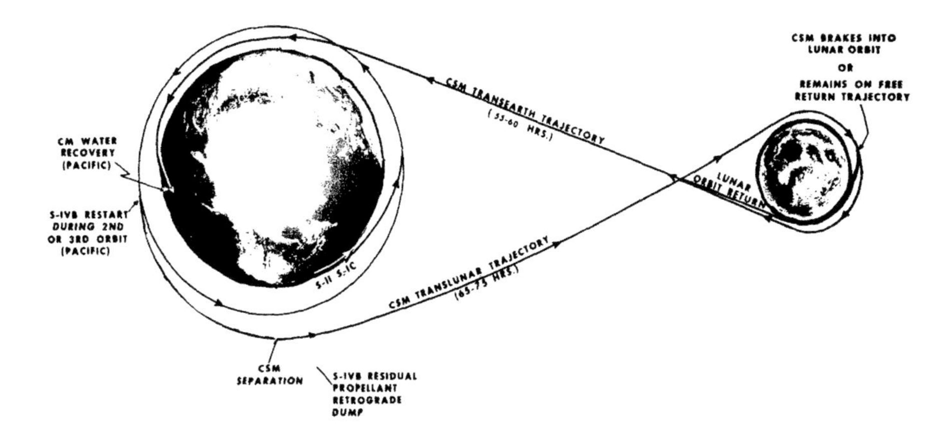

FIGURE 5.38 Free-return trajectory. (Courtesy of NASA.)

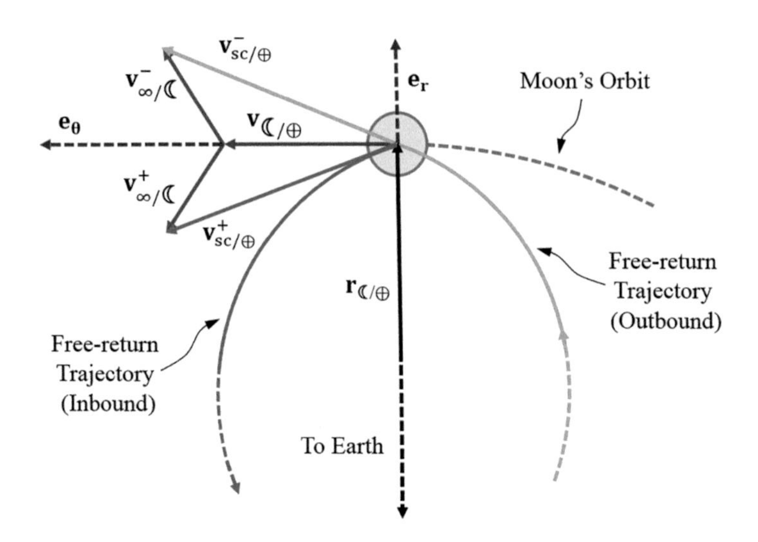

FIGURE 5.39 Geometry of a free-return trajectory for Earth-Moon transfers.

Figure 5.39 shows a more detailed geometry of a free-return trajectory for an Earth-Moon example.

Here, the outbound orbit (in green) is chosen in such a way that the spacecraft performs a lunar flyby resulting in a post-flyby orbit (in blue) that has the same $\mathbf{e}_\theta$ velocity component, but opposite $\mathbf{e}_r$ velocity component, resulting in a "free" return to Earth without the use of any Δv.

Thus, for an orbital transfer to be a free-return trajectory, the pre-flyby and post-flyby radial and transversal geocentric velocity components must satisfy the following relationship

$$\begin{aligned} v_r^+ &= -v_r^- \\ v_\theta^+ &= v_\theta^- \end{aligned} \tag{5.189}$$

where the "r" and "θ" subscripts represent the radial and transverse components, respectively.

It is important to note that a Hohmann transfer generally does not result in a free-return trajectory since the above conditions can only be satisfied for orbital transfers with apoapsis that goes beyond the targeted orbit. Hence, Lambert's problem must be first solved, and then patched conics must be implemented to ensure that the conditions in Equation (5.189) are met. To do so, we must first compute the velocity of the secondary body with respect to the primary body. In case of the Earth and the Moon, the velocity of the Moon with respect to the Earth, $\mathbf{v}_{☾/\oplus}$, has a magnitude of approximately 1.022 km/s. This can be computed by either assuming that the secondary body orbits the primary in a circular orbit – if appropriate, such as in the case of the Earth and the Moon – or using prescribed ephemeris data (see Sections 4.5–4.8).

For the analysis presented here, we will use the Earth and the Moon as primary and secondary bodies, respectively, and we will assume that their orbits are circular.

For the next step, we need to know the departure orbit's radius at Earth, $r_{p/\oplus}$. From the solution of Lambert's problem, we compute the pre-flyby transversal velocity component, v_θ^- at arrival. Next, we use the value of v_r^- – still coming from the same solution to Lambert's problem – such that the Earth-centered pre-flyby spacecraft velocity is

$$v_{sc/\oplus}^- = \sqrt{\left(v_\theta^-\right)^2 + \left(v_r^-\right)^2} \tag{5.190}$$

which means that the spacecraft has a pre-flyby velocity with respect to the Moon of

$$v_{\infty/☾}^- = \sqrt{\left(v_{☾/\oplus} - v_\theta^-\right)^2 + \left(v_r^-\right)^2} \tag{5.191}$$

where ☾ is the symbol used to denote the Moon. Then, the semimajor axis of the trajectory can be computed as

$$a_{FR} = \frac{r_{☾/\oplus}\mu_\oplus}{2\mu_\oplus - r_{☾/\oplus}v_{\infty/☾}^-} \tag{5.192}$$

where the subscript "FR" stands for "free-return" and $r_{☾/\oplus}$ is the orbital distance between the Earth and the Moon, or 384,400 km if assuming circular orbits. Therefore, the eccentricity of the free-return trajectory is

$$e_{FR} = 1 - \frac{r_{p/\oplus}}{a_{FR}} \tag{5.193}$$

Now, we can compute the turn angle of the lunar flyby using the geometry in Figure 5.39 – refer to Section 5.12 for additional details on gravity assists – so that

$$\sin\left(\frac{\delta}{2}\right) = \frac{v_r^-}{v_{\infty/☾}^-} \tag{5.194}$$

We can then combine Equations (5.178) and (5.179) to compute the required flyby perilune, $r_{p/☾}$ as

$$r_{p/☾} = \frac{\mu_☾}{\left(v_\infty^-\right)^2}\left(\frac{1}{\sin\left(\frac{\delta}{2}\right)} - 1\right) \tag{5.195}$$

Then, we can compute the eccentricity of the spacecraft with respect to the Moon during its flyby maneuver as

$$e_{sc/☾} = \frac{1}{\sin\left(\frac{\delta}{2}\right)} \tag{5.196}$$

and its semimajor axis as

$$a_{sc/☾} = \frac{r_{p/☾}}{1 - e_{sc/☾}} \tag{5.197}$$

Since the conditions of Equation (5.189) will generally not be met by a generic Earth-Moon transfer, we need to iterate on the cislunar trajectory that connects the spacecraft from departure to its lunar flyby. In other words, we need to solve Lambert's problem using different parameters until the free-return trajectory conditions are met. Based on the mission requirements, we can choose said parameters for these iterations to be one or more of the following parameters while leaving the rest fixed:

1. Departure true anomaly: this is usually tied to launch time and location requirements.
2. Time of flight (TOF): while robotic missions can afford longer times of flight, missions involving a crew generally tend to minimize time of flight.
3. Flyby altitude, which is limited by the physical size of the secondary body.

Mission designers usually use a combination of the above variables – and potentially others – in order to determine the free-return trajectory that best fits their mission. Figure 5.40 shows qualitatively how various Earth-Moon trajectories would look like for a fixed departure location – and having common line of apsides (LOA) – but

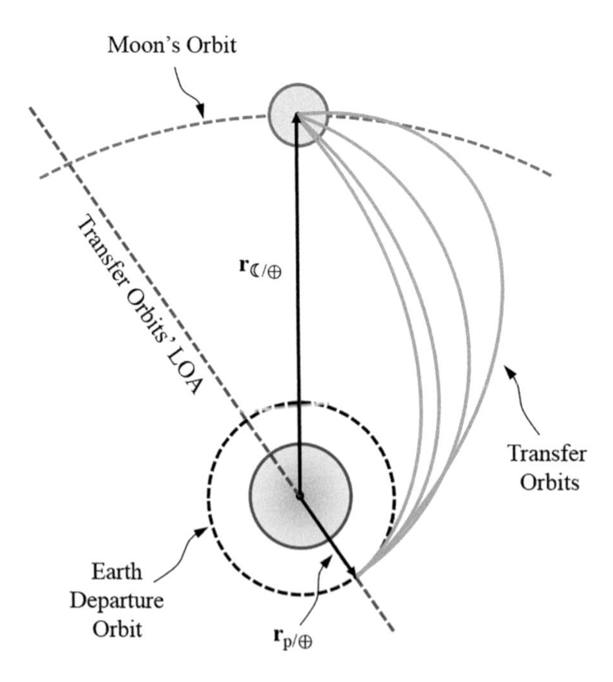

FIGURE 5.40 Earth-Moon trajectories for fixed departure location but varying time of flight.

varying the time of flight (assuming that the Moon is appropriately aligned for each different transfer). Note that of these orbits, not all will result in a free-return trajectory.

If we did not set a prescribed TOF for the free-return trajectory, then we can compute the geocentric eccentric anomaly, E_{FR2}, at which the flyby occurs as

$$e_{FR} \cos E_{FR2} = 1 - \frac{r_{☾/\oplus}}{a_{FR}} \tag{5.198}$$

keeping in mind that there is no need to perform a quadrant correction for E_{FR} since it is always in the outbound direction. We can use this, in combination with the departure eccentric anomaly, E_{FR1} – we can compute this from the Earth departure true anomaly using Equation (2.91) – to compute the TOF of the free-return trajectory with Kepler's time equation as

$$TOF = \sqrt{\frac{a_{FR}^3}{\mu_\oplus}}\left[E_{FR2} - E_{FR1} + -e_{FR}\left(\cos E_{FR2} - \cos E_{FR1}\right)\right] \tag{5.199}$$

The Apollo program made use of free-return trajectories for the first few missions, namely Apollo 8, 10, and 11. This was done so that in case of main engine failure, the reaction control system (RCS) engines could be used to fine-tune the approach for a safe return to Earth. For the missions after Apollo 12, a mid-course correction changed this approach to an orbit that did not lead to a free-return trajectory. This would enable the Apollo spacecraft to approach the Moon in such a manner so as to provide better landing options within the constraints of their propellant supply. However, Apollo 13 suffered a malfunction that was caused by an explosion and rupture of one of the oxygen tanks in the service module. Because of this, the spacecraft had to perform a series of maneuvers that put it onto a return trajectory back to Earth, as shown in Figure 5.41.

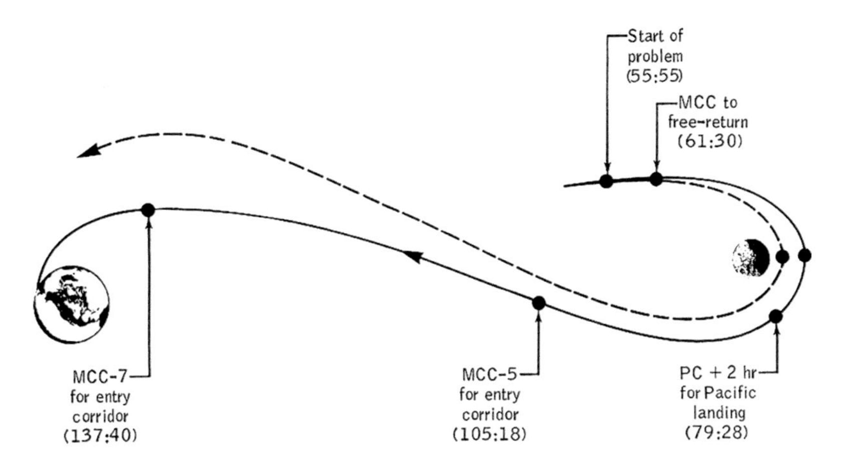

FIGURE 5.41 Apollo 13's trajectory as seen from an Earth-Moon synodic rotating reference frame. (Courtesy of NASA.)

Other kinds of Earth-Moon free-return trajectories exist. They can be classified based on two parameters: near (N) or far (F) side lunar passage and prograde (P) or retrograde (R) orbital transfer, as seen from the perspective of an Earth-Moon synodic reference frame. Each combination leads to a different kind of free-return trajectory: far side prograde (a), far side retrograde (b), near side prograde (c), far side retrograde (d).

Exploration Mission 1 (EM-1) also known as Artemis 1[14] also used a free-return trajectory similar to the one shown in Figure 5.38. Artemis 1 was the first in a series of increasingly complex missions to the Moon that aimed at enabling human exploration of both the Moon, and eventually Mars, including the assembly and usage of the Lunar Gateway.

Free-return trajectories are not limited to celestial bodies in orbit around their barycenter, like the Earth and Moon. Free-return trajectories can be computed and have been proposed for interplanetary missions, such as missions to Mars [Drake et al., 2010]. In fact, a free-return transfer orbit from Earth to Mars could be considered for crewed missions. For example, an interplanetary Earth-Mars Hohmann transfer orbit can be modified to be made into a free-return trajectory. This means that an approximately 260-day (0.71 years) transfer to Mars would result into a free-return from Mars to Earth in approximately 1.5 years. Of course, faster transfers can be found albeit at the cost of Δv. Alternative free-return options for Earth-Mars missions exist. In fact, there are "free-return" orbits that take exactly 2 or 3 years to return to Earth without the use Mars's gravitational assists – thus not making such orbits into free-return trajectories in the sense that we have discussed here, using flybys. These are heliocentric transfer orbits between Earth and Mars that have periods of 1 or 1.5 years, respectively. If aborted, a mission would need to wait two orbits in order to reencounter the Earth in the same orbital location where they originally departed.

5.14 ATMOSPHERIC INTERACTIONS

The majority of celestial bodies in the Solar System does not have an atmosphere. However, some of the most scientifically interesting celestial bodies, such as Mars, Venus, and Titan, have atmospheres that can make missions targeting their surfaces more challenging. Furthermore, maintaining an orbit that is affected by an atmosphere results in additional use of propellant. For example, the altitude of the ISS needs to be increased every few months or the space station would slowly reenter into the Earth's atmosphere.

As we describe in the following sections, while atmospheric interactions pose a risk to spacecraft arriving from interplanetary space – and thus at relatively high speeds – some of these phenomena can be used to facilitate certain aspects of a mission, such as reduce a spacecraft orbital energy in order to be captured by the celestial body. This is the case for aerocapture and aerobraking maneuvers. We will also analyze entry, descent, and landing (EDL) maneuvers, and will consider the heating a spacecraft would experience, and the use of unpowered vs. powered descents.

[14] https://www.nasa.gov/artemis-1

Assuming that gravity ($\mathbf{F}_{grav}$) and atmospheric drag ($\mathbf{F}_{drag}$) are the only forces acting on the spacecraft, the equations of motion become

$$\ddot{\mathbf{r}} = -\frac{\mu \mathbf{r}}{r^3} - \frac{1}{2}\left(\frac{C_D A}{m}\right)\rho v_{rel}\mathbf{v}_{rel} \tag{5.200}$$

where $\mathbf{r}$ is the position of the spacecraft with respect to the celestial body, μ is the gravitational parameter of the celestial body, C_D is the drag coefficient, A is the projected spacecraft area in the direction of motion, m is the mass of the spacecraft, and ρ is the atmospheric density. The reciprocal of the terms in parenthesis is the so-called spacecraft ballistic coefficient BC, i.e., $BC = m/C_D A$, since these are all parameters that depend on the size, mass, and shape of the spacecraft. $\mathbf{v}_{rel}$ is the difference between the velocity of the spacecraft and the velocity of the atmosphere, $\mathbf{v}_{atm}$, or

$$\mathbf{v}_{rel} = \mathbf{v} - \mathbf{v}_{atm} = \mathbf{v} - \boldsymbol{\omega}_B \times \mathbf{r} \tag{5.201}$$

where $\boldsymbol{\omega}_B$ is the angular rotation of the celestial body since the atmosphere rotates with the body. Figure 5.42 shows the geometry of a spacecraft under the influence of a celestial body's gravitational and drag forces, where the gravitational ($\mathbf{F}_{grav}$) and drag ($\mathbf{F}_{drag}$) forces are labeled along the spacecraft's osculating orbit (in green), the celestial body (in red), and its atmosphere (in blue). Note that, generally, $\mathbf{v}_{rel}$ is not purely along the transverse direction, although its radial component is usually small, especially for low eccentricity orbits.

Because of the presence of drag, the energy of the orbit is no longer constant. This means that also other orbital elements change in time, such as semimajor axis

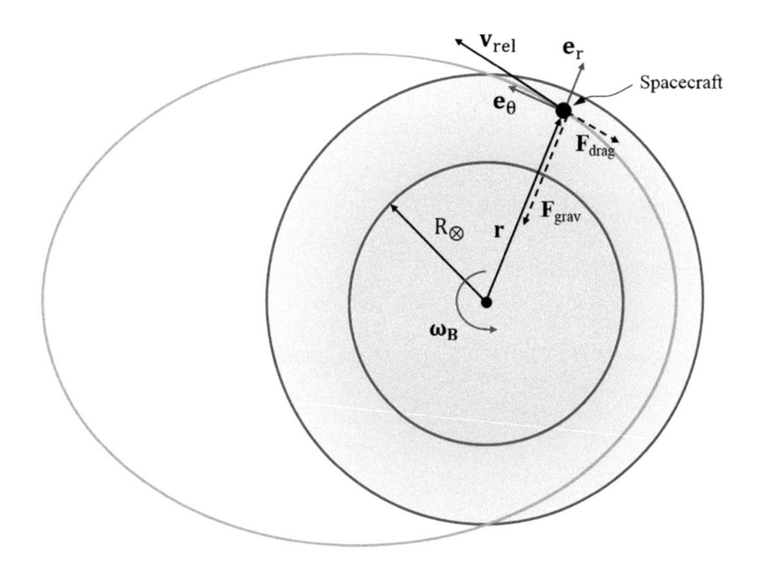

FIGURE 5.42 Geometry of a spacecraft under the influence of a celestial body's gravitational and drag forces.

and eccentricity. Starting from the *vis viva* equation (Equation 2.50), taking a time derivative, and solving for the rate of change of semimajor axis results in

$$\frac{da}{dt} = \frac{2a^2}{\mu}\frac{dE}{dt} \tag{5.202}$$

where dE/dt is the energy dissipation due to atmospheric drag. Solving Equation (2.34) for eccentricity and taking a time derivative gives

$$\frac{de}{dt} = \frac{\mu}{2e}\left(\frac{2ah\dot{h} - h^2\dot{a}}{a^2}\right) \tag{5.203}$$

where the dotted notation is used to denote time derivative, as usual. Now, considering the work – change in energy – done by the drag force, we have

$$\frac{dE}{dt} = \mathbf{F}_{drag} \cdot \mathbf{v}_{rel} \tag{5.204}$$

where $\mathbf{F}_{drag}$ is the force of drag. Since the velocity of the spacecraft is constrained to the orbital plane by definition, in cylindrical coordinates it is $\mathbf{v} = \dot{r}\mathbf{e}_r + r\dot{\theta}\mathbf{e}_\theta$, while the force of drag is $\mathbf{F}_{drag} = F_r\mathbf{e}_r + F_\theta\mathbf{e}_\theta + F_n\mathbf{e}_n$. Ignoring the velocity of the atmosphere, we can simplify Equation (5.203) as

$$\frac{dE}{dt} = F_r\dot{r} + F_\theta r\dot{\theta} \tag{5.205}$$

where here we used the fact that $F_n = 0$ since drag can only act in the plane of motion. Similarly, we can derive an equation for the rate of change of angular momentum from its definition, $\mathbf{h} = \mathbf{r} \times \mathbf{v}$, as

$$\frac{d\mathbf{h}}{dt} = rF_\theta\mathbf{e}_n - rF_n\mathbf{e}_\theta \tag{5.206}$$

from which, using $F_n = 0$, we get

$$\frac{dh}{dt} = rF_\theta \tag{5.207}$$

Plugging Equation (5.205) into Equation (5.202) and simplifying yields

$$\frac{da}{dt} = \frac{2a^2}{\mu}\left(F_r\frac{\mu e\sin\theta}{h} + F_\theta\frac{h}{r}\right) \tag{5.208}$$

Similarly, we can simplify Equation (5.203) to get

$$\frac{de}{dt} = \frac{\mu}{e}\left[\left(\frac{hr^2 - aph}{ar}\right)F_\theta - eh\sin\theta F_r\right] \tag{5.209}$$

If we assume that the spacecraft is located in a circular orbit, then the radial component of drag can be dropped, further simplifying Equations (5.208) and (5.209) with

$$F_r = 0$$

$$F_\theta = -\frac{1}{2}\left(\frac{C_D A}{m}\right)\rho v_{rel}^2 \tag{5.210}$$

which gives a rate of change for semimajor axis

$$\frac{da}{dt} = -\frac{ha^2}{\mu r}\left(\frac{C_D A}{m}\right)\rho v_{rel}^2 \tag{5.211}$$

and for the rate of change of eccentricity gives

$$\frac{de}{dt} = -\frac{\mu}{2e}\left(\frac{hr^2 - aph}{ar}\right)\left(\frac{C_D A}{m}\right)\rho v_{rel}^2 \tag{5.212}$$

Integrating Equations (5.208) and (5.209) or – if assuming the orbit is nearly circular – Equations (5.211) and (5.212) over one orbital period with the substitution $h = r^2\dot{\theta}$ reduced the expressions to an average change in semimajor axis and eccentricity.

In the next sections, we will analyze the use of the atmosphere of celestial bodies for interplanetary missions, including orbiters and landers, and how such atmospheric interactions make these missions challenging.

5.14.1 Aerocapture

An aerocapture is a kind of maneuver that allows a spacecraft to be captured by the gravity of the arrival body after only one pass through that body's atmosphere. In order for the spacecraft to be captured from its arrival hyperbolic orbit ($e > 1$) to an elliptical or circular orbit ($e < 1$), the amount of energy that the spacecraft must dissipate over time into that celestial body's atmosphere (as given in Equation 5.204) must be sufficient to drop the pre-aerocapture hyperbolic (and thus positive) energy to a post-aerocapture elliptical (and thus negative) energy. An aerocapture maneuver requires the spacecraft to dive deep into the celestial body's atmosphere – a walk-in maneuver – and then perform another maneuver at apoapsis in order to raise its periapsis above the atmosphere of the celestial body and thus be on a stable orbit that is no longer significantly affected by atmospheric drag.

Aerocapture maneuvers have not been used to date due to their high risk – the spacecraft only has one pass to decrease its orbital energy enough to be captured. However, this type of maneuver is fast, requiring only one pass to complete the capture, as opposed to aerobraking, which requires multiple passes – we will discuss aerobraking in the next section. The right part of Figure 5.43 shows the main phases of an aerocapture maneuver along with listing pro's and con's associated to it.

One the main challenges of atmospheric effects is estimating the drag force on the spacecraft and other heating effects due to atmospheric interactions. One of the

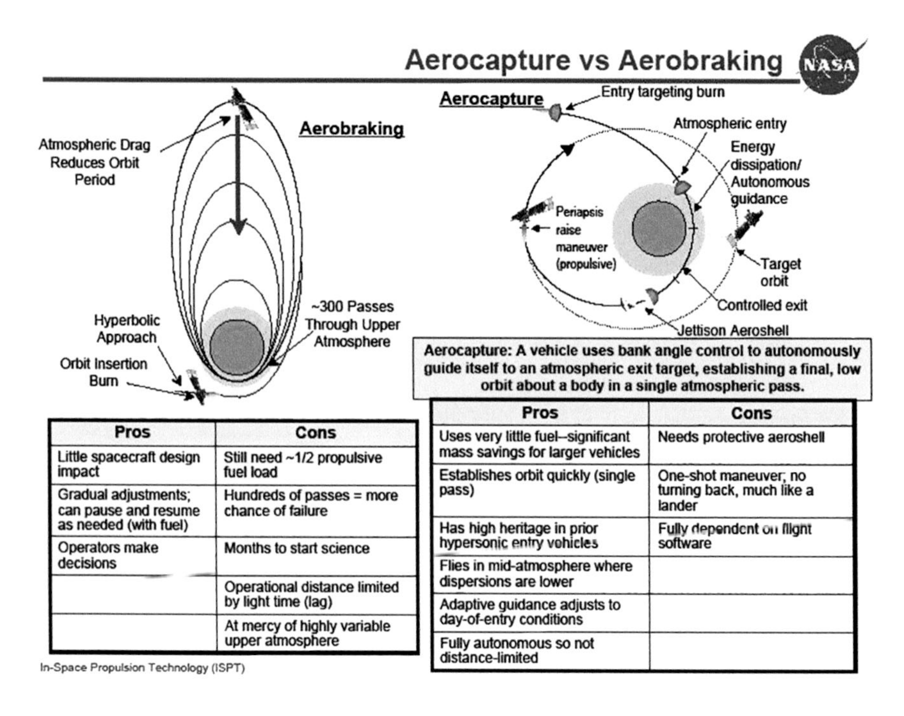

Pros	Cons
Little spacecraft design impact	Still need ~1/2 propulsive fuel load
Gradual adjustments; can pause and resume as needed (with fuel)	Hundreds of passes = more chance of failure
Operators make decisions	Months to start science
	Operational distance limited by light time (lag)
	At mercy of highly variable upper atmosphere

Pros	Cons
Uses very little fuel–significant mass savings for larger vehicles	Needs protective aeroshell
Establishes orbit quickly (single pass)	One-shot maneuver, no turning back, much like a lander
Has high heritage in prior hypersonic entry vehicles	Fully dependent on flight software
Flies in mid-atmosphere where dispersions are lower	
Adaptive guidance adjusts to day-of-entry conditions	
Fully autonomous so not distance-limited	

FIGURE 5.43 Main differences between aerobraking and aerocapture maneuvers. (Courtesy of NASA.)

variables that plays a main role in this is atmospheric density, ρ. A simple atmospheric model that is generally used is given by modeling ρ as follows

$$\rho = \rho_0 e^{\left(\frac{-r+R_\otimes+h}{H_0}\right)} \tag{5.213}$$

where ρ_0 is the reference density, r is the radial position of the spacecraft, $R_\otimes$ is the radius of the celestial body, h is the reference height, and H_0 is the scale height. Equation (5.213) models atmospheric density as an exponential function that decays as altitude increases. While this is mostly accurate when compared to the real density distribution of a body, it is often necessary to chain together multiple of these simply models to have a better representation of atmospheric density. However, this model does not account for seasonal variations, weather, or other local phenomena, such as effects due to time of the day.

Let's consider a sample mission as we've seen in Section 5.7, where a spacecraft is performing an interplanetary Hohmann transfer from Earth to Mars with arrival v_∞ of 2.64479 km/s. For this spacecraft, let's assume that it has $C_D = 2.2$, $A = 23\,m^2$, and $m = 1{,}100\,kg$, resulting in $BC = 21.7391\ kg/m^2$. Using an atmospheric model like the one described by Equation (5.213) and spacecraft parameters given in Table 5.15, we can integrate Equation (5.200) assuming that the spacecraft is targeting a periapse altitude of 78 km – higher values of periapsis do not result into a capture.

TABLE 5.15
Spacecraft and atmospheric density parameters used for a Martian mission example

Density Model	Range	Values
Reference density	0–25 km	0.0525 kg/m^3
	25–125 km	0.0159 kg/m^3
Scale height	0–25 km	11.049 km
	25–125 km	7.295 km
Ballistic coefficient	N/A	21.7391 kg/m^2

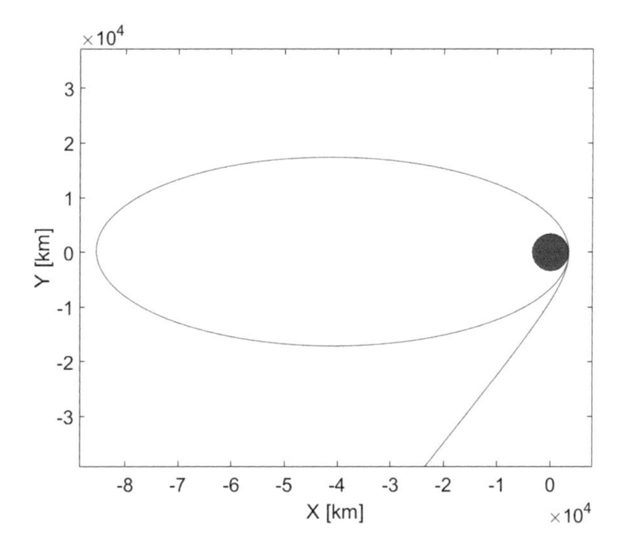

FIGURE 5.44 Aerocapture maneuver for a Martian mission targeting a periapsis altitude of 78 km.

Doing so results in the aerocapture orbit shown in Figure 5.44, which sets the spacecraft on a 85,600 × 76 km altitude orbit after the spacecraft reaches apoapsis – the periapsis also decreases, albeit by a very small amount.

After the spacecraft has been captured, a prograde Δv maneuver of 12 m/s at apoapsis is sufficient to raise the periapsis to a 200 km altitude – high enough that atmospheric effects are much less significant. This maneuver is called a walk-out maneuver. Table 5.16 summarizes the pre- and post-aerocapture orbital parameters for this example.

Figure 5.45 shows the change of altitude, velocity, energy, and dynamic pressure over time. The effects of the aerocapture maneuver are visible especially in the abrupt change to energy, going from positive to negative, and dynamic pressure.

Had the spacecraft completed its capture at Mars using solely a propulsive maneuver at a periapsis of 200 km, it would have required a Δv of 1.063 km/s.

TABLE 5.16
Orbital parameters before and after the aerocapture maneuver

	Semimajor Axis [km]	Eccentricity [N/A]	Apoapse Altitude [km]	Periapse Altitude [km]	Energy [km²/s²]
Before	–6,154	1.564	N/A	78.00	3.497
After	44,529	0.92205	85,600	77.00	–0.4830

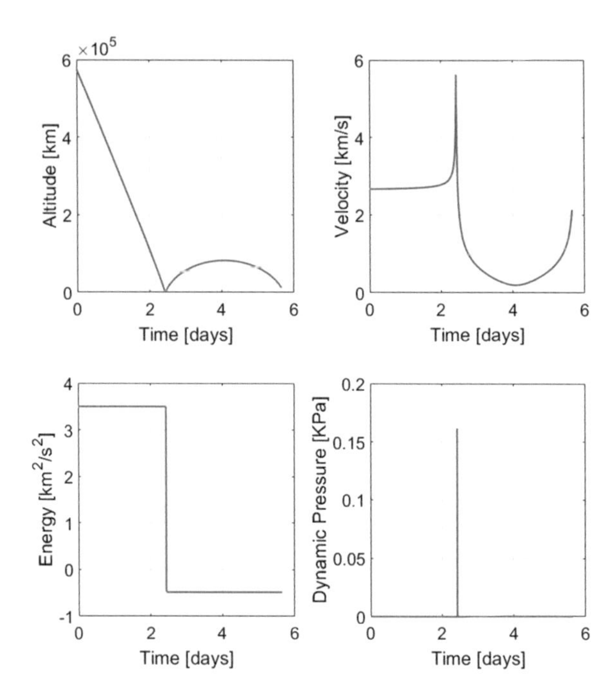

FIGURE 5.45 Altitude, velocity, energy, and dynamic pressure over time due to aerocapture for a Martian mission.

Thus, while aerocapture maneuvers require essentially no propellant except for the small periapse raise maneuver, they require the spacecraft to be equipped with appropriate heat shielding systems in order to survive the extreme heating caused by the drag forces during the spacecraft's atmospheric pass. Additionally, there is some level of uncertainty in the atmospheric model of Mars – or any other celestial body for that matter – which could lead to the aerocapture maneuver to have different results based on the season, local weather (especially winds), and time of the day, which are all factors that can change the amount of drag that the spacecraft experiences. Too much drag could result in an orbit that crashes onto the surface, while too little drag could result in the spacecraft not being captured.

An alternative maneuver that has been used in past missions and has lower – although not inexistent – risks is aerobraking.

5.14.2 Aerobraking

Aerobraking maneuvers are comprised by a sequence of atmospheric passes that slowly lower the orbital energy – and thus apoapsis – of a spacecraft that is already on an elliptical orbit. The left part of Figure 5.43 shows the main phases of an aerobraking maneuver along with a list of pro's and con's. The first atmospheric pass is usually used to calibrate the atmospheric model that mission designers will use to predict the drag that the spacecraft is expected to encounter on subsequent passes. Just like in aerocaptures, aerobraking maneuvers begin with a walk-in maneuver where a spacecraft – usually on a highly elliptical orbit – propulsively lowers its periapsis to an altitude at which atmospheric drag is significant. It is important to note that while aerocapture maneuvers would be used by spacecraft arriving from interplanetary space – and thus arriving at relatively high speeds – for aerobraking maneuvers, the spacecraft already completed its interplanetary arrival with a propulsive orbit insertion. Because of this, generally the atmospheric passes that a spacecraft experiences during aerobraking happen at significantly lower speeds and lower heating is experienced by the onboard systems, thus not requiring the spacecraft to be equipped with a heat shield. Once the desired apoapse altitude is reached, the spacecraft performs a walk-out maneuver to raise its periapsis above the celestial body's atmosphere.

Let's consider an example of a Mars-orbiting spacecraft that needs to reduce its apoapsis, starting from an orbit of 20,000 × 110 km. If we can integrate Equation (5.200) for 15 days using the same atmospheric model and ballistic coefficient listed in Table 5.15, we obtain a series of 57 aerobraking orbits as shown in Figure 5.46. Table 5.17 shows how the spacecraft orbit changes due to the aerobraking maneuver.

After the aerobraking maneuvers, a walk-out maneuver of 18 m/s is required at apoapsis for the spacecraft to raise its periapse altitude to 200 km. If the same final orbit were to be achieved using propulsive maneuvers alone, a Δv of 1.412 km/s would be needed. It is possible to combine aerobraking with Δv maneuvers, primarily to

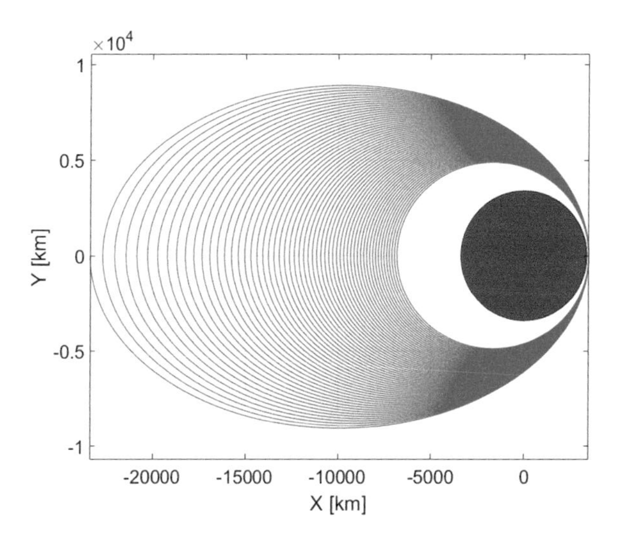

FIGURE 5.46 Aerobraking orbits for a Martian mission.

TABLE 5.17
Orbital parameters before and after the aerobraking maneuver

	Semimajor Axis [km]	Eccentricity [N/A]	Apoapse Altitude [km]	Periapse Altitude [km]	Energy [km^2/s^2]
Before	13,449	0.7395	20,000	110.0	−1.601
After	5,092	0.3123	3,289	107.9	−4.227

reduce the amount of time taken to lower the apoapsis or to perform other orbital adjustments, such as inclination changes.

Figure 5.47 shows the changes in altitude, velocity, energy, and dynamic pressure over the course of the atmospheric passes of the aerobraking maneuver. It should be noted that these changes are not as abrupt as those happening during aerocapture maneuvers, also resulting in significantly less heating effect on the spacecraft components, as seen in the next section.

Figure 5.48 shows the changes to semimajor axis, eccentricity, apoapse and periapse altitudes over the course of the aerobraking maneuvers. Each "step" on the graphs corresponds to an atmospheric pass.

Past missions that have used aerobraking to adjust their orbit include NASA's Magellan spacecraft, aimed at exploring and mapping Venus, and a series of Mars missions, such as Mars Global Surveyor, Mars Odyssey, and Mars Reconnaissance Orbiter, among others.

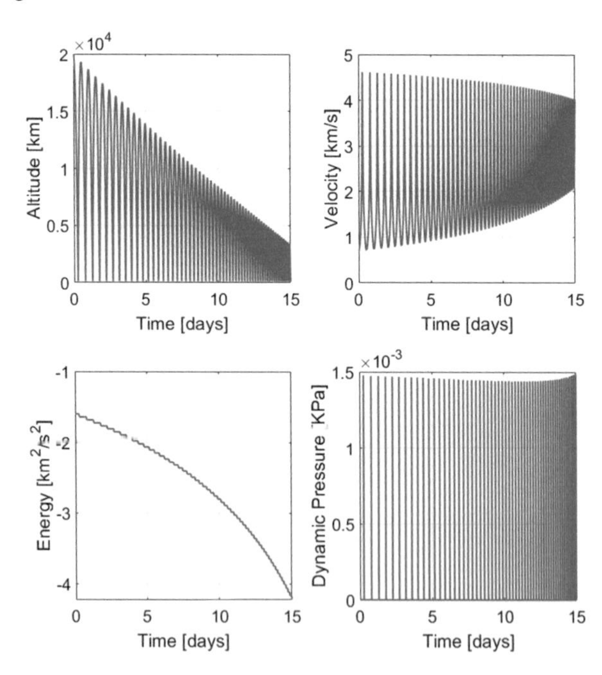

FIGURE 5.47 Altitude, velocity, energy, and dynamic pressure over time due to aerobraking for a Martian mission.

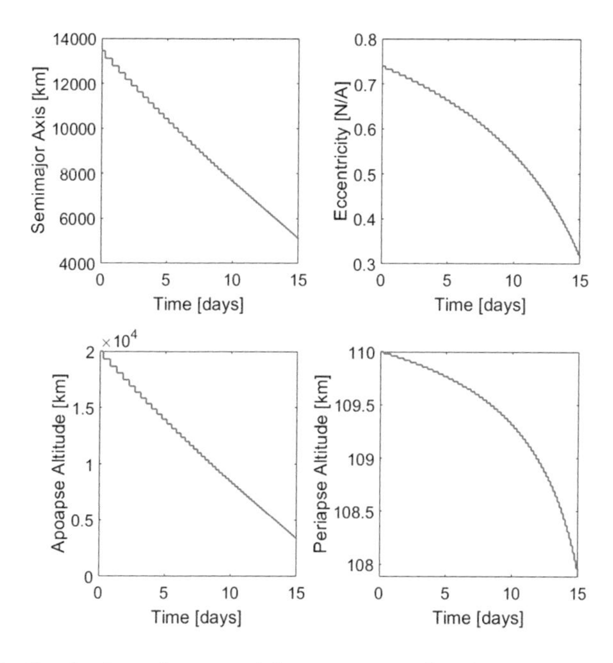

FIGURE 5.48 Semimajor axis, eccentricity, apoapse, and periapse altitudes over time due to aerobraking for a Martian mission.

In summer of 1993, NASA's Magellan spacecraft was the first planetary spacecraft to use aerobraking while in orbit around Venus.[15] Its success cleared the way for the use of aerobraking by Mars Global Surveyor which, in turn, paved the way for Mars Odyssey and Mars Reconnaissance Orbiter to use the same technique. Magellan was not originally designed to operate in a planetary atmosphere, but in order to acquire higher resolution images to map the Venusian surface, its orbit needed to be decreased. There were three main constraints on Magellan's aerobraking maneuver: temperature constraints on the high gain antenna and solar panels, timing constraints – a maximum of 80 days to complete the maneuver – and power constraints since the final orbital period had to be greater than 94 minutes in order for the spacecraft solar panels to be exposed to the Sun for a sufficient amount of time. The aerobraking maneuver performed by Magellan changed its orbit from an elliptic 280 × 8,500 km orbit to a near-circular 197 × 541 km orbit, reducing the orbital period from 3.22 to 1.57 hours – just above 94 minutes. The process took a total of 730 orbits in 70 days, which resulted in a total of 1.219 km/s Δv savings. The aerobraking process had four phases, which were combined with some propulsive maneuvers for station-keeping needs and orbital adjustments: a walk-in phase (4 days), a reduction of periapse with repeated drag passes to correctly target the desired periapsis and calibrate the atmospheric effects that would be expected on subsequent passes, an increase of periapse to keep dynamic pressure within spacecraft tolerance, and a final circularization phase to raise its periapse (walk-out maneuver).

[15] https://solarsystem.nasa.gov/missions/magellan/in-depth/

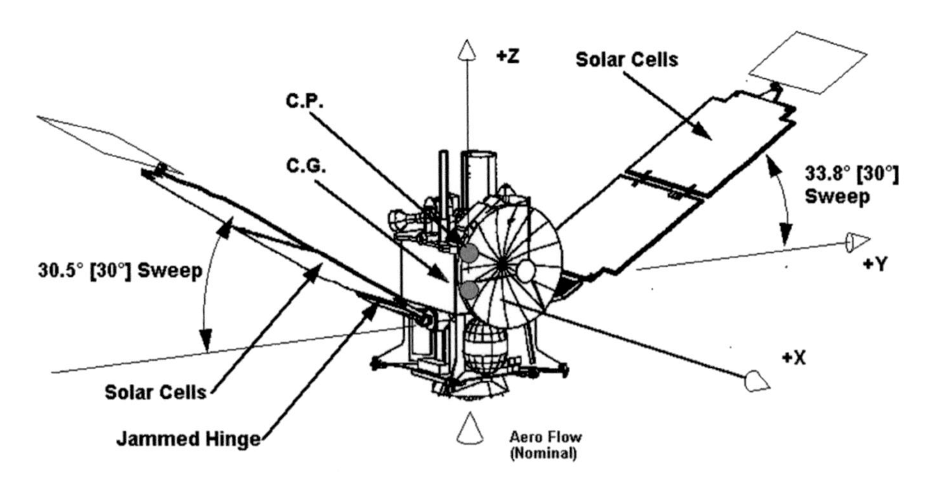

FIGURE 5.49 Mars Global Surveyor. (Courtesy of NASA.)

To estimate the orbital perturbations on Magellan, mission designers and mission planners relied on the use of a 21×21 JPL-MGN05 gravity model, used the gravitational perturbations of the Sun, the Moon, and the planets – modeled as point masses – and the relativistic effects due to the Sun's gravitational field. The atmospheric effects of Venus were modeled as an exponential model, as given by Equation (5.213), with a reference density that was computed based on tracking data that was acquired during the mission.

NASA's Mars Global Surveyor,[16] which is depicted in Figure 5.49, was launched in on November 7, 1996, and arrived at Mars on September 12, 1997. MGS performed a propulsive orbit insertion to be placed into a highly elliptical capture orbit about Mars. To establish the required mapping orbit, the spacecraft supplemented its propulsive capabilities by aerobraking, as planned by the mission designers. However, following launch, the aerobraking plans for the mission had to be reviewed in order to accommodate a failure that occurred during the solar array deployment sequence. In fact, the inboard hinge of one of the two spacecraft's solar arrays failed to properly latch, causing one of the solar panels to be angled differently than the other, as shown in Figure 5.49. This meant that the ground crew had to devise a new plan to ensure that the spacecraft would still be able to perform the planned aerobraking maneuvers with its new configuration. Additionally, it was originally thought that the new constraints on aerobraking would be due to the torque that the unlatched solar panel would be able to withstand during atmospheric passes. However, testing revealed that despite the incorrect solar panel deployment position, the most strict requirement was still based on the thermal capabilities of the spacecraft components, rather than due to mechanical holding torque limitations.

Figures 5.50 and 5.51 show the aerobraking phases that were planned for the spacecraft and the resulting orbital period reduction (due to one of these phases). The ground crew was still able to follow this baseline architecture, although modifications

[16] https://mars.nasa.gov/mgs/sci/aerobrake/SFMech.html

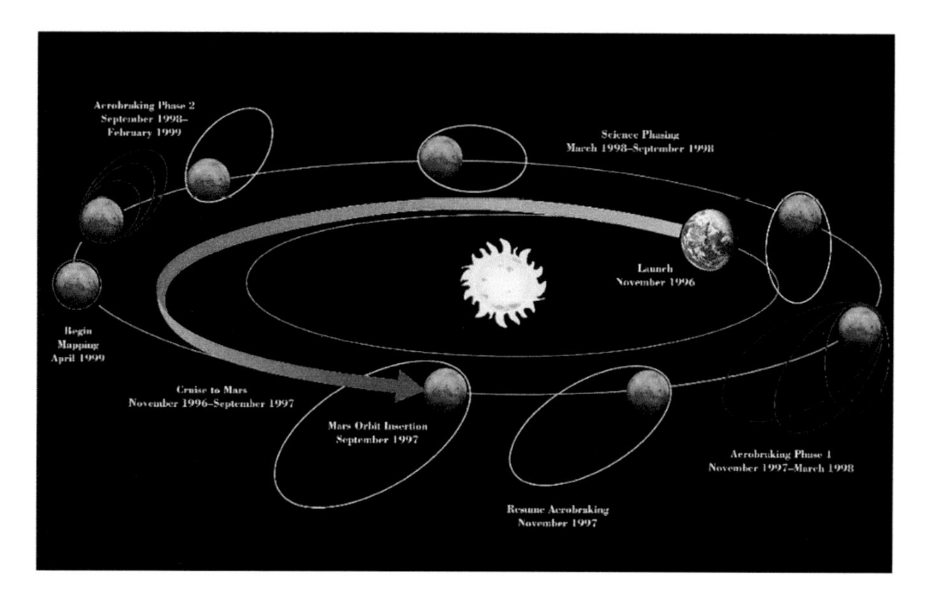

FIGURE 5.50 Main aerobraking phases of the Mars Global Surveyor. (Courtesy of NASA.)

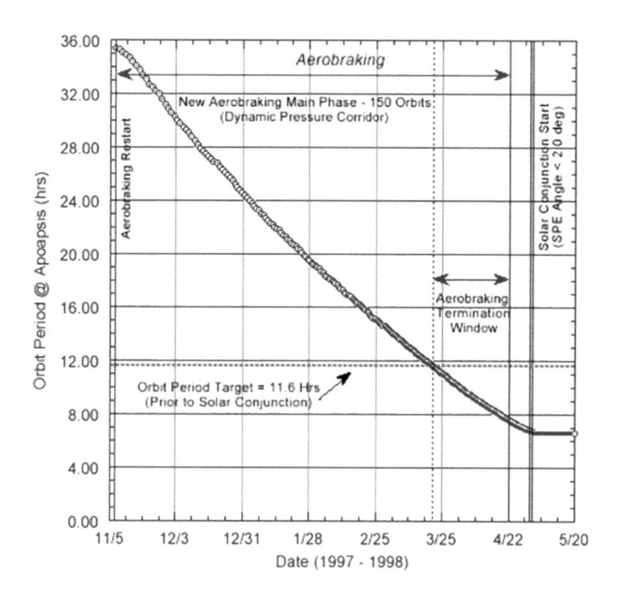

FIGURE 5.51 Mars Global Surveyor's orbital period reduction during one of its aerobraking phases. (Courtesy of NASA.)

to the original plans had to be made, mostly due to the fact that MGS did not experience the average aerodynamic pressure of 0.2 Pa that mission designers were expecting. Similarly to MGS, NASA's Mars Odyssey also made use of aerobraking maneuvers to lower its original capture orbit – a highly elliptical Mars orbit – from a period of 14 hours to a 1.9-hour orbital period over the course of 332 aerobraking passes, which took 76 days to complete.

5.14.3 Entry, Descent, and Landing (EDL)

In this section, we will discuss the challenges associated with entry, descent, and landing (EDL) into celestial bodies that have atmospheres.

There exist two main types of entry and descent capabilities: direct insertion and indirect insertion. The former puts the spacecraft onto a (re)entry[17] trajectory directly from interplanetary space, i.e., from a hyperbolic trajectory, resulting in higher entry speeds – and thus higher heating – but needing low or no Δv, while the latter makes first use of a propulsive orbit insertion and then lowers its periapsis enough to initiate the (re)entry trajectory, resulting in lower entry speeds and heating, but requiring the significant use of Δv.

In order to estimate the required heat shield and other EDL systems needed for a mission, heating effects must be computed. The dynamic pressure q is computed as

$$q = \frac{1}{2}\rho v_{rel}^2 \tag{5.214}$$

where q has units of mPa if keeping velocity in terms of km/s, as we usually do. The heat flux experienced by the spacecraft during its entry and descent phase can be computed as

$$\dot{Q} = \frac{1}{2}\rho v_{rel}^3 \tag{5.215}$$

where $\dot{Q}$ has units of kg/s^3 or, as more commonly seen, units of W/m^2.

The average spacecraft surface temperature can be computed using Stefan-Boltzmann law

$$T = \left(\frac{\dot{Q}}{\sigma}\right)^{1/4} \tag{5.216}$$

where σ is Stefan-Boltzmann constant and is equal to 5.670374×10^{-8} Wm^{-2}K^4.

Figures 5.46 and 5.48 show the dynamic pressure for the aerocapture and aerobraking examples we presented in the last two sections, respectively. Note how this quantity spikes suddenly for aerocapture maneuvers while it has periodic (but smaller) spikes for aerobraking maneuvers. Figures 5.52 and 5.53 show the heat flux and temperature experienced by the aerocapture and aerobraking examples. The temperature maxima are 1,700°C and 315°C, respectively. While 315°C is still a very high temperature to endure for a spacecraft – our example has a rather "aggressive" aerobraking strategy trying to minimize time – it most certainly does not require the

---

[17] An entry trajectory implies that the spacecraft has not encountered the body yet (e.g. entry at Mars from a ballistic Earth-Mars trajectory), while reentry implies that a spacecraft is returning to that body (e.g. a spacecraft returning to Earth).

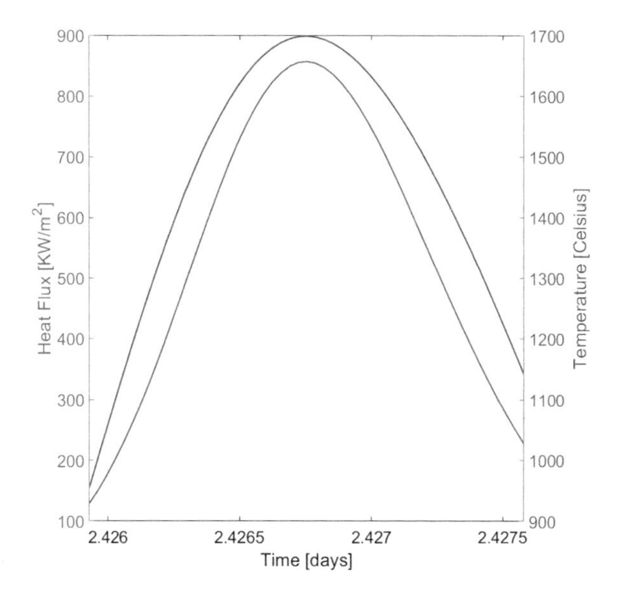

FIGURE 5.52 Heat flux and temperature vs. time for a Martian aerocapture maneuver.

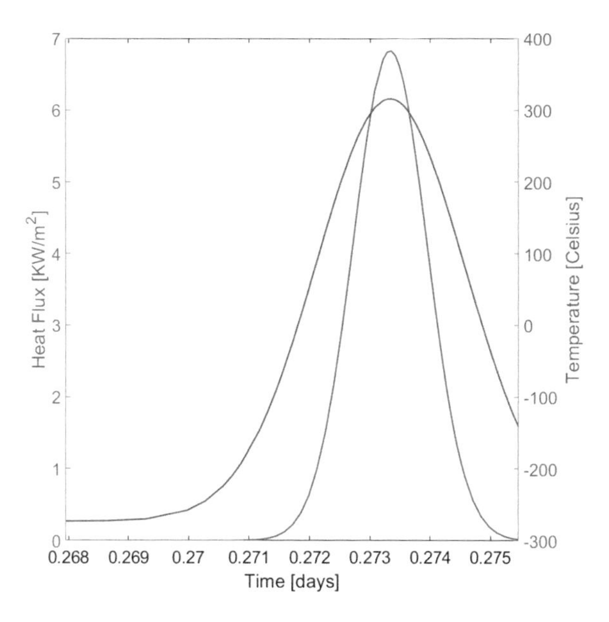

FIGURE 5.53 Heat flux and temperature vs. time for a Martian aerobraking maneuver.

same heat shielding capabilities that an aerocapture maneuver would. It should also be noted that using Δv in combination with these maneuvers can further reduce the heating experienced by the spacecraft. However, if the final objective is to land on the surface, the use of Δv needs to be accurately distributed over the course of each pass so that the final landing location remains the one that was originally intended.

Several spacecraft have performed EDL maneuvers, requiring the following systems:

1. an aeroshell, an aerodynamic cover that protects the lander/rover from the aerodynamic forces and gives the entire spacecraft the geometry required to produce the lift and drag needed for a successful EDL sequence; this component is also generally equipped with a heat shield, which is used to protect the spacecraft from extremely high temperatures;
2. parachutes, used to decelerate from large orbital speeds on the order of kilometers per second to hundreds of meters per second (or less); different types of parachutes exist and are mainly distinguished between their operating speed (hypersonic vs. supersonic vs. subsonic);
3. airbags and/or rocket assisted descent (RAD) motors, depending on whether the spacecraft needs to use propulsive maneuvers during the EDL sequence (through the use of a sky crane, for example) and/or airbags;
4. lander structure, a lightweight yet robust structure that houses and protects the lander or rover from mechanical vibrations and impacts;
5. guidance and navigation systems, such as Radar or LiDAR, used to measure the altitude and other atmospheric parameters.

Considering the aerobraking example from the previous section, if we let the spacecraft continue its aerobraking sequence until it reaches the surface, a total of 77 passes are required in 17.2 days, during which the spacecraft experiences a peak heating of approximately 1,600°C during its final approach and landing. Figure 5.54 shows the entire aerobraking sequence until landing, which is denoted by a green dot on the surface of Mars.

Some of the most notable missions that have performed EDL at interplanetary destinations include all of the Martian landers and rovers, from Viking 1 in 1976 to current and future Mars mission, and Venus landers, such as the Venera probes, as discussed in Chapter 1. Among these, Vikings 1 and 2, which were the first missions to touch the Martian surface on the western slope of Chryse Planitia and Utopia Planitia, respectively, took photographs of the Martian surface and performed a series of biology experiments designed to look for possible signs of life. The results showed unexpected chemical activity in the Martian regolith, but provided no clear evidence for the presence of living microorganisms on Mars.

Other planetary probes include Galileo's probe at Jupiter, Cassini's Huygens lander at Titan (one of the moons of Saturn), Near-Earth Asteroid Rendezvous – Shoemaker (NEAR Shoemaker), a robotic space probe designed to study the near-Earth asteroid Eros – it was the first spacecraft to successfully orbit an asteroid and also land on it – and Rosetta's Philae lander, the first probe to land on a comet.

There exist a multitude of different EDL techniques, which depend on several factors, such as the targeted celestial body, the targeted landing location, and the mass of the spacecraft. For example, the EDL sequence of the Mars Exploration Rovers (MER), Spirit and Opportunity (each approximately 190 kg), made use of airbags for the final landing procedure, while the Mars Science Laboratory, Curiosity (900 kg) made use of a propulsive system, the so-called sky crane as shown in Figure 5.55.

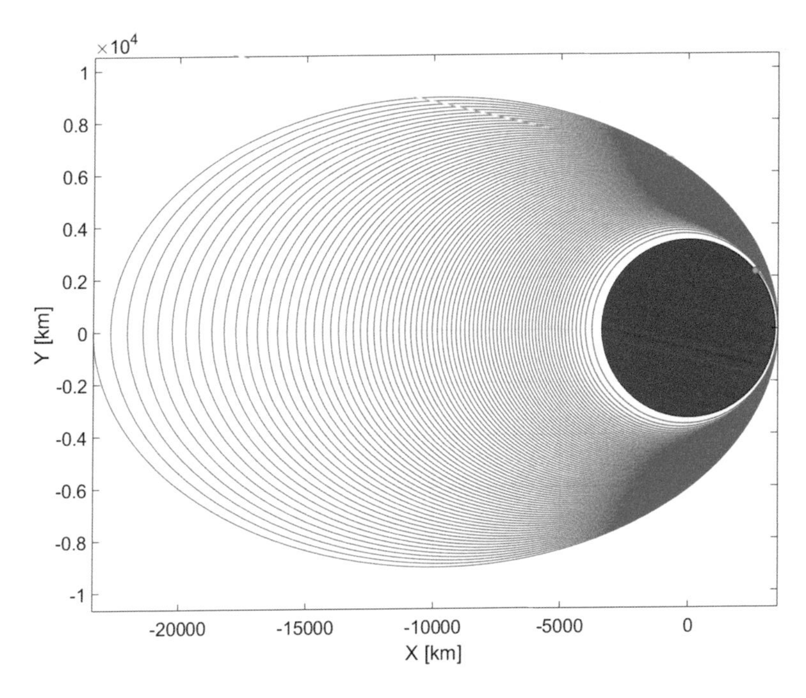

FIGURE 5.54 Aerobraking orbits for a Martian mission.

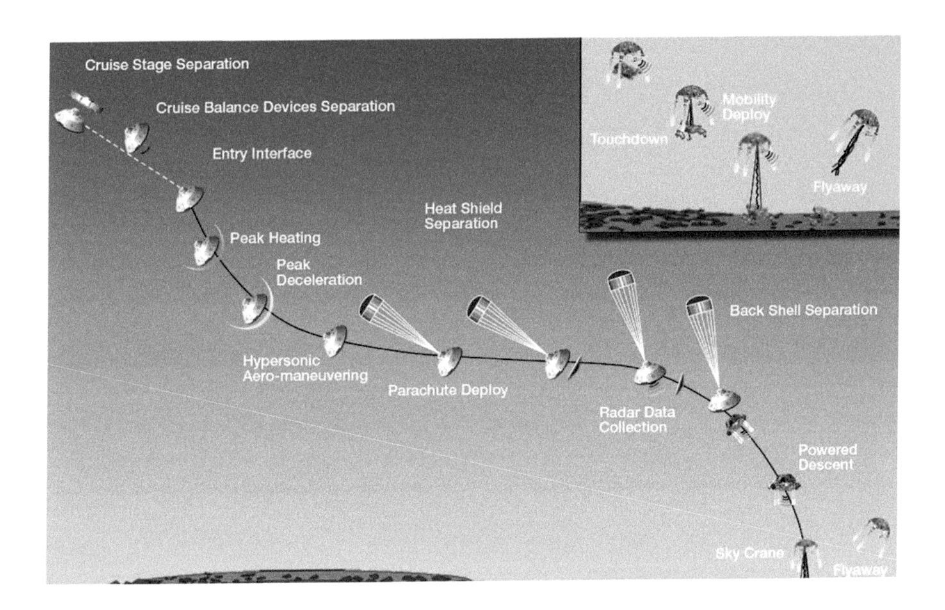

FIGURE 5.55 Entry, descent, and landing sequence of Mars Science Laboratory. (Curiosity; Courtesy of NASA.)

While airbags require no propellant and were used for many Martian lander and rover missions such as MER, they are not capable of completely slowing down and protect larger payloads like Curiosity or Perseverance.

A few are also the robotic missions that returned to Earth, such as Genesis and Stardust.

Stardust was launched on February 7, 1999, with the objective to collect dust samples from the coma of comet Wild 2 and cosmic dust to be returned to Earth. Stardust was the first sample-return mission that provided scientists with the ability to study a comet and cosmic particles first hand. The trajectory of Stardust is shown in Figure 5.56.

Genesis was a sample-return mission aimed at collecting and returning particles coming from the solar wind. This mission was the first mission that was planned to return material from outer space since the Apollo program, and the first mission to perform a sample return beyond the Moon's orbit. Genesis was launched on August 8, 2001, and it was placed in a quasi-halo orbit as shown previously in Figure 3.50. Genesis was planned to reenter Earth's atmosphere and safely land at the end of its mission. However, on September 8, 2004, the capsule crash-landed in Utah's desert due to a design flaw preventing the correct deployment of the spacecraft's drogue parachute. Although the crash contaminated a significant portion of its payload, some samples were still recovered and then analyzed, as originally intended. Since a regular parachute landing was thought to be too dangerous for the payload's integrity, a so-called mid-air retrieval was planned. This consisted in the capsule deploying a drogue parachute to slow down from hypersonic speeds, and then deploy a large parafoil to further slow the spacecraft's descent into the Earth's atmosphere, leaving the capsule in a stable yet slow descent. A pair of helicopters was then supposed to

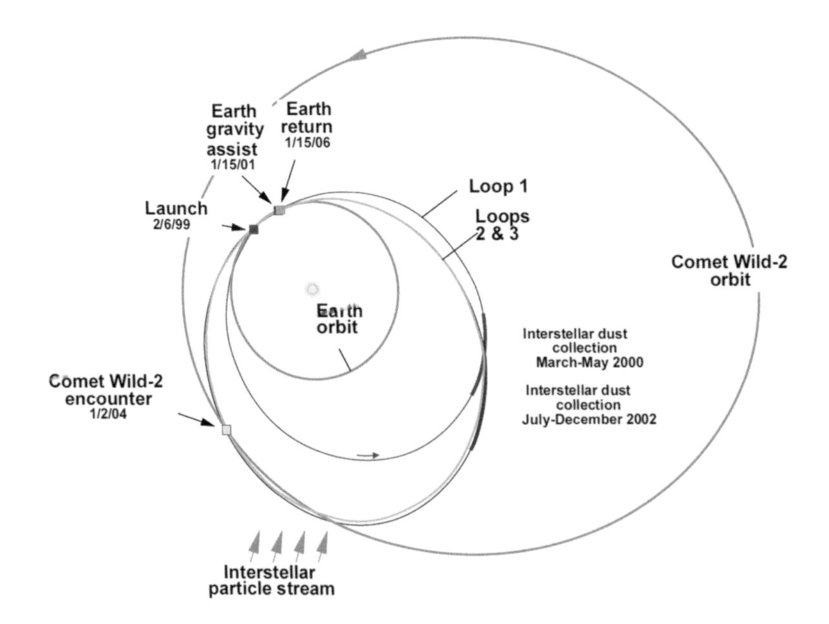

FIGURE 5.56 Trajectory of Stardust. (Courtesy of NASA-JPL/Caltech.)

catch the capsule by its parachute with a 5-m hook – certainly making for one of the most unorthodox landing procedures. Unfortunately, the capsule never deployed its parachute, causing the aforementioned crash-landing. Later, an investigation revealed that an accelerometer (also known as G-switch) was mounted backward.[18] This sensor was meant to release the parachute at 3g through the expected 30g deceleration. However, due to the incorrect installation of this sensor, this never happened, causing the capsule to free-fall for the majority of its descent.[19] It should be noted that the same parachute concept was adopted Stardust, which landed successfully.

In addition to robotic missions, several crewed missions that returned to – and thus reentered at – Earth took place, although none from interplanetary space. Such missions include missions from the early stages of human spaceflight, such as the Gemini program and the Apollo program, to more modern missions such as the Space Shuttle program, and currently operating crewed vehicles such as the Soyuz and SpaceX's Dragon capsules. While the Space Shuttle was piloted somewhat like a glider during the landing phase with the objective of landing on a runway, most capsule designs such as the Apollo crew capsule and Soyuz utilize parachutes and perform water landings, or splashdowns. Notable is the "land landing" of Dragon, which is performed with a combination of parachutes and rocket motors.

5.14.3.1 Precision Landing

One possible requirement for landing missions is that they land near a predetermined target. For example, NASA's Surveyor 3 landed on the Moon in 1967. Thirty-one months later, the Apollo 12 lunar module landed approximately 160 m from the Surveyor 3 spacecraft. This was done in order to allow for the study of the long-term effects of the lunar environment on various parts of the Surveyor spacecraft. This precision landing (referred to in the Apollo program as pinpoint landings) was accomplished due to several factors, including using the technique of pilotage, using fixed references on the lunar surface and real-time orbit determination and navigation from both the Earth and from the command/service module (CSM) [Woods, 2008, p. 226; NASA SP-4029].

In addition to wanting to land within a short distance of another spacecraft, it may be desirable to land close to a region with some sort of scientific interest. Apollo 11's primary goal was to land on the Moon. To do this, there were many things that needed to be learned, and a sequence of maneuvers that were devised to ensure that the astronauts in the lunar module would be able to have enough reaction time for each phase of the landing procedure (shown in Figure 5.57). Mission planners chose a very flat area of the Moon, which meant that a pinpoint landing could be relaxed. Apollo 11 landed 7.4 km downrange of its original target. In addition to Apollo 12's pinpoint landing, Apollo 14 ended up as the most accurate landing, missing its target by around 45 m. Apollo 15 was the first Apollo lander to carry the lunar rover. It undocked from the CSM at a lower altitude, flew a steeper trajectory, and landed

[18] Michael Ryschkewitsch, NASA investigation board chair on Genesis, noted that none of NASA's review procedures had highlighted a mistake, also adding that "it would be very easy to mix this up."

[19] An accelerometer installed backwards is similar to the original event that inspired Edward A. Murphy Jr. to formulate the well-known Murphy's Law: "Anything that can go wrong will go wrong."

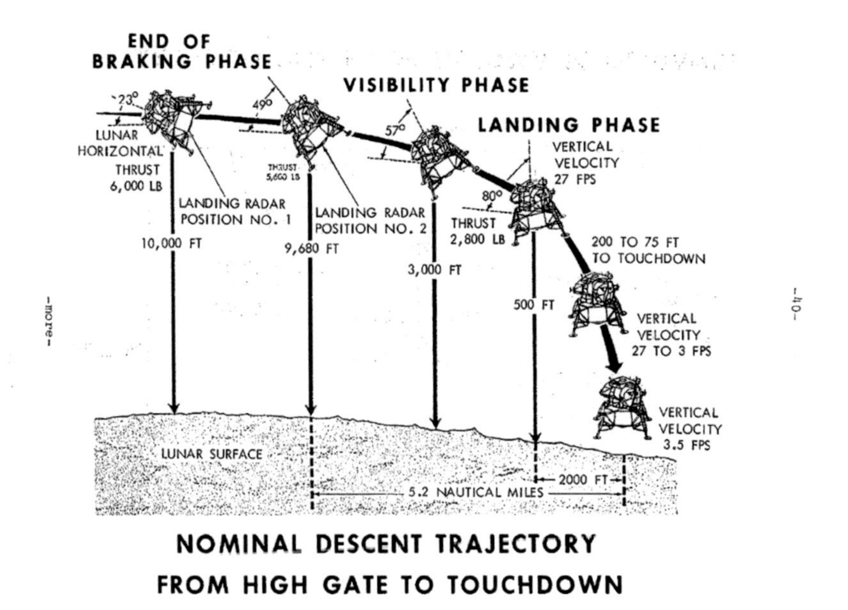

FIGURE 5.57 Nominal Apollo descent trajectory. (Courtesy of NASA.)

550 m from its target. Apollo 16 had a 6-hour delay leaving orbit and still landed within 212 m of its target. The final mission, Apollo 17, landed 200 m east of its target. Considering that the view of the lunar module crew was poor, this was a phenomenal feat of piloting skills! [Orloff and Harland, 2006]. Table 5.18 shows a summary of the Apollo landing precision for all of their missions.

Future missions back to the Moon and on to other planets will no doubt require similar or greater precision. Today, probably the most accurate pinpoint landing feat is the landing of the Space Shuttle. This spacecraft starts at orbital velocity, deorbiting through a complex atmospheric maneuver, and lands within a small target area (essentially a runway). This system has substantially more tracking and ground control, as well as better pilot visibility, than spacecraft landing on planets.

TABLE 5.18
Summary of the lunar landing accuracy of the each Apollo mission [Orloff and Harland, 2006]

Mission	Distance from Targeted Landing Site	Notes
Apollo 11	7.4 km	First lunar landing
Apollo 12	160 m	
Apollo 13	N/A	Did not land
Apollo 14	45 m	
Apollo 15	550 m	First lander with rover
Apollo 16	212 m	
Apollo 17	200 m	

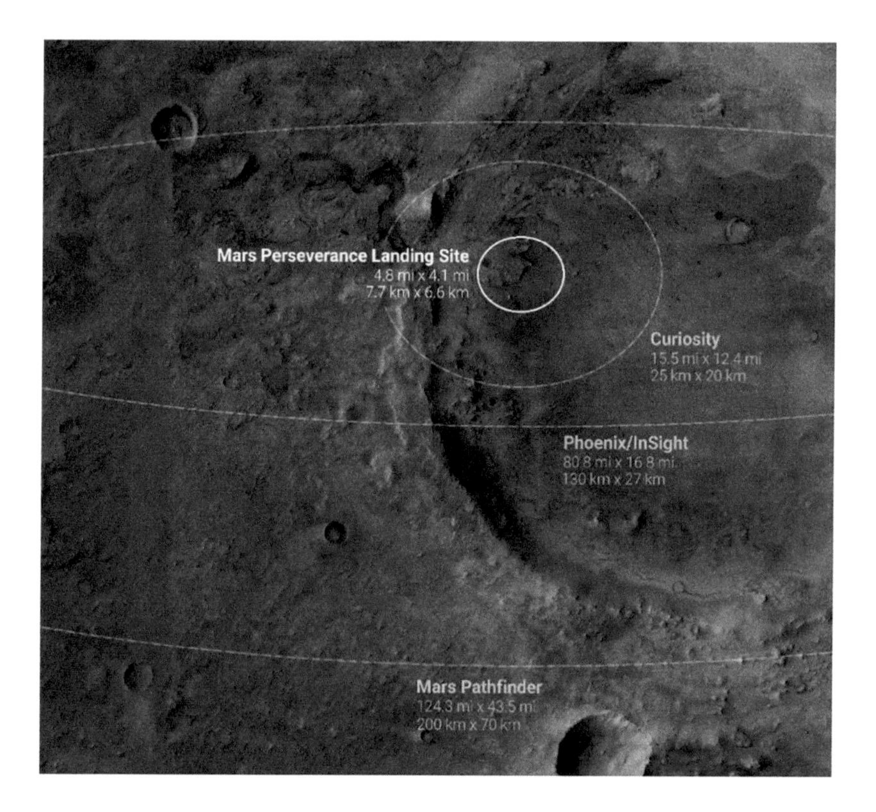

FIGURE 5.58 Landing ellipses for five NASA missions. (Courtesy of NASA-JPL/Caltech.)

Future infrastructures, such as a base on the Moon or another planet, will likely have some of these systems available to future interplanetary pilots.

Autonomous landing precision for robotic missions has increased over time, as seen in Figure 5.58, where landing ellipses for five NASA missions to Mars are shown. Landing ellipses are calculated by mission designers and represent the region within which a spacecraft is expected to land based (up to 3σ accuracy) on its arrival trajectory as it approaches the planet, also known as B-plane targeting, as discussed in more detail in Chapter 6. This figure illustrates how spacecraft landings on Mars have significantly increased in precision since NASA's first Mars landing of the Viking probes in 1976 which, for reference, had landing ellipse sizes of 280×100 km (174×62 mi).

5.15 ROUNDTRIP TRAJECTORY DESIGN

In this section, we want to briefly describe the concept of roundtrips applied to interplanetary trajectories. While the vast majority of robotic missions are one-way missions to their destinations, there have been a few that have returned to Earth to deliver the samples they collected during their time in space, such as the Genesis and Stardust missions, which we previously described. These kinds of robotic missions are known as sample-return missions. On the other hand, crewed missions have all been roundtrip missions – at least for now – requiring the crew to return to Earth after their time spent in space. The main difference between robotic sample-return missions and crewed roundtrip missions is considerations regarding the time of flight. In the next sections, we will describe these considerations, along with presenting a type of recurring roundtrip orbits, called cycler orbits.

5.15.1 Robotic Sample-Return Missions

Most robotic missions are one-way missions to their destinations. This means that the spacecraft must be equipped with appropriate scientific instrumentation capable of performing all of the sampling and experimentations for the duration of the mission, turning spacecraft into deep space laboratories. While this approach has led to countless scientific discoveries, scientists would be able to obtain much more significant results if they were able to analyze samples using the laboratories that we possess here on Earth. While we have a few examples of samples returned from space, such as the Genesis and Stardust missions and the Apollo missions, most samples we get from space are fallen meteorites that are pieces of asteroids or, in rare cases, believed to be chunks of other celestial bodies, such as Mars, that over the course of many years made their way to Earth.

Robotic sample-return missions have the objective to travel to a given destination, collect samples, and return them to Earth. The main challenge with sample-return missions is that they require both an outbound and inbound trajectory, resulting in significantly more Δv and time of flight, double those of a one-way trip at best – although they can be even higher, depending on the case.

Let's consider the Earth-Mars example that we discussed in Section 5.6. If a mission wanted to make use of a Hohmann transfer to both travel from Earth to Mars (outbound trajectory, ~9 months), and return to Earth (inbound trajectory, also ~9 months assuming a Hohmann transfer), then they would need to wait at Mars until the next available launch window opens to transfer back to Earth, or approximately 15 months. This mission would therefore last approximately 900 days. Figure 5.59 shows an example of the resulting interplanetary transfers between Earth and Mars for this example, which is sometimes referred to as a long-stay roundtrip mission. In this figure, the orbits of Earth and Mars are blue and red, respectively, while the outbound and inbound Hohmann transfers are green and purple, respectively.

One of the major difficulties of robotic sample-return missions is that surface operations could be slow and complex, and thus require a longer stay time on Mars. In fact, while sample collection has been completed by multiple spacecraft in past and current missions, taking said samples, and stowing them safely into a return vehicle would be a rather difficult operation which would need to be performed autonomously. For our Earth-Mars example, adding a synodic period (~26 months) to the stay time would give the opportunity to still use the same return trajectory while giving the spacecraft additional surface time, if needed.

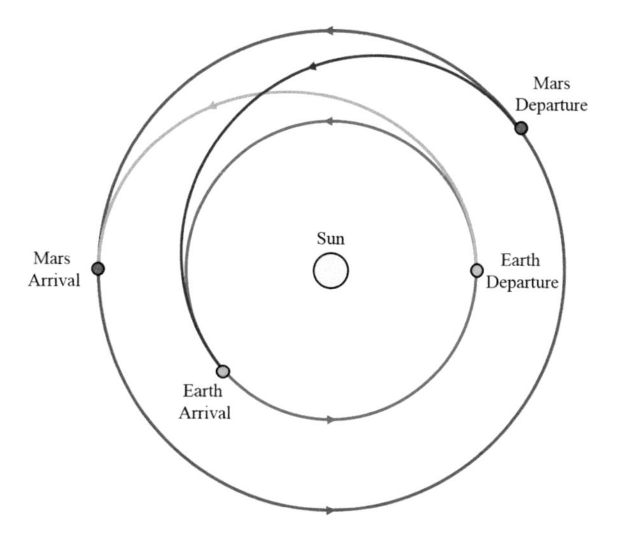

FIGURE 5.59 Example of a long-stay Earth-Mars roundtrip mission.

5.15.2 Human Roundtrip Missions

While robotic missions can afford longer times of flight, crewed missions should try to limit the amount of time that astronauts spend in interplanetary space. In fact, crews can suffer physiological and psychological side effects from interplanetary missions, including bone and muscle loss due to microgravity, increased risk of cancer due to radiation, and isolation to name a few.

As humanity reaches for further destinations, the time it would take a crew to return to Earth drastically increases, from a few hours for ISS missions, a few days for lunar missions, to multiple months, or even years for missions to Mars or asteroids. However, the energy (and thus Δv) needed to reach these destinations does not increase as drastically. For example, the Δv needed for an optimal two-burn transfer from Earth to the Moon, including descending to the lunar surface, is comparable to the Δv needed to reach an orbit around Mars.

While increases in Δv imply an increase in propellant mass following the rocket equation, Equation (5.1), crew resources, such as oxygen, water, food, and other supplies, increase linearly with the time spent in space. Therefore, finding a Δv-optimal trajectory – such as a Hohmann transfer from Earth to Mars – results in a low propellant mass, but a longer time of flight means an increase in crew supplies, which might result in an overall launch mass that would be higher than that of a faster transfer that needs more Δv.

Let's consider the same Earth-Mars roundtrip example from before. If we wanted to minimize the total mission time for the health of the crew, alternative roundtrip trajectories exist. For example, a short-stay mission would have the crew spend 30 days on the Martian surface, while the outbound and inbound transfers would

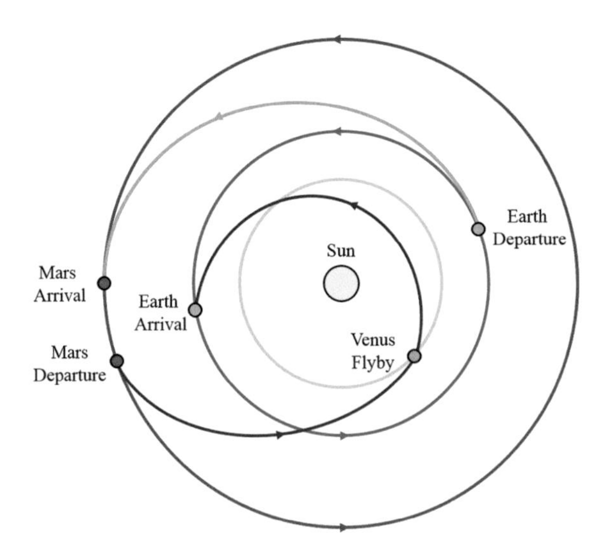

FIGURE 5.60 Example of a short-stay Earth-Mars roundtrip mission.

take approximately 7 and 9.5 months, respectively, for a total of about 1.5-year Mars mission, as shown in Figure 5.60. This mission makes use of a Venus flyby on the inbound trajectory, which means that the transfer vehicle would need to be appropriately equipped to withstand the increased solar radiation from the vicinity to the Sun.

If one wanted to minimize the transfer time while keeping the time spent at Mars on the order of multiple months, a fast transfer option could be used. This makes use of higher energy transfers between Earth and Mars, resulting in more Δv, but lower time of flight. Such a transfer would require an outbound transfer of ~5 months, a stay time of ~20 months, and a return transfer of ~3.5 months. Figure 5.61 shows this type of transfer.

Table 5.19 shows a summary of the Earth-Mars roundtrip missions that we discussed in this and the previous sections using times of flight computed using planetary ephemeris data during the 2014–2016 Earth-Mars synodic period. Similar launch and arrival opportunities repeat every Earth-Mars synodic period.

5.15.3 Cycler Orbits

A cycler orbit is a type of transfer orbit that periodically encounters two celestial bodies at regular intervals. In theory, a cycler orbit could be used to establish a "taxi" transport system between two celestial bodies. As we will discuss in this section, Earth-Moon and Earth-Mars cyclers are the most studied and perhaps interesting uses of cycler orbits.

Cycler orbits make use of free-return trajectories in a way that allows the cycler spacecraft to keep going between two celestial bodies with the use of minimal Δv

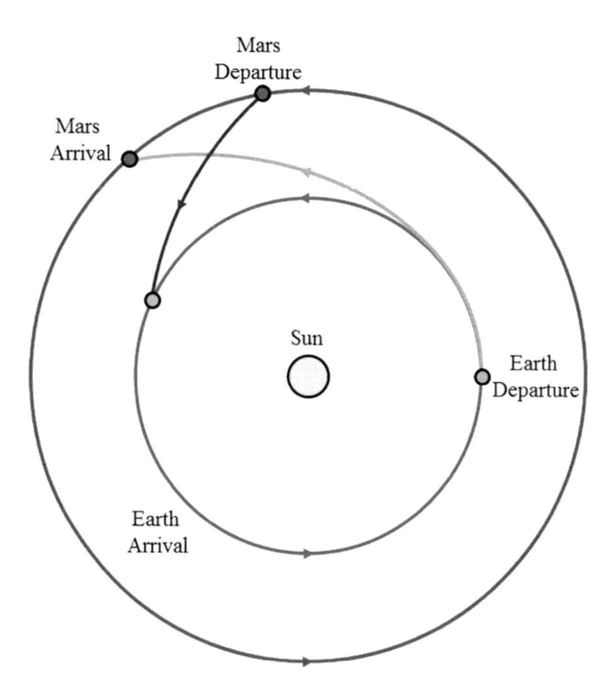

FIGURE 5.61 Example of a fast transfer Earth-Mars roundtrip mission.

TABLE 5.19
Summary of times (days) for various types of Earth-Mars roundtrip missions (Courtesy of NASA)

Type	Long Stay	Short Stay	Fast Transfer
Outbound transfer	224 days	224 days	150 days
Stay time	458 days	30 days	619 days
Inbound transfer	237 days	291 days	110 days
Total time	919 days	545 days	879 days

maneuvers – sometimes even no maneuvers at all. However, due to their prolonged flight durations, cyclers need to be equipped with heavy shielding against cosmic radiation and solar storms. Payloads that need to be shuttled between celestial bodies would be launched from a celestial body, rendezvous and dock with the cycler, and then detach from it once in proximity to the target celestial body. This concept of operations requires payloads to perform hyperbolic rendezvous maneuvers, which are considered to be extremely dangerous. In fact, if rendezvous and/or docking fails, the cycler orbit would not be able to slow down significantly to reattempt the maneuver, or it would otherwise miss its rendezvous window at the targeted celestial body.

The first idea of a cycler for interplanetary travel was an Earth-Venus cycler devised by Hollister [1969], where he suggested that such a cycler could be used for both communication and transportation purposes. Buzz Aldrin, the second man to walk on the surface of the Moon, was the first to present an Earth-Mars cycler – or simply Mars cycler – in 1985, which was based on his earlier lunar cycler research. Aldrin proposed a Mars cycler corresponding to a single Earth-Mars synodic period cycler [Aldrin, 1985], which is known as the Aldrin cycler. An Aldrin cycler is a special case of a Mars cycler that has an orbital period equal to the Earth-Mars synodic period (2.1351 years). In fact, an Aldrin cycler that is launched from Earth into a heliocentric elliptical orbit travels to and intercepts Mars in 146 days, continues orbiting the Sun for approximately 16 months beyond Mars's orbit, crosses Mars's orbit again – although Mars is not there at the time of crossing – and returns to Earth 146 days later, in the same orbital position from which it left Earth, requiring no additional Δv from its original Earth departure [Aldrin, 1985]. Since Earth and Mars have orbital periods around the Sun of 1 and 1.881 years, respectively, Earth completes approximately 15 orbits in the same time that Mars completes 8 orbits around the Sun. Because of this, Mars cyclers with a time per cycle equal to seven times the Earth-Mars synodic period (14.94 years $\approx$ 15 years), return to Earth in approximately the same position from which they departed originally. Figure 5.62 shows the geometry of a generic Earth-Mars cycler: Earth's orbit is in blue, Mars's orbit is in red, and the cycler's orbit is in green. Also, the cycler's orbit line of apsides (LOA) is labeled.

Each encounter of Earth and Mars is labeled with E for Earth and M for Mars, respectively. Here, we ignored the gravitational effects of Earth and Mars here, although, in reality, each Earth and Mars flyby needs to carefully planned to guarantee that the cycler stays on course once it returns to interplanetary space.

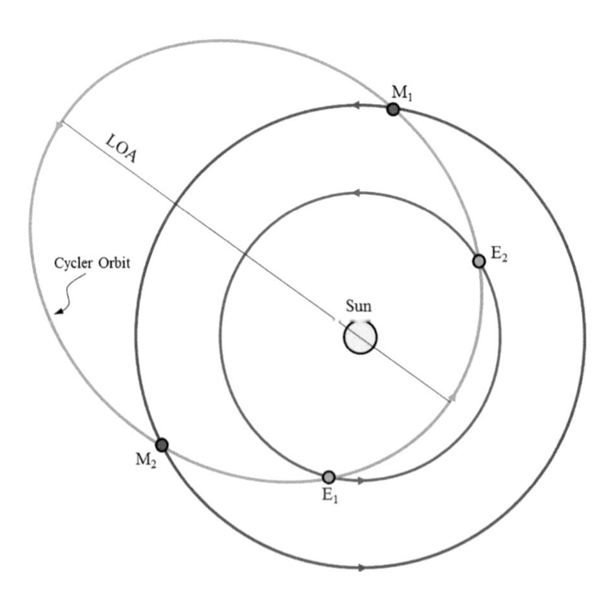

FIGURE 5.62 A single Earth-Mars cycler.

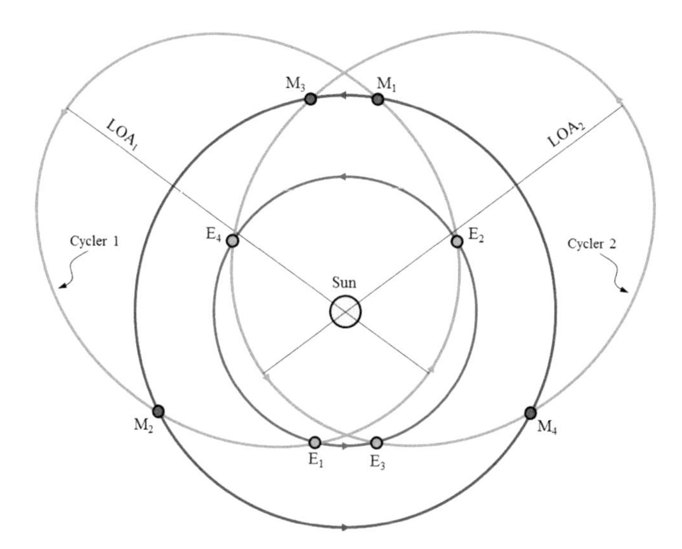

FIGURE 5.63 A double Earth-Mars cycler configuration.

Alternatively, the cycler could remain far enough from the sphere of influence of the target to minimize third-body gravitational effects. Additionally, a propellant-less interplanetary cycler is unlikely possible due to the fact that the orbits of the planets are not perfectly circular or coplanar. However, the amount of Δv needed to correct for these approximations are minimal when compared to the propulsive maneuvers needed to go between Earth and Mars, as we've seen in Sections 5.6 and 5.7.

For long-duration Mars missions, two separate cyclers, as shown in Figure 5.63, could be used in order to provide outbound and inbound trips at the same time: while one cycler is directed to Mars, the other would be returning to Earth, making exchanges of crew members, cargo resupplies, and deployment of scientific instruments faster and more reliable.

Lunar cyclers have also been proposed for Earth-Moon resupply missions and for other operations [Uphoff and Crouch, 1991]. Specifically, lunar cyclers could be used to resupply the Lunar Gateway, for example, or for delivering and returning crews between Earth and the Moon during the Artemis program. Furthermore, lunar cyclers could be used also to help assemble an interplanetary spacecraft that would then escape the Earth-Moon system to target another celestial body in the Solar System, such as an asteroid, Mars, or beyond [Genova and Aldrin, 2015].

5.16 LOW-THRUST TRAJECTORIES

The analysis of low-thrust orbit transfers has the advantage that for certain cases simplifications and assumptions can be made so as to decrease the reliance on numerical techniques. For example, if the force of thrust is purely in the radial direction or in the direction of motion, we can derive rather accurate analytical approximations to these problems – we'll look at these special cases at the end of this section. However, solutions to low-thrust – and thus continuous-thrust – trajectories generally require the use of sophisticated optimization methods and numerical techniques which are widely discussed in current literature.

5.16.1 Electric Propulsion

In the 1960s, many studied the utility of a solar-electric transfer vehicle. This spacecraft would have used solar-electric panels to power an electric (ion or arc-jet) engine. The thrust level developed for such an orbital transfer vehicle was very small (on the order of thousandths to hundredths of a g). The typical orbit transfer took a very long time, with the transfer trajectory resembling a spiral. Dickerson and Smith [1967] derived the necessary conditions required for optimal solar-electric powered flight, using the calculus of variations techniques from classical optimization theory. Olgevie [1968] proposed developing a solar-electric propulsion planetary orbiter spacecraft. He found that the use of this type of spacecraft (based on 1960s technology) could deliver to Mars a significantly larger payload than conventional rockets. Additionally, using a chemical rocket to escape Earth, he found that there was only a 15% increase in transfer time from Earth to Mars.

In the 1970s, NASA embarked on their Solar-Electric Propulsion Stage (SEPS) program. The significant problem associated with the trajectory optimization problem, as found by Olgevie et al. [1975], was that of maintaining the optimal path, while pointing the solar arrays (within tolerances) at the Sun. In the late 1980s, the U.S. Air Force began a program entitled ELectric Insertion Transfer Experiment (ELITE), in which the objective was to build, test, and fly a solar-electric orbit transfer and orbit maneuvering vehicle, as a precursor to an operational electric orbit transfer vehicle [Avila, 1992]. Both of these programs have faced the problem of an extended duration in a transfer orbit. This is one of the many factors that folds into the design trade studies.

The first flight of an electric low-thrust orbit transfer system was NASA's Deep Space 1 mission [Rayman et al., 1999]. This mission was launched in October 1998, and was part of NASA's New Millennium Program. After the technology demonstration was completed, the spacecraft encountered the asteroid Braille [Rayman, 2003]. The primary mission ended in September 1999, but an extended mission to the comet Borelly. The mission finally ended in December 2001, with the propulsion system accumulating 16,265 hours of operation, and expended 73.4 kg of propellant. The total Δv delivered by this electric propulsion system was approximately 4.3 km/s. The spacecraft is now still in orbit around the Sun.

Leveraging on the success of the Deep Space 1 mission, NASA has undertaken a low-thrust mission, named Dawn, to two protoplanets, Vesta and Ceres. The nominal trajectory of this spacecraft was shown previously in Figure 1.4. This mission began in 2007, and continued until 2018, when the extended mission ended due to the depletion of the spacecraft's hydrazine reserves. Dawn performed its first orbit insertion around the protoplanet Vesta on July 16, 2011, where it stayed for approximately 14 months to survey the celestial body. In 2012, it left Vesta and on March 6, 2015, it arrived at Ceres where it performed its second extraterrestrial orbit insertion. Although the mission has ended, the spacecraft is currently in orbit around Ceres in a stable orbit. Dawn is the first spacecraft to have orbited two different celestial bodies in our Solar System during the same mission, the first to have visited Vesta and Ceres, and the first to orbit a dwarf planet. These "records" would not have been possible without the use of electric propulsion which, despite the longer-than-usual interplanetary transfer times, delivered an amount of Δv that chemical propulsion systems would not have been able provide for the same amount of launch mass.

There are other missions (current and planned) that use electric propulsion systems under development, and research continues into trajectory optimization techniques for electric propulsion systems. While this technology is promising for many missions, there exist constraints that should be considered, such as eclipsing by a planet (which can cause a solar-panel-powered propulsion system to shut down). This is more common while in orbit around a planet, rather when in orbit around the Sun. Solar propulsion systems have other technical challenges, such degradation of the performance of solar panels; simple or complex models (depending on what is known about the behavior of the solar panels or the fidelity needs) can be developed and included in simulations in order to account for this and other effects on the trajectory of the spacecraft.

Electric propulsion finds use in many interplanetary and Earth-orbiting missions. One example, which covers both of these cases, is that of orbit raising maneuvers. These are continuous-thrust maneuvers that aim at reaching a prescribed final orbit, whether it be GEO or an Earth escape trajectory.

5.16.2 Solar Sails

A new technology that is under study that would provide a propellant-less propulsion system is the use of a solar sail to capture small solar radiation pressure. As the Sun emits light, photons that hit a high-reflectivity surface, such as a mirror, need to bounce back, changing their linear momentum. This change in momentum gives the object being hit an equal and opposite change in momentum. In fact, while massless, photons have a linear momentum which is equal to their energy divided by the speed of light. The combination of all of the photons together hitting a mirror acts somewhat like a wind hitting the sail of a boat that is soaring across the water. Therefore, a large solar sail could, in theory, provide a propulsive force to a spacecraft. Much like a sail on a sailboat, this solar sail could be reoriented to direct this propulsive force in various directions, as shown in Figure 5.64, where α represents the angle of incidence of the sail with respect to its orientation relative to the Sun.

We begin our analysis with Einstein's most famous equation from his theory of relativity, which states

$$E = mc^2 \tag{5.217}$$

where E is energy, m is mass, and c is the speed of light, which is approximately 2.998×10^8 m/s. Rewriting this equation in terms of linear momentum we get

$$mc = \frac{E}{c} \tag{5.218}$$

where mc is linear momentum.

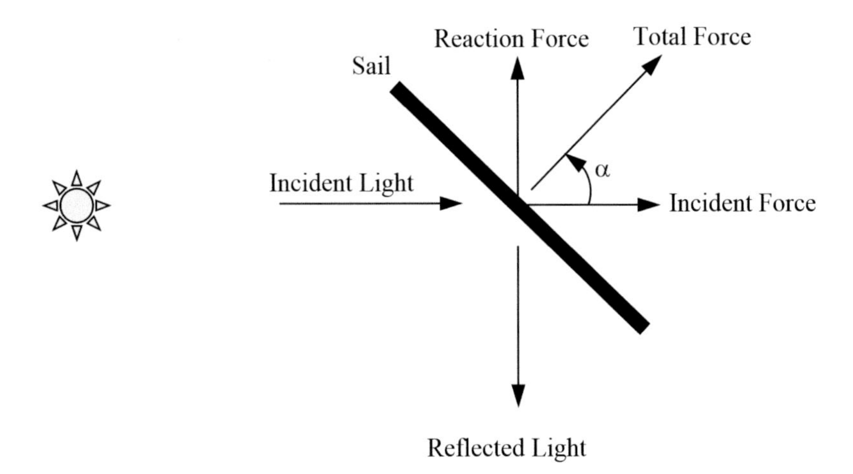

FIGURE 5.64 Solar sail geometry.

We also need to know the solar flux (SF), which, for spacecraft at Earth-like distance from the Sun, can be approximated as [Wertz, 1978]

$$SF = \frac{1358}{1.004 + 0.0334 \cos D_{aphelion}} \frac{W}{m^2} \tag{5.219}$$

where $D_{aphelion}$ is

$$D_{aphelion} = \frac{2\pi}{365} \Delta days_{aphelion} \tag{5.220}$$

where $\Delta days_{aphelion}$ are the number of days from aphelion for a given year – generally aphelion occurs on July 4. In Equation (5.220), the number 365 needs to be substituted with 366 for leap years. Additionally, the average value of the solar flux at Earth is 1,367 W/m^2. If a value of SF needs to be computed for a different celestial body, Equation (5.219) can still be used, keeping in mind that solar flux decreases as a function of the reciprocal of the distance squared. For example, at Mars (1.524 times further from the Sun than the Earth), the solar flux on average is approximately $1/(1.524)^2 = 0.4306$ times that of the Earth, or 589 W/m^2.

The change in linear momentum as given by Equation (5.218) due to solar flux as given in Equation (5.219) is therefore

$$p_{SRP} = \frac{SF}{c} \tag{5.221}$$

where the subscript SRP stands for solar radiation pressure (SRP). At Earth's distance, this value is approximately 4.560×10^{-6} N/m^2. Consequently, the force due to SRP on a spacecraft is

$$\mathbf{F}_{SRP} = -p_{SRP} C_R A_S \mathbf{n} \tag{5.222}$$

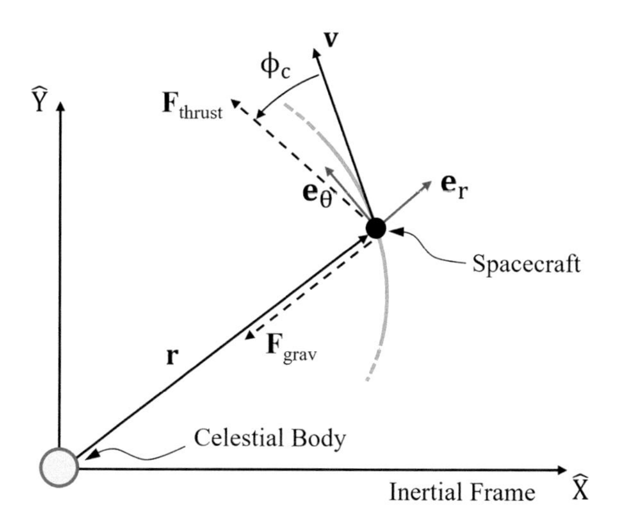

FIGURE 5.65 Diagram of a spacecraft under the influence of gravity and a generic force of thrust.

where A_S is the area facing the Sun, **n** is the vector normal to the solar sail's surface as shown in Figure 5.65, and C_R is a reflectivity coefficient ranging from 0 to 2. A value of 0 indicates that the object is translucent (no momentum change due to SRP), a value indicates of 1 that an object absorbs SRP acting as a black body, and a value of 2 indicates that an object reflects the incoming radiation like a perfectly reflective mirror perpendicular to the Sun. Solar sails are made of materials such that C_R is as close to 2 as possible, although it is a value that is very difficult to compute for any given spacecraft – including solar sails – similar to drag coefficients.

Lastly, in order to compute the acceleration due to SRP, we simply divide the force given by Equation (5.222) by the mass of the spacecraft

$$\mathbf{a}_{SRP} = \frac{\mathbf{F}_{SRP}}{m} \tag{5.223}$$

In a heliocentric reference frame, the equations of motion can be developed, resulting in the perturbation acceleration due to solar radiation pressure becoming

$$\ddot{\mathbf{r}} = -\frac{\mu}{r^3}\mathbf{r} + \mathbf{a}_{SRP} \tag{5.224}$$

While solar sails do not require propellant to function, they produce extremely small levels of thrust, and thus large sails would be required even for a very small payload to be thrusted from a given location to its destination in a reasonable amount of time. For example, a 1 km^2 square solar sail would be able to produce approximately only 4.56 N of thrust at Earth's distance from the Sun.

An excellent detailed description of the dynamics of solar sails can be found in McInnes [1999].

5.16.3 Low-Thrust Missions

Let's assume that a spacecraft is using a continuous, low-thrust propulsion system to change its orbit. This differs from the impulsive maneuver approximation that we use previously, since here, the propulsive maneuver is a significant percentage of the entire transfer – sometimes the entire time of flight – while in the latter, the duration of the propulsive maneuver is infinitesimally small when compared to the total duration of the transfer.

When discussing low-thrust maneuvers, we often refer to accumulated velocity change as the total velocity imparted by the thrust during the time of flight of a given thrusting maneuver. The accumulated velocity change helps us determine how much propellant the spacecraft needs for the maneuver, according to Equation (5.1), and is computed with the following expression

$$\Delta v_{acc} = \frac{a_{LT}}{\dot{m}_{spec}} \ln\left(1 + \dot{m}_{spec} TOF\right) \tag{5.225}$$

where the subscript LT stands for "low-thrust" and thus a_{LT} is the initial acceleration due to thrust, and $\dot{m}_{spec}$ is the specific mass flow rate, which is equal to $\dot{m}/m_0$, with m_0 being the starting mass of the spacecraft.

Let's now derive the equations of motion of a spacecraft under the influence of gravity ($\mathbf{F}_{grav}$) and thrust ($\mathbf{F}_{thrust}$), as shown in Figure 5.65.

Considering Equation (2.13), the acceleration of a spacecraft in the inertial frame, we can rewrite it in cylindrical coordinates using $\mathbf{e}_r$, $\mathbf{e}_\theta$, and $\mathbf{e}_n = \mathbf{e}_r \times \mathbf{e}_\theta$ as

$$\frac{d^2\mathbf{r}^{(I)}}{dt^2} = \left(\ddot{r} - r\dot{\theta}^2\right)\mathbf{e}_r + \left(r\ddot{\theta} + 2\dot{r}\dot{\theta}\right)\mathbf{e}_\theta + 0\,\mathbf{e}_n \tag{5.226}$$

Considering the thrust only acting in the orbital plane of motion of the spacecraft, we can write the acceleration due to low-thrust as

$$\mathbf{a}_{LT} = a_{LT,r}\,\mathbf{e}_r + a_{LT,\theta}\,\mathbf{e}_\theta \tag{5.227}$$

Since the only forces acting on the spacecraft are those of gravity and low-thrust, the overall acceleration on the spacecraft must be

$$\frac{d^2\mathbf{r}^{(I)}}{dt^2} = \left(a_{LT,r} - \frac{\mu}{r^2}\right)\mathbf{e}_r + a_{LT,\theta}\,\mathbf{e}_\theta \tag{5.228}$$

We can thus combine Equations (5.226) and (5.228) and equate terms in the radial and transverse directions to get the following system of equations

$$\begin{aligned} \ddot{r} - r\dot{\theta}^2 &= a_{LT,r} - \frac{\mu}{r^2} \\ r\ddot{\theta} + 2\dot{r}\dot{\theta} &= a_{LT,\theta} \end{aligned} \tag{5.229}$$

which represent the equations of motion of the spacecraft. Recognizing that the term $r\ddot{\theta} + 2\dot{r}\dot{\theta}$ can be written as

$$r\ddot{\theta} + 2\dot{r}\dot{\theta} = \frac{d}{dt}\left(r^2\dot{\theta}\right) \quad (5.230)$$

we can simplify Equations (5.229) as

$$\begin{aligned} \ddot{r} - r\dot{\theta}^2 &= a_{LT,r} - \frac{\mu}{r^2} \\ \frac{d}{dt}\left(r^2\dot{\theta}\right) &= a_{LT,\theta} \end{aligned} \quad (5.231)$$

It is sometimes convenient to rewrite Equation (5.231) in terms of the radial and transverse velocity components, v_r and v_θ, such that

$$v_r = \dot{r} \quad \text{and} \quad v_\theta = r\dot{\theta} \quad (5.232)$$

so that, after some algebraic manipulations, Equation (5.231) becomes

$$\begin{aligned} \dot{v}_r &= \frac{v_s^2}{r} - \frac{\mu}{r^2} + \frac{a_{LT}\sin\phi_c}{1 + \dot{m}_{spec}t} \\ \dot{v}_\theta &= -\frac{v_r v_\theta}{r} + \frac{a_{LT}\cos\phi_c}{1 + \dot{m}_{spec}t} \end{aligned} \quad (5.233)$$

where here we included specific mass flow rate, which is assumed to be constant, resulting in a linear decrease of spacecraft mass, and thus a non-constant acceleration due to thrust. Additionally, the angle φ_c is a control angle, which is measured from the velocity vector to the thrust vector, as shown in Figure 5.65. Writing Equation (5.233) using this control angle φ_c is used to appropriately set up optimal control problems where the φ_c is an unknown function of time that solves the optimal control problem. The formulation given by Equation (5.233) is also used by Bryson and Ho [1975] to derive optimal control laws for low-thrust trajectories, Additional resources on this optimization problem include von Stryk and Bulrich [1992], among others. Here, we will present a brief introduction on this matter, with a focus on the problem setup. In fact, there has been a significant amount of work in developing methods to solve the optimal low-thrust orbit transfer problem, including several kinds on solution methods.

The study of optimal orbit transfers dates back to the 1920s with the initial work of Hohmann [1925]. From there, a hiatus in the work occurred until the 1950s and 1960s. Lawden [1963] developed the theory of primer vector to examine the optimality of impulsive transfers, and this work is still used today. Recent work has involved using multiple burn transfers, i.e., a number of burns greater than two.

The fundamental form of general optimization problems (one which includes low-thrust trajectory optimization) involving non-linear differential equations is

$$\frac{d\mathbf{X}}{dt} = f\left[\mathbf{X}(t), \mathbf{u}(t), t\right]; \mathbf{X}(t_o), t_o \text{ given}; t_o \leq t \leq t_f, \tag{5.234}$$

given initial conditions $\mathbf{X}(t_0) = \mathbf{X}_0$ at some initial time, t_0, with some final value $\mathbf{X}(t_f)$ at some given or fixed time, t_f, such that $t_f - t_0$ is the time of flight, TOF, of interest. The state vector $\mathbf{X}(t)$ is an n-dimensional vector, and the control vector, $\mathbf{u}(t)$ is a k-dimensional vector. The goal is to find a control function $\mathbf{u}(t)$ that minimizes a cost function,

$$J\left[\mathbf{u}(t)\right] = \Phi\left(\mathbf{X}(t_f), t_f\right) \tag{5.235}$$

with the inequality constraint,

$$C(\mathbf{u}, t) \leq \mathbf{0} \tag{5.236}$$

and the terminal boundary conditions

$$\boldsymbol{\Psi}\left[\mathbf{X}(t_f)\right] = \mathbf{0} \tag{5.237}$$

where $\mathbf{C}(\mathbf{u}, t)$ is a constraint vector of dimension $k < m-1$, and $\boldsymbol{\Psi}$ is a constraint vector of dimension $q < n-1$. For low-thrust optimization, our state vector can include the position and velocity vectors and the mass. Constraints can include such things as maximum thrust level, bounds on pointing angles, and eclipsing geometry characteristics.

Optimization schemes can be categorized into two types. The first is the "indirect" method, and involves deriving the necessary conditions for optimality using the calculus of variations techniques. This problem is then solved numerically, which introduces discretization of the equations. The second method is known as the "direct" method. In this method, the optimal control problem is transformed into a non-linear programming formulation. This section discusses both methods and shows how they can be applied to optimal interplanetary trajectories.

Many authors have used different types of optimization schemes to solve the optimal trajectory transfer problem. Lawden's primer vector is limited, in that it is valid for impulsive approximations only. Shooting methods have been used over the years to solve any number of two-point boundary value problems [Osbourne, 1969]. However, the solution of optimal orbit transfer problems by this method is hampered by two problems. (1) the instability of the initial value problem for the system of differential equations, and (2) the requirement of "good" starting values of the parameters in the integration. As the solution is extremely sensitive to the initial guesses on the costate variables, even with the best of these approximate methods, errors in the final solution build up quickly and the true optimal solution is missed. Handelsman [1966] and Pines [1964] both found that guessing of the initial values of the costate variables may be eliminated by the use of the initial values from impulsive trajectories to initiate iteration for non-impulsive thrust. Igarashi and Spencer [2005] found that evolutionary algorithms can be used to help refine the guess of initial conditions for the adjoint variables.

The indirect method is the classical Mayer problem in the calculus of variation. The necessary conditions to minimize the cost function are best expressed in terms of the auxiliary functions and the Hamiltonian. The indirect method using the calculus of variations, the first-order necessary conditions for an optimal trajectory state that there exists a vector of adjoint variables, $\boldsymbol{\lambda}(t)$, such that

$$H(t) = \boldsymbol{\lambda}^T(t) f(\mathbf{X}, \mathbf{u}, t) \tag{5.238}$$

where $\boldsymbol{\lambda}$ is the adjoint vector, and the superscript T represents the transpose of the vector.

The first-order necessary conditions on the adjoint vector [Bryson and Ho, 1975] are

$$\frac{d\boldsymbol{\lambda}(t)}{dt} = -\left(\frac{\partial H}{\partial \mathbf{X}}\right)^T \tag{5.239}$$

Thus, in addition to the equations of motion, Equation (5.234), Equation (5.239) must be integrated simultaneously.

The set of adjoint equations, expressed in the **r**, **v**, m state representation, as

$$\frac{d\boldsymbol{\lambda}_r(t)}{dt} = -\left(\frac{\partial g}{\partial r}\right)^T \boldsymbol{\lambda}_v(t) \tag{5.240}$$

$$\frac{d\boldsymbol{\lambda}_v(t)}{dt} = -\boldsymbol{\lambda}_r(t) \tag{5.241}$$

$$\frac{d\lambda_m(t)}{dt} = -\lambda_v(t)\frac{T}{m^2(t)} \tag{5.242}$$

where $\boldsymbol{\lambda}_r(t)$ is the adjoint vector for $\mathbf{r}(t)$, $\boldsymbol{\lambda}_v(t)$ is the adjoint vector for $\mathbf{v}(t)$, and $\lambda_m(t)$ is the adjoint scalar for m(t). The scalar $\lambda_v(t) = \sqrt{\boldsymbol{\lambda}_v(t) \cdot \boldsymbol{\lambda}_v(t)}$.

The optimal control is found by minimizing the Hamiltonian with respect to the control vector, **u,**

$$\frac{\partial H}{\partial u} = \boldsymbol{\lambda}^T \frac{\partial f}{\partial u} = 0 \tag{5.243}$$

and the calculus of variation's Legendre-Clebsch condition $\frac{\partial^2 H}{\partial u^2} \geq 0$ (which means that this is a positive semidefinite matrix).

Using a variation on Pontryagin's maximum principle, coupled with a numerical search, for the Hamiltonian to be minimized with respect to the control vector, $\hat{\mathbf{u}}(t)$, the optimal control vector is

$$\hat{u}(t) = -\frac{\lambda_V(t)}{\lambda_V(t)} \tag{5.244}$$

Primer vector uses the adjoint vector or costate variables associated with the velocity vector on an optimal trajectory. The primer vector points in the exact opposite direction of the adjoint variable associated with the velocity vector. This is the optimal pointing direction. The behavior of these variables helps determine the optimality of the solution obtained in the optimal trajectory problem. The impulsive transfers used by Lawden [1963] use infinite thrust and zero burn time. Many analytical approximations use this impulsive assumption, since for short duration burns, and high-thrust engines, the analytical solution compares favorably with the exact solution.

Inserting this into the expanded form of the Hamiltonian, H, for the constant thrust, T, becomes

$$H = \lambda_r^T(t)v(t) + \lambda_v^T(t)g(r,t) - \left(\frac{\lambda_v(t)}{m(t)} + \frac{\lambda_m(t)}{c}\right)T \tag{5.245}$$

Previous authors [Zondervan, 1983] have called the function,

$$K(t) = \frac{\lambda_v(t)}{m(t)} + \frac{\lambda_m(t)}{c} \tag{5.246}$$

the switching function, which determines whether the thrust is *on* or *off*. If $K > 0$, the thrust is on; if $K < 0$, the thrust is off, and the instantaneous time that $K = 0$, the thrust is turned on or off.

Using many of the optimization methods previously mentioned, optimal solutions can be found. The numerical integration is relatively straightforward, but the determination of the initial conditions of the costate variables is where the problem lies.

In the direct method, the infinite dimensional control problem is converted into a mathematical programming problem. This procedure can best be described as direct transcription (first coined by Canon [1970]), or through collocation, which is fitting the results to mathematical curves. Direct transcription discretizes the problem with high order integration methods, while collocation represents the state and control variables with parameterized functions, commonly piecewise polynomials.

Enright and Conway [1991, 1992] developed a direct transcription method to find the optimal solution to the finite-thrust spacecraft trajectory problem, including a low-thrust escape spiral, a three-burn rendezvous, and a low-thrust transfer to the Moon. In addition to the many direct transcription applications, many collocation methods have also been implemented. Russell and Shampine [1972] used a piecewise polynomial function as a method for solving two-point boundary problems. These collocation methods are a type of integration method, shown to be equivalent to implicit Runge-Kutta methods. Hargraves and Paris [1987] found that these methods were easy to program, and showed them to be very efficient for several sample problems. However, in a further study by Enright and Conway [1991], they developed a collocation method to solve the low-thrust optimal transfer problem and showed that this method was a specific implementation of the general direct transcription method, and that there were deficiencies in the accuracy of their solution procedure.

Each of these authors made assumptions and approximations that allowed for simple solutions for their individual problems. Although every case was not necessarily an optimal solution, their results closely matched the optimal solutions and are useful for their individual research goals.

In the next two sections, we will present derivations and approximations for two particular kinds of low-thrust maneuvers that make use of radial-only and tangential-only thrust, respectively, which allow us to significantly simplify the low-thrust problems and obtain analytic expressions for each.

5.16.3.1 Radial Thrust

As seen in Section 5.16.2, solar sails are capable of producing an optimal level of thrust if they point perpendicular to the Sun, thus producing primarily a radially outward thrust from the Sun, or in the $\mathbf{e}_r$ direction. In this section, we analyze the special low-thrust case of a spacecraft that is subjected to the force of gravity and a purely radially outward thrust, whether it be from solar radiation pressure or, for example, a thruster. Refer to Figure 5.65, assuming that thrust is only pointed in the radial direction.

Considering radial-only thrust, $a_{LT,\theta} = 0$, the equations of motion given by Equation (5.231) simplify to

$$\begin{aligned} \ddot{r} - r\dot{\theta}^2 &= a_{LT,r} - \frac{\mu}{r^2} \\ \frac{d}{dt}\left(r^2\dot{\theta}\right) &= 0 \end{aligned} \tag{5.247}$$

If we integrate the second part of Equation (5.247), we obtain the angular momentum h as the constant of integration. Assuming that the spacecraft starts from a circular orbit with radius r_0, we can use the angular momentum to write $\dot{\theta}$ as

$$\dot{\theta} = \frac{h}{r^2} = \frac{\sqrt{\mu r_0}}{r^2} \tag{5.248}$$

Substituting Equation (5.248) into (5.247) and rewriting $\ddot{r}$ as $\frac{d}{dt}(\dot{r})^2$, the first equation becomes

$$\frac{1}{2}\frac{d}{dt}(\dot{r})^2 - \frac{\mu r_0}{r^3} + \frac{\mu}{r^2} = a_{LT,r} \tag{5.249}$$

and integrating with respect to time gives

$$(\dot{r})^2 = 2\int_{r_0}^{r}\left(\frac{\mu r_0}{\xi^3} - \frac{\mu}{\xi^2} + a_{LT,r}\right)d\xi \tag{5.250}$$

where ξ is used a dummy variable for the purpose of integration and we assumed that the acceleration due to thrust is constant – which would not be true for a solar sail since thrust would decrease as the spacecraft travels further from the Sun. The result of the integration of Equation (5.250) gives

$$(\dot{r})^2 = (r - r_0)\left[2a_{LT,r} - \frac{\mu}{r_0 r^2}(r - r_0)\right] \tag{5.251}$$

If we are interested in seeing when the spacecraft will escape, then we can combine Equation (5.251) with the *vis viva* equation (Equation 2.50) and the fact that $v^2 = (\dot{r})^2 + (r\dot{\theta})^2$ to get

$$r_{esc} = r_0\left(1 + \frac{\mu}{2r_0 a_{LT,r}}\right) \tag{5.252}$$

where r_{esc} is the radius at which the spacecraft escapes. Furthermore, substituting Equation (5.252) into Equation (5.251) gives

$$(\dot{r})^2 = \frac{2a_{LT,r}}{r^2}(r - r_0)\left[r^2 - (r_e - r_0)(r - r_0)\right] \tag{5.253}$$

from which it can be deduced that in order for $\dot{r}$ to remain positive – and thus for the spacecraft to increase its orbit to escape – the following condition must hold true

$$r_{esc} \le 5r_0 \tag{5.254}$$

In other words, the thrust in the radial direction must be sufficient to allow the spacecraft to escape, or combining Equations (5.252) and (5.254)

$$a_{LT,r} \ge \frac{\mu}{8r_0} \tag{5.255}$$

which is equal to 1/8 of the gravitational acceleration of the spacecraft at its starting radius. This means that, for $a_{LT,r} = \frac{\mu}{8r_0}$, the spacecraft is barely able to escape the system.

For example, let's consider a spacecraft that is launched into a heliocentric orbit similar to that of the Earth, i.e. with $r_0 = 1$ AU and $v_0 = 29.78$ km/s. We are interested in seeing how far (r_{esc}) and how long (t_{esc}) it needs to travel before it reaches escape velocity with respect to the Solar System, assuming that it outputs constant thrust in the radial direction only. We then numerically integrate the equations of motion Equation (5.247) for various levels of $a_{LT,r}$ until the spacecraft reaches a parabolic orbit. Specifically, looking at values of radial thrust that go from 1/8 to 1/4 of the starting gravitational acceleration, the resulting low-thrust orbits are shown in Figure 5.66, where we plotted six of these orbits along with the starting orbit. The orbits are plotted until the escape condition – orbital energy becoming zero – is reached.

Table 5.20 shows the corresponding escape radii and times of flight for the orbits shown in Figure 5.67.

Computing the escape radii and times for all levels of thrust from 1/8 to 1 of the starting gravitational acceleration, we can plot the graph shown in Figure 5.67.

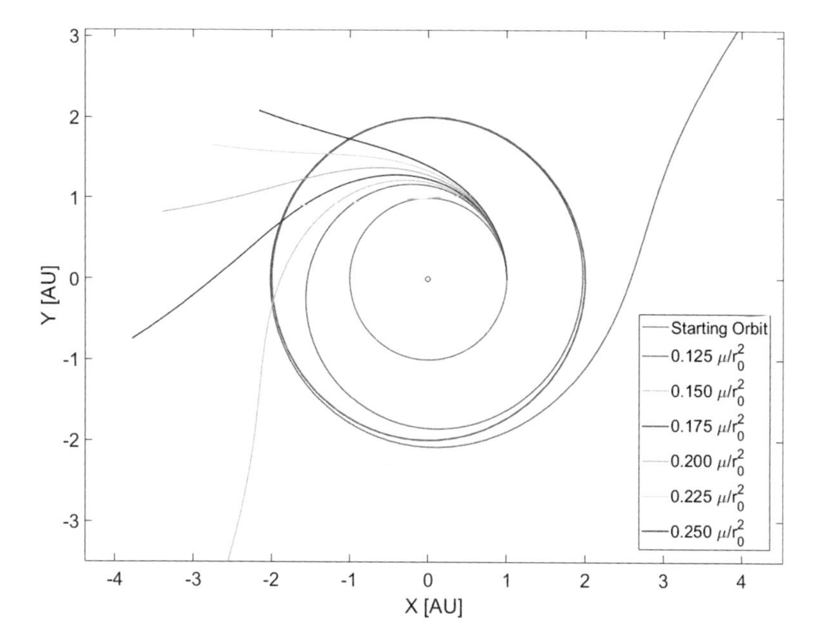

FIGURE 5.66 Plot of six low-thrust orbits that escape the Solar System.

TABLE 5.20
Radial thrust example for various levels of thrust

Acceleration [μ/r_0^2]	r_{esc} [AU]	t_{esc} [Years]
0.125	4.968	4.118
0.150	4.333	1.883
0.175	3.857	1.383
0.200	3.450	1.109
0.225	3.222	0.9308
0.250	3.000	0.8034

As expected, the values of r_{esc} and t_{esc} decrease as the thrust level increases. Note that here we assumed that the spacecraft outputs a constant level of thrust for the entire time it takes from start to escape conditions. Implementing thrusting and coasting arcs would result in different results.

If the spacecraft does not have sufficient radial thrust, the spacecraft is unable to reach escape conditions and leave the system. In our example, if we set a level of thrust equal to $a_{LT,r} = \frac{\mu}{10r_0}$, we obtain the orbit that is shown in Figure 5.68, where it can be seen that the orbit of the spacecraft oscillates between the starting radius – here 1 AU – and a maximum value of 1.382 AU. Here, the orbit was propagated for a total of 10 years.

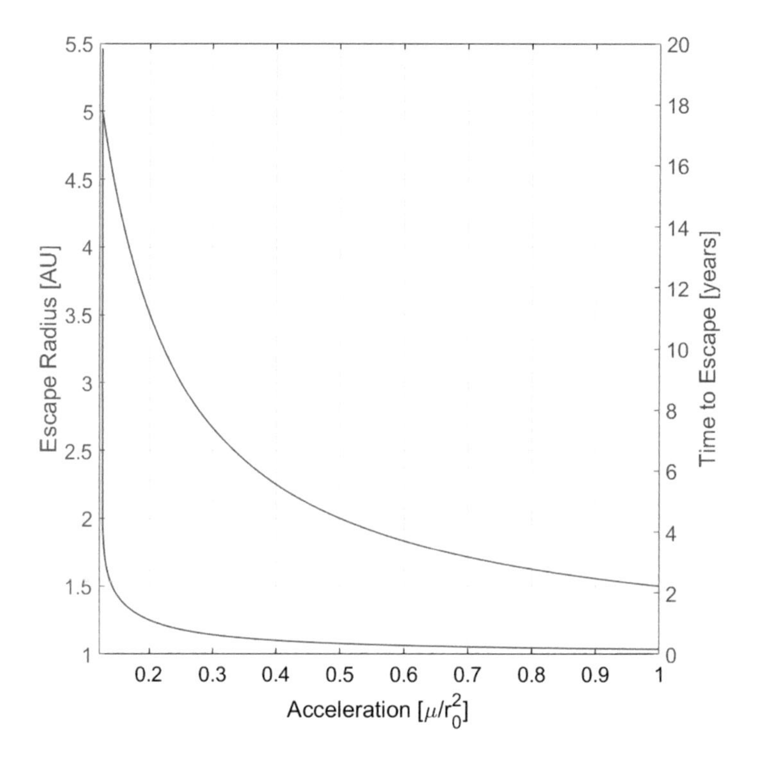

FIGURE 5.67 Plot of the escape radii and times for radial thrust.

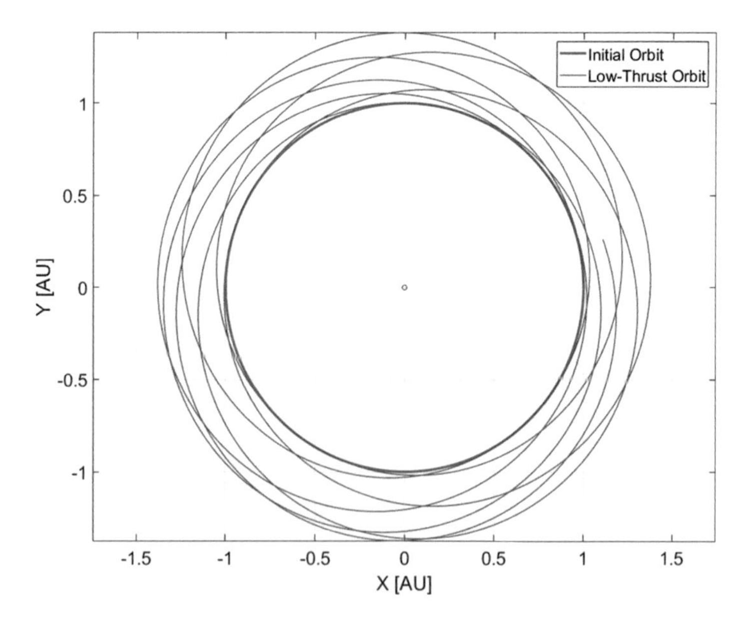

FIGURE 5.68 Plot of a radial thrust orbit without enough thrust to escape.

In this section, we assumed that the acceleration due to thrust is constant. However, this would not be true for a solar sail that is always pointing away from the Sun, for example, requiring to plug Equation (5.223) into Equation (5.250) – keeping in mind that spacecraft mass is constant for a solar sail – and perform the integration accordingly. Alternatively, one can integrate the equations of motion (Equation 5.247) numerically using any numerical methods, such as the ones described in Section 2.8.

5.16.3.2 Tangential Thrust

Let's now consider the thrust force to be always in the direction of motion of the spacecraft – or always tangent to its path – as shown in Figure 5.69.

Here, $\mathbf{e}_N$ – not to be confused with $\mathbf{e}_n$, the direction normal to the orbital plane – is a unit vector that always points in the direction normal to the tangential direction, $\mathbf{e}_T$, and that lies on the orbital plane of motion of the spacecraft. Additionally, φ is an angle that is measured positive counterclockwise from a given reference direction, which is generally the x-axis of an inertial frame.

Along the path, the differential length ds is given by

$$ds = \sqrt{(dr)^2 + (rd\theta)^2} \tag{5.256}$$

so that

$$\frac{ds}{dt} = \sqrt{\left(\frac{dr}{dt}\right)^2 + \left(r\frac{d\theta}{dt}\right)^2} \tag{5.257}$$

and thus, we can write the velocity vector in terms of ds/dt simply as

$$\mathbf{v} = \frac{ds}{dt}\mathbf{e}_T \tag{5.258}$$

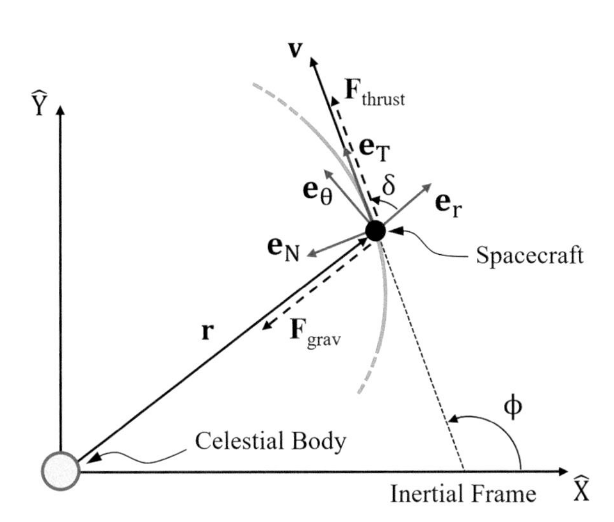

FIGURE 5.69 Diagram of a spacecraft under the influence of gravity and tangential thrust.

Also, we can write the acceleration as the time derivative of velocity, which, using the quantities described here, becomes

$$\frac{d\mathbf{v}^{(I)}}{dt} = \dot{v}\mathbf{e}_T + \dot{\phi}\mathbf{e}_n \times v\mathbf{e}_T \tag{5.259}$$

Using the definition of radius of curvature ρ

$$\rho = \frac{d\phi}{ds} \tag{5.260}$$

we can simplify Equation (5.259) to

$$\frac{d\mathbf{v}^{(I)}}{dt} = v\frac{dv}{ds}\mathbf{e}_T + \frac{v^2}{\rho}\mathbf{e}_N \tag{5.261}$$

from which we can write the square of the acceleration, a^2, as

$$a^2 = \left(\frac{dv}{dt}\right)^2 + \frac{v^4}{\rho^2} \tag{5.262}$$

Solving Equation (5.262) for the reciprocal of the radius of curvature leads to

$$\frac{1}{\rho} = \frac{1}{v^2}\sqrt{a^2 - \left(\frac{dv}{dt}\right)^2} \tag{5.263}$$

Using the definitions of acceleration and velocity along the path, we can also write this quantity as

$$\frac{1}{\rho} = \frac{1 - \left(\frac{dr}{ds}\right)^2 - r\frac{d^2r}{ds^2}}{r\sqrt{1 - \left(\frac{dr}{ds}\right)^2}} \tag{5.264}$$

In terms of tangential and normal coordinates, the acceleration vector can be written as

$$\mathbf{a} = \frac{dv}{dt}\mathbf{e}_T + \frac{v^2}{\rho}\mathbf{e}_N \tag{5.265}$$

where we have assumed that no out-of-plane acceleration occurs and we have included the acceleration due to gravity and tangential thrust on the right-hand side of the equation.

$$\mathbf{a} = a_{LT,T}\,\mathbf{e}_T - \frac{\mu}{r^2}\mathbf{e}_r \tag{5.266}$$

Note that in Equation (5.266), we have a mixture of cylindrical and normal-tangential unit vectors. In order to ensure that all terms of Equations (5.213) and (5.214) are written in the same coordinate frame, we introduce an angle δ, as shown in Figure 5.69. This angle δ is called the flight-direction angle, and it is measured positive counterclockwise from $\mathbf{e}_r$ to $\mathbf{e}_T$. Using angle δ, we can write the acceleration components in Equation (5.265) as

$$\begin{aligned} \frac{dv}{dt} &= a_{LT,T} - \frac{\mu}{r^2}\cos\delta \\ \frac{v^2}{\rho} &= \frac{\mu}{r^2}\sin\delta \end{aligned} \tag{5.267}$$

where cosδ and sinδ can be written as

$$\begin{aligned} \cos\delta &= \frac{dr}{ds} \\ \sin\delta &= r\frac{d\theta}{ds} \end{aligned} \tag{5.268}$$

Substituting Equation (5.268) into Equation (5.267) yields

$$\begin{aligned} v\frac{dv}{ds} &= a_{LT,T} - \frac{\mu}{r^2}\frac{dr}{ds} \\ \frac{v^2}{\rho} &= \frac{\mu}{r^2}\frac{d\theta}{ds} \end{aligned} \tag{5.269}$$

where the term 1/ρ can be substituted using Equation (5.264) and, simplifying, results in the following equations of motion

$$\begin{aligned} &\frac{1}{2}\frac{d\left(v^2\right)}{ds} + \frac{\mu}{r^2}\frac{dr}{ds} = a_{LT,T} \\ &rv^2\frac{d^2r}{ds^2} + \left(v^2 - \frac{\mu}{r}\right)\left[\left(\frac{dr}{ds}\right)^2 - 1\right] = 0 \end{aligned} \tag{5.270}$$

which are sometimes referred to as Bernoulli's equations. Assuming that the starting orbit is circular, the initial conditions are given by

$$\begin{aligned} r(t_0) &= r_0 \\ v^2(t_0) &= \frac{\mu}{r_0} \\ \left.\frac{dr}{ds}\right|_{t=t_0} &= 0 \end{aligned} \tag{5.271}$$

Assuming that $a_{LT,T}$ is constant, Equation (5.270) can be integrated using the initial conditions provided by Equation (5.271), from which we obtain

$$v^2 = 2 s a_{LT,T} + \mu\left(\frac{2}{r} - \frac{1}{r_0}\right) \tag{5.272}$$

where s is the length of the path traversed by the spacecraft.

If $a_{LT,T} \ll \frac{\mu}{r^2}$, which is often case for missions using continuous-thrust maneuvers, then $\frac{d^2r}{ds^2} \approx 0$ always, meaning that the orbit remains approximately circular throughout the duration of the maneuver and thus $v \approx \sqrt{\frac{\mu}{r}}$. As a result, Equation (5.272) can be rearranged and approximated as

$$r = \frac{\mu r_0}{\mu - 2 s r_0 a_{LT,T}} \tag{5.273}$$

Since the spacecraft escapes when the speed reaches the value of $\sqrt{2\mu/r}$, the second part of Equation (5.270) requires that

$$2r\frac{d^2r}{ds^2} = 1 - \left(\frac{dr}{ds}\right)^2 \tag{5.274}$$

Thus, it can be shown that the escape conditions for a spacecraft subjected to a small tangential constant thrust are

$$\begin{aligned} s_{esc} &= \frac{v_0^2}{2 a_{LT,T}}\left(1 - \frac{1}{v_0}\sqrt[4]{20 a_{LT,T}^2 r_0^2}\right) \\ r_{esc} &= \frac{r_0 v_0}{\sqrt[4]{20 a_{LT,T}^2 r_0^2}} \end{aligned} \tag{5.275}$$

Also, the time required to escape, t_{esc}, can be computed by integrating $dt = ds/v$ with the expression for v given by Equation (5.272), giving

$$t_{esc} - t_0 = \frac{v_0}{a_{LT,T}}\left[1 - \left(\frac{20 a_{LT,T}^2 r_0^2}{v_0^4}\right)^{\frac{1}{8}}\right] \tag{5.276}$$

Additionally, the number of revolutions required to escape, N_{esc}, can be computed with as follows

$$N_{esc} = \frac{1}{2\pi}\int_0^{s_{esc}} \frac{ds}{r} = \frac{v_0^2}{8\pi r_0 a_{LT,T}}\left(1 - \frac{\sqrt{20} a_{LT,T} r_0}{v_0^2}\right) \tag{5.277}$$

TABLE 5.21
Tangential thrust example for various levels of thrust

Acceleration [μ/r_0^2]	s_{esc} [AU]	r_{esc} [AU]	t_{esc} [Years]	N_{esc}
1/8	1.009	1.337	0.1723	0.1404
1/100	39.43	4.729	8.597	3.801
1/1,000	466.6	14.95	118.0	39.61

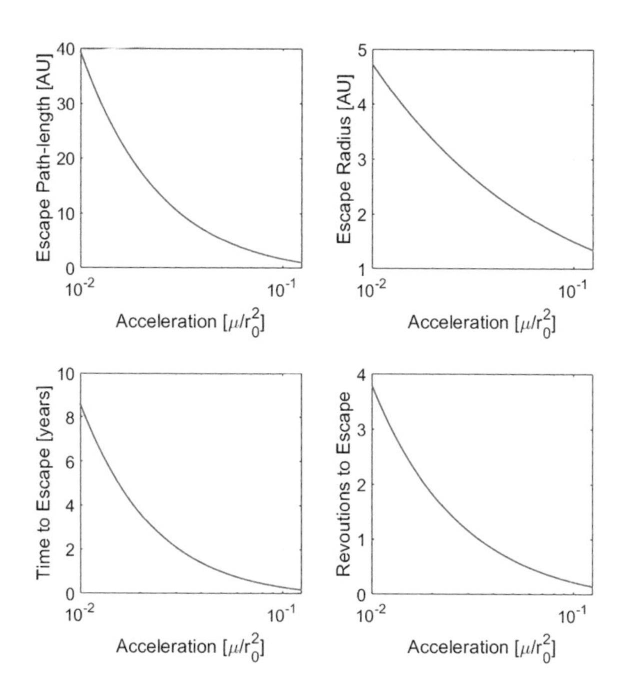

FIGURE 5.70 Tangential thrust example for various levels of thrust.

Note that, unlike the case of radial-only thrust, here there is no minimal amount of thrust required to escape. Of course, larger amounts of thrust will require less time to escape. If we consider the same example as before ($r_0 = 1$ AU and $v_0 = 29.78$ km/s), using the same levels of thrust but in the tangential direction, we obtain the results listed in Table 5.21. Figure 5.70 shows the various values of escape path-length, radius, time to escape, and revolutions to escape for values of acceleration between 1/100 and 1/8 of μ/r_0^2.

PROBLEMS

1. A spacecraft is launched into a circular orbit around the Earth at an altitude of 450 km. The upper stage of the launch vehicle initiates a Hohmann transfer to GEO through a GTO. How much Δv is the launch vehicle expected

to perform (Δv_1) and how much Δv is the spacecraft's propulsion system expected to deliver in order to perform the adequate orbit insertion at GEO (Δv_2)? How long does the transfer take?

2. A space vehicle in a circular orbit at an altitude of 300 km above Mars's surface executes a Hohmann transfer. The total Δv available is 0.35 km/s. This means that this Δv needs to be split between Δv_1 (to initiate the transfer) and Δv_2 (to complete the transfer). What is the *maximum* altitude of the final orbit that this spacecraft can reach with the allotted Δv? How long does this transfer take?
3. A spacecraft completes an interplanetary Hohmann transfer from Earth to Venus. The crew wants to come back to Earth.
    a. How long do they have to wait before Earth and Venus align themselves to allow for a Hohmann transfer from Venus to Earth?
    b. What is the angle between Earth and Venus at departure?
    c. What is the angle between Earth and Venus at arrival?
4. For a Hohmann transfer from Earth to Venus, plot C3 vs. time for a window that begins 15 days before the optimal Hohmann geometry and ends 15 days after the Hohmann alignment. Increment launch date by 1 day, and use the time of flight of 146 days. Keep the date of arrival constant.
5. Starting with Equations (5.12)–(5.15) and the definitions of R and S, prove Equation (5.16). *Hint*: an equation that you will find useful in this derivation is that of angular momentum h written as a function of apoapse radius r_a and periapse radius r_p only, i.e. $h = \sqrt{2\mu}\sqrt{\frac{r_a r_p}{r_a + r_p}}$
6. Compute the Δv and TOF for a bi-elliptic heliocentric transfer between Earth (1 AU) and Uranus (19.19 AU) using $S = 2$. How do Δv and TOF for this transfer compare with those of a Hohmann transfer?
7. A resupply of potatoes for Mark Watney is currently orbiting Mars in a circular orbit with altitude of 2,000 km (green orbit in Figure 5.71). The resupply vehicle performs a retrograde Δv that puts it on an elliptical orbit (blue orbit) intersecting Mars at an angle $\alpha = 30°$ from the line of apsides (LOA) of the new orbit, as shown in Figure 5.71. Calculate the Δv (in km/s) required to perform this orbital maneuver.
8. A spacecraft is orbiting Earth with semimajor axis $a_1 = 15{,}000$ km, and eccentricity $e_1 = 0.45$.
    a. Calculate the flight-path angle ϕ_1 when the spacecraft reaches a true anomaly of $\theta_1 = 258°$.
    b. Calculate the Δv needed to circularize the orbit at $\theta_1 = 258°$ where the radius is r_1.
    c. How much has the line of apsides rotated from the initial elliptical orbit to the final circular orbit?
9. Calculate the Δv required to change the inclination of an Earth orbit at apoapsis, using the following data: $a = 13{,}952$ km, $e = 0.47$, $i_1 = 28.5°$, $i_2 = 63.4°$, $\omega = 0°$. Note that these orbital elements define an orbit with peri apsis at the ascending node.

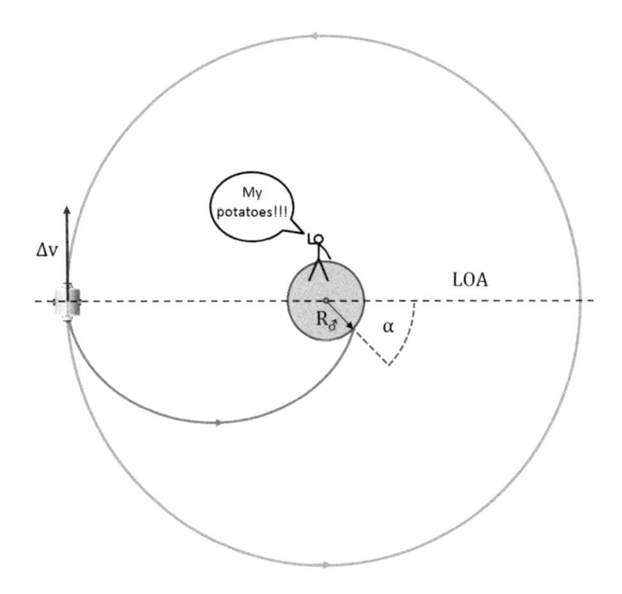

FIGURE 5.71 Mark Watney, visibly angry.

10. Consider a spacecraft in GEO. It is desired to shift its longitude by 12° westward using three revolutions of its phasing orbit. Calculate the orbital period of the phasing orbit, the total TOF, and Δv for such phasing maneuver.
11. In this problem use canonical units such that $\mu_{\odot} = 4\pi^2 \, LU^3/TU^2$, where 1 LU = 1 AU and 1 TU = 1 year. *Planet Express* has been tasked with delivering a shipment of buggalo[20] milk from Mars to a colony of Amphibiosans orbiting Jupiter. Assume all transfers in this problem are prograde and that Mars and Jupiter are in circular orbits around the Sun. Also, ignore the gravity of Mars and Jupiter.
    a. Determine the type of orbit (i.e., conic section) for the heliocentric cruise required to transfer from Mars to Jupiter provided that the transfer angle $\Delta\theta$ is 192.558° and the time of flight (TOF) is 1.4 years.
    b. Treating the transfer angle $\Delta\theta$ as a variable, determine the range of values for the speed v_1 (in LU/TU) that the *Planet Express* Ship ("Old Bessie") would require for a *minimum-energy transfer ellipse* as it enters the transfer orbit at Mars.
12. A spacecraft is transferring from Earth to Mars. Assume a change in true anomaly of 120° (true anomaly from Earth departure to Mars arrival). Find the heliocentric transfer orbit characteristics (a and e) assuming a 300-, 200-, 100-, 50-, and 25-day transfer. Be sure to state all of your assumptions. Make sure to compute the minimum transfer time you can have and still have an elliptical transfer to ensure that you are solving Lambert's equation using the correct conic type.

[20] (From Futurama) Buggalos are a type of creature that are half beetle, half cow. They are thought to be native to Mars as Native Martians have used them to fly on.

13. Consider Lambert's problem. Determine the expression for the parabolic transfer time t_p, i.e., the transfer time between P_1 and P_2 assuming that the orbit is a parabola, or $t_p = \frac{\sqrt{2}}{3\sqrt{\mu}}\left[s^{3/2} - \text{sgn}(\Delta\theta)(s-c)^{3/2}\right]$. Start with Lambert's equation for an elliptical orbit, $\sqrt{\mu}(t_2 - t_1) = a^{3/2}\left[\alpha - \beta - (\sin\alpha - \sin\beta)\right]$ and proceed to the limit as $a \to \infty$. Be sure to account for the two cases, $\Delta\theta \le \pi$ and $\Delta\theta > \pi$.
14. Show that $p_m = \frac{2(s-r_1)(s-r_2)}{c} = \frac{r_1 r_2(1-\cos\Delta\theta)}{c}$
15. At time $t_0 = 0$, the International Space Station (ISS) is ejecting a CubeSat through its CubeSat deployer mechanism. Assume the ISS is in a geocentric circular orbit with period of 92 minutes. With respect to the Cartesian Hill-Clohessy-Wiltshire (HCW) frame of reference, the CubeSat is ejected with a relative velocity of $\delta\mathbf{v}(t_0) = 5\mathbf{i} + 10\mathbf{j} + 0\mathbf{k}$ m/s
    a. Using the HCW relative motion equations, compute how far away (in km) the satellite is from the ISS at $t = 46$ minutes.
    b. What happens to the relative distance as time increases? Does this make physical sense? Briefly explain why or why not.
16. During the Apollo program and during the lunar orbit rendezvous, the lunar module launched from the Moon and would rendezvous with the command module. Had the lunar module been able to get into orbit, but was not able to get to the command module, the command module could have maneuvered to rendezvous with the lunar module.
    a. Assume that the command module is in a 100 km altitude circular orbit about the Moon. The lunar module is a circular orbit 100 km behind and 10 km below the command module. What is the Δv necessary for the lunar module to rendezvous with the command in 1 hour?
    b. If the command module had to go after the lunar module in its orbit, what Δv would be needed to rendezvous in 1 hour (assuming the same relative positions in part (a))?
17. Compute the Δv_1, Δv_2, and Δv_{total} (in km/s) and TOF (in years) required to perform an interplanetary Hohmann transfer from Earth to Jupiter. Assume that you start from a circular LEO having altitude of 300 km and you are targeting a circular orbit around Jupiter having altitude of 4,300 km. How could one lower the Δv requirements for a mission targeting Jupiter? Explain, referring to maneuvers discussed in this chapter.
18. Compute the Δv_1, Δv_2, and Δv_{total} (in km/s) and TOF (in years) required to perform an interplanetary Hohmann transfer from Earth to Venus. Assume that you start from a circular LEO having altitude of 300 km and you are targeting a circular orbit around Venus having altitude of 500 km.
19. Create a porkchop plot for Earth-Venus transfers for the following sets of dates
    - Departure dates: 01 Sep 2052 through 01 Mar 2054
    - Arrival dates: 01 Sep 2053 through 01 Jul 2054

using 1-day time steps. Use planetary ephemerides provided either by JPL Horizons or MATLAB's *planetaryEphemeris* function. Assuming departure and arrival planetary altitudes of 400 and 500 km, respectively, compute and plot a porkchop plot for total mission Δv. What is the minimum total Δv? For which departure and arrival dates does it occur? Is the transfer type I or type II?

20. Recreate the porkchop plot shown in Figure 5.22. Use similar contour levels for C_3 at launch, arrival v_∞, and TOF.
21. Assume that a spacecraft performs an Earth-Venus interplanetary Hohmann transfer – assume circular coplanar orbits for simplicity. However, at arrival, instead of performing an orbit insertion, the spacecraft does a lit-side leading edge flyby of Venus. Compute and plot the turn angle, hyperbolic flyby eccentricity, impact parameter, and post-flyby aphelion as a function of flyby altitude, from a minimum altitude of 200 km to a maximum altitude of 10,000 km.
22. For a hyperbolic gravity assist (GA) of a target planet "$\otimes$" with gravitational parameter $\mu_\otimes$ and physical radius $R_\otimes$, the periapsis of the hyperbolic GA trajectory has eccentricity e, periapse radius r_p, and the inbound and outbound velocities at the sphere of influence of the planet are $\mathbf{v}_\infty^-$ and $\mathbf{v}_\infty^+$, respectively. Determine the values of r_p and magnitude of $\mathbf{v}_\infty^-$ that yield the *maximum* possible magnitude of the gravity assist Δv_{GA}. Express your answers for r_p and v_∞^- in terms of $\mu_\otimes$ and $R_\otimes$ (noting that $r_p \geq R_\otimes$). Then, determine the *maximum* value of Δv_{GA} in terms of $\mu_\otimes$ and $R_\otimes$. Lastly, determine the numerical values for the corresponding turn angle δ and eccentricity e.
23. A spacecraft is in orbit around Mars and is beginning to aerobrake. Initially, it starts out in an orbit with periapse and apoapse altitudes of 140 and 1,000 km, respectively. Use the following atmospheric and spacecraft physical characteristics

Density Model	**Values**
Reference density	30 kg/km^3
Reference altitude	140 km
Scale height	10 km
Drag coefficient	2.2
Projected area	23 m^2
Mass	1,100 kg

   a. Propagate the orbit and estimate how long it takes to reduce the apoapse to 140 km altitude.
   b. If a propulsive maneuver of 1 m/s is added at apoapse on the first pass, how long does it take to reduce the apoapse to 140 km?
   c. If a propulsive maneuver of 10 m/s is added at apoapse on the first pass, how long does it take to reduce the apoapse to 140 km?

24. A spacecraft is entering the Martian atmosphere, on its way to land. At 112,000 m altitude, it is traveling at 5,588 m/s.

a. Using the atmospheric drag parameters in the table below, integrate the equations of motion and estimate the velocity 7 minutes later. Be sure to state your assumptions.

Density Model	**Range**	**Values**
Reference density	0–25 km	0.0525 kg/m³
	25–125 km	0.0159 kg/m³
Scale height	0–25 km	11.049 km
	25–125 km	7.295 km
Ballistic coefficient	N/A	65 kg/m²

b. Provide three plots: altitude vs. time, heating vs. time, and dynamic pressure vs. time.

25. Design a roundtrip trajectory for an Earth-Mars crewed mission during the 2050s, such that a crew is given a stay time between 50 and 100 days. What types of outbound and inbound transfers would you choose? Explain why, considering the time the crew spends in space and ensuring that your total mission Δv is within a reasonable value. Use JPL's planetary ephemeris data to obtain the position and velocities of Earth and Mars. (Just like real mission design, there is no unique answer for this problem)
26. Compute the semimajor axis, eccentricity, radii of perihelion and aphelion for the Aldrin cycler discussed in Section 5.15.3. For this problem, assume that the orbits of Earth and Mars are coplanar and circular. Furthermore, assume that Earth and Mars orbit the Sun in exactly 1 year and 1.875 years, respectively. This means that the cycler would encounter Earth on successive passes that are 1/7 of an orbit (or 51.4°) apart. Lastly, assume that Mars flybys do not provide any changes in the heliocentric orbit of the cycler, so that only Earth flybys can provide the necessary orbital changes to maintain the cycler's orbit periodicity.

    Hint: you will need to solve a multi-rev Lambert's problem to find the semimajor axis of the cycler orbit.
27. Compute the minimum radial-only acceleration needed to transfer a spacecraft from a circular LEO with 400 km altitude to GEO. Then, using the acceleration value you computed, find the time at which the spacecraft would reach GEO.
28. Assume that a spacecraft starts from a circular heliocentric orbit equal to that of the Earth ($r_0 = 1$ AU). The vehicle turns on its thrusters, producing a constant tangential acceleration of 1/32 μ/r_0. How long will the spacecraft take to escape the Solar System? Compute this value by integrating the equations of motion and also by using the approximations given in Section 5.16.3.2. How do the values compare to each other?

6 Navigation and Targeting

This chapter discusses methods and techniques to take measurements via radio signals and/or optical techniques to estimate the spacecraft's position and velocity. This chapter also discusses the mathematics and methods of tracking and estimation of spacecraft states, which can then be used to correct the ongoing process of going from one point to another.

6.1 TRACKING SOURCES AND METHODS

While it is important to know where the spacecraft is estimated to be, what is more important to know is where the spacecraft actually is. While the trajectory design process gives us a theoretical trajectory, various things can happen that prevent the spacecraft from flying that exact trajectory. Some of these deviations between planned and actual trajectories include mismodeling of the performance of the spacecraft maneuvers and mismodeling of gravitational and non-gravitational forces. Thus, it is necessary to determine the location of the spacecraft, compare it to its theoretical location, and correct the trajectory as necessary. Figure 6.1 shows a diagram

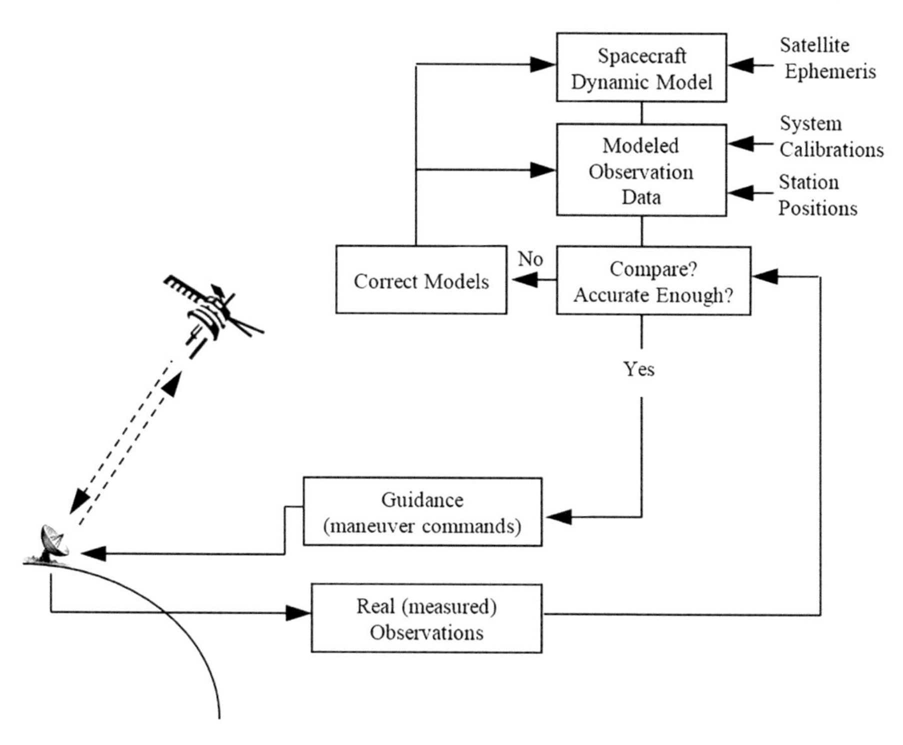

FIGURE 6.1 Navigation process.

DOI: 10.1201/9781003165071-6

summarizing the main steps required for correct navigation. We will discuss these important steps throughout this chapter.

Once a spacecraft has moved out of the vicinity of the Earth, common methods used to track and perform orbit determination for spacecraft in Earth orbit generally no longer become feasible. For spacecraft in Earth orbit, various space surveillance networks throughout the world perform the day-to-day tracking and cataloging space objects are in place. Navigating an object in Earth orbit is accomplished using a dense amount of various tracking data, which can be used in a variety of ways. Contrarily, in deep space, the Deep Space Networks are used to track objects. This data is generally sparse, with large gaps in the tracking data. The data is also generally one-dimensional (range and range rate), with the goal of determining a three-dimensional state.

Coordinate frames and orbital element sets used for Earth-orbiting objects are no longer usable. Different coordinate frames and orbital descriptors are used for interplanetary space, as we discussed in Chapter 4. Generally, tracking methods for objects outside of Earth orbit are either secondary uses of communications systems or satellite-based instrumentation.

6.1.1 Deep Space Network

NASA's Deep Space Network (DSN) is the communications infrastructure used for interplanetary spacecraft missions. Run by NASA's Jet Propulsion Laboratory, the DSN consists of three Earth-based stations located in Madrid, Spain, Canberra, Australia, and Goldstone, California. At each of these stations, there are several different antennas consisting of one 34-m diameter high-efficiency antenna and one 34-m beam waveguide antenna (three at the Goldstone site), one 26-m antenna and one 70-m antenna. These large dishes give the DSN the power to communicate with spacecraft on interplanetary missions throughout the Solar System. The locations of the DSN antennas are found in Table 6.1. A comprehensive description of the DSN is found in NASA's technical report "DSN Telecommunications Link Design Handbook" [Chang, 2018]. Some pictures from the three DSN sites are shown in Figure 6.2.

The prime purpose of the DSN is to communicate with spacecraft at interplanetary distances. However, a secondary use for the communication signal is for tracking of spacecraft. This is done by determining round-trip travel time between the ground and the spacecraft, which provides information on range to the spacecraft, and the Doppler shift in the frequency of the signal due the spacecraft moving toward or away from the ground station, which provides information on range rate.

There are numerous methods in use to determine the position and velocity of an orbiting spacecraft including two-way and three-way range, two-way and three-way Doppler, and very long baseline interferometry (VLBI). Two-way and three-way range measurements are simple measurements of distance based on the round-trip light time of a signal sent to the spacecraft where in the three-way technique, the signal returns to a second ground station. The two-way and three-way Doppler are techniques that interpret the phase shift of a signal sent to the spacecraft to determine the speed and direction of the spacecraft, where the three-way version again returns to a second ground station. VLBI enlists the services of two stations at very large

TABLE 6.1
DSN antenna locations [Chang, 2018]

Name	Antenna Size (m)	Latitude (deg)	Longitude (deg)	Geocentric Radius (km)
DSS-13	34	35.06602	243.2055	6,372.125
DSS-14	70	35.24435	243.1105	6,371.993
DSS-15	34	35.24031	243.1128	6,371.967
DSS-16	26	35.16018	243.1264	6,371.966
DSS-24	34	35.15853	243.1252	6,371.974
DSS-25	34	35.15626	243.1246	6,371.983
DSS-26	34	35.15434	243.127	6,371.992
DSS-27	34	35.05715	243.2233	6,372.110
DSS-34	34	−35.217	148.982	6,371.694
DSS-43	70	−35.2209	148.9813	6,371.689
DSS-45	34	−35.217	148.9777	6,371.676
DSS-46	26	−35.2235	148.9831	6,371.676
DSS-54	34	40.23577	355.7459	6,370.025
DSS-55	34	40.23445	355.7474	6,370.008
DSS-63	70	40.24136	355.752	6,370.051
DSS-65	34	40.23736	355.7493	6,370.022
DSS-66	26	40.24012	355.7486	6,370.037

FIGURE 6.2 Deep Space Network: Goldstone (top left), Canberra (top right), and Madrid (bottom). (Courtesy of NASA/JPL-Caltech.)

distances apart on the Earth surface to give a delta one-way range (DOR) measurement normally expressed by time in seconds [Altunin et al., 2000]. This time represents the difference in time it takes a signal to travel from the spacecraft to the two ground stations, which is a concept that will be explained later.

The European Space Agency and the Chinese National Space Agency have additional Deep Space Networks that operate similarly to NASA's DSN.

6.1.2 Radiometric Measurements

To track a spacecraft beyond Earth orbit, one technique is to use the communications signal from the spacecraft. Various types of this radiometric data include Doppler, range, and delta-differential one-way range (delta-DOR).

First, a spacecraft is generally moving away or toward a transmitter or receiver. The signal that is broadcast (as part of the communications system) initially has a frequency. Due to the Doppler shift of the signal, if the spacecraft is moving toward the ground station, the frequency of the signal increases. If it is moving away from the ground station, the frequency of the signal decreases.

Second, the round-trip travel time of the signal can also be measured, and with an accurate time difference measurement, the range (radial distance) can easily be found. For added precision, modeling or measuring the decrease in the speed of light as the signal passes through the atmosphere can be included. Accurate measurement devices, such as clocks, need to be included in the system. Information on the hardware systems used is found in Delva et al. [2017] and Duchayne et al. [2009], among others.

Third, the delta-differential one-way range (delta-DOR) measures the plane-of-sky angle between the spacecraft and a baseline between two tracking stations. These two tracking stations are separated by some distance, and a preliminary study found that a larger baseline can improve the accuracy [Haeberle et al., 2004].

Any combination of these methods can be used. We will now discuss the mathematics of various observation types and show how they can be used to determine the orbit of a spacecraft in interplanetary space.

6.1.2.1 Doppler

The base equation for a Doppler measurement at time, t, according to the geometrical elements of the observables is

$$f_R(t) - \frac{2f_T}{cT}\left[\rho(t) - \rho(t - T)\right] \tag{6.1}$$

where T is the step size, c is the speed of light, and f_T is the transmitter frequency.

6.1.2.2 Range

The second measurement type is the two-way range. The range is a vector quantity that represents the difference between the spacecraft position and the ground station position. The range defined in Cartesian coordinates is given by

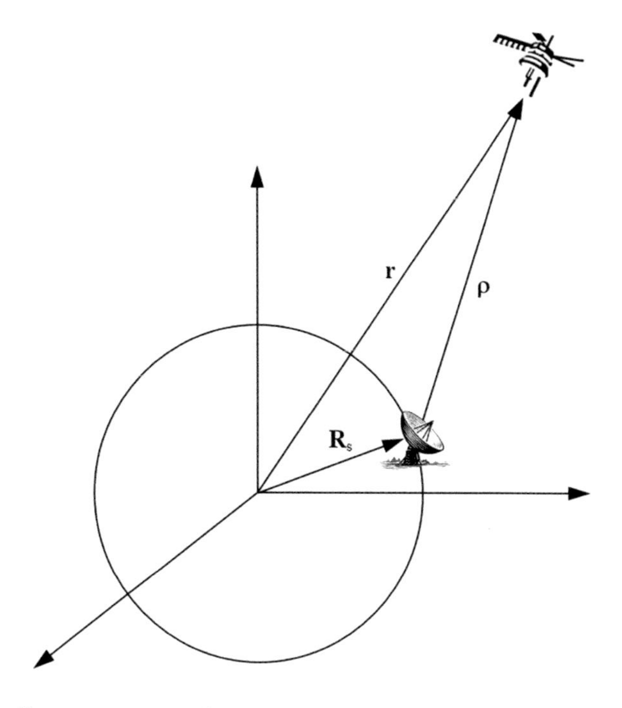

FIGURE 6.3 Range measurement.

$$\rho = \mathbf{r} - \mathbf{R}_s = \begin{Bmatrix} x - x_s \\ y - y_s \\ z - z_s \end{Bmatrix} \tag{6.2}$$

where $(x, y, z)^T$ are the Cartesian elements of the spacecraft in an inertial frame and $(x_s, y_s, z_s)^T$ are the Cartesian elements of the station in the same frame. The magnitude of the range is

$$\rho = \sqrt{(x - x_s)^2 + (y - y_s)^2 + (z - z_s)^2} \tag{6.3}$$

The geometry is shown in Figure 6.3.

6.1.3 Delta-Differential One-Way Range (Delta-DOR)

The final measurement type need is the DOR measurement. The DOR measurement is the time difference in seconds it takes for a signal sent by a spacecraft to reach two different target ground stations. A visual of a DOR measurement system is represented in Figure 6.4 to show how the algebraic representation of the time can be inferred through its geometry.

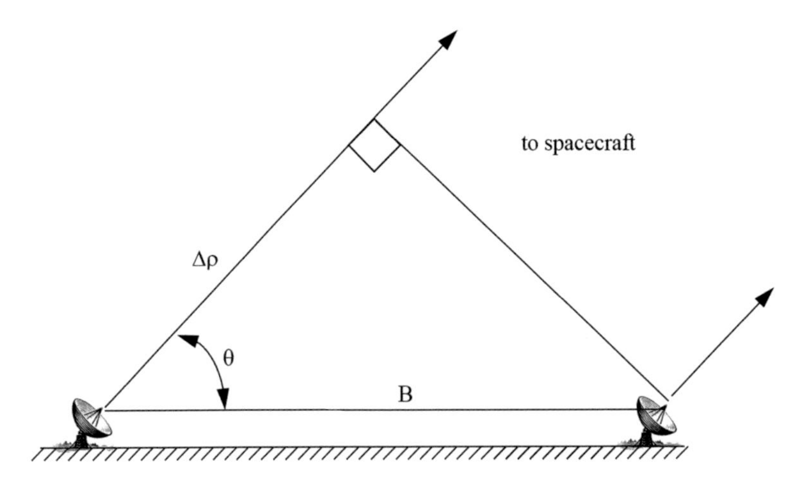

FIGURE 6.4 VLBI geometry for observing distant spacecraft.

Here, B represents the baseline vector between the two ground stations and $\Delta\rho$ is the difference in one-way range. Using the geometry of Figure 6.4, an equation for the delta one-way range is derived as

$$\Delta\rho = B\cos\theta \tag{6.4}$$

The right-hand side of the equation can be represented by vectors in the sense that

$$\Delta\rho = \mathbf{B}\cdot\hat{\mathbf{s}} \tag{6.5}$$

where $\hat{\mathbf{s}}$ is the unit vector in the direction of the spacecraft. In other words, the differenced range is the projection of the baseline in the direction of the spacecraft. In order for this to be true, it must be assumed that the ground station-to-spacecraft direction is approximately the same for both stations.

Physically, the extra distance traveled can be represented by the time it takes to travel that extra distance, τ_g, multiplied by the speed of the signal

$$\Delta\rho = \tau_g c \tag{6.6}$$

This allows for a representation of the measurement in terms of physical parameters that can be differentiated

$$\tau_g = \frac{1}{c}\left(\mathbf{B}\cdot\hat{\mathbf{s}}\right) = \frac{1}{c}\left[\frac{x}{r}B_x + \frac{y}{r}B_y + \frac{z}{r}B_z\right] \tag{6.7}$$

where $r = \sqrt{x^2 + y^2 + z^2}$ is the magnitude of the position vector to the spacecraft, and $\left(B_x, B_y, B_z\right)^T$ are the Cartesian components of the baseline vector **B**.

Using a classical Kalman filter requires, in addition to the dynamical equations of motion and the equations for the observation-state relationship, the partial derivatives of each measurement type with respect to the state vector. In this case, the state vector is comprised of 12 elements evaluated at the initial epoch: the position of the spacecraft (3), velocity of the spacecraft (3), position of the tracking station (3), and velocity of the tracking station (3). For two-way Doppler and range, the partials associated with the state elements of the ground stations not used are set equal to zero.

To find the partials, one needs to first look at the Cartesian elements of the range. Taking the derivative with respect to the first spacecraft state variable, x, results in

$$\frac{\partial \rho}{\partial x} = \frac{x - x_s}{\sqrt{(x - x_s)^2 + (y - y_s)^2 + (z - z_s)^2}} \tag{6.8}$$

Similarly, the partials for the y and z state variables are:

$$\frac{\partial \rho}{\partial y} = \frac{y - y_s}{\sqrt{(x - x_s)^2 + (y - y_s)^2 + (z - z_s)^2}} \tag{6.9}$$

$$\frac{\partial \rho}{\partial z} = \frac{z - z_s}{\sqrt{(x - x_s)^2 + (y - y_s)^2 + (z - z_s)^2}} \tag{6.10}$$

From the above equations, one can see that the state partials vector for the position variables is equal to the unit vector of the range, $\hat{\rho}$. The velocity terms do not appear in the equation for the range and therefore, their partials are zero. This allows for a much simpler analysis for range measurements. A similar analysis to that for the spacecraft state variables can be performed to show that the state partials vector for the ground stations is also the unit vector of the range, except it is negative. Therefore, finding the range unit vector gives the partials of the state vectors

$$\frac{\partial \rho}{\partial \mathbf{r}_{s/c}} = \frac{\rho}{\rho} = \hat{\rho} \tag{6.11}$$

$$\frac{\partial \rho}{\partial \mathbf{r}_s} = -\frac{\rho}{\rho} = -\hat{\rho} \tag{6.12}$$

These partials are multiplied be the state transition matrix of spacecraft states and the Earth rotation matrix of the ground station states, respectively, to get the final partials to be used.

Second, the partials for the Doppler measurements need to be derived. From Equations (6.11) and (6.12), it is known that the derivative of the range is its unit vector, and because there is a linear relation between f_R and ρ (Equation 6.1), the range partials are applied here. Thus, the partial vector for the Doppler observable is

$$\frac{\partial f_R}{\partial \mathbf{r}_{s/c}} = \frac{2f_T}{cT}\left(\hat{\rho}_k - \hat{\rho}_{k-1}\right) \tag{6.13}$$

This can also be applied to the station locations where the partial vector for the range is the negative of its unit vector

$$\frac{\partial f_R}{\partial \mathbf{r}_s} = -\frac{2f_T}{cT}\left(\hat{\rho}_k - \hat{\rho}_{k-1}\right) \tag{6.14}$$

The state transition matrix associated with the spacecraft and the rotation matrix associated with the stations are then used to transition the Cartesian elements at the current time back to the initial epoch. Note that for the simulated data white noise is added to the data with a $1-\sigma$ strength of σ_r.

Now that there is a representation for the time difference, it must be differentiated to get the partials starting with the spacecraft state variables. The first partial derivative is with respect to x

$$\frac{\partial \tau_g}{\partial x} = \frac{1}{c}\left[\frac{B_x}{\sqrt{x^2+y^2+z^2}} - \frac{x\left(xB_x+yB_y+zB_z\right)}{\sqrt{\left(x^2+y^2+z^2\right)^3}}\right] \tag{6.15}$$

Rearranging this equation, the partial becomes

$$\frac{\partial \tau_g}{\partial x} = \frac{1}{rc}\left[B_x - \frac{x\left(xB_x+yB_y+zB_z\right)}{r^2}\right] \tag{6.16}$$

This same process can be repeated for the y and z partials to get

$$\frac{\partial \tau_g}{\partial y} = \frac{1}{rc}\left[B_y - \frac{y\left(xB_x+yB_y+zB_z\right)}{r^2}\right] \tag{6.17}$$

$$\frac{\partial \tau_g}{\partial z} = \frac{1}{rc}\left[B_z - \frac{z\left(xB_x+yB_y+zB_z\right)}{r^2}\right] \tag{6.18}$$

Now that the partials for the position have been found, it is necessary to look at the other components of the state vector – the velocity components. Conveniently, there are no velocity components in the DOR measurement equation. This means that all of the velocity partials are zero.

The partials now must be found for the ground stations. First, the station elements need to be linked to the DOR equation. The station vectors are represented by

$$l_i = (l_{ix}, l_{iy}, l_{iz})^T, \text{ for } i - 1, ? \tag{6.19}$$

In the DOR equation, the station state vectors are represented in the baseline vector term

$$\mathbf{B} = \mathbf{l}_2 - \mathbf{l}_1 \tag{6.20}$$

Going back to the DOR equation, the chain rule can be used to generalize the partial. With respect to the x-direction, the DOR partial becomes

$$\frac{\partial \tau_g}{\partial l_{ix}} = \frac{\partial \tau_g}{\partial B_x} \frac{\partial B_x}{\partial l_{ix}} \tag{6.21}$$

This allows the partial to be taken with respect to the baseline instead of directly with respect to the state variables

$$\frac{\partial \tau_g}{\partial B_x} = \frac{1}{c} \frac{x}{r} \tag{6.22}$$

Converting back to the state partials the equation becomes

$$\frac{\partial \tau_g}{\partial l_{ix}} = \frac{1}{c} \frac{x}{r} \frac{\partial B_x}{\partial l_{ix}} \tag{6.23}$$

This analysis matches for the other state variables as well. Looking at Equation (6.7), the derivative of the baseline is equal to one for station number two and negative one for station number one, so the state vector partials become

$$\begin{aligned} \frac{\partial \tau_g}{\partial l_{1x}} &= -\frac{1}{c}\frac{x}{r} & \frac{\partial \tau_g}{\partial l_{2x}} &= \frac{1}{c}\frac{x}{r} \\ \frac{\partial \tau_g}{\partial l_{1x}} &= -\frac{1}{c}\frac{y}{r} & \frac{\partial \tau_g}{\partial l_{2x}} &= \frac{1}{c}\frac{y}{r} \\ \frac{\partial \tau_g}{\partial l_{1x}} &= -\frac{1}{c}\frac{z}{r} & \frac{\partial \tau_g}{\partial l_{2x}} &= \frac{1}{c}\frac{z}{r} \end{aligned} \tag{6.24}$$

These are all the DOR partials needed for the analysis. As with the other measurements, the state transition matrices are used to map the partials back to the initial epoch. For the analysis of the true satellite, a data noise term, σ_d, can be added to simulate actual conditions for the signal propagation.

6.2 OPTICAL NAVIGATION

Optical navigation (opnav) is a method of using optical devices on-board spacecraft to image a Solar System body with a star background. This is used, either alone or with other radiometric data, to determine a navigation solution to improve the knowledge of the location of the body relative to the star background. This form of data provides the best improvement in spacecraft ephemeris when the target body's position is not known well [Martin-Mur et al., 2006]. The first use of spacecraft optical navigation was demonstrated for the Mariner 6 and 7 missions in 1967. Opnav tools and techniques have been improved over the years, and optical navigation continues to be a cornerstone of the navigation system of many interplanetary spacecraft. This data source was used extensively for missions to the outer planets, as well as for precision applications for missions to Mars, and the accumulation of this data has helped improve information about the ephemerides of various bodies.

Probably the biggest benefit of optical navigation data is the improved accuracy contribution to the location of small bodies in the Solar System, when compared to the use of only ground-based measurement data. Asteroid ephemerides from ground-based measurements only are accurate to tens to hundreds of kilometers, while the error for comets is on the order of thousands of kilometers. The combined use of optical navigation data with ground-based measurements has made the various Galileo spacecraft encounters with the Jovian moons Gaspra and Ida possible. The Near-Earth Asteroid Rendezvous (NEAR) spacecraft's rendezvous with 433 Eros was also made possible thanks to opnav, Stardust and Deep Impact would not have been possible if opnav had not been used, and, more recently, OSIRIS-REx has made significant use of opnav [Mastrodemos et al., 2005; McCarthy et al., 2022; Miller et al., 1995].

When optical navigation first began, the data was collected using cameras whose primary function was scientific data collection. As technology has improved, cameras and their associated support hardware have been developed to be made lighter, require less power, and be less expensive, which will allow inclusion of dedicated optical navigation systems on developing spacecraft missions. The predicted reference location and actual reference location of a target is measured with respect to a reference star, as seen in Figure 6.5.

Optical navigation uses many of the same techniques used in classical photographic astrometry, but uses the spacecraft pictures of objects in the foreground with a star background. Measuring the image coordinates of the stars and other objects allows estimates of the angular positions of the objects to be made. This is very useful for determining coordinate frames in the B-plane (as we saw in Section 4.2.7) and

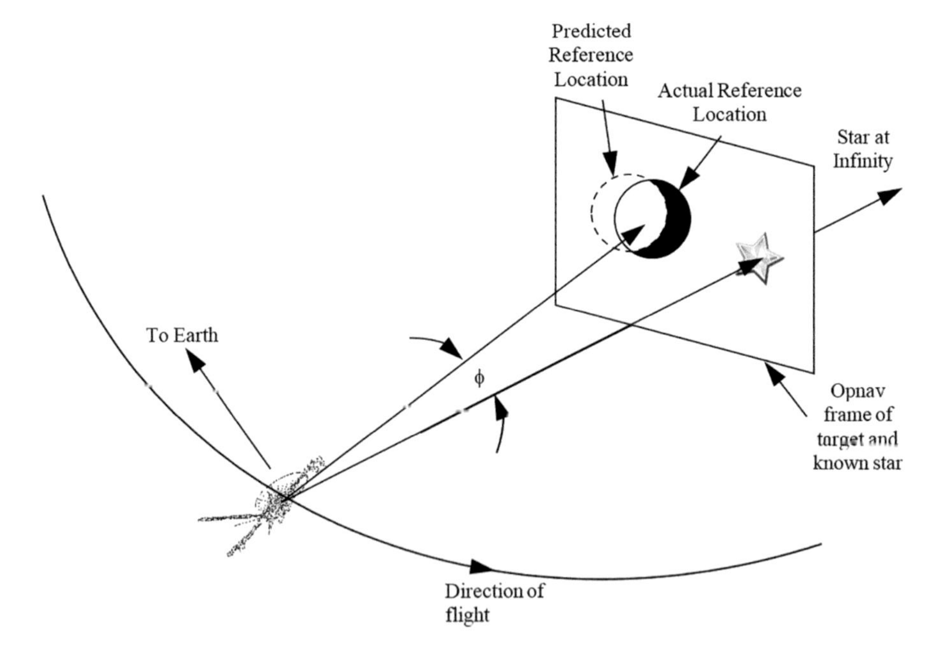

FIGURE 6.5 Optical navigation schematic.

can be used to determine the orbits of satellites. This method requires an accurate star catalog and sufficient knowledge of the characteristics of the spacecraft camera(s) used for these measurements. These methods have been used over the history of interplanetary flight, both with ground control and autonomous operations. With advances in more powerful on-board computing capabilities on spacecraft, more autonomous optical navigation techniques are trending toward more spacecraft, ranging from small, CubeSat-sized spacecraft to large flagship missions. Details on the mechanics of this technique are detailed further in references [Bhaskaran, 2020; Bhaskaran at al., 1996; Owen, 2003; Owen at al., 2008].

6.3 DETERMINISTIC ORBIT DETERMINATION

Perfect measurements and models do not exist – all have noise and errors that cause the orbit determination solution to be an approximation rather than an exact solution. However, Josiah Gibbs developed a deterministic method (known as Gibbs' method) that can be used as a starting place for finding the first estimate for an approximate solution of an orbit.

To apply Gibbs' method, we need to know three position vectors at three different times. One crucial characteristic of these vectors is that all three vectors are in the same plane. We see these three vectors, $\mathbf{r}_1$, $\mathbf{r}_2$, and $\mathbf{r}_3$, in Figure 6.6.

Since these vectors are coplanar, any third vector can be written as a linear combination of the other two,

$$c_1\mathbf{r}_1 + c_2\mathbf{r}_2 + c_3\mathbf{r}_3 = 0 \tag{6.25}$$

where all constants c_1, c_2, and c_3 are not equal to zero. In the perifocal reference frame, each position vector can be written in polar coordinates as

$$\mathbf{r} = r\left(\cos\theta\hat{\mathbf{p}} + \sin\theta\hat{\mathbf{q}}\right) \tag{6.26}$$

and a common eccentricity vector,

$$\mathbf{e} = \mathbf{e}\hat{\mathbf{p}} \tag{6.27}$$

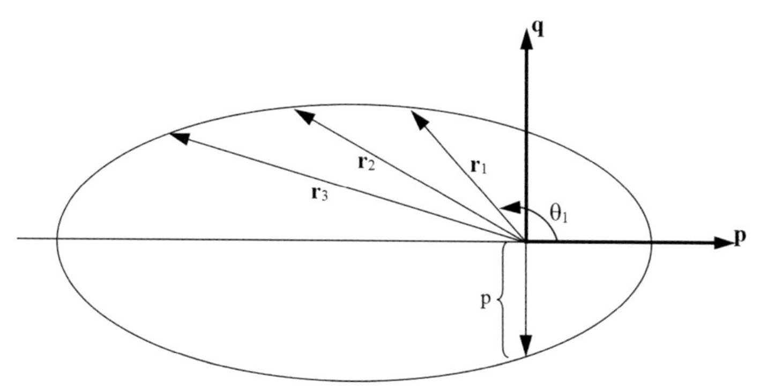

FIGURE 6.6 Coplanar position vectors.

Taking the dot product of Equation (6.27) with (6.26) using each of the three position vectors, Equation (6.28) shows that,

$$\mathbf{e} \cdot \mathbf{r} = \mathrm{re}\cos\theta = \frac{\mathrm{pe}\cos\theta}{1+\mathrm{e}\cos\theta} \tag{6.28}$$

With a little algebraic manipulation, Equation (6.28) can be rewritten as

$$\mathbf{e} \cdot \mathbf{r} = \mathrm{p} - \mathrm{r} \tag{6.29}$$

Taking the dot product of the eccentricity vector, **e**, with Equation (6.25) and applying the relationship in Equation (6.29) yields

$$\begin{aligned} \mathbf{e} \cdot \left(c_1\mathbf{r}_1 + c_2\mathbf{r}_2 + c_3\mathbf{r}_3\right) &= c_1\mathbf{e} \cdot \mathbf{r}_1 + c_2\mathbf{e} \cdot \mathbf{r}_2 + c_3\mathbf{e} \cdot \mathbf{r}_3 \\ &= c_1\left(\mathrm{p} - \mathrm{r}_1\right) + c_2\left(\mathrm{p} - \mathrm{r}_2\right) + c_3\left(\mathrm{p} - \mathrm{r}_3\right) = 0 \end{aligned} \tag{6.30}$$

Taking the cross-product of Equation (6.25) with $\mathbf{r}_1$, $\mathbf{r}_2$, and $\mathbf{r}_3$,

$$\left(c_1\mathbf{r}_1 + c_2\mathbf{r}_2 + c_3\mathbf{r}_3\right) \times \mathbf{r}_1 = 0 = c_2\mathbf{r}_2 \times \mathbf{r}_1 + c_3\mathbf{r}_3 \times \mathbf{r}_1 \tag{6.31}$$

$$\left(c_1\mathbf{r}_1 + c_2\mathbf{r}_2 + c_3\mathbf{r}_3\right) \times \mathbf{r}_2 = 0 = c_1\mathbf{r}_1 \times \mathbf{r}_2 + c_3\mathbf{r}_3 \times \mathbf{r}_2 \tag{6.32}$$

$$\left(c_1\mathbf{r}_1 + c_2\mathbf{r}_2 + c_3\mathbf{r}_3\right) \times \mathbf{r}_3 = 0 = c_1\mathbf{r}_1 \times \mathbf{r}_3 + c_2\mathbf{r}_2 \times \mathbf{r}_3 \tag{6.33}$$

Using the vector identity, $\mathbf{a} \times \mathbf{b} = -\mathbf{b} \times \mathbf{a}$, Equations (6.31) through (6.33) can be rewritten as

$$c_2\mathbf{r}_1 \times \mathbf{r}_2 = c_3\mathbf{r}_3 \times \mathbf{r}_1 \tag{6.34}$$

$$c_1\mathbf{r}_1 \times \mathbf{r}_2 = c_3\mathbf{r}_2 \times \mathbf{r}_3 \tag{6.35}$$

$$c_1\mathbf{r}_3 \times \mathbf{r}_1 = c_2\mathbf{r}_2 \times \mathbf{r}_3 \tag{6.36}$$

Multiplying Equation (6.30) by $\mathbf{r}_3 \times \mathbf{r}_1$, yields

$$c_1\left(\mathrm{p} - \mathrm{r}_1\right)\mathbf{r}_3 \times \mathbf{r}_1 + c_2\left(\mathrm{p} - \mathrm{r}_2\right)\mathbf{r}_3 \times \mathbf{r}_1 + c_3\left(\mathrm{p} - \mathrm{r}_3\right)\mathbf{r}_3 \times \mathbf{r}_1 = 0 \tag{6.37}$$

and using the identities in Equations (6.34) and (6.36), Equation (6.30) becomes

$$c_2\left(\mathrm{p} - \mathrm{r}_1\right)\mathbf{r}_2 \times \mathbf{r}_3 + c_2\left(\mathrm{p} - \mathrm{r}_2\right)\mathbf{r}_3 \times \mathbf{r}_1 + c_2\left(\mathrm{p} - \mathrm{r}_3\right)\mathbf{r}_1 \times \mathbf{r}_2 = 0 \tag{6.38}$$

Dividing by c_2 (since $c_2 \neq 0$), and grouping terms, Equation (6.38) becomes

$$\mathrm{p}\left(\mathbf{r}_2 \times \mathbf{r}_3 + \mathbf{r}_3 \times \mathbf{r}_1 + \mathbf{r}_1 \times \mathbf{r}_2\right) = \mathrm{r}_1\mathbf{r}_2 \times \mathbf{r}_3 + \mathrm{r}_2\mathbf{r}_3 \times \mathbf{r}_1 + \mathrm{r}_3\mathbf{r}_1 \times \mathbf{r}_2 \tag{6.39}$$

Defining the right-hand side of Equation (6.39) as **N** and the coefficient of p as **D**,

$$\mathrm{p}\mathbf{D} = \mathbf{N} \tag{6.40}$$

The vectors **N** and **D** have the same direction (they both point out of the orbital plane, **w**), so we can remove the vector notation and the semilatus rectum, p, becomes

$$p = \frac{N}{D} \tag{6.41}$$

In the orthogonal perifocal reference frame,

$$\hat{\mathbf{q}} = \hat{\mathbf{w}} \times \hat{\mathbf{p}} \tag{6.42}$$

The vector **N** can be written as

$$\mathbf{N} = N\hat{\mathbf{w}} \tag{6.43}$$

Solving Equation (6.43) for $\hat{\mathbf{w}}$ and inserting the definition of $\hat{\mathbf{p}}$ found by rearranging Equation (6.27) into Equation (6.42),

$$\hat{\mathbf{q}} = \frac{\mathbf{N}}{N} \times \frac{\mathbf{e}}{e} \tag{6.44}$$

Inserting the definition of **N** used in Equation (6.40) and rearranging Equation (6.44),

$$Ne\hat{\mathbf{q}} = r_1(\mathbf{r}_2 \times \mathbf{r}_3) \times \mathbf{e} + r_2(\mathbf{r}_3 \times \mathbf{r}_1) \times \mathbf{e} + \mathbf{r}_3(\mathbf{r}_1 \times \mathbf{r}_2) \times \mathbf{e} \tag{6.45}$$

We can use the vector identity,

$$(\mathbf{a} \times \mathbf{b}) \times \mathbf{c} = (\mathbf{a} \cdot \mathbf{c})\mathbf{b} - (\mathbf{b} \cdot \mathbf{c})\mathbf{a} \tag{6.46}$$

in Equation (6.45),

$$\begin{aligned} Ne\hat{\mathbf{q}} = {} & r_1(\mathbf{r}_2 \cdot \mathbf{e})\mathbf{r}_3 - r_1(\mathbf{r}_3 \cdot \mathbf{e})\mathbf{r}_2 + r_2(\mathbf{r}_3 \cdot \mathbf{e})\mathbf{r}_1 - r_2(\mathbf{r}_1 \cdot \mathbf{e})\mathbf{r}_3 \\ & + r_3(\mathbf{r}_1 \cdot \mathbf{e})\mathbf{r}_2 - r_3(\mathbf{r}_2 \cdot \mathbf{e})\mathbf{r}_1 \end{aligned} \tag{6.47}$$

From Equation (6.29), using the values of $\mathbf{r}_1$, $\mathbf{r}_2$, and $\mathbf{r}_3$, insertion of this into Equation (6.47) yields,

$$\begin{aligned} Ne\hat{\mathbf{q}} = {} & r_1(p - r_2)\mathbf{r}_3 - r_1(p - r_3)\mathbf{r}_2 + r_2(p - r_3)\mathbf{r}_1 \\ & - r_2(p - r_1)\mathbf{r}_3 + r_3(p - r_1)\mathbf{r}_2 - r_3(p - r_2)\mathbf{r}_1 \end{aligned} \tag{6.48}$$

Grouping these terms,

$$Ne\hat{\mathbf{q}} = p\left[(r_2 - r_3)\mathbf{r}_1 + (r_3 - r_1)\mathbf{r}_2 + (r_1 - r_2)\mathbf{r}_3\right] = p\mathbf{S} \tag{6.49}$$

Here, too, we can drop the vector notation since **S** must be in the $\hat{\mathbf{q}}$ direction, so

$$Ne = pS \tag{6.50}$$

Inserting Equation (6.41) into Equation (6.50), we can solve for the eccentricity,

$$e = \frac{S}{D} \tag{6.51}$$

Since we know p and e, we can find the semimajor axis,

$$a = \frac{p}{1-e^2} \tag{6.52}$$

From our earlier definition of the eccentricity vector given by Equation (2.27),

$$\frac{d\mathbf{r}}{dt} \times \mathbf{h} = \mathbf{v} \times \mathbf{h} = \mu\left(\frac{\mathbf{r}}{r} + \mathbf{e}\right) \tag{6.53}$$

Taking the cross-product of the angular momentum, **h**, with Equation (6.53),

$$\mathbf{h} \times (\mathbf{v} \times \mathbf{h}) = \mu\left(\frac{\mathbf{h} \times \mathbf{r}}{r} + \mathbf{h} \times \mathbf{e}\right) \tag{6.54}$$

Using the vector identity in Equation (6.46) for the left-hand side of Equation (6.54),

$$\underbrace{(\mathbf{h} \cdot \mathbf{h})}_{h^2}\mathbf{v} - \underbrace{(\mathbf{h} \cdot \mathbf{v})}_{0}\mathbf{v} = \mu\left(\frac{\mathbf{h} \times \mathbf{r}}{r} + \mathbf{h} \times \mathbf{e}\right) \tag{6.55}$$

We know that the angular momentum vector points out of the orbital plane,

$$\mathbf{h} = h\mathbf{w} \tag{6.56}$$

and the eccentricity vector definition in Equation (6.27), solving for the velocity vector in Equation (6.55),

$$\mathbf{v} = \frac{\mu}{h^2}\left(\frac{\mathbf{h} \times \mathbf{r}}{r} + \mathbf{h} \times \mathbf{e}\right) \tag{6.57}$$

Simplifying, Equation (6.57) becomes

$$\mathbf{v} = \frac{\mu}{h}\left(\frac{\mathbf{w} \times \mathbf{r}}{r} + e\hat{\mathbf{q}}\right) \tag{6.58}$$

In Equation (2.30), we defined the angular momentum as $h = \sqrt{\mu p}$, so using the definition for p in Equation (6.40) and the definition for eccentricity in Equation (6.51), the velocity vector becomes

$$\mathbf{v} = \frac{1}{r}\sqrt{\frac{\mu}{ND}}\mathbf{D} \times \mathbf{r} + \sqrt{\frac{\mu}{ND}}\mathbf{S} \tag{6.59}$$

To find the velocity at any of the position vectors $\mathbf{r}_1$, $\mathbf{r}_2$, and $\mathbf{r}_3$, we simply use the corresponding position vector $\mathbf{r}_1$, $\mathbf{r}_2$, and $\mathbf{r}_3$ in Equation (6.59).

While this whole process seems complex, there are very few computations that are needed. First, we need to find the magnitudes of all position vectors. Second, we find the cross-products of one position vector with another, and use the definitions for the vectors **S**, **D**, and **N** to easily compute the eccentricity, semilatus rectum, and semimajor axis. One more cross-product between **D** and a position vector, along with some more algebraic computations, yields the velocity vector when the object is at that corresponding position vector. With any pair of position and velocity, we can then compute all of the classical orbital elements (or other orbital elements of interest), as we discussed in Chapter 2.

So far, we have assumed that we have perfect knowledge of the position vectors and can compute perfect knowledge of the velocity vectors. However, actual measurements are subject to various error sources, so we can use vectors found through Gibbs' method for preliminary orbit determination and to produce an initial estimate for a statistical analysis to process the large amount of measurement data. Additional descriptions and applications of this method can be found in Bate et al. [2020, pp. 109–113], among others.

6.4 STATISTICAL ORBIT DETERMINATION

One outcome of using any of the various tracking tools is an abundance of tracking data. Whether the data is range and range rate found from signal travel time and Doppler shift information found from radiometric sources (radar and/or communication signal), or angle and angle-rate information found from telescopes, this data must be analyzed. While it was shown with Gibbs' method that only three position vectors were needed, in reality a large amount of data needs to be compiled and analyzed to find the true position of the spacecraft.

The initial conditions for the nominal orbit are changed by a user-defined amount representing the initial knowledge of the spacecraft position. Both these orbits are illustrated in Figure 6.7 with $\Delta\mathbf{X}$ representing the differential spacecraft state. The filter processes simulated measurements to estimate values of $\Delta\mathbf{X}(t_0)$ at the initial epoch t_0.

Having a trajectory path to track, we can simulate the process as explained in the flowchart in Figure 6.8. The time loop starts after initializing various filter variables

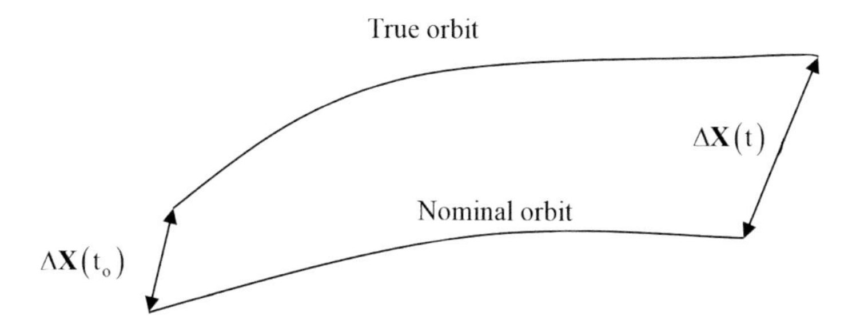

FIGURE 6.7 Description of true and nominal orbits.

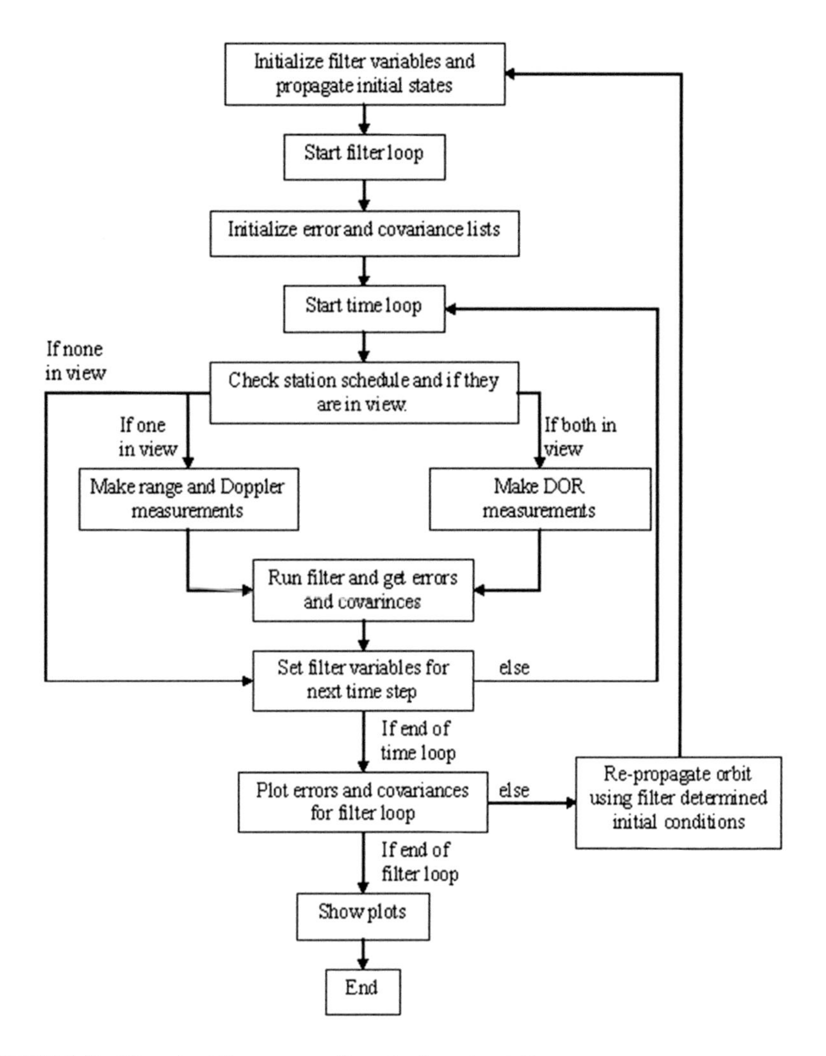

FIGURE 6.8 Flowchart for spacecraft navigation simulation.

and output lists. At each time step, first we check to see if the station schedule permits us to determine if any measurements should be taken. Applying a pre-chosen minimum time increment between measurements prevents measurements that are taken too close to each other and may not be distinguishable from each other. Next, we determine if the stations are in view of the spacecraft. If both are in view and a pre-established schedule calls for a DOR measurement, a DOR measurement is taken. If one station is in view and the schedule calls for a Doppler or range measurement, these measurements are taken. Next, we use the relationships between the actual measurements of the true and nominal spacecraft trajectories as well as the partial derivatives of the DOR measurement, from Equations (6.8) to (6.10) with respect to the state vector. These values are then fed into an initial state Kalman filter which maps the current conditions back to an original state which is updated every time the filter is called. The filter outputs an error and a covariance associated with the position error of the spacecraft. At the end of the track, these values are plotted and saved.

The user has many options when running this simulation. There are a large number of decisions in the initialization and the data collection process. The initial conditions of the spacecraft orbit can be defined using the Cartesian position and velocity components as well as information on the reference frame and central body of the orbit. The station locations need to be specified by their geodesic coordinates: latitude, longitude, and altitude. In relation to these initial conditions, the uncertainty associated with the initial conditions would need to be defined to create the true spacecraft state. The numerical propagator used to create the spacecraft's trajectory and the initial epoch and time duration would need to be set.

A tracking scheduler needs information about the time intervals between measurement passes, the time interval between individual measurements, and the length of tracking pass. Each individual tracking technique has its own set of these parameters. The tracking techniques, Doppler, range and DOR, can be chosen, providing the ability to assess the effects of individual measurement types on solution accuracy. Each measurement type has a data weight associated with it, again something that can be changed to suit a specific simulation.

Errors in the state variables of the spacecraft and the stations affect the way the filter performs and converges. The process can include the option of allowing choice of evaluating how individual state variables contribute to errors in state system. This can be useful in determining the sensitivity of the filter to specific state variables and allowing states that are of no interest to be ignored. Each state is then given an error to simulate process noise. This is done for the spacecraft by including an acceleration error which is then used to determine process noise on the position and velocity states at each time step. To find the process noise on the position and velocity, the state equation must be looked at first. With noise, the state equations take the form,

$$\left\{\begin{array}{c}\frac{d\mathbf{r}}{dt}\\ \frac{d\mathbf{v}}{dt}\end{array}\right\} = A\left\{\begin{array}{c}\mathbf{r}\\ \mathbf{v}\end{array}\right\} + \left\{\begin{array}{c}0\\ \mathbf{w}\end{array}\right\} \tag{6.60}$$

$$A = \begin{bmatrix} 0 & 1 \\ 0 & 0 \end{bmatrix}$$

where a white noise process $\mathbf{w}$ is driving the acceleration states. Using the state relation

$$\frac{dP}{dt} = AP + PA^T + Q \tag{6.61}$$

and substituting, Equation (6.60) becomes

$$\begin{bmatrix} \frac{dp_{11}}{dt} & \frac{dp_{12}}{dt} \\ \frac{dp_{21}}{dt} & \frac{dp_{22}}{dt} \end{bmatrix} = \begin{bmatrix} 0 & 1 \\ 0 & 0 \end{bmatrix} \begin{bmatrix} p_{11} & p_{12} \\ p_{21} & p_{22} \end{bmatrix} + \begin{bmatrix} p_{11} & p_{12} \\ p_{21} & p_{22} \end{bmatrix} \begin{bmatrix} 0 & 0 \\ 1 & 0 \end{bmatrix} + \begin{bmatrix} 0 & 0 \\ 0 & q \end{bmatrix} \quad (6.62)$$

where q is the user specified strength of the process noise on the acceleration and the p values need to be determined. By multiplying the matrices, Equation (6.62) becomes

$$\begin{bmatrix} \frac{dp_{11}}{dt} & \frac{dp_{12}}{dt} \\ \frac{dp_{21}}{dt} & \frac{dp_{22}}{dt} \end{bmatrix} = \begin{bmatrix} p_{11} & p_{12} \\ 0 & 0 \end{bmatrix} + \begin{bmatrix} p_{11} & 0 \\ p_{21} & 0 \end{bmatrix} + \begin{bmatrix} 0 & 0 \\ 0 & q \end{bmatrix} \quad (6.63)$$

This leads to a set of four differential equations

$$\frac{dp_{11}}{dt} = p_{21} + p_{12} \qquad \frac{dp_{12}}{dt} = p_{22}$$
$$\frac{dp_{21}}{dt} = p_{22} \qquad \frac{dp_{22}}{dt} = q \quad (6.64)$$

Beginning with $\frac{dp_{22}}{dt} = q$, the differential equations can easily be solved to arrive at a solution for the process noise covariance

$$P(t) = \begin{bmatrix} p_{11} & p_{12} \\ p_{21} & p_{22} \end{bmatrix} = \begin{bmatrix} \frac{1}{3}qt^3 & \frac{1}{2}qt^2 \\ \frac{1}{2}qt^2 & qt \end{bmatrix} \quad (6.65)$$

In a 6×6 matrix for the actual spacecraft state variables

$$P(\Delta t) = \begin{bmatrix} \frac{q_x \Delta t^3}{3} & 0 & 0 & \frac{q_x \Delta t^2}{2} & 0 & 0 \\ 0 & \frac{q_y \Delta t^3}{3} & 0 & 0 & \frac{q_y \Delta t^2}{2} & 0 \\ 0 & 0 & \frac{q_z \Delta t^3}{3} & 0 & 0 & \frac{q_z \Delta t^2}{2} \\ \frac{q_x \Delta t^2}{2} & 0 & 0 & q_x t & 0 & 0 \\ 0 & \frac{q_y \Delta t^2}{2} & 0 & 0 & q_y \Delta t & 0 \\ 0 & 0 & \frac{q_z \Delta t^2}{2} & 0 & 0 & q_z \Delta t \end{bmatrix} \tag{6.66}$$

where

$$\Delta t = t - t_{lm} \tag{6.67}$$

and t_{lm} is the time of the last measurement taken. Finally, since this is an epoch state filter, the noise is backdated to the initial epoch using $P(t_0) = \Phi(t_0, t) P(t) \Phi^T(t_0, t)$. The process noise on the station locations is assigned as a constant at the beginning of the simulation and is not affected by time.

6.5 MANEUVER PLANNING AND EXECUTION

With the use of modern propulsion systems, and gravity assists from other planets, it is possible to place Earth-orbiting satellites in interplanetary trajectories using relatively few high-thrust, and low-duration burn maneuvers. These maneuvers are generally accomplished by using a liquid or solid rocket propellant propulsion system, which can produce tremendous thrusts necessary to gain the sufficient change in velocity for an interplanetary transfer orbit. Using accurate force models for the Earth, Moon, Sun, and other planets, as well as various perturbing forces such as solar radiation pressure, solar flux, atmospheric drag, etc., the dynamics of an interplanetary spacecraft trajectory can be modeled, and mission stages can be planned. Some of these mission stages may contain thrusting maneuvers to change the spacecrafts velocity vector to place it into an intersecting trajectory with its destination planet. However, while the force models used to calculate the necessary change in velocities as the location of the burns may represent analytical solutions, they do not take into account various uncertainties within burn and coast segments of an interplanetary trajectory. While these uncertainties may be within design tolerances, and representing very small perturbations from the accurate force models used, when propagated over the long duration of an interplanetary coast stage, they can cause very large deviations from a desired final state, which could lead to a failure of certain mission objectives. Therefore, to obtain the correct final state, mid-course corrections (which were introduced in Section 5.11) can be implemented to counteract these uncertainties which may lead to deviations from a nominal trajectory. Additionally, statistical methods are used in order to predict and estimate how such corrections and/or unaccounted or mismodeled perturbations affect the orbit of the spacecraft over time.

6.5.1 Mid-Course Corrections Determination

Mid-course corrections can be calculated in a similar way as rendezvous or station keeping, in which a very small change in velocity is calculated in order to match the current trajectory to a desired one.

At the basic levels of calculating a mid-course correction, thrusting segments would be assumed to be impulsive, and the orbit of the spacecraft could be modeled analytically, with exact known states. In this scenario, a mid-course correction could simply be viewed as a change in the velocity vector, which may or may not result in a plane change. A diagram of this elementary example can be viewed in Figure 6.9.

However, in real-world applications, orbits cannot be modeled analytically, and must thus rely on numerical methods and proper data acquisition to obtain necessary orbit information. Secondly, thrusting maneuvers are not impulsive, and more importantly, exactly modeling the overall thrust vector during these finite maneuvers can be difficult. This is due to the uncertainties associated with the rocket engine design, force model, time-keeping system, and attitude/position sensors. Each of these components is designed and performs their objectives to certain tolerance, leading to a very small amount of unpredictability to calculating the actual position and velocity of the spacecraft.

Furthermore, orbit determination methods often need to be employed to keep accurate and time changing orbits of the spacecraft (presented as a two-line element, for example). Provided an accurate orbit determination system (which may contain

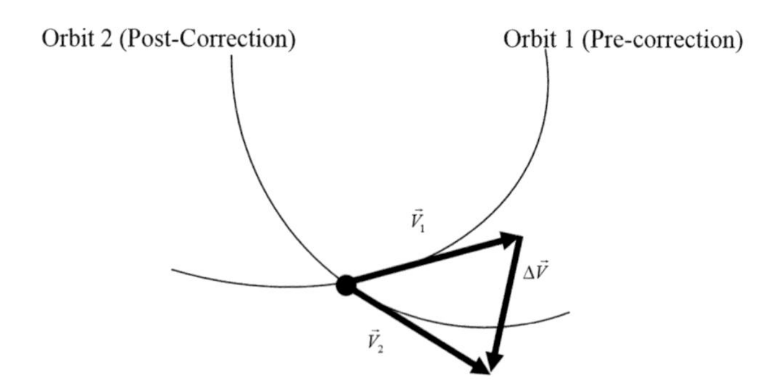

FIGURE 6.9 Basic impulsive course correction.

uncertainties within the data it receives, or the numerical method itself) discrepancies from the modeled position and velocity vectors can be calculated. Should discrepancies be large enough that mission objectives cannot be accomplished with the current satellite state, a mid-course correction can be implemented to get closer to a nominal state. Just as the perturbing forces on low Earth-orbiting satellites require the need for frequent station keeping maneuvers, the vast distances covered by interplanetary spacecraft facilitate the need for mid-course corrections.

An example of how important mid-course corrections are due to various uncertainties can be easily illustrated by looking at a basic Earth-Mars interplanetary Hohmann transfer (ignoring hyperbolic escape from parking orbit). Typical discrepancies in the change in velocity acquired in a thrusting maneuver from an Earth parking orbit to heliocentric orbit are on the range of up to 3%.[1] As we saw in in Chapter 5, solving for the first magnitude of the change in velocity, Δv_1, for a Hohmann transfer can be calculated as

$$\Delta v_1 = \sqrt{2\left(\frac{\mu_\odot}{r_\oplus} - \frac{\mu_\odot}{r_\oplus + r}\right)} - \sqrt{\frac{\mu_\odot}{r_\oplus}} = 2.9446\ \text{km/s} \tag{6.68}$$

Assuming a 0%–3% change in this calculated change in velocity, the resulting radius of the final orbit, r_2, could be calculated as

$$r_2 = \frac{\mu_\odot}{\dfrac{\mu_\odot}{r_\oplus} - \dfrac{\left(\left\{1 + \dfrac{\%Err}{100}\right\}\Delta V_1 + \sqrt{\dfrac{\mu_\odot}{r_\oplus}}\right)^2}{2}} - r_\oplus \tag{6.69}$$

Using Equation (6.69), a plot can be made of how the final orbit radius differs with a 0%–3% error in Δv_1, which can be seen in Figure 6.10. When observing Figure 6.10, it can be seen that the radius calculated in Equation (6.69) can be in excess of up to 3 million kilometers larger than the Martian orbital radius. A discrepancy in a desired position such as this would most certainly lead to problems satisfying mission objectives, and the velocity must be corrected during the coast stage. While real-world interplanetary transfers will be much more complex than a simple Hohmann transfer, it can still be seen how small errors in orbital maneuvers can have drastic impacts on the interplanetary trajectories.

With the understanding of the discrepancies from design trajectories interplanetary spacecraft encounter, NASA and other space agencies have been incorporating mid-course corrections on satellite trajectories since the Apollo missions. In these Apollo missions, using the knowledge gained in the rendezvous of the Gemini missions, mid-course corrections were made to ensure that a rendezvous between a Lunar Module Ascent Stage (LM-AS) crewed vehicle and a command/service module (CSM) could be possible. In these transfers, the LM-AS would be considered an "active vehicle" which would adjust its orbit radius and inclination in several

[1] Mars Pathfinder Navigation, http://science.ksc.nasa.gov/mars/mpf/mpfnavpr.html

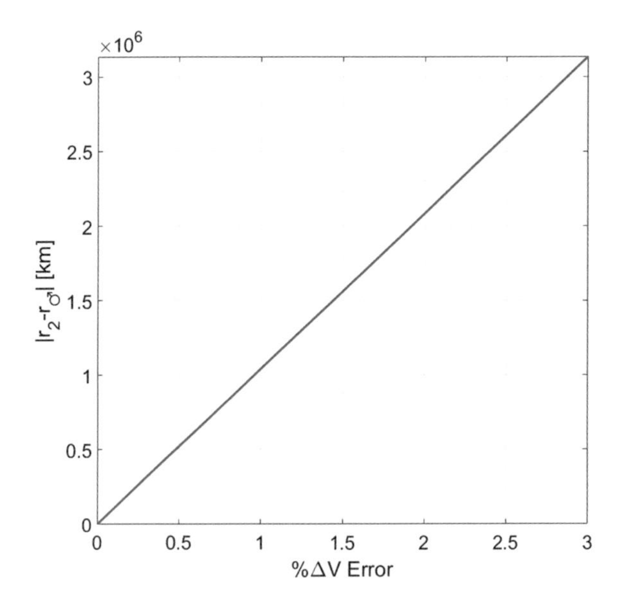

FIGURE 6.10 Plot of percent of discrepancy vs. Δv_1 error in final orbit radius for Earth-Mars Hohmann transfer.

maneuvers to place it in a position to rendezvous with the CSM. Since rendezvous of the two spacecraft was based on tight tolerances for time of rendezvous and relative position to initiate the docking procedure, the burns to place the LM into the proper orbit needed to be exact. Due to this, two mid-course corrections located between the two thrusts (initiation and breaking) producing the orbital intersect trajectory of the two spacecraft. These corrections for a sample coelliptic docking method can be visualized in Figure 6.11. The trajectories in Figure 6.11 outline the launch, Coelliptic Sequence Initiation (CSI), Constant Delta Height (CDH, places the LM in a 15 km height difference relative to the CSM), Terminal Phase Initiation (TPI), mid-course corrections, and Terminal Phase Final (TPF) maneuver sequences.[2]

From the successes of mid-course corrections in the Apollo programs, techniques were improved, and corrections were incorporated into the mission designs of all NASA interplanetary missions. In recent years, there have been several interplanetary missions to Mars, which were all designed to implement mid-course corrections in their trajectories. For example, the Mars Pathfinder missions incorporated four mid-course correction maneuvers, or trajectory correction maneuvers (TCMs) which would ensure the proper insertion into the Martian atmosphere. The first two scheduled maneuvers, TCM-1 and TCM-2, were scheduled for the first 2 months of the interplanetary trajectory, while the second two, TCM-3 and TCM-4, would be scheduled for the end of the interplanetary cruise phase, when the spacecraft is in close proximity to Mars. As a failsafe, an optional fifth TCM was also scheduled to be performed right before insertion into the Martian atmosphere, should the incoming

[2] The Apollo Flight Journal, http://history.nasa.gov/afj/loressay.htm

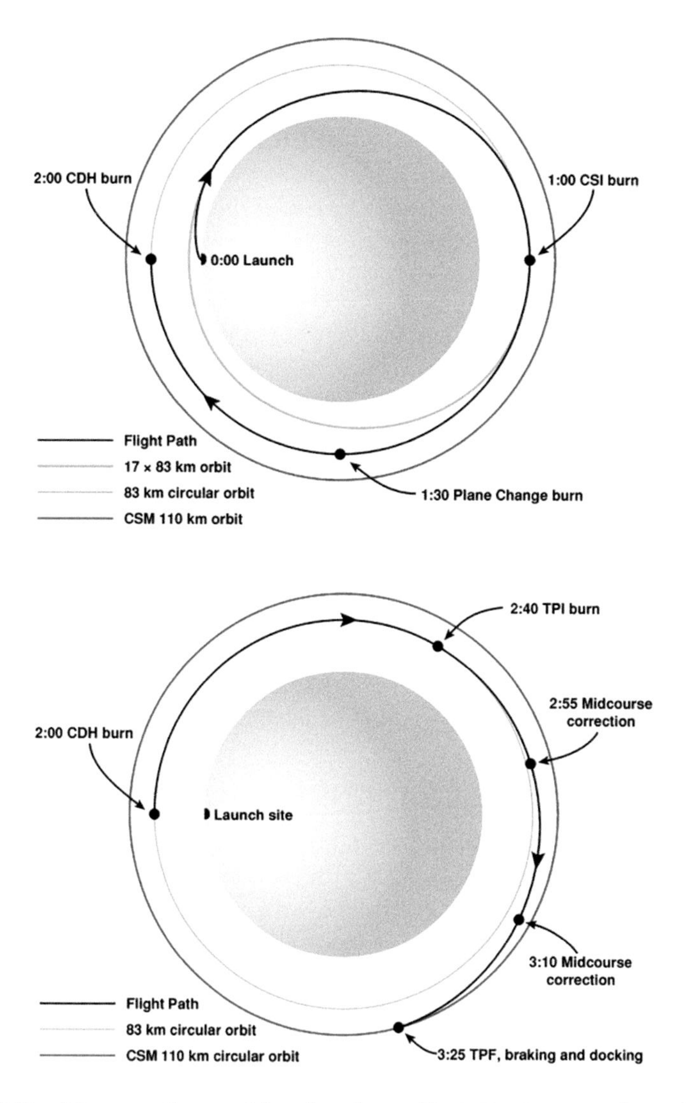

FIGURE 6.11 Maneuver times and locations for coelliptic Rendezvous scheme for Apollo LM and CSM Docking procedure. (Courtesy of NASA.)

trajectory remain unsatisfactory. A schedule of these maneuvers, and their specific objectives can be seen in the previously presented Table 5.14.

In Table 5.14, the TCM change in velocity magnitudes steadily decreases as the spacecraft gets closer to Mars. This behavior is expected, due to the fact that the spacecraft state gets closer and closer to its nominal value with each correction. As for the specific objectives of the corrections, TCM-1 and TCM-2 were designed for Mars B-plane targeting satisfying two design criteria that the probability of an unsterilized launch vehicle upper stage impacting the Martian surface be less than 0.0001%, and the probability that the Pathfinder vehicle impacting the Martian atmosphere at a speed greater than 300 m/s be less than 0.001%.[3] The latter corrections, TCM-3 and TCM-4, were designed to place the spacecraft in a specific incoming trajectory to land at a desired site on the Martian surface. These mid-course correction sequences ensured that the Pathfinder vehicle could be delivered to meet its mission objectives for both orbital insertion and landing.

Therefore, through these two simple examples from the Apollo and Pathfinder missions, it can be seen how necessary and valuable mid-course corrections can be in the planning and modeling of all spacecraft missions. However, with the addition of these correction maneuvers comes the decision of how to implement them. Several advanced methods of astrodynamics are devoted to identifying how to optimize the performance of these corrections, in both propellant use and accuracy. Some of these methods include, but are not limited to the use of dynamic programming, linearizing impulsive corrections, as well as direct and indirect methods of trajectory optimization [Cicolani, 1966; Groves, 1966; Pfeiffer, 1963; Stern and Potter, 1965].

6.5.2 Statistical Methods Applied to Targeting

The maneuver material presented in previous chapters assumed that all maneuvers were perfectly applied – the direction and magnitude of the maneuver were exact. However, physical systems have inherent error in their maneuver execution. To better plan for such errors, statistical methods are applied. One such statistical method is the estimation of error propagation using Monte Carlo simulations.

6.6 MONTE CARLO SIMULATIONS FOR MISSION PLANNING

Monte Carlo methods are a statistical method to assess the range of potential outcomes based on potential variations in the initial parameters used to execute a maneuver. A commonly used error estimation distribution is the normal, or Gaussian, distribution. This normal distribution is found using the probability density function,

$$f(x) = \frac{1}{\sigma\sqrt{2\pi}} e^{-\frac{1}{2}\left(\frac{x-\mu}{\sigma}\right)^2} \tag{6.70}$$

where σ is the standard deviation and μ is the mean – not to be confused with the use of μ to denote the gravitational parameter. This distribution is plotted in Figure 6.12, where various sections of the distribution are shown. These correspond to probability distributions related to the so-called 1σ, 2σ, and 3σ deviations.

For example, if 1,000 different values of x are chosen, approximately 34.1% (341) will lie between the mean and 1σ (the blue region in the figure), about 13.6% (136) of values will lie between 1σ and 2σ (the red region), and about 2.1% (21) will lie between 2σ and 3σ (the green region). Note that within round off error, the percentages in this example approximately sum up to 100%. This is true for either side of the distribution (positive and negative x-values) due to symmetry. This is why 1σ, 2σ, and 3σ accuracies refer to the combined sum of the probabilities on either side of

[3] Mars Pathfinder Navigation: http://science.ksc.nasa.gov/mars/mpf/mpfnavpr.html

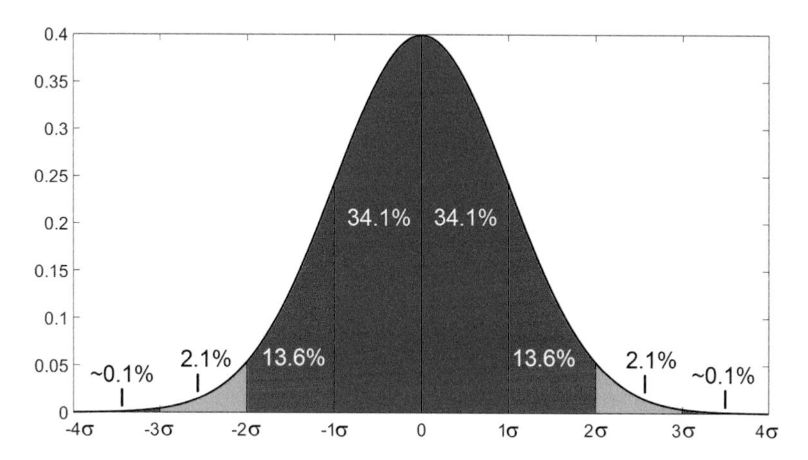

FIGURE 6.12 Normal (or Gaussian) distribution.

the distribution, or approximately 68%, 95%, and 99.7%. This distribution based on standard deviation is also referred to as the "68–95–99.7 rule."

In astrodynamics, 3σ accuracy (or approximately 99.7% accuracy) is used to estimate the effects of orbital maneuvers. This results in propagating an orbit that allows mission designers and mission planners to estimate the position of a spacecraft within a three-dimensional 3σ accuracy ellipsoid and to estimate quantities of trajectory correction propellant. As shown previously in Section 5.14.3.1, Mars landings are predicted using landing ellipses (which are essentially two-dimensional since they are projected onto the surface of Mars) to estimate where spacecraft would land. Here, we present an example of B-plane targeting, which is used in interplanetary missions to predict and estimate the spacecraft trajectory and compare it to the nominal designed trajectory. For the definition of B-plane, refer to Section 4.2.7.

Consider a spacecraft on a hyperbolic orbit with respect to Mars, with nominal hyperbolic eccentricity of 2.0. At the nominal periapse point, the spacecraft will have an altitude of 500 km above the surface of Mars, and the angle between the **T** and **B** vectors is 30°. At the boundary of the Martian sphere of influence (which is approximately 580,000 km from Mars), the velocity of the spacecraft is

$$\mathbf{v}_{\infty} = \mathbf{v}_{\infty,\text{nominal}} + \delta\mathbf{v}_{\infty} \tag{6.71}$$

where $\mathbf{v}_{\infty,\text{ nominal}}$ is the nominal two-dimensional velocity (in the S-B plane) that is on the trajectory plane at the boundary of the sphere of influence and $\delta\mathbf{v}_{\infty}$ is the three-dimensional error velocity at the boundary of the sphere of influence with zero mean and standard deviations σ of 0.1% of the nominal spacecraft speed at the boundary of the sphere of influence.

We want to run a series of Monte Carlo simulations and compute the size of the 3σ error ellipse on the B-plane resulting from the velocity deviations given by Equation (6.71). Here, we will use 100 samples, or trajectories, so that we easily visualize the results, although generally hundreds, thousands, or millions of trajectories are computed for actual interplanetary missions. From this, we can also determine

TABLE 6.2
Nominal hyperbolic trajectory parameters

Parameter	Value	Reference
e	2.0	Given
r_p	3,889.5 km	Given
r_{SOI}	578,420 km	Equation (5.148)
a	−3,889.5 km	Equation (2.37)
$v_{\infty, nominal}$	3.3492 km/s	Equation (2.50)
Δ	6,647.4 km	Equation (4.15)
Δ	60.0°	Equations (5.176) and (5.177)

how many trajectories intersect, and thus impact, Mars by computing their periapse radius and comparing it with the physical radius of Mars. Lastly, we want to compute the turn angle for each trajectory, assuming that no Δv maneuver is applied – thus resulting in gravity assist trajectories. Table 6.2 shows the nominal parameters of the hyperbolic trajectory for this example. Refer to Sections 4.2.7 and 5.12 for B-plane and gravity assist equations, if needed.

The nominal initial conditions of the spacecraft with respect to the **SBR** B-plane frame at the boundary of the sphere of influence are thus

$$\begin{aligned} \mathbf{r} &= -\sqrt{r_{SOI}^2 - \Delta^2}\,\mathbf{S} + \Delta\mathbf{B} + 0\mathbf{R} \\ \mathbf{v} &= v_{\infty,nominal}\,\mathbf{S} + 0\mathbf{B} + 0\mathbf{R} \end{aligned} \tag{6.72}$$

which, for our example, becomes

$$\begin{aligned} \mathbf{r} &= -578{,}390\mathbf{S} + 6{,}647.4\mathbf{B} + 0\mathbf{R} \quad \text{km} \\ \mathbf{v} &= 3.3492\mathbf{S} + 0\mathbf{B} + 0\mathbf{R} \quad \text{km/s} \end{aligned} \tag{6.73}$$

Implementing deviations on the spacecraft velocity as stated above results in a series of randomly generated initial conditions which can then be propagated analytically using the Lagrange coefficients (Section 2.6.5). Figure 6.13 shows the **SB** plane projection of the nominal trajectory (in black) along with the 100 trajectories (in blue) resulting from the Monte Carlo simulation. Figure 6.14 shows the B-plane along with a circle, representing Mars, and where the trajectories pierce through the B-plane. In this figure, the black dot represents the nominal trajectory, the blue crosses are the Monte Carlo simulation trajectories, the blue dot is their average, and the blue ellipse represents the 3σ ellipse based on this particular simulation. Figure 6.15 shows the variation of eccentricity, turn angle δ, semimajor axis, and periapse radius as a result of the v_∞ deviations. Periapse radii smaller than the physical radius of Mars result in trajectories that impact Mars. As we saw in Section 5.14, even periapse radii above Mars's surface can result in impact trajectories due to atmospheric effects. However, here, we did not consider the effects of the Martian atmosphere. Table 6.3 shows numerical results for this example.

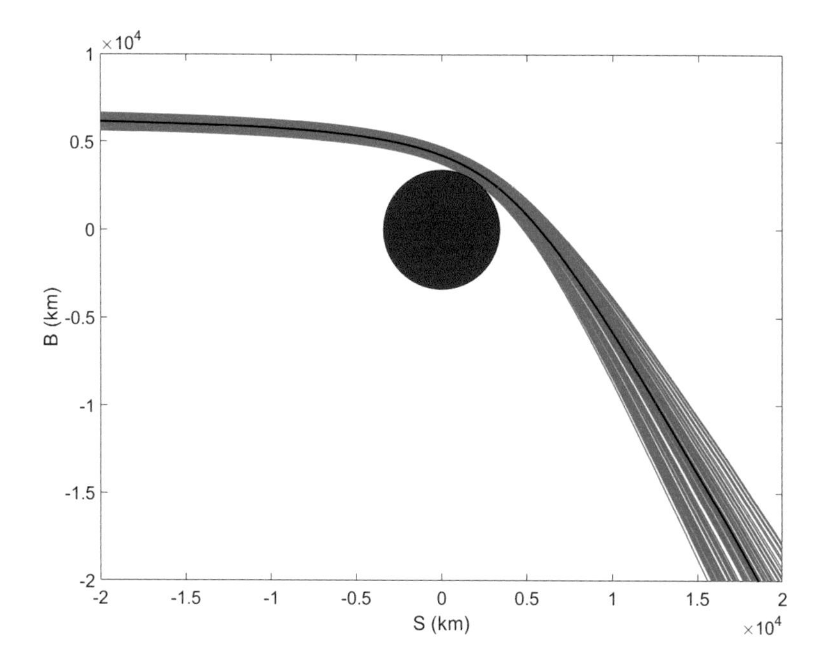

FIGURE 6.13 Monte Carlo simulation of Mars flyby trajectories (**SB** plane).

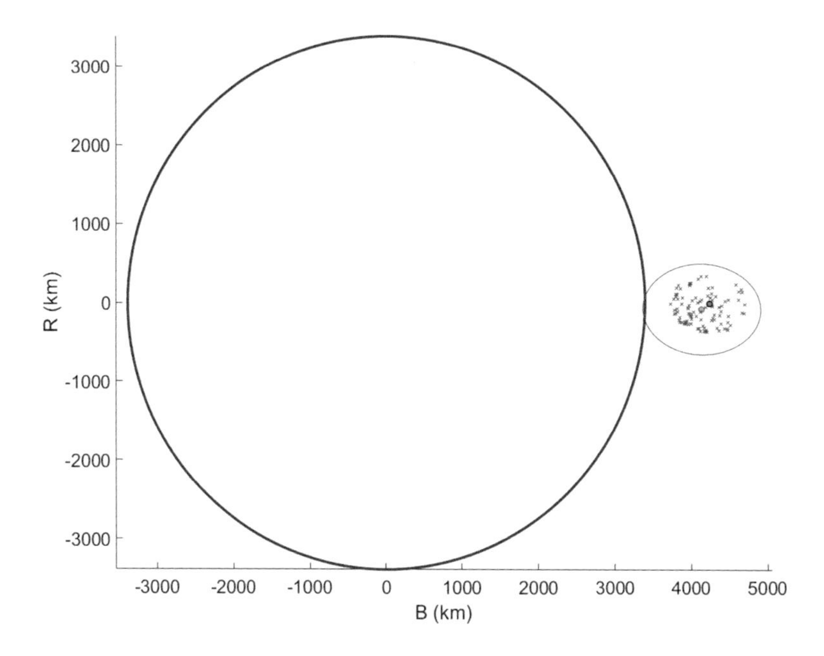

FIGURE 6.14 Monte Carlo simulation of Mars flyby trajectories (B-plane).

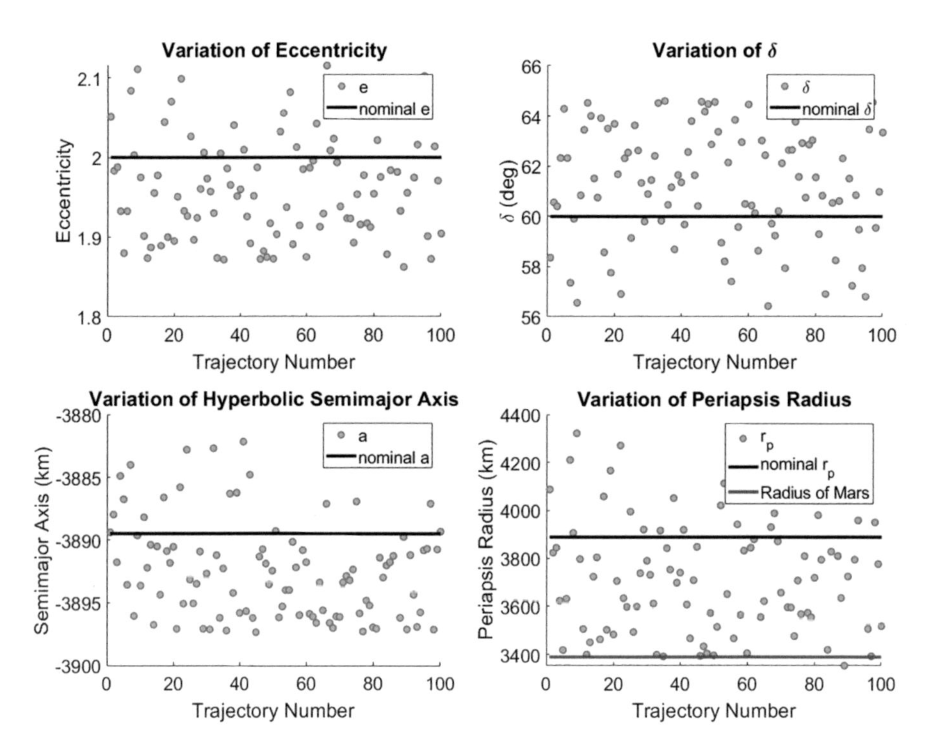

FIGURE 6.15 Variation of orbital parameters from Monte Carlo simulation of Mars flyby trajectories (B-plane).

TABLE 6.3
Numerical results from the Monte Carlo simulation

Parameter	Value
Semimajor axis of the 3σ error ellipse	770.2 km
Semiminor axis of the 3σ error ellipse	577.1 km
Percent of impact trajectories	1%
δ range	(56.42°, 64.96°)

In actual mission design, Monte Carlo simulations are used extensively. Although we only showed a simple example where variables in velocity had a statistical error, in an actual mission design many more variables would be varied. The various terms would each have their own statistical distribution and many possible outcomes would be evaluated. In addition to using this method to determine position and velocity errors, it can also be applied to helping a mission designer assess a DV budget needed to make sure the spacecraft has enough propellant to accomplish its mission. Using modern computational techniques such as parallel processing aids a designer to assess the effects of these variations in performance parameters on the overall mission performance.

PROBLEMS

1. Consider a Mars-orbiting spacecraft in a circular orbit at an altitude of 500 km. The spacecraft initiates a maneuver to land on Mars, targeting a periapsis equal to the radius of Mars at 180° from the location of the maneuver. However, the propulsion system malfunctions, causing the spacecraft to deliver a higher retrograde Δv than expected. It is believed that the extra Δv that the propulsion system delivered is somewhere between 1% and 5% of the nominal Δv. Based on this, what is the range of new landing location angles? What is the new range of times of flight for the spacecraft to hit the Martian surface? For this problem, ignore the atmospheric effects of Mars.
2. Consider Problem 1 from Chapter 5. Assume that the upper stage has a velocity deviation which can be modeled as a normal distribution with zero mean and 0.01% of the spacecraft speed in LEO in all x-, y-, and z-directions. Create a Monte Carlo simulation with these parameters, and propagate the orbit until apoapsis. How does the orbit differ from the nominal GTO? Create scatter plots similar to Figure 6.15 showing the variation of inclination, semimajor axis, and eccentricity.
3. A spacecraft is arriving in the sphere of influence of Jupiter, where it is expected to perform a gravity assist of the planet at an altitude of 1 million km. The spacecraft has a nominal hyperbolic eccentricity of 2.5. Create a Monte Carlo simulation assuming that the spacecraft position and velocity at the sphere of influence can be modeled as provided in Equation (6.72) and that the nominal v_∞ has a deviation that can be represented as a normal distribution with zero mean and 1% of the nominal v_∞ (in all directions). Propagate 1,000 orbits, plot them and create a B-plane analysis similar to Figure 6.14, showing the 3σ uncertainty ellipse resulting from your simulation. How likely is it that the trajectory would impact Jupiter?

APPENDIX A: ASTRONOMICAL CONSTANTS

TABLE A.1
Solar System physical characteristics

Body	Equatorial Radius (km)	Mass ($\times 10^{24}$ kg)	Gravitational Constant (μ) (km^3/s^2)	Semimajor Axis (km)	Eccentricity	Sidereal Orbit Period (Year)
Sun	695,990	1,989,000	1.327×10^{11}	–	–	–
Mercury	2,439.7	0.33022	2.2034×10^4	57,909,308.4	0.2056	0.2408467
Venus	6,051.9	4.8690	3.257×10^5	108,204,088.7	0.0068	0.61519726
Earth	6,378.140	5.9742	3.986×10^5	149,597,800	0.0167	1.0000174
Earth's Moon	1,738	0.073483	4,903	384,400[a]	0.0549	27.45 (days)
Mars	3,397	0.64191	4.305×10^4	227,942,167.9	0.0934	1.8808476
Jupiter	71,492	1,898.8	1.268×10^8	778,327,433.8	0.0483	11.862615
Saturn	60,268	568.50	3.794×10^7	1,426,983,495	0.0560	29.447498
Uranus	25,559	86.625	5.794×10^6	2,870,991,219	0.0461	84.016846
Neptune	24,764	102.78	6.809×10^6	4,497,074,426	0.0097	164.79132

Gravitational constant, $G = 6.67259\ (\pm 0.00030) \times 10^{-11}\ m^3/(kg\ s^2)$.
Astronautical unit (AU) = 1.4959787×10^8 km.
[a] Around the Earth.

TABLE A.2
Orbital Synodic periods (years)

	Mercury	Venus	Earth	Mars	Jupiter	Saturn	Uranus	Neptune
Mercury	–	0.3957	0.3172	0.2762	0.2458	0.2428	0.2415	0.2412
Venus	0.3957	–	1.5988	0.9142	0.6489	0.6283	0.6197	0.6175
Earth	0.3172	1.5988	–	2.1351	1.0921	1.0351	1.0120	1.0061
Mars	0.2762	0.9142	2.1351	–	2.2356	2.0093	1.9241	1.9027
Jupiter	0.2458	0.6489	1.0921	2.2356	–	19.852	13.809	12.780
Saturn	0.2428	0.6283	1.0351	2.0093	19.852	–	45.370	35.873
Uranus	0.2415	0.6197	1.0120	1.9241	13.809	45.370	–	171.37
Neptune	0.2412	0.6175	1.0061	1.9027	12.780	35.873	171.37	–

Periapsis and apoapsis are generic terms for the closest point and furthest point in an orbit, respectively. The prefixes peri- and apo- are also applied to various suffixes representing the astronomical body. These suffixes are the Greek or Roman names of the bodies. The terms pericynthion and apocynthion (related to Earth's Moon) are used for objects not launched from the surface of the Moon and were adopted for use in NASA's Apollo program.

TABLE A.3
Planetary symbols and terminology

Body	Symbol	Apoapsis	Periapsis
Sun	☉	Apohelion	Perihelion
Mercury	☿	Apohermion, Apoherm	Perihermion, Periherm
Venus	♀	Apocytherion, Apocytherean, Apokrition	Pericytherion, Pericytherean, Perikrition
Earth	⊕	Apogee	Perigee
Earth's Moon	☾	Aposelene, Apocynthion, Apolune	Periselene, Pericynthion, Perilune
Mars	♂	Apoareion	Periareion
Jupiter	♃	Apojove, Apozene	Perijove, Perizene
Saturn	♄	Apokrone, Aposaturnium	Perikrone, Perisaturnium
Uranus	⛢	Apouranion	Periuranion
Neptune	♆	Apoposedion	Periposedion
Pluto	⯓ ♇	Apohadion	Perihadion

APPENDIX B: SPHERICAL TRIGONOMETRY

A variant of planar trigonometry replaces the spatial distance on the sides of a triangle with an angular distance. This geometry is known as spherical trigonometry. Common applications of spherical trigonometry include calculating the distance between two points on the sphere, whether that sphere is a spherical planet or a celestial sphere. For example, one is to know the distance between latitude/longitude pair with another latitude/longitude pair, we would use the principles of spherical trigonometry to calculate this distance.

Spherical trigonometry was first used by Greek and Persian mathematicians as a way to compute the passage of time. Figure B.1 represents the geometry of a spherical triangle. In this figure, each side of the spherical triangle is an angular distance

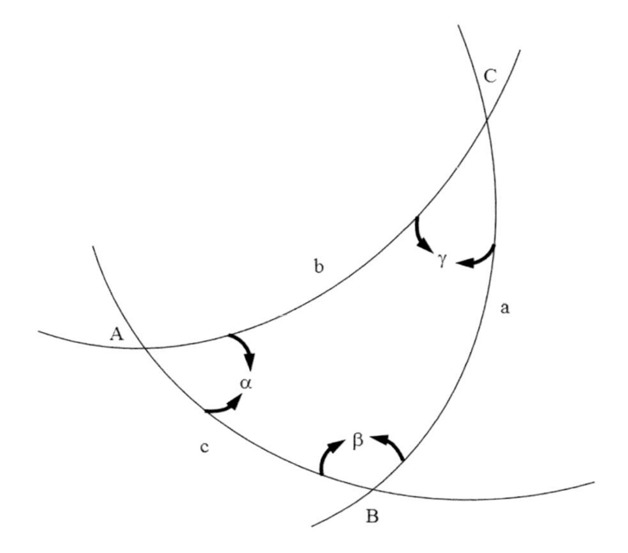

FIGURE B.1 Spherical triangle.

swept out by moving a constant radius vector, centered at the center of a sphere, through an angle. Additional sides of the spherical triangle are created in a similar fashion. At the intersection of two of these legs of the spherical triangle, an angle is formed. The triangle is completed by adding a third side that intersects the other two legs. Mathematical relationships between the angles formed by the legs and the intersection angles can be made. Napier's rules present this relationship. Relationships between the angular distances (sides of the spherical triangle) and the angles between the planes (angles) are shown using the law of sines,

$$\frac{\sin\alpha}{\sin a} = \frac{\sin\beta}{\sin b} = \frac{\sin\gamma}{\sin c} \tag{B.1}$$

and the law of cosines for sides,

$$\begin{aligned}
\cos\alpha &= \cos\beta\cos\gamma + \sin\beta\sin\gamma\cos a \\
\cos\beta &= \cos\gamma\cos\alpha + \sin\gamma\sin\alpha\cos b \\
\cos\gamma &= \cos\alpha\cos\beta + \sin\alpha\sin\beta\cos c
\end{aligned} \tag{B.2}$$

and angles,

$$\begin{aligned}
\cos a &= -\cos b\cos c + \sin b\sin c\cos\alpha \\
\cos b &= -\cos c\cos a + \sin c\sin a\cos\beta \\
\cos c &= -\cos a\cos b + \sin a\sin b\cos\gamma
\end{aligned} \tag{B.3}$$

Note that Napier's rules for spherical triangles are of a similar form to the laws sines and cosines for planar triangles.

Several applications of Napier's rules are used in the development of various orbital relationships that are presented elsewhere in this book.

APPENDIX C: PLANETARY EPHEMERIDES

These data are to be used as described in Section 4.7.1. The reference used for these values is "Keplerian Elements for Approximate Positions of the Major Planets" by E.M. Standish (JPL/Caltech) available from the JPL Solar System Dynamics website (http://ssd.jpl.nasa.gov/).

TABLE C.1A
Keplerian elements and their rates, with respect to the mean ecliptic and equinox of J2000, valid for the time interval 1800 AD–2050 AD

	a (AU) (AU/Cy)	e (rad) (rad/Cy)	i (deg) (deg/Cy)
Mercury	0.38709927	0.20563593	7.00497902
	0.00000037	0.00001906	−0.00594749
Venus	0.72333566	0.00677672	3.39467605
	0.00000390	−0.00004107	−0.00078890
Earth/Moon barycenter	1.00000261	0.01671123	−0.00001531
	0.00000562	−0.00004392	−0.01294668
Mars	1.52371034	0.09339410	1.84969142
	0.00001847	0.00007882	−0.00813131
Jupiter	5.20288700	0.04838624	1.30439695
	−0.00011607	−0.00013253	−0.00183714
Saturn	9.53667594	0.05386179	2.48599187
	−0.00125060	−0.00050991	0.00193609
Uranus	19.18916464	0.04725744	0.77263783
	−0.00196176	−0.00004397	−0.00242939
Neptune	30.06992276	0.00859048	1.77004347
	0.00026291	0.00005105	0.00035372
Pluto	39.48211675	0.24882730	17.14001206
	−0.00031596	0.00005170	0.00004818

TABLE C.1B

Keplerian elements and their rates, with respect to the mean ecliptic and equinox of J2000, valid for the time interval 1800 AD–2050 AD

	L (deg) (deg/Cy)	Longitude of Periapsis (deg) (deg/Cy)	Longitude of the Ascending Node (deg) (deg/Cy)
Mercury	252.25032350	77.45779628	48.33076593
	149,472.67411175	0.16047689	−0.12534081
Venus	181.97909950	131.60246718	76.67984255
	58,517.81538729	0.00268329	−0.27769418
Earth/Moon	100.46457166	102.93768193	0.0
barycenter	35,999.37244981	0.32327364	0.0
Mars	−4.55343205	−23.94362959	49.55953891
	19,140.30268499	0.44441088	−0.29257343
Jupiter	34.39644051	14.72847983	100.47390909
	3034.74612775	0.21252668	0.20469106
Saturn	49.95424423	92.59887831	113.66242448
	1,222.49362201	−0.41897216	−0.28867794
Uranus	313.23810451	170.95427630	74.01692503
	428.48202785	0.40805281	0.04240589
Neptune	−55.12002969	44.96476227	131.78422574
	218.45945325	−0.32241464	−0.00508664
Pluto	238.92903833	224.06891629	110.30393684
	145.20780515	−0.04062942	−0.01183482

TABLE C.2A

Keplerian elements and their rates, with respect to the mean ecliptic and equinox of J2000, valid for the time interval 3000 BC–3000 AD

	a (AU) (AU/Cy)	e (rad) (rad/Cy)	i (deg) (deg/Cy)
Mercury	0.38709843	0.20563661	7.00559432
	0.00000000	0.00002123	−0.00590158
Venus	0.72332102	0.00676399	3.39777545
	−0.00000026	−0.00005107	0.00043494
Earth/Moon Barycenter	1.00000018	0.01673163	−0.00054346
	−0.00000003	−0.00003661	−0.01337178
Mars	1.52371243	0.09336511	1.85181869
	0.00000097	0.00009149	−0.00724757
Jupiter	5.20248019	0.04853590	1.29861416
	−0.00002864	0.00018026	−0.00322699

(Continued)

TABLE C.2A (*Continued*)
Keplerian elements and their rates, with respect to the mean ecliptic and equinox of J2000, valid for the time interval 3000 BC–3000 AD

	a (AU) (AU/Cy)	e (rad) (rad/Cy)	i (deg) (deg/Cy)
Saturn	9.54149883	0.05550825	2.49424102
	−0.00003065	−0.00032044	0.00451969
Uranus	19.18797948	0.04685740	0.77298127
	−0.00020455	−0.00001550	−0.00180155
Neptune	30.06952752	0.00895439	1.77005520
	0.00006447	0.00000818	0.00022400
Pluto	39.48686035	0.24885238	17.14104260
	0.00449751	0.00006016	0.00000501

Note: The computation of M for Jupiter through Pluto must be augmented by the additional terms given in Table C.2c (below).

TABLE C.2B
Keplerian elements and their rates, with respect to the mean ecliptic and equinox of J2000, valid for the time interval 3000 BC–3000 AD

	L (deg) (deg/Cy)	Longitude of Periapsis (deg) (deg/Cy)	Longitude of the Ascending Node (deg) (deg/Cy)
Mercury	252.25166724	77.45771895	48.33961819
	149,472.67486623	0.15940013	−0.12214182
Venus	181.97970850	131.76755713	76.67261496
	58,517.81560260	0.05679648	−0.27274174
Earth/Moon	100.46691572	102.93005885	−5.11260389
Barycenter	35,999.37306329	0.31795260	−0.24123856
Mars	−4.56813164	−23.91744784	49.71320984
	19,140.29934243	0.45223625	−0.26852431
Jupiter	34.33479152	14.27495244	100.29282654
	3,034.90371757	0.18199196	0.13024619
Saturn	50.07571329	92.86136063	113.63998702
	1,222.11494724	0.54179478	−0.25015002
Uranus	314.20276625	172.43404441	73.96250215
	428.49512595	0.09266985	0.05739699
Neptune	304.22289287	46.68158724	131.78635853
	218.46515314	0.01009938	−0.00606302
Pluto	238.96535011	224.09702598	110.30167986
	145.18042903	−0.00968827	−0.00809981

Note: The computation of M for Jupiter through Pluto must be augmented by the additional terms given in Table C.2c (below).

TABLE C.2C
Additional terms which must be added to the computation of M for Jupiter through Pluto, 3000 BC–3000 AD, as described in Section 4.7.1

	b	c	s	f
Jupiter	−0.00012452	0.06064060	−0.35635438	38.35125000
Saturn	0.00025899	−0.13434469	0.87320147	38.35125000
Uranus	0.00058331	−0.97731848	0.17689245	7.67025000
Neptune	−0.00041348	0.68346318	−0.10162547	7.67025000
Pluto	−0.01262724			

APPENDIX D: INITIAL CONDITIONS FOR PERIODIC ORBITS IN THE CIRCULAR RESTRICTED THREE-BODY PROBLEM FOR THE EARTH-MOON SYSTEM

Here we present a series of Cartesian initial conditions (ICs) that can be used to generate periodic orbit in the Earth-Moon system by integrating the initial state $\mathbf{X}_0 = \begin{bmatrix} x_0 & y_0 & z_0 & \dot{x}_0 & \dot{y}_0 & \dot{z}_0 \end{bmatrix}^T$ for period T using the CR3BP equations of motion with the following parameters.

$\mu = 0.012150585609624 04$

1 DU = 389,703 km

1 TU = 382,981 s

All ICs were obtained from JPL's Solar System Dynamics database of periodic orbits (https://ssd.jpl.nasa.gov/tools/periodic_orbits.html) unless otherwise specified.

TABLE D.1
L_1 Lyapunov family initial conditions

x_0	$\dot{y}_0$	T
−0.832238309337782	−0.040586687132343	2.69653738139437
−0.825493941781705	−0.104799117352634	2.72624465817082
−0.817941254632040	−0.185147584643998	2.81000743196363
−0.810845482838303	−0.263423648988838	2.96468162529025
−0.802670575558952	−0.338409540598485	3.22847108381605
−0.791588087402816	−0.403973414773399	3.61934468682740
−0.776967318624104	−0.458023677043333	4.09299106529158
−0.759307393676014	−0.505745128342778	4.59348555733136
−0.739138841121786	−0.552481266182751	5.08520906795249
−0.716739642038189	−0.601693709185475	5.54607828652737

TABLE D.2
L_2 Lyapunov family initial conditions

x_0	$\dot{y}_0$	T
−1.14528182830311	−0.0545770359443779	3.37735005419650
−1.12180424859669	−0.169303380549629	3.41230416949785
−1.09588216566390	−0.292752139547197	3.49578490120742
−1.06995881019479	−0.425986791126913	3.65533876085210
−1.04735418747544	−0.568352349396663	3.91939186782640
−1.03039806902818	−0.715816496175862	4.28425817857587
−1.01881025005046	−0.864592497586580	4.70949751978763
−1.01100780119720	−1.01425353184537	5.15207615828473
−1.00562585324935	−1.16614281806734	5.58263918799237
−1.00179311263346	−1.32171867740340	5.98376818547925

TABLE D.3
L_3 Lyapunov family initial conditions

x_0	$\dot{y}_0$	T
0.980879450369756	0.0491784269918098	6.21841208453133
0.956012157897287	0.100406081064013	6.21847983850141
0.931154894273377	0.152331744016536	6.21859362818550
1.11544844493969	−0.217632294767988	6.21884522623900
1.16504323917404	−0.312378112993135	6.21934895454349
1.21460332261702	−0.405559249223934	6.22004216998744
1.26416217010976	−0.497476480409322	6.22093018914807
1.31375575411501	−0.588414895696382	6.22202075166065
1.36340972692572	−0.678628454736362	6.22332424320227
1.41312387338205	−0.768320536007814	6.22485374251881

TABLE D.4
L_4 Lyapunov family initial conditions with $x_0 = -0.487849413449431$

$\dot{x}_0$	y_0	$\dot{y}_0$	T
0.900672279641547	0.0543416621242159	0.0221880516899251	6.58204851736832
0.938505172001864	0.112809023710120	0.0469037312777928	6.57991147467613
0.976066937147407	0.170085580472898	0.0719018105005575	6.57635726862477
1.01298348622529	0.225749489469015	0.0968976210403637	6.57151166206235
1.04895475650490	0.279497518832409	0.121664470151946	6.56552736313172
1.08376450511285	0.331147275230792	0.146040632657424	6.55856641355020
1.11727319702597	0.380618351865406	0.169924084761900	6.55078682409673
1.14940204637922	0.427904632705515	0.193260672171316	6.54233473633554
1.18011590033953	0.473048098389812	0.216031265854909	6.53334104062054
1.20940972732785	0.516119845080523	0.238241503916451	6.52392046481846

TABLE D.5
L_5 Lyapunov family initial conditions with $x_0 = -0.487849413449431$

$\dot{x}_0$	y_0	$\dot{y}_0$	T
−0.900672279641547	−0.0543416621242159	0.0221880516899251	6.58204851736832
−0.938505172001864	−0.112809023710120	0.0469037312777928	6.57991147467613
−0.976066937147407	−0.170085580472898	0.0719018105005575	6.57635726862477
−1.01298348622529	−0.225749489469015	0.0968976210403637	6.57151166206235
−1.04895475650490	−0.279497518832409	0.121664470151946	6.56552736313172
−1.08376450511285	−0.331147275230792	0.146040632657424	6.55856641355020
−1.11727319702597	−0.380618351865406	0.169924084761900	6.55078682409673
−1.14940204637922	−0.427904632705515	0.193260672171316	6.54233473633554
−1.18011590033953	−0.473048098389812	0.216031265854909	6.53334104062054
−1.20940972732785	−0.516119845080523	0.238241503916451	6.52392046481846

TABLE D.6
Northern L_1 halo family initial conditions

x_0	z_0	$\dot{y}_0$	T
−0.824425578101826	0.0623292756428522	−0.173521194438447	2.76578525934204
−0.831192933194168	0.121301232534929	−0.236017313116492	2.78524853011638
−0.847985699641184	0.172936321552328	−0.2637627216628	2.58628320934705
−0.896599395437523	0.199288220251230	−0.191388835192734	1.94949884636495
−0.867665590841194	0.188268675777110	−0.243797091812538	2.28867357323915
−0.926937785485373	0.302069508023322	−0.0821934389885669	2.20148577185167
−0.888314401342313	0.416447217421474	−0.111547740752567	2.59859148738831
−0.836155609768091	0.518045968621904	−0.160472627550042	2.79595529570443
−0.775170814044179	0.608209906824177	−0.219778187609515	2.90461301602606
−0.707199299355286	0.687977147308974	−0.286716024291547	2.97028967006645

TABLE D.7
Southern L_1 halo family initial conditions

x_0	z_0	$\dot{y}_0$	T
−0.824425578101826	−0.0623292756428522	−0.173521194438447	2.76578525934204
−0.831192933194168	−0.121301232534929	−0.236017313116492	2.78524853011638
−0.847985699641184	−0.172936321552328	−0.263762721629628	2.58628320934705
−0.896599395437523	−0.199288220251230	.191388835192734	1.94949884636495
−0.867665590841194	−0.188268675777110	−0.243797091812538	2.28867357323915
−0.926937785485373	−0.302069508023322	−0.0821934389885669	2.20148577185167
−0.888314401342313	−0.416447217421474	−0.111547740752567	2.59859148738831
−0.836155609768091	−0.518045968621904	−0.160472627550042	2.79595529570443
−0.775170814044179	−0.608209906824177	−0.219778187609515	2.90461301602606
−0.707199299355286	−0.687977147308974	−0.286716024291547	2.97028967006645

TABLE D.8
Northern L_2 halo family initial conditions

x_0	z_0	$\dot{y}_0$	T
−1.18074709536068	0.0124391738744698	0.156747918155089	3.41426583322763
−1.18011462822169	0.0279521327394916	0.160227193856732	3.40912588132459
−1.17906239499376	0.0420474898439647	0.165319774318642	3.40096618037991
−1.17762311755614	0.0550965326640780	0.171267683869400	3.39033688720297
−0.990267369727388	0.123498353971951	0.0182653180572176	0.835392824593204
−1.17369190510765	0.0787138475958238	0.183811891758638	3.36289674952148
−1.17119937495413	0.0897224172953401	0.189981443360360	3.34600644789538
−1.16846438262244	0.0999595200000499	0.195691188265634	3.32754539164282
−0.993184321735992	0.140539375079398	0.0289781370907505	0.997960661356441
−0.994540582683005	0.145521144186807	0.0333838924060610	1.04702966715675

TABLE D.9
Southern L_2 halo family initial conditions

x_0	z_0	$\dot{y}_0$	T
−1.18074709536068	−0.0124391738744698	0.156747918155089	3.41426583322763
−1.18011462822169	−0.0279521327394916	0.160227193856732	3.40912588132459
−1.17906239499376	−0.0420474898439647	0.165319774318642	3.40096618037991
−1.17762311755614	−0.0550965326640780	0.171267683869400	3.39033688720297
−0.990267369727388	−0.123498353971951	0.0182653180572176	0.835392824593204
−1.17369190510765	−0.0787138475958238	0.183811891758638	3.36289674952148
−1.17119937495413	−0.0897224172953401	0.189981443360360	3.34600644789538
−1.16846438262244	−0.0999595200000499	0.195691188265634	3.32754539164282
−0.993184321735992	−0.140539375079398	0.0289781370907505	0.997960661356441
−0.994540582683005	−0.145521144186807	0.0333838924060610	1.04702966715675

TABLE D.10
Distant retrograde family initial conditions

x_0	$\dot{y}_0$	T
−0.977263585255518	0.156747918155089	3.41426583322763
−0.970694815322703	0.160227193856732	3.40912588132459
−0.954605049237006	0.165319774318642	3.40096618037991
−0.898335354870926	0.171267683869400	3.39033688720297
−0.805553576909219	0.0182653180572176	0.835392824593204
−0.716350027362627	0.183811891758638	3.36289674952148
−0.624586649586040	0.189981443360360	3.34600644789538
−0.533200080131239	0.195691188265634	3.32754539164282
−0.445826141634795	0.0289781370907505	0.997960661356441
−0.364941466327420	0.0333838924060610	1.04702966715675

APPENDIX E: INITIAL CONDITIONS FOR PERIODIC ORBITS IN THE CIRCULAR RESTRICTED THREE-BODY PROBLEM FOR THE MARS-PHOBOS SYSTEM

Here we present a series of Cartesian initial conditions that can be used to generate periodic orbit in the Mars-Phobos system by integrating the initial state $\mathbf{X}_0 = \begin{bmatrix} x_0 & y_0 & z_0 & \dot{x}_0 & \dot{y}_0 & \dot{z}_0 \end{bmatrix}^T$ for period T using the CR3BP equations of motion with the following parameters [Conte, 2019]

$\mu = 1.660513\text{E-}08$ 1 DU = 9,376 km 1 TU = 4,387.42 s

TABLE E.1
Distant retrograde family initial conditions

x_0	$\dot{y}_0$	T
−1.0015998	−5.1545257E-03	2.2895915
−1.0028797	−6.5371177E-03	4.3149486
−1.0041596	−8.6397483E-03	5.4051710
−1.0054394	−1.0999270E-02	5.8551253
−1.0067193	−1.3453458E-02	6.0491296
−1.0079991	−1.5949485E-02	6.1427703
−1.0092790	−1.8464451E-02	6.1928042
−1.0105589	−2.0989299E-02	6.2217169
−1.0118387	−2.3519602E-02	6.2395029
−1.0131186	−2.6053102E-02	6.2509613
−1.0143985	−2.8587404E-02	6.2588075
−1.0156783	−3.1121443E-02	6.2644638
−1.0169582	−3.3656003E-02	6.2683095
−1.0182381	−3.6189882E-02	6.2711970
−1.0195179	−3.8722962E-02	6.2733889
−1.0207978	−4.1255104E-02	6.2750615
−1.0220776	−4.3786185E-02	6.2763530
−1.0233575	−4.6316710E-02	6.2772431
−1.0246374	−4.8845349E-02	6.2780779
−1.0259172	−5.1372291E-02	6.2789589
−1.0271971	−5.3899203E-02	6.2792866
−1.0284770	−5.6423736E-02	6.2798477
−1.0297568	−5.8947014E-02	6.2802644
−1.0310367	−6.1468952E-02	6.2806086
−1.0319966	−6.3359314E-02	6.2808749

References

Abilleira, F., "Broken-Plane Maneuver Applications for Earth to Mars Trajectories", 20th International Symposium on Space Flight Dynamics, Annapolis, Maryland, September 24–28, 2007.

Aldrin, B., "Cyclic Trajectory Concepts", Meeting with Jet Propulsion Laboratory, October 1985.

Alfano, S., "Low-Thrust Orbit Transfer", M.S. Thesis, Air Force Institute of Technology, AFIT/GA/AA/82D-2, December 1982.

Altunin, V., K. Miller, D. Murphy, J. Smith, and R. Wietfeldt, "Space Very Long Baseline Interferometry (SVLBI) Mission Operations", NASA Technical Report TMO Progress Report 42–142, NASA Jet Propulsion Laboratory, Pasadena, CA, 2000.

Anderson, G.M., and E.A. Smith, "A Combined Gradient/Neighboring Extremal Algorithm for the Calculation of Optimal Transfer Trajectories between Noncoplanar Orbits Using a Constant Low Thrust Rocket", *Journal of the Astronautical Sciences*, Vol. XXIII, No. 3, pp. 225–239, July–September 1975.

Astronomical Almanac. Annual Printing, U.S. Government Printing Office, Washington D.C., 1984.

Astronomical Almanac. Annual Printing, U.S. Government Printing Office, Washington, D.C., 2006.

Auwers, A., "Fundamental-Catalog für die Zonen-Beobachtungen am Nördlichen Himmel", Publ. d Astron. Gesellschaft, No. 14, 1879.

Auwers, A., "Mittlere Örter von 83 südlichen Sternen für 1875.0", Publ. d. Astron. Gesellschaft, No. 17, 1883.

Avila, E.R., "ELITE Program Overview", AIAA 92–1559-CP, AIAA Space Programs and Technologies Conference, Huntsville, AL, March 1992.

Barden, B.T., "Using Stable Manifolds to Generate Transfers in the Circular Restricted Problem of Three Bodies", M.S. Thesis, School of Aeronautics and Astronautics, Purdue Univ., West Lafayette, IN, December 1994.

Barrar, R.B., "An Analytic Proof that the Hohmann-Type Transfer is the True Minimum Two-Impulse transfer", *Acta Astronautica*, Vol. 9, No. 1, pp. 1–11, 1963.

Bate, R.R., D.D. Mueller, and J.E. White, "*Fundamentals of Astrodynamics*", 1st ed., Dover Publications, New York, 1971.

Bate, R.R., D.D. Mueller, and J.E. White, "*Fundamentals of Astrodynamics*", 2nd ed., Dover Publications, New York, 2020.

Battin, R.H., "*An Introduction to the Mathematics and Methods of Astrodynamics*", 1st ed., AIAA Education Series, New York, 1987.

Battin, R.H., "*An Introduction to the Mathematics and Methods of Astrodynamics*", Rev. ed., AIAA Education Series, Reston, VA, 1999.

Belbruno, E.A., and J.P. Carrico, "Calculation of Weak Stability Boundary Ballistic Lunar Transfer Trajectories", AIAA/AAS Astrodynamics Specialist Conference, Denver, CO, August 14–17, 2000.

Belbruno, E.A., and J.K. Miller, "A Ballistic Lunar Capture Trajectory for the Japanese Spacecraft Hiten", Jet Propulsion Laboratory, IOM 312/90.41371-EAB, 1990.

Belbruno, E.A., and J.K. Miller, "Sun-Perturbed Earth-to-Moon Transfers with Ballistic Capture", *Journal of Guidance, Control, and Dynamics*, Vol. 16, No. 4, pp. 770–775, 1993.

Bell, E.T., "*Men of Mathematics*", Simon and Schuster, New York, 1937.

Bezrouk, C., and J. Parker, "Long Duration Stability of Distant Retrograde Orbits", AIAA/AAS Astrodynamics Specialist Conference, San Diego, CA, August 9–13, 2014.

Bhaskaran, S., "Autonomous Optical-Only Navigation for Deep Space Missions", AIAA ASCEND 2020, November 2, 2020. https://doi.org/10.2514/6.2020-4139.

Bhaskaran, S., J.E. Riedel, and S.P. Synnott, "Autonomous Optical Navigation for interplanetary missions", Proceedings of SPIE 2810, Space Sciencecraft Control and Tracking in the New Millennium, October 28, 1996. https://doi.org/10.1117/12.255151.

BIPM (Bureau international des poids et mesures; International Bureau of Weights and Measures), "*SI Brochure*", 8th ed., p. 112, 2006.

BIPM (Bureau international des poids et mesures; International Bureau of Weights and Measures), "*BIPM Annual Report on Time Activities*", Vol. 10, 2015.

Brooks, C.G., J.M. Grimwood, and L.S. Swenson, "Chariots for Apollo: A History of Manned Lunar Spacecraft", NASA SP-4205, 1979.

Broucke, R. "Stability of Periodic Orbits in the Elliptic, Restricted Three-Body Problem", *AIAA Journal*, Vol. 7, No. 6, pp. 1003–1009, 1969.

Broucke, R.A., "On the History of the Slingshot Effect and Cometary Orbits", *Advances in the Astronautical Sciences*, Vol. 109, pp. 1927–1939, 2001.

Brouwer, D., and G.M. Clemence, "*Methods of Celestial Mechanics*", Academic Press, New York and London, 1961.

Brusch, R.G., and T.L. Vincent, "Low-Thrust, Minimum Fuel, Orbit Transfers", *Astronautica Acta*, Vol. 16, No. 2, pp. 65–73, 1971.

Brussels, D.M., "Copernicus Earns Papal Blessing", *New Scientist*, 1993.

Bryson, A.E., and Y. Ho, "*Applied Optimal Control: Optimization, Estimation, and Control*", Hemisphere Publishing, New York, 1975.

Canon, M.D., "*Theory of Optimal Control and Mathematical Programming*", McGraw-Hill, New York, 1970.

Capitaine, N., J. Vondrak, M. Soffel, D. McCarthy, G. Petit, J. Kovalevsky, B. Guinot, K. Seidelmann, J. Bangert, M. Standish, P. Wallace, A.M. Gontier, "Proceedings of the IERS Workshop on the Implementation of the New IAU resolutions", IERS Technical Note Number 29, Frankfurt am Main, 2002.

Chaisson, E.a.M., Steve, "*Astronomy: A Beginner's Guide to the Universe*", 5th ed., Pearson Education Inc., Upper Saddle River, NJ, 2007.

Chang, C. "DSN Telecommunications Link Design Handbook", NASA Technical Report 810–005, Rev. E, 301, Rev. C Coverage and Geometry, 2018, http://deepspace.jpl.nasa.gov/dsndocs/810-005/.

Chao, C.C., "*Applied Orbit Perturbation and Maintenance*", Aerospace Press and American Institute of Aeronautics and Astronautics, El Segundo, CA and Reston, VA, 2005.

Chicone, C., "*Ordinary Differential Equations with Applications*", Springer-Verlag, New York, 1999.

Chobotov, V.A., "*Orbital Mechanics*", 3rd ed., American Institute of Aeronautics and Astronautics Education Series, Reston, VA, 2002.

Cicolani, L.S., "Linear Theory of Impulsive Velocity Corrections for Space Mission Guidance" NASA Technical Report TN D-3365, 1966.

Clohessy, W.H., and R.S. Wiltshire, "Terminal Guidance System for Satellite Rendezvous", *Journal of Aerospace Science*, Vol. 27, No. 9, pp. 653–658, 1960.

Condon, G.L. and J. Williams, "Asteroid Redirect Crewed MissionNominal Design and Performance", SpaceOps 2014 Conference, 2014–3563, Pasadena, CA, May 5–9, 2014.

Conley, C., "Low Energy Transit orbits in the Restricted Three-Body Problem", *SIAM Journal on Applied Mathematics*, Vol. 16, No. 4, pp. 732–746, 1968.

Conte, D. "Semi-analytical Solutions for Proximity Operations in the Circular Restricted Three-Body Problem", Doctoral Dissertation, Department of Aerospace Engineering, The Pennsylvania State University, University Park, PA, August 2019.

Conte, D., M. Di Carlo, K. Ho, D.B. Spencer, and M. Vasile, "Earth-Mars Transfers through Moon Distant Retrograde Orbits", *Acta Astronautica*, Vo. 143, pp. 372–379, 2017.

Conte, D., and D.B. Spencer, "Mission Analysis for Earth to Mars-Phobos Distant Retrograde Orbits", *Acta Astronautica*, Vol. 151, pp. 761–771, 2018.

Copernicus, N., "*On the Revolutions of Heavenly Spheres*", reprinted by the Running Press Book Publishers, 2002.

Cowell, P.H., and A.C.D. Crommelin, "The Orbit of Jupiter's Eighth Satellite", *Monthly Notices of the Royal Astronomical Society*, Vol. 68, pp. 576–581, 1908.

Crone, G, A. Elfving, T. Passvogel, G. Pilbratt, and J. Tauber, "Unveiling the Universe. Two Missions to Unlock the Secrets of the Cold Cosmos", *ESA Bulletin*, Vol. 128, pp. 10–17, 2006.

Curtis, H.D., "*Orbital Mechanics for Engineering Students*", Elsevier Butterworth-Heinemann, Burlington, MA, 2019.

D'Amario, L.A., D.V. Byrnes, R.E. Diehl, L.E. Bright, and A.A. Wolf, "Preliminary Design for a Proposed Saturn Mission with a Second Galileo Spacecraft", *Journal of the Astronautical Sciences*, Vol. 37, 1989.

Darling, D., "*Gravity's Arc*", John Wiley & Sons, Hoboken, NJ, 2006.

Delva, P., A. Hees, P. Wolf, "Clocks in Space for Tests of Fundamental Physics", *Space Science Reviews*, Vol. 212, No. 3–4, pp. 1385–1421, 2017.

Diacu, F., "The Solution of the N-body Problem", *The Mathematical Intelligencer*, Vol. 18, pp. 66–70, 1996.

Dickerson, W.D., and D.B. Smith, "Trajectory Optimization for Solar-Electric Powered Vehicles", AIAA 67–583-CP, AIAA Guidance, Control and Flight Dynamics Conference, Huntsville, AL, August 1967.

Diehl, R., E. Belbruno, D. Bender, M. Myers, and D. Stetson, "Low Launch-Energy Trajectories to the Outer Solar System via Venus and Earth Gravity-Assist Flybys", AAS 87–419, AAS/AIAA Astrodynamics Conference, Kalispell, MT, January 1988.

Drake, B.G., S.J. Hoffman, and D.W. Beaty, "Human exploration of Mars, Design Reference Architecture 5.0," 2010 IEEE Aerospace Conference, Big Sky, MT, March 6–13, 2010.

Duchayne, L., F. Mercier, and P. Wolf, "Orbit Determination for Next Generation Space Clocks", *Journal of Astronomy and Astrophysics*, Vol. 504, No. 2, pp. 653–661, 2009.

Dunham, D.W., and D.P. Muhonen, "Tables of Libration-Point Parameters for Selected Solar System Objects," *Journal of the Astronautical Sciences*, Vol. 49, No. 1, pp.197–217, 2001.

Edelbaum, T.N., "Propulsion Requirements for Controllable Satellites", *ARS Journal*, pp. 1079–1089, 1961.

Edelbaum, T.N., "How Many Impulses?" *Astronautics and Aeronautics Journal*, pp. 64–69, 1967.

Edelbaum, T.N., L. Sackett, H. Malchow, "Optimal Low Thrust Geocentric Transfer", AIAA 73–1074-CP, AIAA 10th Electric Propulsion Conference, Lake Tahoe, NV, October–November 1973.

Eichhorn, H., "*Astronomy of Star Positions*", Frederick Ungar Publishing Co., New York, 1974.

Enright, P.J., and B.A. Conway, "Optimal Finite-Thrust Spacecraft Trajectories Using Collocation and Nonlinear Programming", *Journal of Guidance, Control, and Dynamics*, Vol. 14, No. 5, pp. 981–985, 1991.

Enright, P.J., and B.A. Conway, "Discrete Approximations to Optimal Trajectories Using Direct Transcription and Nonlinear Programming", *Journal of Guidance, Control, and Dynamics*, Vol. 15, No. 4, pp. 994–1002, 1992.

ESA, "*ESA Science and Technology – Gaia*", http://sci.esa.int/gaia/, Accessed: 2020-08-15.

Farquhar, R.W., "The Control and Use of Libration-Point Satellites", NASA Technical Report TR R-346, 1970.

Farquhar, R.W., "The Flight of ISEE-3/ICE: Origins, Mission History, and a Legacy", AIAA-98–4464, AIAA/AAS Astrodynamics Conference, Boston, MA, August 10–12, 1998.

Farquhar, R., D. Dunham, Y. Guo, and J. McAdams, "Utilization of Libration Points for Human Exploration in the Sun-Earth-Moon System and Beyond", *Acta Astronautica*, Vol. 55, pp. 687–700, 2004.

Farquhar, R.W., and A.A. Kamel, "Quasi-Periodic Orbits About the Translunar Libration Point", *Celestial Mechanics*, Vol. 7, pp. 458–473, 1973.

Farquhar, R.W., D. Muhonen, and D. Richardson, "Mission Design for a Halo Orbiter of the Earth", *Journal of Spacecraft and Rockets*, Vol 14, No. 3, pp. 170–177, 1977.

Fricke, W., H. Schwan, T. Lederle, U. Bastian, R. Bien, G. Burkhardt, B. Du Mont, R. Hering, R. Jährling, H. Jahreiß, S. Röser, H.M. Schwerdtfeger, H.G. Walter, "Fifth Fundamental Catalog (FK5), Part I: The Basic Fundamental Stars", Veröffentlichungen Astronomisches Rechen-Institut Heidelberg, No. 32. G. Braum, Karlsruhe, 1988.

Fricke, W., H. Schwan, T. Lederle, U. Bastian, R. Bien, G. Burkhardt, B. Du Mont, R. Hering, R. Jährling, H. Jahreiß, S. Röser, H.M. Schwerdtfeger, H.G. Walter, "Fifth Fundamental Catalog (FK5), Part II: The FK5 Extension - New Fundamental Stars", Veröffentlichungen Astronomisches Rechen-Institut Heidelberg, No. 33. G. Braum, Karlsruhe, 1991.

Galilei, G., "*Dialogues Concerning Two New Sciences*" (original edition, 1638), Hawking, S. (ed.), Running Press, Philadelphia, 2002.

Genova, A.L, and B. Aldrin, "A Free-Return Earth-Moon Cycler Orbit for an Interplanetary Cruise Ship," AAS/AIAA Astrodynamics Specialist Conference, Vail, CO, August 9–13, 2015.

Giorgini, J.D., D.K. Yeomans, A.B. Chamberlin, P.W. Chodas, R.A. Jacobson, M.S. Keesey, J.H. Lieske, S.J. Ostro, E.M. Standish, and R.N. Wimberly, "JPL's On-Line Solar System Data Service", *Bulletin of the American Astronomical Society*, Vol 28, No. 3, p. 1158, 1996.

Gómez, G., W.S. Koon, M.W. Lo, J.E. Marsden, J. Masdemont, and S.D. Ross, "Invariant Manifolds, the Spatial Three-Body Problem and Space Mission Design", AAS 01–301, AAS/AIAA Astrodynamics Specialists Conference, Quebec, Canada, July 30–August 2, 2001a.

Gómez, G., J. Llibre, R. Martinez, and C. Simó, "Dynamics and Mission Design Near Libration Points", Vol. 1, *Fundamentals: The Case of Collinear Points*, 2001b.

Gómez, G., and J. Masdemont, "Some Zero Cost Transfers Between Libration Point Orbits", AAS 00–177, AAS/AIAA Spaceflight Mechanics Meeting, Santa Barbara, CA, 2000.

Gooding, R.H., "A Procedure for the Solution of Lambert's Orbital Boundary-Value Problem", *Celestial Mechanics and Dynamical Astronomy*, Vol. 48, No. 2, pp. 145–165 1990.

Groves, R.T., "*Injection and Midcourse Correction Analysis for the Galactic Probe*", NASA Goddard Spaceflight Center, Greenbelt, MD, 1966.

Hacker, B.C., and J.M. Grimwood, "On the Shoulders of Titans: A History of Project Gemini", NASA SP-4203, 1977.

Haeberle, D.W., D.B. Spencer, and T.A. Ely, "Preliminary Investigation of Interplanetary Navigation Using Large Antenna Arrays with Varied Baselines", AIAA-2004–4744, AIAA/AAS Astrodynamics Specialists Conference, Provence, RI, August 16–19, 2004.

Handelsman, M., "Optimal Free-Space Fixed-Thrust Trajectories Using Impulsive Trajectories as Starting Interatives", *AIAA Journal*, Vol. 4, No. 6, pp. 1077–1082, 1966.

Hargraves, C.R., and S.W. Paris, "Direct Trajectory Optimization Using Nonlinear Programming and Collocation", *Journal of Guidance, Control and Dynamics*, Vol 10, No. 4, July–August 1987.

Hawking, S.W., "*On the Shoulders of Giants: The Great Works of Physics and Astronomy*", Running Press. XIII, Philadelphia, PA, 2002.

Hazelrigg, G.A., Jr., "The Use of Green's Theorem to Find Globally Optimal Solutions to a Class of Impulsive Transfers", AAS 68–092, AAS/AIAA Astrodynamics Specialist Conference, Jackson, WY, September 1968.

Hénon, M. "Vertical Stability of Periodic Orbits in the Restricted Problem", *Journal of Astronomy and Astrophysics*, Vol. 28, pp. 415–426, 1973.

Herrick, S., *Astrodynamics: Orbit Determination, Space Navigation, Celestial Mechanics*, Van Nostrand Reinhold Company, the University of California, 1971.

Hill, G.W., "Researches in Lunar Theory", *American Journal of Mathematics*, Vol. 1, pp. 5–26, 1878.

Hirani, A.N., and R.P. Russell, "Approximations of Distant Retrograde Orbits for Mission Design", *Advances in the Astronautical Sciences*, Vol. 124, pp. 273–288, 2006.

Ho, K., O.L. de Weck, J.A. Hoffman, and R. Shishko, "Dynamic Modeling and Optimization for Space Logistics Using Time-expanded Networks", *Acta Astronautica*, Vol. 105, No. 2, pp. 428–443, 2014.

Hohmann, W., "Die Erreichbarkeit der Himmelshkörper" ("The Attainability of Heavenly Bodies"), Munich, 1925. NASA's Technical Translation: NASA-TR-R-36, 1960.

Hollister, W.M., "Periodic Orbits for Interplanetary Flight", *Journal of Spacecraft and Rockets*, Vol. 6, No. 4, pp. 366–369, 1969.

Holzinger, M.J., C.C. Chow, and P. Garretson, "A Primer on Cislunar Space", AFRL 2021–1271, 2021.

Howell, K.C., "Three-Dimensional, Periodic Halo Orbits in the Restricted Three-Body Problem", Doctoral Dissertation, Department of Aeronautics and Astronautics, Stanford University, 1983a.

Howell, K.C., "Effects of Eccentricity on Halo Orbits in the Restricted Three-Body Problem", AAS 83–335, AAS/AIAA Astrodynamics Specialist Conference, Lake Placid, NY, August 1983b.

Howell, K.C., "Families of Orbits in the Vicinity of the Collinear Libration Points", *Journal of the Astronautical Sciences*, Vol. 49, No. 1, pp. 107–125, 2001.

Howell, K.C., B.T. Barden, and M. Lo, "Application of Dynamical Systems Theory to Trajectory Design for a Libration Point Mission", *Journal of the Astronautical Sciences*, Vol. 45, No. 2, pp. 161–178, 1997.

Howell, K.C., B.G. Marchand, and M.W. Lo, "Temporary Satellite Capture of Short-Period Jupiter Family Comets from the Perspective of Dynamical Systems", *Journal of the Astronautical Science*, Vol. 49, No. 4, pp. 539–557, 2001.

Howell, K.C., and H.J. Pernicka, "Numerical Determination of Lissajous Trajectories in the Restricted Three-Body Problem", *Celestial Mechanics*, Vol. 41, pp. 107–124, 1988.

Hughes, D.W., B.D. Yallop, and C.Y. Hohenkerk, "The Equation of Time", *Monthly Notices of the Royal Astronomical Society*, Vol. 238, pp. 1529–1535, 1989.

Igarashi, J., and D.B. Spencer, "Optimal Continuous Thrust Orbit Transfer Using Evolutionary Algorithms", *Journal of Guidance, Control and Dynamics*, Vol. 28, No. 3, pp. 547–549, 2005.

International Astronomical Union (IAU), "IAU 2006 Resolution B3- English Version", XXVIth International Astronomical Union General Assembly, 2006.

Jacobi, C.G.J., "Sur le Movement D'un Point et Sur Un Cas Particulier du Problème des Trois Corps", *Comptes Rendus de l'Académie des Sciences de Paris*, Vol. 3, pp. 59–61, 1836.

Jaffe, L.D., and R.J. Parks, "Surveyor 1 Mission Report", JPL/Caltech, TR 32–1023, Pts. 1, 2, and 3, Pasadena, CA, August 1966.

Jah, M., "*Derivation of the B-plane (Body Plane) and Its Associated Parameters*", Jet Propulsion Lab, 2002.

Jesick, M., M. Wong, S. Wagner, J. Kangas, and G. Kruizinga, "Mars 2020 Trajectory Correction Maneuver Design," *Journal of Spacecraft and Rockets*, Vol. 59, No. 4, 2022.

Johnson, N.L., "*Handbook of Soviet Lunar and Planetary Exploration*", Univelt, 1979.

Kalman, R.E., "A New Approach to Linear Filtering and Prediction Problems", *Journal of Basic Engineering*, Vol. 82, No. 1, pp. 35–45, 1960.

Kaplan, G.H., "The IAU Resolutions on Astronomical Reference Systems, Time Scales, and Earth Rotation Models: Explanation and Implementation", U.S. Naval Observatory Circular, No. 179, U.S. Naval Observatory, Washington, DC, 2005.

Kaula, W.M., "*Theory of Satellite Geodesy*", Dover Publications, Blaisdell, 1966.

Kennedy, J.F., "Message to the Congress on Urgent National Needs", Speech before Joint Session of U.S. Congress, May 25, 1961.

Kennedy, J.F., "Address at Rice University on the Nation's Space Effort", Speech at Rice University, September 12, 1962.

Kim, M., and C. Hall, "Lyapunov and Halo Orbits About L2", *Advances in the Astronautical Sciences*, Vol. 109, pp. 349–366, 2002.

Koon, W.S., M. Lo, J. Marsden, and S. Ross, "Heteroclinic Orbits between Periodic Orbits and Resonanace Transitions in Celestial Mechanics", *Chaos*, Vol. 10, No. 2, June 2000.

Koon, W.S., M.W. Lo, J.E. Marsden, and S.D. Ross, "Low Energy Transfer to the Moon", *Celestial Mechanics and Dynamical Astronomy*, Vol. 81, No. 1–2, pp. 63–73, 2001.

Koon, W.S., M.W. Lo, J.E. Marsden, and S.D. Ross, "Constructing a Low Energy Transfer between Jovian Moons", *Contemporary Mathematics*, Vol. 292, pp. 129–145, 2002.

Kreyszig, E., "*Advanced Engineering Mathematics*", 10th ed., John Wiley & Sons, 2021.

Labunsky, A.V., O.V. Papkov, and K.G. Sukhanov, "*Multiple Gravity Assist Interplanetary Trajectories*", Gordon and Breach Science Publishers, Amsterdam, 1998.

Lam, T., and G.J. Wahiffen, "Exploration of distant retrograde orbits around Europa" AAS 05–110, AAS/AIAA Space Flight Mechanics Meeting, Copper Mountain, CO, January 2005.

Lawden, D.F., "*Optimal Trajectories for Space Navigation*", Butterworths Mathematical Texts, Butterworthsm, London, 1963.

Llibre, J., R. Martinez, and C. Simó, "Transversality of the Invariant Manifolds Associated to the Lyapunov Family of Periodic Orbits near L2 in the Restricted Three-Body Problem", *Journal of Differential Equations*, Vol. 58, No. 1, pp. 104–156, 1985.

Lo, M.W., "Libration Point Trajectory Design", *Numerical Algorithms*, Vol. 14, pp. 153–164, 1997.

Lo, M.W., and S.D. Ross, "SURFing the Solar System: Invariant Manifolds and the Dynamics of the Solar System", JPL IOM 312/97, 1997.

Lo, M.W., and S.D. Ross, "Low Energy Interplanetary Transfers Using the Invariant Manifolds of L1, L2 and Halo Orbits", AAS 98–136, AAS/AIAA Astrodynamics Specialist Conference, Monterey, CA, February 1998.

Lo, M.W., and S.D. Ross, "The Lunar L1 Gateway: Portal to the Stars and Beyond," AIAA 2001–4768, AIAA Space 2001 Conference, Albuquerque, New Mexico, August 2001.

Lo, M.W., B.G. Williams, W.E. Bollman, D. Han, Y. Hahn, J.L. Bell, E.A. Hirst, R.A. Corwin, P.E. Hong, K.C. Howell, B. Barden, and R. Wilson, "Genesis Mission Design", *Journal of the Astronautical Sciences*, Vol. 49, No. 1, pp. 169–184, 2001.

Marec, J.P., "*Optimal Space Trajectories*", 1st ed., Elsevier, Amsterdam, 1979.

Martin-Mur, T.J., D.S. Abraham, D. Berry, S. Bhaskaran, R.J. Cesarone, and L.J. Wood, "The JPL Roadmap for Deep Space Navigation", *Advances in the Astronautical Sciences*, Vol. 124, pp. 1925–1932, 2006.

Mastrodemos, N., D.G. Kubitschek, and S.P. Synnott, "Autonomous Navigation for the Deep Impact Mission Encounter with Comet Tempel 1", Russell, C.T. (eds.) *Deep Impact Mission: Looking Beneath the Surface of a Cometary Nucleus*. Springer, Dordrecht, 2005. https://doi.org/10.1007/1-4020-4163-2_4.

McCarthy, D., "*IERS conventions (2000)*", IERS Technical Note, International Earth Rotation Service, 2000.

McCarthy, L.K., C.D. Adam, J.M. Leonard, P.G. Antresian, D. Nelson, E. Sahr, J. Pelgrift, E.J. Lessac-Chenen, J. Geeraert, and D. Lauretta, "OSIRIS-REx Landmark Optical Navigation Performance During Orbital and Close Proximity Operations at Asteroid Bennu", AIAA SciTech 2022 Forum, San Diego, CA & Virtual, January 3–7, 2022.

McCarthy, D., and P.K. Seidelmann, "*Time – From Earth Rotation to Atomic Physics*", Wiley-VCH Verlag, Weinheim, 2009.

McGehee, R.P., "Some Homoclinic Orbits for the Restricted Three-Body Problem," Doctoral Dissertation, University of Wisconsin, 1969.

McInnes, C.R., "*Solar Sailing: Technology, Dynamics and Mission Applications*", Praxis Publishing Ltd., Chichester, UK, 1999.

Miller, J.K., B.G. Williams, W.E. Bollman, and R.P. Davis, "Navigation of the Near Earth Asteroid Rendezvous (NEAR) Mission", AAS/AIAA Spaceflight Mechanics Meeting, Albuquerque, NM, February 1995.

Moulton, F.R., "*An Introduction to Celestial Mechanics*", Dover Publications, Mineola, NY, 1970 (republication of 2nd edition, Macmillan Company, 1914).

NASA, "*Solar and Heliospheric Observatory*", https://sohowww.nascom.nasa.gov/, Accessed: 2019-02-15.

Newton, I., "*Philosophiæ Naturalis Principia Mathematica*", London, 1687.

O'Brien, F., "*The Apollo Flight Journal*", http://history.nasa.gov/afj/loressay.htm, 2000

Olgevie, R.E., P.D. Andrews, and T.P. Jasper, "Attitude Control Requirements for an Earth Orbital Solar Electric Propulsion Stage," AIAA 75–353, AIAA 11th Electric Propulsion Conference, New Orleans, LA, March 1975.

Orloff, R.W., and D.M. Harland, "*Apollo: The Definitive Sourcebook*", Praxis Publishing Ltd., Chichester, UK, 2006.

Osbourne, M.R., "On Shooting Methods of Boundary Value Problems," *Journal of Mathematical Analysis and Applications*, Vol. 27, No. 2, pp. 1969.

Owen, W.M. Jr., "Interplanetary Optical Navigation 101", NASA/Jet Pro-pulsion Laboratory, in Space Mission Challenges for Information Technology SMC-IT 2003, Pasadena, CA, July 13–16, 2003. http://hdl.handle.net/2014/38410.

Owen, W.M. Jr., T.C. Duxbury, C.H. Acton, Jr., S.P. Synnott, J.E. Riedel, and S. Bhaskaran, "A Brief History of Optical Navigation at JPL", *Advances in the Astronautical Sciences*, Vol. 131, pp. 329–348, 2008.

Palmore, J.I., "An Elementary Proof of the Optimality of Hohmann Transfers", *Journal of Guidance, Control and Dynamics*, Vol. 7, No. 5, pp. 629–630, 1984.

Park, R.S., W.M. Folkner, J.G. Williams, and D.H. Boggs, "The JPL Planetary and Lunar Ephemerides DR440 and DE441", *Advances in the Astronautical Sciences*, Vol. 161, No. 3, p. 105, 2021.

Parker, T.S., and L.O. Chua, "*Practical Numerical Algorithms for Chaotic Systems*", Springer-Verlag, New York, 1989.

Perko, L., "*Differential Equations and Dynamical Systems*", 3rd ed., Springer-Verlag, New York, 2000.

Peters, J., "Neuer Fundamentalkatalog des Berliner Astronomischen Jahrbuchs nach den Grundlagen von A. Auwers für die Epochen 1875 und 1900", Veröffentl. D. Königl. Astron. Rechen-Instituts zu Berlin, No. 33, 1907.

Pfeiffer, C.G., "*A Dynamic Programming Analysis of Multiple Guidance Corrections of a Trajectory*", NASA Jet Propulsion Laboratory, Pasadena, CA. 1963.

Pines, S., "Constants of the Motion for Optimum Thrust Trajectories in a Central Force Field", *AIAA Journal*, Vol. 3, No. 11, pp. 2010–2014, 1964.

Poincaré, H., "*Les Méthodes Nouvelles da la Mécanique Céleste*", Gauthier-Villars, Paris, 1892–1899. Reprinted by Dover, New York, 1957, NASA TTF-450, 1967.

Prussing, J.E., "Simple Proof of the Global Optimality of the Hohmann Transfer", *Journal of Guidance, Control and Dynamics*, Vol. 15, No. 4, pp. 1037–1038, 1992.

Prussing, J.E., and B.A. Conway, "*Orbital Mechanics*", Oxford University Press, New York, 2012.

Rayman, M.D., "The Successful Conclusion of the Deep Space 1 Mission: Important Results without a Flashy Title", *Space Technology*, Vol. 23, No. 2–3, p. 185, 2003.

Rayman, M.D., P. Varghese, D.H. Lehman, and L.L. Livesay, "Results from the Deep Space 1 Technology Validation Mission", IAA-99-IAA.11.2.01, 50th International Astronautical Congress, Amsterdam, October 4–8, 1999.

Redding, D.C., "Optimal Low-Thrust Transfers to Geosynchronous Orbit," Stanford University Guidance and Control Laboratory Report SUDAAR 539, September 1983.

Richardson, D.L., and N.D. Cary, "A Uniformly Valid Solution for Motion About the Interior Libration Point of the Perturbed Elliptic-Restricted Three-Body Problem", AAS 75-021, AIAA/AAS Astrodynamics Conference, Bahamas, July 1975.

Roy, A.E, "*Orbital Motion*", 2nd edition, Adam Holger Ltd., Bristol, 4th Edition, CRC Press, 2005.

Russell, R.D., and L.F. Shampine, "A Collocation Method for Boundary Value Problems", *Numerical Mathematics*, Vol. 19, pp. 1–28, 1972.

Sauer, C.G., "*Optimization of a Solar-Electric Propulsion Planetary Orbiter Spacecraft*," AAS 68-104, AAS/AIAA Astrodynamics Conference, Jackson, WY, September 1968.

Seeber, G, "*Satellite Geodesy*", 2nd ed., de Gruyter, Berlin, 2003.

Seidelmann, P.K., "*Explanatory Supplement to the Astronomical Ephermeris and the American Ephemeris and Nautical Almanac*", University Science Books, Mill Valley, CA, 1992.

Seitz, F., "*The Science Matrix: The Journey, Travails, Triumphs*", Springer-Verlag, New York, 1992.

Siddiqi, A.A. "Deep Space Chronicle: A Chronology of Deep Space and Planetary Probes 1958–2000", NASA Monographs in Aerospace History No. 24, 2017.

Simpson, D.G., "An Alternative Lunar Ephemeris Model for On-Board Flight Software Use", Proceedings of the 1999 NASA/GSFC Flight Mechanics Symposium, pp. 175–184, 1999.

Spencer, D.B., "The Gravitational Influences of a Fourth Body on Periodic Halo Orbits", M.S. Thesis, School of Aeronautics and Astronautics, Purdue University, 1985.

SP-4029, "*Apollo by the Numbers: A Statistical Reference*", NASA, 2001, http://history.nasa.gov/SP-4029/Apollo_00_Welcome.htm ny Richard Orloff.

Standish, E.M., and J.G. Williams, "*Keplerian Elements for Approximate Position of the Major Planets*", NASA JPL, 2006.

Stern, R.G., and J.E. Potter, "*Optimization of Midcourse Velocity Corrections*", Experimental Astronomy Laboratory, Massachusetts Institute of Technology, Cambridge, MT, 1965.

Stewart, D.J., and R.G. Melton, "Approximate Analytic Representation for Low-Thrust Trajectories", AAS 91–512, AAS/AIAA Astrodynamics Conference, Durango, CO, August 1991.

Swenson, L.S., Jr., J.M. Grimwood, and C.C. Alexander, "This New Ocean: A History of Project Mercury", NASA SP-4201, 1989.

Szebehely, V.G., "*Theory of Orbits: The Restricted Problem of Three Bodies*", Academic Press, New York, 1967.

Szebehely, V.G., "*Adventures in Celestial Mechanics: A First Course in the Theory of Orbits*", 1st ed., University of Texas Press, Austin, TX, 1989.

Tauber, J.A., and J. Clavel, "The Planck Mission", The Extragalactic Infrared Background and its cosmological Implications, IAU Symposium, Vol. 204, 2001.

Tisserand, F.F., "*Traité de la Mécanique Céleste*", Gauthier-Villars, Paris, 1889.

Uphoff, C., and M.A. Crouch, "Lunar Cycler Orbits with Alternating Semi-Monthly Transfer Windows," A93–17901 05–13, Proceedings of the 1st AAS/AIAA Annual Spaceflight Mechanics Meeting, Houston, TX, February 11–13, 1991.

Vallado, D.A., "*Fundamentals of Astrodynamics and Applications*", 4th ed., Microcosm Press, Kluwer Academic Publishers, El Segundo, CA, 2013.

Vaughan, R.M., P.H. Kallemeyn, D.A. Spencer, and R.D. Braun, "*Navigation Flight Operations for Mars Pathfinder*", NASA Technical Report 20040110335, 2004.

Villac, B.F., and D.J. Scheeres, "Escaping Trajectories in the Hill Three-Body Problem and Applications", *Journal of Guidance, Control, and Dynamics*, Vol. 26, No. 2, pp. 224–232, 2003.

von Stryk, O., and R. Bulirsch, "Direct and Indirect Methods for Trajectory Optimization", *Annals of Operations Research*, Vol. 37, pp. 357–373, 1992.

Wallace, M.S., J. S. Parker, N.J. Strange, and D. Grebow, "Orbital operations for Phobos and Deimos exploration", AIAA 2012–5067, AIAA/AAS Astrodynamics Specialist Conference, Minneapolis, MN, August 13–16, 2012.

Walter, H.G., and O.J. Sovers, "*Astronomy of Fundamental Catalogues*", Springer-Verlag, Berlin Heidelberg, 2000.

Wertz, J.R., "*Spacecraft Attitude Determination and Control*", D. Reidel Publishing Company, Dordrecht, Holland, 1978.

Whitley, R., and R. Martinez, "Options for Staging Orbits in Cislunar Space", 2016 IEEE Aerospace Conference, Big Sky, MT, March 5–12, 2016.

Wiesel, W.E., "*Spaceflight Dynamics*", 2nd ed., McGraw-Hill, New York, 1997.

Wiggins, S., "*Introduction to Applied Nonlinear Dynamical Systems and Chaos*", Springer-Verlag, New York, 1990.

Wilson, R.S., and K.E. Williams, "Genesis Trajectory and Maneuver Design Strategies during Early Flight," AAS 03–202, AAS/AIAA Space Flight Mechanics Meeting, Ponce, Puerto Rico, February 9–13, 2003.

Woods, W.D., "*How Apollo Flew to the Moon*", Praxis Publishing Ltd., Chichester, UK, 2008.

Woolley, R., and A. Nicholas, "SEP Mission Design for Mars Orbiters", *Advances in the Astronautical Sciences*, Vol. 156, pp. 865–884, 2014.

Yamato, H., "Trajectory Design Methods for Restricted Problems of Three Bodies with Perturbations," Doctoral Dissertation, Department of Aerospace Engineering, The Pennsylvania State University, University Park, PA, August 2003.

Yamato, H., and D.B. Spencer, "Orbit Transfer via Tube Jumping in Restricted Problems of Four Bodies", *Journal of Spacecraft and Rockets*, Vol. 42, No. 2, pp. 321–328, 2005.

Zondervan, K.P., "Optimal Low Thrust, Three Burn Orbit Transfers with Large Plane Changes", Doctoral Dissertation, California Institute of Technology, May 1983.

Index

Printed in the United States
by Baker & Taylor Publisher Services